Heat and Mass Transfer

Series editors

D. Mewes, Hannover, Germany

F. Mayinger, München, Germany

Heat and mass transfer occur in coupled form in most production processes and chemical-engineering applications of a physical, chemical, biological or medical nature. Very often they are associated with boiling, condensation and combustion processes and also with fluids and their flow fields. Hence rheological behavior and dissipative heating also play a role.

The increasing interplay of experimental research and computer-assisted evaluation and analysis methods has led to new results, which not only confirm empirical representations and their physical interpretation but, in addition, extend their previously limited applications significantly.

The series covers all fields of heat and mass transfer, presenting the interrelationships between scientific foundations, experimental techniques, model-based analysis of results and their transfer to technological applications. The authors are all leading experts in their fields.

Heat and Mass Transfer addresses professionals and researchers, students and teachers alike. It aims to provide both basic knowledge and practical solutions, while also fostering discussion and drawing attention to the synergies that are essential to start new research projects.

More information about this series at http://www.springer.com/series/4247

Christo Boyadjiev · Maria Doichinova
Boyan Boyadjiev · Petya Popova-Krumova

Modeling of Column Apparatus Processes

Second Edition

 Springer

Christo Boyadjiev
Institute of Chemical Engineering (IChE)
Bulgarian Academy of Sciences (BAS)
Sofia
Bulgaria

Maria Doichinova
Institute of Chemical Engineering (IChE)
Bulgarian Academy of Sciences (BAS)
Sofia
Bulgaria

Boyan Boyadjiev
ChemEng Ltd.
Sofia
Bulgaria

Petya Popova-Krumova
Institute of Chemical Engineering (IChE)
Bulgarian Academy of Sciences (BAS)
Sofia
Bulgaria

ISSN 1860-4846   ISSN 1860-4854   (electronic)
Heat and Mass Transfer
ISBN 978-3-030-07905-5   ISBN 978-3-319-89966-4   (eBook)
https://doi.org/10.1007/978-3-319-89966-4

1st edition: © Springer International Publishing Switzerland 2016
2nd edition: © Springer International Publishing AG, part of Springer Nature 2018
Softcover re-print of the Hardcover 2nd edition 2018

Printed on acid-free paper

This Springer imprint is published by the registered company Springer International Publishing AG part of Springer Nature
The registered company address is: Gewerbestrasse 11, 6330 Cham, Switzerland

*The **logic** and the **intuition** are the **foundation** of the human knowledge and the science*

This book is dedicated to our Institute of Chemical Engineering Bulgarian Academy of Sciences

Christo Boyadjiev
Maria Doichinova
Boyan Boyadjiev
Petya Popova-Krumova

Preface

The complex processes in the column apparatuses have a combination of hydro-dynamic processes, convective and diffusive mass (heat) transfer processes, and chemical reactions between the reagents (components of the phases).

The fundamental problem in the column apparatuses modeling is a result of the complicated hydrodynamic behavior in the columns, and thus, the velocity distributions in the phases and the interphase boundaries in the columns are unknown. This problem can be circumvented if the phase boundary processes are replaced by equivalent volumetric processes in the phases and the unknown distributions of the phase velocities are replaced by the average velocity of the cross section of the columns.

The column apparatuses are possible to be modeled, using a new approach on the base of the physical approximations of the mechanics of continua, where the mathematical point is equivalent to a small (elementary) physical volume, which is sufficiently small with respect to the apparatus volume, but at the same time sufficiently large with respect to the intermolecular volumes in the medium.

The mathematical models of the processes in the column apparatuses, in the physical approximations of the mechanics of continua, will be the mass balances in the phase volumes (phase parts in the elementary volume) between the convective mass transfer (as a result of the fluid motions), the diffusive mass transfer (as a result of the concentration gradients) and the volume mass sources (sinks) (as a result of chemical reactions or interphase mass transfer). In the case of balance between these three effects, the mass transfer process is stationary, or in the opposite case, the process is non-stationary. These convection–diffusion-type models permit to be made a qualitative analysis of the processes (models). On this base, it is possible to obtain the main, small, and slight physical effects (mathematical operators). The slight effects (operator) must be rejected because they are not experimentally observable. As a result, the process mechanism identification is possible to be made. These models permit to determinate the mass transfer resistances in the gas and liquid phases and the optimal dispersion system finding in gas absorption (gas–liquid drops or liquid–gas bubbles). The convection–diffusion models are a base of the average-concentration models, which allow a quantitative analysis of the processes in column apparatuses.

The convection–diffusion models are possible to be used for qualitative analysis only, because the velocity distribution functions are unknown and cannot be obtained. The problem can be avoided by the average-concentration models, where the average values of the velocity and concentration over the cross-sectional area of the column are used; i.e., the medium elementary volume (in the physical approximations of the mechanics of continua) will be equivalent to a small cylinder with a real radius and a height, which is sufficiently small with respect to the column height and at the same time sufficiently large with respect to the inter-molecular distances in the medium.

The convection–diffusion models and average-concentration models are used for the qualitative and quantitative analyses of the processes in single phase (simple and complex chemical reactions), two phase (absorption, adsorption, and catalytic processes), and three phase (two-phase absorbent processes and absorption–adsorption processes).

In many cases, the computer modeling of the processes in column apparatuses, on the base of a new approach, using the convection–diffusion-type model and average-concentration-type model, does not allow a direct use of the MATLAB. In these cases, it is necessary to create combinations of MATLAB with appropriate algorithms.

Practically, the new type models are characterized by the presence of small parameters at the highest derivatives. As a result, the use of the conventional software for solving of the model differential equations is difficult. This difficulty may be eliminated by an appropriate combination of MATLAB and perturbation method.

In the cases of countercurrent gas–liquid or liquid–liquid processes, the mass transfer process models are presented in two-coordinate systems, because in one-coordinate system one of the equations has not a solution by reason of the negative equation Laplacian value. A combination of an iterative algorithm and MATLAB must be used for the solutions of the equations set in different coordinate systems.

In the cases of a non-stationary adsorption in gas–solid systems, the presence of mobile (gas) and immobile (solid) phases in the conditions of longtime processes leads to the non-stationary process in the immobile phase and stationary process in the mobile phase, practically. As a result, different coordinate systems must be used in the gas- and solid-phase models. A combination of a multi-step algorithms and MATLAB must be used for the solutions of the equations set in different coordinate systems.

The convection–diffusion and average-concentration models, where the radial velocity component is equal to zero in the cases of a constant axial velocity radial non-uniformity along the column height, are presented. There are many cases of industrial columns, where these conditions are not satisfied. The models of chemical, absorption, adsorption, and catalytic processes in the cases of an axial modification of the radial non-uniformity of the axial velocity component, where the radial velocity component is not equal to zero, are presented too.

The absorption process intensification of an average soluble gas is realized with a new patent in two-zone column, where the upper zone is physical absorption in a gas–liquid drops system (intensification of the gas-phase mass transfer) and the lower zone is a physical absorption in liquid–gas bubbles system (intensification of the liquid-phase mass transfer).

The solid fuel combustion in the thermal power plants, which use sulfur-rich fuels, poses the problem of sulfur dioxide removal from the waste gases. This problem is complicated by the fact that it is required to purify huge amounts of gas with low sulfur dioxide concentration. The huge gas amounts need big apparatus size, which is possible to be decreased if the removal process rate is maximized.

The purification of large amounts of waste gases from combustion plants used countercurrent absorbers, where the gas velocity (as a result and absorbers diameter too) is limited by the rate of the absorbent drops fall in an immobile gas medium. A new patent avoids this disadvantage, where co-current sulfur dioxide absorption is realized.

A new patent solves the problem of absorbent regeneration, using two-step process—physical absorption of sulfur dioxide by water and adsorption of sulfur dioxide from the water solution by synthetic anionite particles. The adsorbent regeneration is made by ammonium hydroxide solution. The obtained ammonium sulfite solution is used (after reaction with nitric acid) for concentrated sulfur dioxide and ammonium nitrate solution production.

The Introduction (Chap. 1) concerns linear mass transfer theory (model theories, boundary layer theory, two-phase boundary layers), mass transfer in countercurrent flows (velocity and concentration distribution, comparison analysis), nonlinear mass transfer (influence on the hydrodynamics, boundary conditions, boundary layer theory, Marangoni effect), interphase mass transfer resistances (film theory and boundary layer theory approximations), three-phase mass transfer processes (physical, hydrodynamic, and interphase mass transfer models, absorption mechanism and kinetics).

Part I focuses on the convection–diffusion-type models for qualitative analysis of the column apparatus processes. Chapter 2 presents one-phase chemical processes in column reactors (simple and complex chemical reaction kinetics), approximations of the model (short-column and high-column model, effect of the chemical reaction rate), effect of the radial non-uniformity of the velocity distribution (conversion degree, concentration distribution, influence of the velocity radial non-uniformity shape, scale effect, back mass transfer mechanism), examples (effect of the tangential flow, simultaneous mass and heat transfer processes, circulation zones in column apparatuses, mass transfer in one-phase countercurrent flow). Chapter 3 presents the convection–diffusion-type models of two-phase processes (physical and chemical absorption, physical and chemical adsorption, catalytic processes in the cases of physical and chemical adsorption mechanism), examples (airlift reactors, airlift photobioreactor, moisture adsorption). Chapter 4 presents models of three-phase processes in the cases of two-phase absorbent processes (physical and chemical absorption) and absorption–adsorption processes (physical and chemical adsorption mechanism).

Part II addresses the average-concentration-type models for quantitative analysis of the column apparatuses processes. Chapter 5 presents the average-concentration-type models of the column reactors in the cases of simple and complex chemical reactions (effect of the velocity radial non-uniformity, model parameter identification) and as an example the modeling of a non-isothermal chemical reactor. Chapter 6 presents the interphase mass transfer models of the physical and chemical absorption, physical and chemical adsorption, catalytic processes in the cases of physical and chemical adsorption mechanism and as examples airlift reactor modeling and moisture adsorption modeling.

Part III addresses the calculation problems in the convection–diffusion-type models and average-concentration-type models. Chapter 7 presents the perturbation method approach for the solution of the equations in the convection–diffusion models and average-concentration models. Chapter 8 presents the solutions of two-coordinate systems' problems in the models of the countercurrent absorption processes. Chapter 9 presents the multi-step modeling algorithms in the case of a longtime non-stationary adsorption process, when the interphase gas–solid mass transfer is stationary.

Part IV concerns the convection–diffusion and average-concentration models of the industrial processes, in the cases of an axial modification of the radial non-uniformity of the axial velocity component. Chapters 10–14 present industrial column models of the chemical, absorption, adsorption, and catalytic processes.

Part V concerns the models of the processes, which participate in different patents, related to the waste gas purification from sulfur dioxide in column apparatuses. Chapter 15 presents the modeling of the gas absorption of an average soluble gas in a bizonal absorption apparatus. Chapter 16 presents the processes modeling of an absorption–adsorption method for waste gas purification from sulfur dioxide, where the first step is a physical absorption of sulfur dioxide by water and the second step is a chemical adsorption of sulfur dioxide in the water solution by synthetic anionite. After the sulfur dioxide saturation of the synthetic anionite particles, the adsorbent regeneration is possible to be carried out by water solution of ammonium hydroxide. Chapter 17 presents the processes modeling in a co-current apparatus, where the gas velocity is 4–5 times greater than that of the countercurrent apparatus, which are used in the practice.

Part VI (Chap. 18) concerns the conclusions.

Sofia, Bulgaria   Christo Boyadjiev
                                                                                              Maria Doichinova
                                                                                              Boyan Boyadjiev
                                                                                         Petya Popova-Krumova

Contents

1   **Introduction** 1
   1.1   Linear Mass Transfer Theory 5
      1.1.1   Model Theories 6
      1.1.2   Boundary Layer Theory 8
      1.1.3   Two-Phase Boundary Layers 9
   1.2   Mass Transfer in Countercurrent Flows 12
      1.2.1   Velocity Distribution 13
      1.2.2   Concentration Distribution 16
      1.2.3   Comparison Between Co-current
           and Countercurrent Flows 17
   1.3   Nonlinear Mass Transfer 18
      1.3.1   Influence of the Intensive Interphase Mass
           Transfer on the Hydrodynamics 19
      1.3.2   Boundary Conditions of the Nonlinear Mass
           Transfer Problem 22
      1.3.3   Nonlinear Mass Transfer in the Boundary Layer 24
      1.3.4   Two-Phase Systems 28
      1.3.5   Nonlinear Mass Transfer and Marangoni Effect 37
   1.4   Interphase Mass Transfer Resistances 44
      1.4.1   Film Theory Approximation 45
      1.4.2   Boundary Layer Theory Approximation 45
      1.4.3   Interphase Mass Transfer Intensification 47
   1.5   Three-Phase Mass Transfer Processes 47
      1.5.1   Physical Model 48
      1.5.2   Hydrodynamic Model 49
      1.5.3   Interphase Mass Transfer Model 50
      1.5.4   Absorption Kinetics 51

	1.5.5	Absorption Mechanism	52
	1.5.6	Absorption of Highly Soluble Gases	54
	1.5.7	Absorption of Low Soluble Gases	56
1.6	Conclusions		57
References			58

Part I   Qualitative Analysis of Column Apparatuses Processes

2	**One-Phase Processes**		**67**
	2.1	Column Chemical Reactor	67
		2.1.1 Convection–Diffusion-Type Model	68
		2.1.2 Complex Chemical Reaction Kinetics	70
		2.1.3 Two Components Chemical Reaction	72
		2.1.4 Comparison Qualitative Analysis	73
		2.1.5 Pseudo-First-Order Reactions	73
		2.1.6 Similarity Conditions	74
	2.2	Model Approximations	74
		2.2.1 Short Columns Model	74
		2.2.2 High-Column Model	75
		2.2.3 Effect of the Chemical Reaction Rate	75
		2.2.4 Convection Types Models	76
	2.3	Effect of the Radial Non-uniformity of the Velocity Distribution	76
		2.3.1 Conversion Degree	76
		2.3.2 Concentration Distribution	77
		2.3.3 Influence of the Velocity Radial Non-uniformity Shape	80
		2.3.4 Scale Effect	86
		2.3.5 On the "Back Mixing" Effect	86
	2.4	Examples	93
		2.4.1 Effect of the Tangential Flow	93
		2.4.2 Simultaneous Mass and Heat Transfer Processes	96
		2.4.3 Circulation Zones in Column Apparatuses	98
		2.4.4 Mass Transfer in One-Phase Countercurrent Flow	106
	References		112
3	**Two-Phase Processes**		**115**
	3.1	Absorption Processes	116
		3.1.1 Physical Absorption	117
		3.1.2 Chemical Absorption	120
	3.2	Adsorption Processes	125
		3.2.1 Physical Adsorption	126
		3.2.2 Chemical Adsorption	129

3.3   Catalytic Processes 133
      3.3.1   Physical Adsorption Mechanism 134
      3.3.2   Chemical Adsorption Mechanism 140
3.4   Examples 143
      3.4.1   Airlift Reactor 144
      3.4.2   Airlift Photo-bioreactor 149
      3.4.3   Moisture Adsorption 152
References 156

4   Three-Phase Processes 159
4.1   Two-Phase Absorbent Processes 160
      4.1.1 $CaCO_3/H_2O$ *Absorbent* 160
      4.1.2 $Ca(OH)_2/H_2O$ *Absorbent* 162
4.2   Absorption–Adsorption Processes 164
      4.2.1   Physical Adsorption Mechanism 164
      4.2.2   Chemical Adsorption Mechanism 167
4.3   Three-Phase Catalytic Process 173
References 174

Part II   Quantitative Analysis of Column Apparatuses Processes

5   Column Reactors Modeling 179
5.1   Simple Chemical Reactions 180
      5.1.1   Average-Concentration Model 180
      5.1.2   Effect of the Velocity Radial Non-uniformity 183
      5.1.3   Model Parameter Identification 183
5.2   Complex Chemical Reaction 185
5.3   Examples 189
      5.3.1   Non-isothermal Chemical Reactors 189
References 191

6   Interphase Mass Transfer Process Modeling 193
6.1   Absorption Process Modeling 194
      6.1.1   Physical Absorption 194
      6.1.2   Chemical Absorption 200
6.2   Adsorption Process Modeling 205
      6.2.1   Physical Adsorption 205
      6.2.2   Chemical Adsorption 210
6.3   Catalytic Process Modeling 214
      6.3.1   Physical Adsorption Mechanism 214
      6.3.2   Chemical Adsorption Mechanism 222

6.4   Examples 227
        6.4.1   Airlift Reactor Modeling 227
        6.4.2   Moisture Adsorption Modeling 229
        6.4.3   Three-Phase Process Modeling 233
    References 233

Part III   Computer Calculation Problems

7   Perturbation Method Approach 237
    7.1   Perturbation Method 237
    7.2   Convection–Diffusion-Type Models 239
        7.2.1   Short Columns Model........................ 239
        7.2.2   Calculation Problem 240
        7.2.3   Concentration Distributions 242
    7.3   Average-Concentration Models 243
        7.3.1   Calculation Problem 245
        7.3.2   Average-Concentration Distributions 247
        7.3.3   Parameter Identification...................... 248
    References 250

8   Two-Coordinate Systems Problem 251
    8.1   Convection–Diffusion-Type Model..................... 251
        8.1.1   Calculation Problem 252
        8.1.2   Concentration Distributions 253
        8.1.3   Absorption Process Efficiency 255
    8.2   Average-Concentration Model 255
        8.2.1   Calculation Problem 256
    References 257

9   Multi-step Modeling Algorithms 259
    9.1   Convection–Diffusion-Type Model..................... 259
        9.1.1   Calculation Problem 260
        9.1.2   Concentration Distributions 262
        9.1.3   Adsorption Process Efficiency 264
    9.2   Average-Concentration Model 265
        9.2.1   Model Equations Solution 268
        9.2.2   Parameter Identification...................... 269
    References 270

Part IV   Modeling of Processes in Industrial Column Apparatuses

10 Industrial Column Chemical Reactors 273
    10.1   Effect of the Axial Modification of the Radial
        Non-uniformity of the Axial Velocity Component........... 274
        10.1.1 Model Equation Solution 276

        10.1.2   Parameter Identification.......................... 277

        10.1.3   Influence of the Model Parameter 278

   10.2   Effect of the Radial Velocity Component 279

        10.2.1   Axial and Radial Velocities...................... 281

        10.2.2   Concentration Distributions 282

        10.2.3   Average-Concentration Model 284

        10.2.4   Parameter Identification......................... 287

        10.2.5   Influence of the Model Parameters............... 288

   References 288

11   **Industrial Co-current Column Absorber**..................... 291

   11.1   Effect of the Axial Modification of the Radial
        Non-uniformity of the Axial Velocity Component........... 293

        11.1.1   Model Equations Solution 294

        11.1.2   Parameter Identification......................... 295

        11.1.3   Chemical Absorption 298

        11.1.4   Physical Absorption of Highly Soluble Gas
               $(\omega = Da = 0)$ 301

        11.1.5   Physical Absorption of Lightly Soluble Gas
               $(\omega^{-1} = 0)$ 303

   11.2   Effect of the Radial Velocity Component 305

        11.2.1   Convection–Diffusion Model 305

        11.2.2   Axial and Radial Velocity Components 306

        11.2.3   Convective Type Model 307

        11.2.4   Average-Concentration Model 308

        11.2.5   Physical Absorption of Highly Soluble Gas 310

        11.2.6   Physical Absorption of Lightly Soluble Gas........ 314

        11.2.7   Physical Absorption of Average Soluble Gas 317

        11.2.8   Chemical Absorption 320

   References 322

12   **Industrial Counter-current Column Absorber** 323

   12.1   Effect of the Axial Modification of the Radial
        Non-uniformity of the Velocity 325

        12.1.1   Physical Absorption of the Average Soluble
               Gases.................................... 326

        12.1.2   Physical Absorption of Highly Soluble Gas
               $(\omega = Da = 0)$ 327

        12.1.3   Physical Absorption of Lightly Soluble Gas
               $(\omega^{-1} = 0, Da = 0)$ 330

        12.1.4   Parameter Identification in the Cases of Average
               Soluble Gases 332

        12.1.5   Chemical Absorption 332

12.2   Effect of the Radial Component of the Velocity   333
        12.2.1   Axial and Radial Velocity Components   337
        12.2.2   Convective-Type Model   337
        12.2.3   Average-Concentration Model   337
        12.2.4   Physical Absorption of Highly Soluble Gas   339
        12.2.5   Physical Absorption of Lightly Soluble Gas   340
        12.2.6   Physical Absorption of Average Soluble Gas   340
        12.2.7   Chemical Absorption   341
References   342

13 **Industrial Column Adsorber**   343
    13.1   Effect of the Axial Modification of the Radial
            Non-uniformity of the Axial Velocity Component   343
            13.1.1   Physical Adsorption Process   343
            13.1.2   Modeling of the Industrial Column Adsorbers   344
            13.1.3   Model Equations Solution   345
            13.1.4   Concentration Distributions   347
            13.1.5   Parameter Identification   351
            13.1.6   Chemical Adsorption Process   351
            13.1.7   Influence of the Model Parameters   355
    13.2   Effect of the Radial Velocity Component   358
            13.2.1   Physical Adsorption Process   358
            13.2.2   Axial and Radial Velocities   362
            13.2.3   Chemical Adsorption Process   364
    References   364

14 **Industrial Column Catalytic Reactors**   367
    14.1   Effect of the Axial Modification of the Radial
            Non-uniformity of the Axial Velocity Component   367
            14.1.1   Physical Adsorption Mechanism   367
            14.1.2   Modeling of the Industrial Column Catalytic
                      Reactors   370
            14.1.3   Model Equations Solution   371
            14.1.4   Parameter Identification   379
            14.1.5   Chemical Adsorption Mechanism   380
            14.1.6   Convection-Type and Average-Concentration
                      Models   380
            14.1.7   Industrial Column Modeling   381
            14.1.8   Convective Model Equations Solution   382
            14.1.9   Average-Concentration Model Equations
                      Solution   384
            14.1.10   Parameter Identification   384
            14.1.11   Influence of the Model Parameters   386

14.2 Effect of the Radial Velocity Component 388
 14.2.1 Physical Adsorption Mechanism 390
 14.2.2 Axial and Radial Velocities 397
 14.2.3 Convection-Type Model Equations Solution 398
 14.2.4 Chemical Adsorption Mechanism 399
References 399

Part V Waste Gas Purification in Column Apparatises

15 Bizonal Absorption Apparatus 405
15.1 Absorption Column 405
 15.1.1 Physical Absorption Modeling in the Upper
  Zone 407
 15.1.2 Chemical Absorption Modeling in the Lower
  Zone 408
 15.1.3 Generalized (Dimensionless) Variables Model 409
 15.1.4 Industrial Conditions 410
15.2 Algorithm for Model Equations Solution 411
 15.2.1 Upper Zone Model 411
 15.2.2 Numerical Results 412
References 415

16 Absorption–Adsorption Method 417
16.1 Two-Step Absorption–Adsorption Method 418
16.2 Absorption and Adsorption Modeling 420
 16.2.1 Generalized Analysis 421
16.3 Average-Concentration Model 423
 16.3.1 Generalized Analysis 425
 16.3.2 Algorithm of the Solution 427
 16.3.3 Parameter Identification 427
16.4 An Absorption–Adsorption Apparatus 428
16.5 Absorption–Adsorption Process Modeling 430
 16.5.1 Generalized Analysis 431
 16.5.2 Algorithm of the Solution 433
 16.5.3 Parameter Identification 433
References 434

17 Co-current Apparatus 437
17.1 Co-current Absorber 437
 17.1.1 Use of the Co-current Absorber 440
17.2 Convection–Diffusion Type of Model 441
 17.2.1 Generalized Analysis 442
 17.2.2 Concentration Distributions 443
 17.2.3 Absorption Degree 444

17.3   Average-Concentration Model   445
       17.3.1   Parameter Identification   448
References   449

Part VI   Book Conclusions

18   Conclusion   453

Abstract

In the book is presented a new approach for the modeling of chemical, mass, and heat interphase transfer processes in industrial column apparatuses. The base of the new type models is the physical approximations of the mechanics of continua, where the mathematical point is equivalent to a small (elementary) physical volume, which is sufficiently small with respect to the apparatus volume, but at the same time sufficiently large with respect to the intermolecular volumes in the medium.

The mathematical models of the processes in the column apparatuses, in the physical approximations of the mechanics of continua, are the mass balances in the phase volumes (phase parts in the elementary volume) between the convective mass transfer (as a result of the fluid motions), the diffusive mass transfer (as a result of the concentration gradients) and the volume mass sources (sinks) (as a result of chemical reactions or interphase mass transfer). In the case of balance between these three effects, the mass transfer process is stationary. In the opposite case, the process is non-stationary. These convection–diffusion-type models permit to be made a qualitative analysis of the processes (models). On this base, it is possible to obtain the main, small, and slight physical effects (mathematical operators). The slight effects (operator) must be rejected because they are not experimentally observable. As a result, the process mechanism identification is possible to be made. These models permit to determinate the mass transfer resistances in the gas and liquid phases and to find the optimal dispersion system in gas absorption (gas–liquid drops or liquid–gas bubbles). The convection–diffusion models are a base of the average-concentration models, which allow a quantitative analysis of the processes in column apparatuses.

The convection–diffusion models are used for qualitative analysis only, because the velocity distribution functions are unknown and cannot be obtained. The problem is avoided by the average-concentration models, where the average values of the velocity and concentration over the cross-sectional area of the column are used; i.e., the medium elementary volume (in the physical approximations of the mechanics of continua) is equivalent to a small cylinder with a real radius and a height, which is sufficiently small with respect to the column height and at the same time sufficiently large with respect to the intermolecular distances in the medium.

The convection–diffusion models and average-concentration models are used for the qualitative and quantitative analyses of the processes in single phase (simple and complex chemical reactions), two phase (absorption, adsorption, and catalytic processes), and three phase (two-phase absorbent processes and absorption–adsorption processes).

In many cases, the computer modeling of the processes in column apparatuses, on the base of a new approach, using the convection–diffusion-type model and average-concentration-type model, does not allow a direct use of the MATLAB. In these cases is necessary to create combinations of MATLAB with appropriate algorithms.

Practically, the new type models are characterized by the presence of small parameters at the highest derivatives. As a result, the use of the conventional software for solving of the model differential equations is difficult. This difficulty is eliminated by an appropriate combination of MATLAB and perturbation method.

In the cases of countercurrent gas–liquid or liquid–liquid processes, the mass transfer process models are presented in two-coordinate systems, because in one-coordinate system one of the equations does not has a solution by reason of the negative equation Laplacian value. A combination of an iterative algorithm and MATLAB is used for the solutions of the equations set in different coordinate systems.

In the cases of a non-stationary adsorption in gas–solid systems, the presence of mobile (gas) and immobile (solid) phases in the conditions of longtime processes leads to the non-stationary process in the immobile phase and stationary process in the mobile phase, practically. As a result, different coordinate systems in the gas- and solid-phase models must be used. A combination of a multi-step algorithms and MATLAB is used for the solutions of the equations set in different coordinate systems.

The convection–diffusion and average-concentration models, where the radial velocity component is equal to zero in the cases of a constant axial velocity radial non-uniformity along the column height, are presented. There are many cases of industrial columns, where these conditions are not satisfied. The models of chemical, absorption, adsorption, and catalytic processes in the cases of an axial modification of the radial non-uniformity of the axial velocity component, where the radial velocity component is not equal to zero, are presented too.

The absorption process intensification of an average soluble gas is realized with a new patent in two-zone column, where the upper zone is physical absorption in a gas–liquid drops system (intensification of the gas-phase mass transfer) and the lower zone is a physical absorption in liquid–gas bubbles system (intensification of the liquid-phase mass transfer).

The solid fuel combustion in the thermal power plants, which use sulfur-rich fuels, poses the problem of sulfur dioxide removal from the waste gases. This problem is complicated by the fact that it is required to purify huge amounts of gas with low sulfur dioxide concentration. The huge gas amounts need big apparatus size, which is possible to be decreased if the removal process rate is maximized.

The purification of large amounts of waste gases from combustion plants used countercurrent absorbers, where the gas velocity (as a result and absorbers diameter too) is limited by the rate of the absorbent drops fall in an immobile gas medium. A new patent avoids this disadvantage, where co-current sulfur dioxide absorption is realized.

A new patent solves the problem of absorbent regeneration, using two-step process—physical absorption of sulfur dioxide by water and adsorption of sulfur dioxide from the water solution by synthetic anionite particles. The adsorbent regeneration is made by ammonium hydroxide solution. The obtained ammonium sulfite solution is used (after reaction with nitric acid) for concentrated sulfur dioxide and ammonium nitrate solution production.

Christo Boyadjiev
Maria Doichinova
Boyan Boyadjiev
Petya Popova-Krumova

Chapter 1
Introduction

The logic and the intuition are the foundation of the human knowledge and the science. In the mathematics, the intuitions are the axioms (unconditional statements that cannot be proven), while the logic is the theorems (logical consequences of the axioms). The proportion between the logic and the intuition is different in the different sciences. In the mathematics, the logic predominates. In the natural sciences (physics, chemistry, and biology), the role of the intuition increases, but the "axioms" are not always unconditional. In the humanities, the role of the logic decreases.

In the chemical engineering, which combines chemistry, physics, and mathematics, several "axioms" are used that are conditional. For example, the linear relationship (Stokes postulate) between the stress and the shear rate (deformation velocity) is the basis of the hydrodynamic models of the movement of the Newtonian fluids, while the linear relationship between the mass (heat) flux and the concentration (temperature) gradient (Fick and Fourier laws) is used in the construction of models of mass (heat) transfer. The non-equilibrium thermodynamics summarizes these "axioms" in one—"Onsager's principle of linearity," according to which "the rate with which a system is approaching its thermodynamic equilibrium depends linearly on the deviation of the system from the thermodynamic equilibrium."

The non-equilibrium thermodynamics indicates the direction of change of the complex systems to achieve thermodynamic equilibrium, but cannot show the way (mechanism) of their amendment, i.e., the constituent simple processes and interactions between them. As a result, the rate of complex processes cannot be determined. This problem is solved by determining the complicated process mechanism, formulating the simple process models by the "principle of linearity" and inputting the interactions between the simple processes in the complex process, i.e., creating the complex process model.

© Springer International Publishing AG, part of Springer Nature 2018
C. Boyadjiev et al., *Modeling of Column Apparatus Processes,*
Heat and Mass Transfer, https://doi.org/10.1007/978-3-319-89966-4_1

The chemical engineering processes are realized in one-, two-, and three-phase systems (gas–liquid–solid). They are a result from the reactions, i.e., processes of disappearance or creation of any substance. The reactions are associated with a particular phase and can be homogeneous (occurring in volume of the phase) or heterogeneous (occurring at the interface with another phase). Homogeneous reactions are usually chemical, while heterogeneous reactions may be chemical, catalytic, and adsorption. Heterogeneous reaction is the interphase mass transfer too, where on the interphase boundary the substance disappears (created) in one phase and creates (disappears) in the other phase.

The reactions deviate the system from the thermodynamic equilibrium and the processes for its recovery start. The determination of this process rate is a main problem of the chemical engineering, in which aims are to create, design, and manage high-speed processes in small-size units. This gives a reason to use the thermodynamics laws of the irreversible processes, as mathematical structures in the construction of the simple process models, described by extensive and intensive variables (in the case of a merger of two identical systems, the extensive variables doubled their values, but the intensive variables retain their values).

The kinetics of the irreversible processes uses the mathematical structures, arising from Onsager's "principle of the linearity" [1]. According to him, the average values of the derivatives at the time of the extensive variables depend linearly on the average deviations of the conjugate intensive variables from their equilibrium states (values). This principle is valid near the equilibrium, and the coefficients of proportionality are kinetic (rate) constants.

According to the linearity principle, the derivative of the mass m (kg mol) at the time $J_0 = \frac{dm}{dt}$ (kg mol s^{-1}) depends linearly on the deviation from the thermodynamic equilibrium Δc (kg mol m^{-3}) of the concentrations in two-phase volumes or in one phase and the phase boundary, i.e.

$$J_0 = k_0 \Delta c, \qquad (1.0.1)$$

where k_0 (m^3 s^{-1}) is a proportionality coefficient.

Consider a system, which contains two identical volumes in a single phase $v_1 = v_2 = v$ (m^3). The system contains a substance, whose masses m_i (kg mol) and concentrations $c_i = \frac{m_i}{v_i}$ (kg mol m^{-3}), $i = 1, 2$ are different in two volumes; i.e., the system is not in thermodynamic equilibrium. Assuming for definiteness, $c_1 - c_2 > 0, i = 1, 2$. As a result, a mass transfer of the substance starts from the volume v_1 in the volume v_2, to reach the equilibrium $c_1 - c_2 = 0$. According to the linearity principle, the rate of mass transfer between the two volumes J_0 (kg moll s^{-1}) can be represented as

$$J_0 = \frac{dm_1}{dt} = -\frac{dm_2}{dt} = k_0(c_1 - c_2), \qquad (1.0.2)$$

where k_0 (kg mol^{-1} m^3 s^{-1}) is a coefficient of proportionality. If we substitute the masses with the concentrations $m_i = v_i c_i$, $i = 1, 2$, the law of mass transfer rate in a phase between two points with different concentrations J (kg mol m^{-3} s^{-1}) is

$$J = \frac{dc_1}{dt} = -\frac{dc_2}{dt} = k(c_1 - c_2), \tag{1.0.3}$$

where k (s^{-1}) is a rate coefficient. This equation is possible to present the interphase mass transfer rate in the cases of adsorption or catalytic process, where c_1 is a substance concentration in gas (liquid) phase, while c_2 is the substance concentration in the gas (liquid) part of the solid (adsorbent or catalyst capillaries) phase.

In cases, wherein the volumes are different phases (e.g., 1 is a gas phase and 2 is a liquid phase), the law of thermodynamic equilibrium has the form $c_1 - \chi c_2 = 0$, i.e., the law of Henry, and χ is Henry's number. If $c_1 - \chi c_2 > 0$, the mass transfer is from phase 1 to phase 2 and the law on the interphase mass transfer rate is

$$J = k(c_1 - \chi c_2), \tag{1.0.4}$$

where k (s^{-1}) is the interphase mass transfer rate coefficient.

On the surface between two phases (gas–liquid, liquid–liquid) of practice, thermodynamic equilibrium is established instantly, i.e., $c_1^* - \chi c_2^* = 0$, where c_i^*, $i = 1, 2$ are the equilibrium interphase concentrations. Thus, the rate of mass transfer may also be expressed by the rate of the mass transfer in two phases

$$J = k_1(c_1 - c_1^*) = k_2(c_2^* - \chi c_2), \tag{1.0.5}$$

where k_i, $i = 1, 2$ (s^{-1}) are mass transfer rate coefficients.

Onsager' principle of the linearity represents the thermodynamic approximation of the mathematical description of the irreversible processes kinetics, but does not show the way to reach the equilibrium, i.e., the process mechanism; therefore, the proportionality coefficient is unknown. Obviously, this "thermodynamic level" does not allow actual quantitative description of the kinetics of irreversible processes in chemical engineering and must be used the next level of detail of description, so-called "hydrodynamic level."

The hydrodynamic level uses the approximations of the mechanics of continua, where the mathematical point is equivalent to an elementary physical volume, which is sufficiently small with respect to the apparatus volume, but at the same time sufficiently large with respect to the intermolecular volumes in the medium. In this level, the molecules are not visible, as is done in the next level of detail of Boltzmann.

The column apparatuses are the main devices for separation and chemical process realization in chemical, power, biotechnological, and other industries [2, 3]. Different types of column apparatuses are plate columns, packed bed columns, bubble columns, trickle columns, catalyst bed columns, etc.

The plate columns are used for multistage separation process realization, especially for multicomponent distillation processes. The modeling of distillation processes in plate columns uses the "stage-by-stage calculation" approach (from plate to plate) [2–5]. The model equations relevant to each plate are mass and heat balance equations, where the parameters of the mass and heat transfer kinetics are replaced by efficiency coefficients. To this end, the thermodynamic liquid–vapor equilibrium at each plate has to be calculated, too. All these parameters depend on the types and concentrations of the components in both the liquid and the vapor phases. The use of iterative methods to solve the model equations at each plate requires very effective thermodynamic methods allowing repeated calculations of the liquid–vapor equilibrium. The calculation method for the model equation solutions uses the method of quasi-linearization. As a result, the models of distillation plate columns contain many linear equations, but every equation contains few variables. This leads to an equations set with scarce matrixes, and the use of special mathematical software to solve them is necessary. Obviously, the modeling problems of the distillation plate columns are more of a thermodynamic and mathematical nature without involving a hydrodynamic background. The solutions of such problems are commonly performed by specially developed codes (e.g., ChemCad).

The processes in column apparatuses (except for the plate columns) are realized in one, two, or three phases. The gas phase moves among the columns as a stream of bubbles. The liquid-phase presence in the column is as droplets, films, and jets. The solid-phase forms are packed beds, catalyst particles, or slurries ($CaCO_3/H_2O$ suspension).

The mass transfer processes are the base of the separation and chemical processes in chemical, power, biotechnological, and other industries [2, 3]. The models of the mass transfer processes are possible to be created on the basis of the physical approximations of the mechanics of continua.

The big part of the industrial mass transfer processes is realized in one-, two-, or three-phase systems as a result of volume (homogeneous) and surface (heterogeneous) reactions, i.e., mass creation (disappearance) of the reagents (phase components) in the elementary volumes in the phases or on the its interphase surfaces. As a result, the reactions are mass sources (sinks) in the volume (homogeneous chemical reactions) and on the surface (interphase mass transfer in the cases of absorption, adsorption and catalytic reactions) of the elementary phase volumes.

The volume reactions lead to different concentrations of the reagents in the phase volumes, and as a result, two mass transfer processes are realized—convective transfer (caused by the movement of the phases) and diffusion transfer (caused by the concentration gradients in the phases). The mass transfer models are a mass balance in the phases, where components are convective transfer, diffusion transfer, and volume reactions (volume mass sources or sinks). The surface reactions participate as mass sources or sink in the boundary conditions of the model equations. The models of this complex process are possible to be created on the basis of the mass transfer theory as a main approach to quantitative description of processes in the chemical engineering.

The main process in column apparatuses is mass transfer (complicated with volume reaction) of a component of the moving fluid. The quantitative description of this process is possible if the axial distribution of the average concentration over the cross-sectional area of the column is known. This concentration is possible to be obtained as a solution of the mass transfer model equations in the column apparatuses [6]. In many cases of small concentration gradients, it is possible to use linear mass transfer theory models.

1.1 Linear Mass Transfer Theory

A balance of the convective and the diffusive transfer determines the mass transfer in a moving fluid. If the fields of velocity and concentration are marked by $\mathbf{u}(x, y, z)$ and $c(x, y, z)$, the mass flux ($\mathbf{j}$) through an unit surface of a given elementary volume is the sum of the convective and the diffusive fluxes:

$$\mathbf{j} = \mathbf{u}c - D\,\mathbf{grad}\,c, \tag{1.1.1}$$

where D is the diffusivity.

In the cases of stationary processes and absence of a volume source (sink) of a substance, the material balance of the substance in the elementary volume is obtained through integration of the flow over the whole surface of this volume:

$$\operatorname{div}\mathbf{j} = 0. \tag{1.1.2}$$

From (1.1.1) and (1.1.2), the following is directly obtained:

$$(\mathbf{u}, \mathbf{grad})c = D\nabla^2 c, \tag{1.1.3}$$

where $(\mathbf{u}, \mathbf{grad})$ is the scalar product of the vectors $\mathbf{u}$ and $\mathbf{grad}$.

From (1.1.3), one possible problem may be formulated for a mass transfer in a two-dimensional area with dimensions l and h, where $y = 0$ is the surface, through which the mass transfer with another phase (solid, liquid, gas) takes place:

$$u\frac{\partial c}{\partial x} + v\frac{\partial c}{\partial y} = D\left(\frac{\partial^2 c}{\partial x^2} + \frac{\partial^2 c}{\partial y^2}\right);$$

$$x = 0, \quad y > 0, \quad c = c_0; \quad x = l, \quad 0 \le y < h, \quad c = c^*; \tag{1.1.4}$$

$$y = 0, \quad 0 \le x < l, \quad c = c^*; \quad y = h, \quad c = c_0.$$

From (1.1.4), the rate of mass transfer J could be determined through the coefficient of mass transfer (k) and the local mass flux i:

$$J = k(c^* - c_0) = \frac{1}{l} \int_0^l i \, dx, \quad i = -D \left(\frac{\partial c}{\partial y} \right)_{y=0}. \tag{1.1.5}$$

As a final result, the Sherwood number can be determined:

$$\text{Sh} = \frac{kl}{D} = -\frac{1}{c^* - c_0} \int_0^l \left(\frac{\partial c}{\partial y} \right)_{y=0} dx. \tag{1.1.6}$$

From (1.1.5) and (1.1.6), it is evident that for determination of the mass transfer rate, it is necessary to find the mass transfer coefficient or the Sherwood number, i.e., to solve the problem (1.1.4). A basic difficulty in the solution is the velocity determination by solving the system of nonlinear equations of Navier–Stokes [7]. This difficulty is avoided in some model theories of mass transfer.

1.1.1   Model Theories

The fundamental difficulties of the mathematical description of the mass transfer processes are related with the necessity of foreseeing the exceptionally complicated hydrodynamic conditions under which these processes go off. Unfortunately, many mass transfer theories replace the concrete mass transfer conditions with gratuitous hydrodynamic models.

The first mass transfer theory is Nernst's *film theory* [8]. According to it, the mass transfer is a result of stationary diffusion through immovable fluid film with constant thickness h. This theory is an approximation of the linear mass transfer theory (1.1.4) if film theory conditions are introduced:

$$u = v = 0, \quad h \ll l. \tag{1.1.7}$$

In this way from (1.1.4), the following is obtained:

$$\frac{\partial^2 c}{\partial y^2} = 0; \quad y = 0, \quad c = c^*; \quad y = h, \quad c = c_0, \tag{1.1.8}$$

i.e.

$$c = \frac{c_0 - c^*}{h} y + c^*, \quad k = \frac{D}{h}. \tag{1.1.9}$$

The basic disadvantages of this theory are the linear dependence of k on D, which has not been confirmed experimentally, and the unknown thickness of the film h, which does not allow theoretical determination of the mass transfer coefficient. In spite of that, some prerequisites and consequences of the theory are still valid. Examples of that are the assumptions that mass transfer takes place in a thin layer at the phase boundary, the existence of a thermodynamic equilibrium at the interphase, as well as the basic consequence of the theory regarding the additivity of the diffusion resistances [9, 10]. Similar theories are proposed for gas–liquid and liquid–liquid systems by Langmuir [11], Lewis and Whitman [12].

Another approximation is used in Higbie's *penetration theory* [13] and in some related theories, where it is assumed that the mass transfer is non-stationary in a coordinate system, which moves with velocity u_0:

$$\frac{\partial c}{\partial \tau} = D \frac{\partial^2 c}{\partial y^2}, \quad \tau = \frac{x}{u_0}. \tag{1.1.10}$$

This case is equivalent to mass transfer in a thin layer δ with a constant fluid velocity u_0:

$$u = u_0, \quad v = 0, \quad \delta \ll h < l. \tag{1.1.11}$$

As a result, from (1.1.4), the following is obtained:

$$u_0 \frac{\partial c}{\partial x} = D \frac{\partial^2 c}{\partial y^2}; \tag{1.1.12}$$

$$x = 0, \quad c = c_0; \quad y = 0, \quad c = c^*; \quad y \to \infty, \quad c = c_0.$$

The solution of (1.1.12) is obtained through the Green function [6, 9, 10]:

$$c = c_0 + (c^* - c_0)\mathrm{erfc}\, y \sqrt{\frac{u_0}{4Dx}}, \tag{1.1.13}$$

i.e.

$$\mathrm{Sh} = \frac{2}{\sqrt{\pi}} \mathrm{Pe}^{1/2}, \quad \mathrm{Pe} = \frac{u_0 l}{D}. \tag{1.1.14}$$

In fact, the velocity in the thin layer is usually variable [9, 10, 14, 15] and (1.1.14) presents the zero approximation in the solution of the problem only.

The first attempt to introduce the flow hydrodynamics near solid interface is made by Prandtl [16] and Taylor [17]. They suppose that in turbulent conditions, a laminar flow (like Kouett's flow) exists near a solid interface.

The *surface renewal theory*, proposed by Kishinevsky [18, 19] and Danckwerts [20], has the biggest development. The base of this theory is the idea about replacement of the liquid "elements" at the interface with the liquid "elements" from the volume, as a result of the turbulent stirring. The stay residence time at the interface of each "element" $\Delta \tau$ is a constant (Kishinevsky) or a spectrum of values (Danckwerts). The parameter $\Delta \tau$ cannot be calculated from the theory and must be obtained using experimental data. It replaces turbulent pulsation fading law in the viscous sub-layer in the turbulent mass transfer theory. In the time period $\Delta\tau$, the mass transfer rate is defined by the non-stationary diffusion.

There are different variants of the penetration and renewal theories, as the film-penetration model of Toor and Marchelo [21], Ruckenstein [22–24] et al. Unfortunately, the introduction of the non-stationary mechanism in the model theories has no clear physical basis. The presence of parameters calculated on the bases of experimental data leads to a good agreement with the experimental results, but gives not possibility for prediction of the process behavior in new conditions.

The theoretical analysis of the turbulent mass transfer shows that the calculation of the mass transfer rate is possible if the turbulent pulsation fading law in the viscous sub-layer $(D_{turb} \sim y^n)$ is known. Different values of n ($n = 2$ [25], $n = 3$ [26–29], $n = 4$ [30–34], $n = 5$ [35]) are proposed in the literature. Obviously, accumulation of additional experimental data is necessary.

Several other model theories of linear mass transfer exist that do not differ principally from the ones discussed above, and have the same disadvantage of insufficient physical background of their basic prerequisites. In this sense, the theory of mass transfer in the approximations of the boundary layer has the best physical justification.

1.1.2 Boundary Layer Theory

The interphase mass transfer in gas (liquid)–solid systems [7, 36, 37] is realized through an immobile phase boundary. A potential flow with a constant velocity u_0 on a semi-infinite flat plate is discussed below. In this case, from the boundary layer approximation [7], the following is directly obtained:

$$u\frac{\partial u}{\partial x} + v\frac{\partial u}{\partial y} = v\frac{\partial^2 u}{\partial y^2}, \quad \frac{\partial u}{\partial x} + \frac{\partial v}{\partial y} = 0, \quad u\frac{\partial c}{\partial x} + v\frac{\partial c}{\partial y} = D\frac{\partial^2 c}{\partial y^2};$$

$$x = 0, \quad u = u_0, \quad c = c_0; \quad y = 0, \quad u = 0, \quad v = 0, \quad c = c^*; \qquad (1.1.15)$$

$$y \to \infty, \quad u = u_0, \quad c = c_0,$$

where the boundary conditions express a thermodynamic equilibrium at the phase boundary ($y = 0$) and, depending on the sign of the difference ($c^* - c_0$), a process of solution or crystallization takes place. The problem (1.1.15) has a solution if the following similarity variables are used:

$$u = 0.5u_0\varepsilon\varphi', \quad v = 0.5\left(\frac{u_0 v}{x}\right)^{0.5}(\eta\varphi' - \varphi), \quad c = c_0 + (c^* - c_0)\psi,$$

$$y = \eta\left(\frac{u_0}{4Dx}\right)^{-0.5}, \quad \varepsilon = Sc^{0.5}, \quad \varphi = \varphi(\eta), \quad \psi = \psi(\eta). \qquad (1.1.16)$$

The introduction of (1.1.16) into (1.1.15) leads to:

$$\varphi''' + \varepsilon^{-1}\varphi\varphi'' = 0, \quad \varphi'' + \varepsilon\varphi\psi' = 0,$$
$$\varphi(0) = 0, \quad \varphi'(0) = 0, \quad \psi(0) = 1, \quad \varphi'(\infty) = 2\varepsilon^{-1}, \quad \psi(\infty) = 0.$$
(1.1.17)

The solution of (1.1.17) is obtained [36] through the Blasius function $f(z)$:

$$\varphi(\eta) = f(z), \quad z = \frac{2}{\varepsilon}\eta, \quad \psi(\eta) = 1 - \frac{1}{\varphi}\int_0^z E(\varepsilon, p)\,\mathrm{d}p,$$

$$E(\varepsilon, p) = \exp\left[-\frac{\varepsilon^2}{2}\int_0^p f(s)\,\mathrm{d}s\right],$$
(1.1.18)

$$\varphi = \int_0^\infty E(\varepsilon, p)\,\mathrm{d}p \approx \begin{cases} 3,01 Sc^{-0.35} & -\text{ for gases} \\ 3,12 Sc^{-0.34} & -\text{ for liquids} \end{cases},$$

where the function of Blasius is the solution of the problem:

$$2f''' + ff'' = 0, \quad f(0) = 0, \quad f'(0) = 0, \quad f''(0) = 0.33205 \quad (1.1.19)$$

and its values are given in [38].

The introduction of (1.1.18) in (1.1.6) allows the Sherwood number to be determined:

$$\mathrm{Sh} = \frac{kL}{D} = -\mathrm{Pe}^{0.5}\psi'(0) \approx \frac{2}{3}\sqrt{Re}\sqrt[3]{Sc}. \quad (1.1.20)$$

1.1.3 Two-Phase Boundary Layers

The mass transfer in gas–liquid and liquid–liquid systems is realized at a moving phase boundary [6]. In the approximations of the boundary layer theory, the problem has the form:

$$u_j\frac{\partial u_j}{\partial x} + v_j\frac{\partial u_j}{\partial y} = v_j\frac{\partial^2 u_j}{\partial y^2}, \quad \frac{\partial u_j}{\partial x} + \frac{\partial v_j}{\partial y} = 0,$$
$$u_j\frac{\partial c_j}{\partial x} + v_j\frac{\partial c_j}{\partial y} = D_j\frac{\partial^2 c_j}{\partial y^2}, \quad j = 1,2,$$
(1.1.21)

with boundary conditions taking into account the continuity of velocity, stress tensor, and mass flux at the phase boundary:

$$x = 0, \quad u_j = u_{j0}, \quad c_j = c_{j0}, \quad j = 1, 2; \quad y = 0, \quad u_1 = u_2,$$

$$\mu_1 \frac{\partial u_1}{\partial y} = \mu_2 \frac{\partial u_2}{\partial y}; \quad c_1 = \chi c_2, \quad D_1 \frac{\partial c_1}{\partial y} = D_2 \frac{\partial c_2}{\partial y}, \quad v_j = 0, \quad j = 1, 2;$$

$$y \to \infty, \quad u_1 = u_{10}, \quad c_1 = c_{10}; \quad y \to -\infty, \quad u_2 = u_{20}, \quad c_2 = c_{20},$$

$$(1.1.22)$$

where the first phase ($j = 1$) is a gas or a liquid, and the second one ($j = 2$)—liquid. At the phase boundary, the existence of phase equilibrium is assumed and χ is the coefficient of Henry (gas–liquid) or the coefficient of separation (liquid–liquid).

The average rate of the mass transfer between the phases is determined in an analogous way, as well as the rate of mass transfer through finding the average of the local mass fluxes:

$$J = K_1(c_{10} - \chi c_{20}) = \frac{1}{L} \int_0^L I_1 \mathrm{d}x = k_1\left(c_{10} - c_1^*\right) = K_2\left(\frac{c_{10}}{\chi} - c_{20}\right)$$

$$(1.1.23)$$

$$= \frac{1}{L} \int_0^L I_2 \mathrm{d}x = k_2\left(c_2^* - c_{20}\right), \quad c_1^* = \chi c_2^*, \quad K_2 = \chi K_1,$$

where K_j ($j = 1, 2$) are the interphase mass transfer coefficients, k_j ($j = 1, 2$)—mass transfer coefficients, and c_1^* and c_2^*—the concentrations in the two phases at the phase boundary ($y = 0$). The local mass fluxes

$$I_j = -D_j\left(\frac{\partial c_j}{\partial y}\right)_{y=0}, \quad j = 1, 2 \qquad (1.1.24)$$

are obtained after the solution of (1.1.21) and (1.1.22).

From (1.1.23) and (1.1.24), the Sherwood numbers are obtained:

$$\mathrm{Sh}_j = \frac{K_j L}{D_j} = \frac{\chi^{j-1}}{c_{10} - \chi c_{20}} \int_0^L \left(\frac{\partial c_j}{\partial y}\right)_{y=0} \mathrm{d}x, \quad j = 1, 2. \qquad (1.1.25)$$

The relation between the interphase mass transfer coefficients and mass transfer coefficients in the phases is directly obtained from (1.1.23):

$$K_1^{-1} = k_1^{-1} + \chi k_2^{-1}; \quad \chi = 0, \quad K_1 = k_1;$$

$$K_2^{-1} = (\chi k_1)^{-1} + k_2^{-1}; \quad \chi \to \infty, \quad K_2 = k_2, \qquad (1.1.26)$$

From (1.1.26), it can be seen that when the interphase mass transfer rate is limited by the diffusion resistance in one of the phases, the interphase mass transfer coefficient is equal to the mass transfer coefficient in this phase.

The problem (1.1.21), (1.1.22) has a solution after introducing the following similarity variables:

$$u_j = 0.5ju_{j0}\varepsilon_j\varphi_j', \quad v_j = (-1)^{j-1}0.5j\left(\frac{u_{j0}v_j}{x}\right)^{0.5}\left(\eta_j\varphi_j' - \varphi_j\right),$$

$$c_j = c_{j0} - (-\chi)^{1-j}(c_{10} - \chi c_{20})\psi_j, \quad \varphi_j = \varphi_j(\eta_j), \quad \psi_j = \psi_j(\eta_j), \tag{1.1.27}$$

$$\eta_j = (-1)^{j-1}y\left(\frac{u_{j0}}{4D_jx}\right)^{0,5}, \quad \varepsilon_j = Sc_j^{0.5}, \quad Sc_j = \frac{v_j}{D_j}, \quad j = 1,2.$$

As a result, the following is directly obtained:

$$\varphi_j''' + j\varepsilon_j^{-1}\varphi_j\varphi_j'' = 0, \quad \psi_j'' + j\varepsilon_j\varphi_j\psi_j' = 0,$$

$$\varphi_j(0) = 0, \quad \varphi_j'(\infty) = \frac{2}{j\varepsilon_j}, \quad \psi_j(\infty) = 0, \quad j = 1,2,$$

$$\varphi_1'(0) = 2\theta_1\frac{\varepsilon_2}{\varepsilon_1}\varphi_2'(0), \quad \varphi_2''(0) = -0,5\,\theta_2\left(\frac{\varepsilon_1}{\varepsilon_2}\right)^2\varphi_1''(0), \tag{1.1.28}$$

$$\psi_1'(0) = \frac{\chi}{\varepsilon_0}\psi_2'(0), \quad \psi_1(0) + \psi_2(0) = 1,$$

$$\theta_1 = \frac{u_{20}}{u_{10}}, \quad \theta_2 = \left(\frac{\mu_1}{\mu_2}\right)\left(\frac{v_1}{v_2}\right)^{-0.5}\left(\frac{u_{10}}{u_{20}}\right)^{1.5}, \quad \varepsilon_0 = \left(\frac{D_2\,u_{20}}{D_1\,u_{10}}\right)^{0.5}.$$

The solution of (1.1.28) allows the determination of the interphase mass transfer rate between two phases with moving phase boundary:

$$Sh_j = -\sqrt{Pe_j}\psi_j'(0), \quad Pe_j = \frac{u_{j0}L}{D_j}, \quad j = 1,2. \tag{1.1.29}$$

The problem (1.1.28) is solved numerically [39], but for the systems gas–liquid an asymptotic solution using the perturbation method is found [40, 41] in a series of the orders of the small parameters θ_1, θ_2 and for $\psi_j'(0)(j = 1,2)$, the following expressions are obtained in first approximation, regarding the small parameters θ_1 and θ_2:

$$\psi_1'(0) = -\frac{2}{\varepsilon_1\varphi_{10}}\frac{1}{1+a} - \frac{2\theta_1}{\alpha\varphi_{10}^2\varepsilon_1}\frac{1}{(1+a)^2} - 8\theta_2\alpha\frac{\varepsilon_2}{\varepsilon_1}\frac{\bar{\varphi}_2}{\varphi_{10}}\frac{a}{(1+a)^2},$$

$$\psi_2'(0) = -\frac{2}{\sqrt{\pi}}\frac{a}{1+a} - \theta_1\frac{2}{\sqrt{\pi}\alpha\varphi_{10}}\frac{a}{(1+a)^2} - 8\theta_2\frac{\alpha\varepsilon_2\bar{\varphi}_2}{\sqrt{\pi}}\frac{a^2}{(1+a)^2}, \tag{1.1.30}$$

where

$$\varphi_{10} \approx \frac{3}{\sqrt[3]{Sc}}, \quad a = \frac{\chi\sqrt{\pi}}{\varepsilon_0\varepsilon_1\varphi_{10}}, \quad \alpha = 0.33205, \quad \bar{\varphi}_2 = \frac{1}{8}\sqrt{\frac{\pi}{Sc_2}}. \quad (1.1.31)$$

In the cases when the interphase mass transfer is limited by the mass transfer in the gas phase $\chi/\varepsilon_0 \to 0$, $a \to 0$ and for the Sherwood number, the following equation may be written:

$$Sh_1 = \sqrt{Pe_1}\left(\frac{2}{\varepsilon_1\varphi_{10}} + \frac{2\theta_1}{\varepsilon_1\alpha\varphi_{10}^2}\right). \quad (1.1.32)$$

The Sherwood number can be determined in a similar way when the interphase mass transfer is limited by the mass transfer in the liquid phase:

$$Sh_2 = \sqrt{Pe_2}\left(\frac{2}{\sqrt{\pi}} + 8\theta_2\frac{\alpha\varepsilon_2\bar{\varphi}_2}{\sqrt{\pi}}\right). \quad (1.1.33)$$

The results obtained for the hydrodynamics and the mass transfer in co-current flows for a gas–liquid system are in a good agreement with the experimental data [39, 42–44].

1.2 Mass Transfer in Countercurrent Flows

The chemical technologies based on countercurrent flows in gas–liquid systems are widely spread in practice. The theoretical analysis of such flows [45] demonstrates that there is a possibility to obtain asymptotic solutions for gas–liquid systems which are in conformance with the experimental data obtained from thermo-anemometrical measurements in the boundary layer. The correctness of the proposed asymptotic method [45] was confirmed by numerical experiments, as a result of the exact solution of the problem by means of numerical simulation [46]. The theoretical analysis of the countercurrent flow shows [47] that it is a non-classical problem of mathematical physics, which is not sufficiently discussed in the literature. A prototype of such problem is the parabolic boundary value problem with changing direction of time [48, 49]. It was shown [47] that this non-classical problem can be described as consisting of several classical problems.

1.2.1 Velocity Distribution

The mathematical description of the countercurrent flow in the approximation of the boundary layer theory has the following form:

$$u_j \frac{\partial u_j}{\partial x} + v_j \frac{\partial u_j}{\partial y} = v_j \frac{\partial^2 u_j}{\partial y^2}, \quad \frac{\partial u_j}{\partial x} + \frac{\partial v_j}{\partial y} = 0, \quad j = 1, 2;$$

$$x = 0, \quad y \geq 0, \quad u_1 = u_1^\infty; \quad x = l, \quad y \leq 0, \quad u_2 = -u_2^\infty;$$

$$y \to \infty, \quad 0 \leq x \leq l, \quad u_1 = u_1^\infty; \quad y \to -\infty, \quad 0 \leq x \leq l, \quad u_2 = -u_2^\infty; \tag{1.2.1}$$

$$y = 0, \quad 0 < x < l, \quad u_1 = u_2, \quad \mu_1 \frac{\partial u_1}{\partial y} = \mu_2 \frac{\partial u_2}{\partial y}, \quad v_1 = v_2 = 0.$$

The problem (1.2.1) can be presented in dimensionless form using two different coordinate systems for the two phases, so that the flow in each phase is oriented to the longitudinal coordinate, and the following dimensionless variables and parameters are introduced:

$$x = lX_1 = l - lX_2, \quad y = \delta_1 Y_1 = -\delta_2 Y_2,$$

$$u_1 = u_1^\infty U_1, \quad v_1 = u_1^\infty \frac{\delta_1}{l} V_1, \quad u_2 = -u_2^\infty U_2, \quad v_2 = -u_2^\infty \frac{\delta_2}{l} V_2,$$

$$\delta_j = \sqrt{\frac{v_j l}{u_j^\infty}}, \quad j = 1, 2, \quad \theta_1 = \frac{u_2^\infty}{u_1^\infty}, \quad \theta_2 = \left(\frac{\rho_1 \mu_1}{\rho_2 \mu_2}\right)^{1/2} \left(\frac{u_1^\infty}{u_2^\infty}\right)^{3/2}. \tag{1.2.2}$$

In the new coordinate systems, the model of countercurrent flows takes the following form:

$$U_j \frac{\partial U_j}{\partial X_j} + V_j \frac{\partial U_j}{\partial Y_j} = \frac{\partial^2 U_j}{\partial Y_j^2}, \quad \frac{\partial U_j}{\partial X_j} + \frac{\partial V_j}{\partial Y_j} = 0;$$

$$X_j = 0, \quad U_j = 1; \quad Y_j \to \infty, \quad U_j = 1; \tag{1.2.3}$$

$$Y_1 = Y_2 = 0, \quad U_1 = -\theta_1 U_2, \quad \theta_2 \frac{\partial U_1}{\partial Y_1} = \frac{\partial U_2}{\partial Y_2}, \quad V_j = 0; \quad j = 1, 2.$$

The problem (1.2.3) cannot be solved directly, because the velocities U_i ($i = 1$, 2) change their directions in the ranges $0 \leq X_i \leq 1, 0 \leq Y_i < \infty$ ($i = 1, 2$). This non-classical problem of mathematical physics can be presented as a classical one after the introduction of the following similarity variables:

$$U_j = f_j', \quad V_j = \frac{1}{2\sqrt{X_j}} \left(\eta_j f_j' - f_j\right), \quad f_j = f_j(\eta_j), \quad \eta_j = \frac{Y_j}{\sqrt{X_j}}. \tag{1.2.4}$$

The substitution of (1.2.4) into (1.2.3) leads to:

$$2f_j''' + f_j f_j'' = 0, \quad f_j(0) = 0, \quad f_j(\infty) = 1, \quad j = 1, 2,$$

$$f_1'(0) = -\theta_1 f_2'(0), \quad \theta_2 \sqrt{\frac{X_2}{X_1}} f_1''(0) = f_2''(0), \quad X_1 + X_2 = 1. \tag{1.2.5}$$

It is obvious from (1.2.5) that the problem (1.2.3) has no solution in similarity variables.

However, the problem (1.2.5) can be solved after the introduction of new parameter $\bar{\theta}_2$ for each $X_1 \in (0, 1)$:

$$\bar{\theta}_2 = \theta_2 \sqrt{\frac{1 - X_1}{X_1}}, \tag{1.2.6}$$

i.e., the problem has a local similarity solution. In this way, the problem (1.2.5) is substituted by several separate problems for each $X_1 \in (0, 1)$.

The solutions of these separate problems can be obtained after the introduction of the function

$$F(a, b) = \int_6^7 (f_1' - 1)^2 d\eta_1 + \int_6^7 (f_2' - 1)^2 d\eta_2, \quad a = f_1'(0), \quad b = f_1''(0). \tag{1.2.7}$$

The solution of (1.2.5) for each $X_1 \in (0, 1)$ is obtained after searching the minimum of the function $F(a, b)$, where at each step of the minimization procedure the boundary problem has to be solved:

$$2f_j''' + f_j f_j'' = 0, \quad f_j(0) = 0, \quad j = 1, 2,$$

$$f_1'(0) = a, \quad f_2'(0) = -\frac{a}{\theta_1}, \quad f_1''(0) = b, \quad f_2''(0) = \bar{\theta}_2 b. \tag{1.2.8}$$

The problem (1.2.8) was solved numerically for countercurrent gas and liquid flows for the following parameter values $\theta_1 = 0.1$ and $\theta_2 = 0.152$. In accordance with the requirement for minimum of $F(a, b)$ in (0.2.7), the boundary conditions a, b and $F(a, b)$ were determined [46].

The energy dissipated in the laminar boundary layer [6, 7] is described for both phases by the equations:

$$e_j = \mu_j \int_0^l \int_0^{(-1)^{j+1}} \left(\frac{\partial u_j}{\partial y}\right)^2 dxdy, \quad E_j = -\int_0^1 \int_0^\infty \left(\frac{\partial U_j}{\partial Y_j}\right)^2 dY_j dX_j,$$

$$E_j = \frac{e_j \sqrt{\frac{u_j^\infty l}{\nu_i}}}{\nu_i \rho_i u_j^{\infty 2}}, \quad j = 1, 2. \tag{1.2.9}$$

For the case of gas–liquid countercurrent flows, the introduction of similarity variables leads to:

$$E_j = \int_0^1 \frac{1}{\sqrt{X_j}} \left[\int_0^\infty \left(f_j''\right)^2 d\eta_j\right] dX_j, \quad j = 1, 2. \tag{1.2.10}$$

In the case of co-current flows, $f_j''^*$ does not depend on X_j and for energy dissipation the following is obtained:

$$E_j^* = 2 \int_0^\infty \left(f_j''^*\right)^2 d\eta_i, \quad i = 1, 2, \tag{1.2.11}$$

where $f_i^*\,(i = 1, 2)$ is the solution of Eq. (1.2.8) with boundary conditions for co-current flows:

$$\theta_1^* = -0.1, \quad \theta_2^* = \theta_2 = 0.152,$$
$$f_1'^*(0) = 0.0908, \quad f_1''^*(0) = 0.32765. \tag{1.2.12}$$

The comparison of the energy dissipated in the laminar boundary layer [6, 7] for the case of gas–liquid countercurrent and co-current flows is shown in Table 1.1. These results show that the energy dissipation for the gas phase in co-current flows is lower that in the countercurrent flows, while in the second (liquid) phase there is no significant change.

Table 1.1 Comparison between countercurrent and co-current flows

$\theta_3 = 0$		$\theta_3 \to \infty$		$\theta_3 = 1$			
J_1	J_1^*	J_2	J_2^*	J_1	J_1^*	J_2	J_2^*
0.554	0.720	4.380	4.822	0.432	0.626	0.432	0.626
A_1	A_1^*	A_2	A_2^*	A_1	A_1^*	A_2	A_2^*
1.06	1.57	739	750	0.82	1.37	72.8	97.3
E_1	E_1^*	E_2	E_2^*	E_1	E_1^*	E_2	E_2^*
0.525	0.458	0.00593	0.00643	0.525	0.458	0.01328	0.00643

1.2.2 Concentration Distribution

The mathematical model of mass transfer in gas–liquid systems with countercurrent flow in a laminar boundary layer with flat phase boundary takes the following form:

$$u_j \frac{\partial c_j}{\partial x} + v_j \frac{\partial c_j}{\partial y} = D_j \frac{\partial^2 c_j}{\partial y^2}, \quad j = 1, 2;$$

$$x = 0, \quad y \geq 0, \quad c_1 = c_1^\infty; \quad x = l, \quad y \leq 0, \quad c_2 = c_2^\infty;$$

$$y \to \infty, \quad 0 \leq x \leq l, \quad c_1 = c_1^\infty; \quad y \to -\infty, \quad 0 \leq x \leq l, \quad c_2 = c_2^\infty;$$

$$y = 0, \quad 0 < x < l, \quad c_1 = \chi c_2, \quad D_1 \frac{\partial c_1}{\partial y} = D_2 \frac{\partial c_2}{\partial y},$$

(1.2.13)

where u_j and v_j ($j = 1, 2$) are the velocity components in the gas and in the liquid phase determined through solving (1.2.8).

Solving problem (1.2.13) should be carried out [50] after the following similarity variables are introduced:

$$\eta_j = (-1)^{j+1} y \sqrt{\frac{u_j^\infty}{v_j l X_j}}, \quad X_1 = \frac{x}{l}, \quad X_2 = \frac{l-x}{l}, \quad X_1 + X_2 = 1,$$

$$u_j = (-1)^{j+1} u_j^\infty f_j', \quad v_j = (-1)^{j+1} \frac{1}{2} \sqrt{\frac{v_j u_j^\infty}{l X_j}} \left(\eta_j f_j' - f_j \right),$$

(1.2.14)

$$f_j = f_j(\eta_j), \quad c_j = c_j^\infty - \chi^{1-j} \left(c_1^\infty - \chi c_2^\infty \right) \psi_j,$$

$$\psi_j = \psi_j(\eta_j), \quad j = 1, 2.$$

The substitution of Eq. (1.2.14) into Eq. (1.2.13) leads to:

$$2f_j''' + f_j f_j'' = 0, \quad 2\psi_j'' + Sc_j f_j \psi_j' = 0, \quad j = 1, 2,$$

$$f_1(0) = 0, \quad f_1'(0) = a, \quad f_1''(0) = b,$$

$$f_2(0) = 0, \quad f_2'(0) = -\frac{a}{\theta_1}, \quad f_2''(0) = \bar{\theta}_2 b,$$

(1.2.15)

$$\psi_1(0) + \psi_2(0) = 1, \quad \bar{\theta}_3 \psi_1'(0) = \psi_2'(0), \quad \psi_j(\infty) = 0, \quad j = 1, 2,$$

where

$$Sc_j = \frac{v_j}{D_j} \quad (j = 1, 2), \quad \theta_1 = \frac{u_2^\infty}{u_1^\infty}, \quad \theta_2 = \left(\frac{\rho_1 \mu_1}{\rho_2 \mu_2} \right)^{\frac{1}{2}} \left(\frac{u_1^\infty}{u_2^\infty} \right)^{\frac{3}{2}},$$

$$\bar{\theta}_2 = \theta_2 \sqrt{\frac{X_2}{X_1}}, \quad \theta_3 = \chi \frac{D_1}{D_2} \sqrt{\frac{u_1^\infty v_2}{u_2^\infty v_1}}, \quad \bar{\theta}_3 = \theta_3 \sqrt{\frac{X_2}{X_1}}.$$

(1.2.16)

The boundary conditions a and b are determined after the minimization of (1.2.7).

It is clearly seen from Eq. (1.2.15) that it is possible to obtain the similarity solution for different values of $X_1 = 1 - X_2$.

The solution of (1.2.15) was carried out at new boundary conditions for ψ_j $(j = 1, 2)$:

$$\psi_1(0) = \alpha, \quad \psi_1'(0) = \beta, \quad \psi_2(0) = 1 - \alpha, \quad \psi_2'(0) = \bar{\theta}_3\beta, \qquad (1.2.17)$$

where α and β are determined for different values of $X_1 = 1 - X_2$ so that the conditions $\psi_j(\infty) = 0$ $(j = 1, 2)$ be fulfilled.

The Sherwood number can be obtained analogously to (1.1.25):

$$\mathrm{Sh}_j = -\sqrt{Re_j} \int\limits_0^1 \frac{\psi_1'(0)}{\sqrt{X_j}}\,dX_j, \quad Re_j = \frac{u_j^\infty L}{\nu_j}, \quad j = 1, 2, \qquad (1.2.18)$$

where the dimensionless diffusion flux has the form:

$$J_j = -\int\limits_0^1 \frac{\psi_j'(0)}{\sqrt{X_j}}\,dX_j, \quad j = 1, 2. \qquad (1.2.19)$$

1.2.3 Comparison Between Co-current and Countercurrent Flows

For the purpose of comparing the mass transfer rates in the countercurrent and co-current flows, Eq. (1.2.15) should be solved using parameter's values corresponding to co-current flow:

$$\theta_1 = -0.1, \quad \theta_2 = 0.152, \quad f_1'(0) = 0.0908, \quad f_1''(0) = 0.37265,$$
$$\bar{\theta}_3 = \theta_3, \quad J_j^* = -2\psi_j'(0), \quad j = 1, 2. \qquad (1.2.20)$$

The results obtained for J_1^* $(j = 1, 2)$ are shown in Table 1.1. The comparison of these results with values corresponding to co-current flow shows that the co-current flow mass transfer rate is higher than in case of countercurrent flow.

The obtained numerical results allow determining the ratio (A) between mass transfer rate (Sh) and corresponding energy dissipation (E) in cases of counter-current and co-current flows:

$$A_i = \frac{Sh_i}{E_i}, \quad A_i^* = \frac{Sh_i^*}{E_i^*}, \quad i = 1, 2. \tag{1.2.21}$$

A comparison between countercurrent and co-current flows is shown in Table 1.1. The data demonstrate higher efficiency of the co-current flow, i.e., higher mass transfer rate, at equal energy losses [50].

The presented theoretical analysis shows that the linear mass transfer theory allows the prediction of the mass transfer resistance distribution in two-phase systems. In the approximations of the diffusion boundary layer theory of the mass transfer, the criterion is the parameter χ/ε_0 (see (1.1.28)); i.e., the process is limited by the mass transfer in the first phase if $\chi/\varepsilon_0 \geq 10^2$. In the cases when $\chi/\varepsilon_0 \leq 10^2$ the process is limited by the mass transfer in the second phase. The diffusion resistances are comparable if $\chi/\varepsilon_0 \sim 1$. In the approximations of the film mass transfer theory, the criterion is χ only (see 1.1.26).

The results obtained up to now represent the linear theory of mass transfer in the approximations of the boundary layer with a flat phase boundary. In an analogous way, these problems can be solved for different forms of the interphase surface (wave, sphere, cylindrical, etc.) in the processes in film flows, droplets, bubbles, jets, etc.

1.3 Nonlinear Mass Transfer

The theoretical analysis of the linear mass transfer shows that this process occurs when the equation of convection–diffusion (1.1.4) is linear; i.e., the velocity (u, v) and the diffusivity (D) do not depend on the concentration (c) of the transferred substance. These conditions are valid for systems, in which the mass transfer does not influence the hydrodynamics and the dependence of the mass flux on the concentration gradient is linear. The linear theory of mass transfer, built on this basis, has two main consequences: *The mass transfer coefficient does not depend on the concentration, and the interphase mass transfer direction does not influence the mass transfer rate.*

Any deviation of experimental data from these two consequences shows the availability of nonlinear effects, which can occur as a result of secondary flows, caused by the mass transfer, or the dependence of viscosity, diffusivity, and density on concentration [51]. The secondary flows can appear as a result of the concentration gradient (nonlinear mass transfer), surface tension gradient (Marangoni effect), density gradient (natural convection), and pressure gradient (Stephan flow). In these conditions, the concentration influences the velocity field and the convection–diffusion equation becomes nonlinear. The secondary flows may influence the mass transfer rate through a change in the velocity field and therefore in the balance between the convective and diffusive transfer in the equation of convection–diffusion. This effect may increase significantly, if as a result of secondary flows the system loses stability and reaches a new state, becoming a self-organizing dissipative structure.

One of the most interesting nonlinear effects arises from the conditions imposed by high concentration gradients. The latter induce secondary flows at the phase boundaries. This effect has been discussed in detail in this book for a large number of systems taken as examples, and it has been termed "nonlinear mass transfer effect."

The modern development of power, chemical, oil processing, food processing, and other industries calls for creation of systems with intensive mass transfer. For this purpose, it is possible to use the mass transfer processes in two-phase systems at large concentration gradients of the transferred substance. Under these conditions, big mass fluxes of substance through the phase boundary induce secondary flows at the interphase surface. This fact leads to a change in the flow hydrodynamics that influences significantly the mechanism and the kinetics of mass transfer.

The analysis of the mechanism and kinetics of the interphase mass transfer in two-phase systems is possible in many cases, if the velocity distribution is determined in the beginning, and then, after its substitution in the equation of convection–diffusion, the rate of interphase mass transfer is found. However, this procedure cannot be used for systems with large concentration gradients due to the flow at the interphase surface induced by the mass transfer. The velocity of this flow is perpendicular to the interphase surface and is directed toward the mass transfer direction. The physical cause of this movement is the mechanical impulse that is transferred from one phase to the other through the particles, responsible for the mass transfer. In the linear theory of mass transfer, this impulse is considered insignificant. However, at big mass fluxes through the interphase surface in cases of intensive mass transfer it should be considered. Since the transferred impulse is proportional to the diffusive flux of particles taking part in the mass transfer, the velocity field close to the interphase boundary depends on the concentration field. The concrete form of this dependency is determined by the system of equations of the joint transfer of mass and quantity of movement, as well as by the boundary conditions connecting the fluxes of mass and quantity of movement at the interphase boundary.

1.3.1 Influence of the Intensive Interphase Mass Transfer on the Hydrodynamics

The velocity of the induced flow at the interphase surface is determined by the hydrodynamic effect of the intensive mass transfer. This effect reflects, first of all, on the boundary conditions to the equations of hydrodynamics and mass transfer; i.e., these equations cannot be solved separately, unlike the case of systems with low rates of the interphase mass transfer. In order to find the connection between the velocity of the induced flow at the phase boundary and the rate of the interphase mass transfer, an example of an isothermal process of transfer of a dissolved substance from phase 1 into phase 2 will be discussed. It is assumed that each of the phases is a two-component mixture (a solution of the substance m in the corresponding solvent) and that the two solvents do not mix with each other. The diffusion flux $\mathbf{j}_m^{(i)}$ of the substance m at each point of the space inside the phase i is determined in the following way:

$$\mathbf{j}_m^{(i)} = M_m c_m^{(i)} \left(\mathbf{v}_m^{(i)} - \mathbf{v}^{(i)} \right), \quad i = 1, 2. \tag{1.3.1}$$

Here, $c_m^{(i)}$ is the molar concentration of the transferred substance in phase i, M_m— the molecular mass of this substance, $\mathbf{v}_m^{(i)}$—the average statistical velocity of movement of particles of the substance m in arbitrary but fixed coordinate system, and $\mathbf{v}^{(i)}$—velocity of the mass center of the whole liquid mixture in the same coordinate system. Velocity $\mathbf{v}^{(i)}$ is defined by the system of equations of hydrodynamics (in the case of laminar flow—the system of equations of Navier–Stokes). Besides, by definition this velocity is connected to the velocities of movement of the mixture components through the relationship:

$$\rho^{(i)} \mathbf{v}^{(i)} = M_m c_m^{(i)} \mathbf{v}_m + M_0 c_0^{(i)} \mathbf{v}_0^{(i)}, \tag{1.3.2}$$

where the variables with subscript (0) refer to the corresponding solvent and $\rho^{(i)} = M_m c_m^{(i)} + M_0 c_0^{(i)}$ is the summarized density of the solution in the phase i (in the general case, this density is a function of the space coordinates and time).

Let us present each of the velocities in Eq. (1.3.2) as a sum of the velocity of movement at the interphase surface $d\mathbf{r}_s/dt$ ($\mathbf{r}_s$—radius vector of an arbitrary point at the phase boundary) and the velocity of movement in regard to this surface $\mathbf{v}_{rk}^{(i)}$ ($k = m, 0$). Due to the fact that the two solvents do not mix, the normal components of velocities $\mathbf{v}_{r0}^{(1)}$ and $\mathbf{v}_{r0}^{(2)}$ at the interphase boundary must be equal to zero. That is why Eq. (1.3.2), being projected in the normal direction to the interphase boundary, has at each point of this boundary the following form:

$$\rho^{(i)} \left(\mathbf{v}_r^{(i)}, \mathbf{n} \right) = M_m c_m^{(i)} \left(\mathbf{v}_{rm}^{(i)}, \mathbf{n} \right). \tag{1.3.3}$$

Analogously, the projection of Eq. (1.3.1) on the normal direction n [taking into account the relationship (1.3.3)] leads to the equation:

$$\left(\mathbf{v}_s^{(i)}, \mathbf{n}\right) = \left(\frac{d\mathbf{r}_s}{dt}, \mathbf{n}\right) + \frac{\left(\mathbf{j}_{ms}^{(i)}, \mathbf{n}\right)}{\rho_s^{(i)} - M_m c_{ms}^{(i)}}. \tag{1.3.4}$$

This equation is correct for each point at the interphase surface. The subscript "s" here means that the corresponding variable refers to the phase boundary. If the form of the surface is described by the equations:

$$y = f_1(x, t), \quad z = f_2(r, t), \quad r = f_3(\theta, t), \tag{1.3.5}$$

where for the first term in the right part of (1.3.4) the following expressions are correct:

$$\left(\frac{d\mathbf{r}_s}{dt}, \mathbf{n}\right) = \begin{cases} 1 + \left(\frac{\partial f_1}{\partial x}\right)^{-1/2} \frac{\partial f_1}{\partial t} & \text{in Decart coordinate system,} \\ 1 + \left(\frac{\partial f_2}{\partial r}\right)^{-1/2} \frac{\partial f_2}{\partial t} & \text{in cylindrical coordinate system,} \\ 1 + \left(\frac{\partial f_3}{f_3 \partial \theta}\right)^{-1/2} \frac{\partial f_3}{\partial t} & \text{in spherical coordinate system.} \end{cases} \tag{1.3.6}$$

From (1.3.4), it follows that in the case of high enough interphase mass transfer rates the hydrodynamic problem has no solution in spite of the problem of convection–diffusion.

In the literature, a lot of systems of practical interest are described, in which intensive mass transfer leads to a significant change in the hydrodynamic conditions of mass transfer. Good examples of such systems are: condensation of vapors on a cooled wall [52, 53], evaporation of liquids from the surface as drops and bubbles [54–58], crystallization and solution of salts [59–63], heat and mass transfer under conditions of intensive injection (suction) of gases trough a porous wall [64–68]. It is necessary to mention that the effects of nonlinear mass transfer that are further discussed are a result of large concentration gradient (in liquids) or the partial pressure (in gases) of the transferred substance; i.e., the mass flux through the phase boundary in these cases is determined by the mass transfer rate. In this sense, they differ from Stephan's flows [69, 70] and the effects of injection or suction of vapors or gases on a solid surface, which are effects from the gradient of the general pressure.

1.3.2 Boundary Conditions of the Nonlinear Mass Transfer Problem

The mathematical formulation of the mass transfer problem, taking into account the influence of mass transfer on the hydrodynamics, was given for the first time in [71–73]. Equation (1.3.4) presents the basic conclusion, where the velocity of the induced flow is determined with the help of the mass flux through the phase boundary. For the concrete cases in (1.3.4), a factual expression of the mass flux should be placed.

In the general case, one of the phases in a two-phase system in Decart coordinates can be discussed [9, 51], where $y = h(x)$ is the phase boundary. Differentiating from (1.3.4), it will be assumed that the interphase surface is constant with time; i.e., wave surfaces and surfaces of growing drops and bubbles will not be considered. The mass flux of the transferred substance at each point of the discussed phase can be expressed by means of the average statistical velocity of this substance (molecules, atoms, ions) v and the mass center velocity of the mixture (phase) particles v_1:

$$\mathbf{j} = Mc(\mathbf{v} - \mathbf{v}_1). \tag{1.3.7}$$

The velocity $\mathbf{v}_1$ should satisfy the hydrodynamic equations and should be connected with the velocities of the mixture (phase) components through the equation:

$$\rho \mathbf{v}_1 = M_0 c_0 \mathbf{v}_0 + Mc\mathbf{v}, \tag{1.3.8}$$

where ρ is the phase (mixture) density, and the subscript (0) marks the phase (mixture) parameters in the absence of a transferred substance. In this way for the density of the discussed phase, it can be written as:

$$\rho = M_0 c_0 + Mc = \rho_0 + Mc. \tag{1.3.9}$$

The projection of the vector equation (1.3.8) on the normal at the interface surface (vector n) gives:

$$\rho^*\left(\mathbf{v}_1^*, \mathbf{n}\right) = Mc^*(\mathbf{v}^*, \mathbf{n}), \tag{1.3.10}$$

where the superscript (*) denotes the value of the function at the phase boundary. In order to obtain (1.3.10), the condition for complete mutual insolubility of both phases is used:

$$(\mathbf{v}_0, \mathbf{n}) = 0. \tag{1.3.11}$$

Equation (1.3.11) expresses the availability of a normal velocity component of the liquid or gas $(\mathbf{v}_1^*, \mathbf{n})$ at the face interphase, which is determined by the diffusion rate $(\mathbf{v}^*, \mathbf{n})$. The velocity of the face interphase has a tangent component only.

The occurrence of an induced flow at the interphase surface creates a convective flow, i.e., the mass flux of the transferred substance through the interphase surface has convective and diffusive components:

$$I = -MD\left(\frac{\partial c}{\partial n}\right)_{y=h} + Mc^*\left(\mathbf{v}_1^*, \mathbf{n}\right), \tag{1.3.12}$$

where $\partial/\partial n$ is the derivative in direction normal to the interphase surface.

The diffusion component may be expressed by means of projection of the vector Eq. (1.3.7) on the normal vector to the surface:

$$(\mathbf{j}^*, \mathbf{n}) = -MD\left(\frac{\partial c}{\partial n}\right)_{y=h} = Mc^*(\mathbf{v}, \mathbf{n}) - Mc^*\left(\mathbf{v}_1^*, \mathbf{n}\right). \tag{1.3.13}$$

From Eqs. (1.3.10)–(1.3.13), it is obtained:

$$I = \rho^*\left(\mathbf{v}_1^*, \mathbf{n}\right) = -\frac{\rho^*}{\rho^* - Mc^*}MD\left(\frac{\partial c}{\partial n}\right)_{y=h}, \tag{1.3.14}$$

where

$$\rho^* = M_0 c_0^* + Mc^* = \rho_0^* + Mc^*. \tag{1.3.15}$$

For small concentrations of the transferred substance

$$\rho_0^* \approx \rho_0. \tag{1.3.16}$$

The expression (1.3.14) may be presented in the form:

$$I = \rho^*\frac{v^* - h'u^*}{\sqrt{1+h'^2}} = \frac{MD\rho^*}{\rho_0^*}\frac{h'\left(\frac{\partial c}{\partial x}\right)_{y=h} - \left(\frac{\partial c}{\partial y}\right)_{y=h}}{\sqrt{1+h'^2}}, \tag{1.3.17}$$

where u^* and v^* are the components of the velocity v_1 along the x- and y-axes correspondingly.

Equation (1.3.17) gives the relation between the gas or liquid velocity at the face interphase and the concentration gradient of the transferred substance and will be further used as a boundary condition for the Navier–Stokes equations. In the approximations of the linear theory of mass transfer, (0.3.17) represents the condition of "non-leakage" through the face interphase:

$$\frac{v^* - h'u^*}{\sqrt{1 + h'^2}} = 0. \tag{1.3.18}$$

The processes of nonlinear mass transfer, heat transfer, and multicomponent mass transfer in gas (liquid)–solid surface systems will be discussed on the example of the longitudinal streaming of a semi-infinite flat plate in the approximations of the boundary layer theory. The nonlinear effect is taken into account through the introduction of the velocity of the induced flow v^* at the interface $y = 0$. This velocity is obtained directly from (1.3.17), if $u^* = 0$ at $h \equiv 0$ is taken into account:

$$v^* = -\frac{MD}{\rho_0^*}\left(\frac{\partial c}{\partial y}\right)_{y=0}. \tag{1.3.19}$$

1.3.3 Nonlinear Mass Transfer in the Boundary Layer

The kinetics of the nonlinear mass transfer in the approximations of the boundary layer [74–76] will be discussed based on the solution of the equations of hydrodynamics and convection–diffusion, with boundary conditions that take into consideration the influence of the mass transfer on the hydrodynamics. In a rectangular coordinate system, where $y = 0$ corresponds to the interphase surface gas (liquid)–solid, the mathematical description of the nonlinear mass transfer has the form:

$$\begin{aligned}
&u\frac{\partial u}{\partial x} + v\frac{\partial u}{\partial y} = v\frac{\partial^2 u}{\partial y^2}, \quad \frac{\partial u}{\partial x} + \frac{\partial v}{\partial y} = 0, \quad u\frac{\partial c}{\partial x} + v\frac{\partial c}{\partial y} = D\frac{\partial^2 c}{\partial y^2}; \\
&x = 0, \quad u = u_0, \quad c = c_0; \\
&y = 0, \quad u = 0, \quad v = -\frac{MD}{\rho_0^*}\frac{\partial c}{\partial y}, \quad c = c^*; \\
&y \to \infty, \quad u = u_0, \quad c = c_0,
\end{aligned} \tag{1.3.20}$$

where a potential flow, with a velocity u_0 along a plate, and a concentration (c_0) of the transferred substance are assumed. As a result of the rapid establishment of thermodynamic equilibrium, the concentration c^* is always constant on the solid surface. The normal component of the velocity at the interphase is determined from Eq. (1.3.19) as a consequence of intensive interphase mass transfer.

The mass transfer rate for a plate of length L could be determined from the average mass flux:

$$J = Mk(c^* - c_0) = \frac{1}{L} \int_0^L I \, dx, \tag{1.3.21}$$

where k is the mass transfer coefficient and I can be expressed from (1.3.17) as follows:

$$I = -\frac{MD\rho^*}{\rho_0^*} \left(\frac{\partial c}{\partial y}\right)_{y=0}. \tag{1.3.22}$$

In order to solve problem (1.3.20), it is necessary to introduce the similarity variables:

$$u = 0.5u_0\varepsilon\varphi', \quad v = 0.5\left(\frac{u_0 v}{x}\right)^{0.5}(\eta\varphi' - \varphi),$$
$$c = c_0 + (c^* - c_0)\psi, \quad y = \eta\left(\frac{u_0}{4Dx}\right)^{-0.5}, \tag{1.3.23}$$

where

$$\varepsilon = Sc^{0.5}, \quad Sc = v/D, \quad \varphi = \varphi(\eta), \quad \psi = \psi(\eta). \tag{1.3.24}$$

As a result of these substitutions, the problem (1.3.20) gets the following form:

$$\varphi''' + \varepsilon^{-1}\varphi\varphi'' = 0, \quad \psi'' + \varepsilon\varphi\psi' = 0, \quad \theta = \frac{M(c^* - c_0)}{\varepsilon\rho_0^*},$$
$$\varphi(0) = \theta\psi'(0), \quad \varphi'(0) = 0, \quad \varphi'(\infty) = 2\varepsilon^{-1}, \tag{1.3.25}$$
$$\psi(0) = 1, \quad \psi(\infty) = 0,$$

where θ is a small parameter that reflects the effect of the nonlinear mass transfer. In the linear theory of the diffusion boundary layer $\theta = 0$.

Considering the new variables and Eq. (1.3.21), the following is obtained:

$$Sh = \frac{kL}{D} = -\frac{\rho^*}{\rho_0^*} Pe^{0.5}\psi'(0), \quad Pe = \frac{u_0 L}{D}. \tag{1.3.26}$$

It is seen from (1.3.26) that mass transfer kinetics is determined by the dimensionless diffusion flux $\psi'(0)$, which can be obtained solving the problem

(1.3.25). The solution has been found utilizing a perturbation method after presenting φ and ψ as a series in power of the small parameter θ [77]:

$$\varphi = \varphi_0 + \theta\varphi_1 + \theta^2\varphi_2 + \cdots, \quad \psi = \psi_0 + \theta\psi_1 + \theta^2\psi_2 + \cdots. \tag{1.3.27}$$

If (1.3.27) is substituted in (1.3.25), a series of boundary problems that have been solved in [74] could be obtained, and for the functions in (1.3.27), the following can be written:

$$\varphi_0(\eta) = f(z), \quad \varphi_1(\eta) = -\frac{2}{\varepsilon\varphi_0}\varphi(z), \quad \varphi_2(\eta) = -\frac{2\varphi_3}{\varphi_0^3}\varphi(z) - \frac{4}{\varepsilon^2\varphi_0^2}\bar{\varphi}(z),$$

$$\psi_0(\eta) = 1 - \frac{1}{\varphi_0}\int_0^z E(\varepsilon,p)\mathrm{d}p, \quad z = \frac{2}{\varepsilon}\eta, E(\varepsilon,p) = \exp\left[-\frac{\varepsilon^2}{2}\int_0^p f(s)\mathrm{d}s\right],$$

$$\psi_1(\eta) = \frac{\varepsilon\varphi_3}{\varphi_0^3}\int_0^z E(\varepsilon,p)\mathrm{d}p - \frac{\varepsilon}{\varphi_0^2}\int_0^z\left[\int_0^p \varphi(s)\mathrm{d}s\right]E(\varepsilon,p)\mathrm{d}p,$$

$$\psi_2(\eta) = \left(-\frac{2\varepsilon^2\varphi_3^2}{\varphi_0^5} + \frac{\varepsilon^2\varphi_{33}}{2\varphi_0^4} + \frac{2\bar{\varphi}_{33}}{\varphi_0^4}\right)\int_0^z E(\varepsilon,p)\mathrm{d}p$$

$$+ \frac{2\varepsilon^2\varphi_3}{\varphi_0^4}\int_0^z\left[\int_0^p \varphi(s)\mathrm{d}s\right]E(\varepsilon,p)\mathrm{d}p - \frac{\varepsilon^2}{2\varphi_0^3}\int_0^z\left[\int_0^p \varphi(s)\mathrm{d}s\right]^2 E(\varepsilon,p)\mathrm{d}p$$

$$- \frac{2}{\varphi_0^3}\int_0^z\left[\int_0^p \bar{\varphi}(s)\mathrm{d}s\right]E(\varepsilon,p)\mathrm{d}p,$$

$$\tag{1.3.28}$$

In (1.3.28), the functions f, φ, and $\bar{\varphi}$ are solutions of the boundary problems:

$$2f''' + ff'' = 0, \quad 2\varphi''' + f\varphi'' + f''\varphi = 0,$$
$$2\bar{\varphi}''' + f\bar{\varphi}'' + f''\bar{\varphi} = \varphi\varphi'';$$
$$f(0) = 0, \quad f'(0) = 0, \quad f'(\infty) = 1, \quad (f''(0) = 0.33205), \tag{1.3.29}$$
$$\varphi(0) = 1, \quad \varphi'(0) = 0, \quad \varphi'(\infty) = 0,$$
$$\bar{\varphi}(0) = 0, \quad \bar{\varphi}'(0) = 0, \quad \bar{\varphi}'(\infty) = 0.$$

In (1.3.28), the parameters $\varphi_0, \varphi_3, \varphi_{33}, \bar{\varphi}_{33}$ are functions of the Schmidt number:

$$\varphi_0 = \int_0^\infty E(\varepsilon, p)dp \approx \begin{cases} 3.01Sc^{-0.35}\text{–for gases,} \\ 3.12Sc^{-0.34} \quad \text{–for liquids,} \end{cases}$$

$$\varphi_3 = \int_0^\infty \left[\int_0^p \varphi(s)ds \right] E(\varepsilon, p)dp \approx \begin{cases} 6.56Sc^{-0.80}\text{–for gases,} \\ 5.08Sc^{-0.67} \quad \text{–for liquids,} \end{cases}$$

$$\varphi_{33} = \int_0^\infty \left[\int_0^p \varphi(s)ds \right]^2 E(\varepsilon, p)dp \approx \begin{cases} 24.0Sc^{-1.3}\text{–for gases,} \\ 12.2Sc^{-1.0} \quad \text{–for liquids,} \end{cases}$$

$$\bar{\varphi}_{33} = \int_0^\infty \left[\int_0^p \bar{\varphi}(s)ds \right] E(\varepsilon, p)dp \approx \begin{cases} 0.326Sc^{-1.63}\text{–for gases,} \\ 0.035Sc^{-1.1} \quad \text{–for liquids.} \end{cases}$$

(1.3.30)

The dimensionless diffusion flux in the Sherwood number (0.3.26) is obtained directly from (1.3.28):

$$\psi'(0) = -\frac{2}{\varepsilon\varphi_0} + \theta\frac{2\varphi_3}{\varphi_0^3} + \theta^2 \left(-\frac{4\varepsilon\varphi_3^2}{\varphi_0^5} + \frac{\varepsilon\varphi_{33}}{\varphi_0^4} + \frac{4\bar{\varphi}_{33}}{\varepsilon\varphi_0^4} \right). \tag{1.3.31}$$

Equation (1.3.31) shows that the precision of this basic result from the asymptotic theory of the diffusion boundary layer significantly depends on θ and ε. If it is necessary to obtain a theoretical result with an error less than 10%, the second-order approximation of the small parameter θ should be smaller than one-tenth of its zero-order approximation, i.e.

$$\left| \theta^2 \left(-\frac{4\varepsilon\varphi_3^2}{\varphi_0^5} + \frac{\varepsilon\varphi_{33}}{\varphi_0^4} + \frac{4\bar{\varphi}_{33}}{\varepsilon\varphi_0^4} \right) \right| < (0.1) \left| -\frac{2}{\varepsilon\varphi_0} \right|. \tag{1.3.32}$$

From (1.3.32) and (1.3.30) follows:

$$\begin{aligned} \varepsilon = 1, \quad &\theta < 0.41; \quad \varepsilon = 2, \quad \theta < 0.23; \\ \varepsilon = 10, \quad &\theta < 0.056; \quad \varepsilon = 20, \quad \theta < 0.025. \end{aligned} \tag{1.3.33}$$

In order to check the precision of the asymptotic theory of nonlinear mass transfer in a diffusion boundary layer, the finite problem (1.3.25) was solved through a numerical method [46].

In Table 1.2, results of the asymptotic theory $\psi'(0)$ are compared with the results of the numerical experiments $\psi_N'(0)$. The missing data in Table 1.2 correspond to the cases when the singular disturbances in the numerical solution of the problem increase. From (1.3.33), it is obvious that these cases go beyond the limits of the accepted precision of the asymptotic theory.

Table 1.2 Comparison of the results of the asymptotic theory $\psi'(0)$ with the results of the numerical experiments $\psi'_N(0)$

θ	$\varepsilon = 1$		$\varepsilon = 2$		$\varepsilon = 10$		$\varepsilon = 20$	
	$-\psi'_N(0)$	$-\psi'(0)$	$-\psi'_N(0)$	$-\psi'(0)$	$-\psi'_N(0)$	$-\psi'(0)$	$-\psi'_N(0)$	$-\psi'(0)$
0.00	0.664	0.664	0.535	0.535	0.314	0.305	0.250	0.246
+0.03	0.650	0.650	0.515	0.516	0.270	0.265	0.190	0.199
−0.03	0.679	0.679	0.553	0.555	0.384	0.365	0.406	0.363
+0.05	0.641	0.641	0.503	0.504	0.248	0.250	0.166	0.205
−0.05	0.689	0.689	0.572	0.570	0.459	0.415	–	0.479
+0.10	0.620	0.620	0.475	0.478	0.207	0.250	–	0.500
−0.10	0.716	0.716	0.616	0.611	–	0.581	–	0.903
+0.20	0.581	0.584	0.429	0.442	0.160	0.418	–	1.229
−0.20	0.779	0.776	0.736	0.707	–	1.080	–	2.325
+0.30	0.548	0.555	0.393	0.425	–	0.808	–	2.868
−0.30	0.855	0.843	0.936	0.822	–	1.800	–	4.512

The obtained results show that the direction of the intensive mass transfer significantly influences the mass transfer kinetics and this cannot be predicted in the approximations of the linear theory ($\theta = 0$). When the mass transfer is directed from the volume toward the phase boundary ($\theta < 0$), the increase of the concentration gradient in the diffusion boundary layer ($c^* - c_0$) leads to an increase in the diffusion mass transfer. In the cases when the mass transfer is directed from the phase boundary toward the volume ($\theta > 0$), the increase of the concentration gradient leads to a decrease in the diffusion mass transfer.

The nonlinear effects in the mass transfer kinetics under conditions of intensive mass transfer occur in a thin layer on the surface of the phase separation [74], which thickness is approximately three times smaller than the one of the diffusion boundary layers. At the boundary of this "layer of nonlinear mass transfer," the character of the nonlinear effect changes; i.e., the local diffusion flux depends on the concentration gradient and on the value of the parameter θ correspondingly. In the "nonlinear mass transfer layer" for $\theta < 0$ ($\theta > 0$), the flux increases (decreases) with the increase of the absolute value of θ, and out of this layer, this dependence turns to the opposite [51].

1.3.4 Two-Phase Systems

The interphase mass transfer in the gas–liquid and the liquid–liquid systems is associated primarily with the industrial absorption and extraction processes. The process intensification through generation of large concentration gradients in the gas and the liquid leads to manifestation of nonlinear effects in the kinetics of the mass transfer in the gas and liquid phases. In this way, the interphase mass transfer in the gas–liquid and the liquid–liquid systems becomes nonlinear.

The industrial gas absorption is most frequently realized in packed bed columns. The sizes of packing particles used being small, the interphase transfer of the absorbed substance is effected through the thin layers bordering the phase boundary between the gas and the liquid. The main change in the absorbed material concentration takes place in these layers, which allows the theoretical analysis of the kinetics of nonlinear interphase mass transfer to be performed making use of the approximation of the diffusion boundary layer.

The kinetics of the nonlinear interphase mass transfer in the cases of a flat phase interface and co-current movement of the gas and the liquid [9, 31] will be discussed. If the gas and the liquid are designated as a first and a second phase, respectively, Eq. (0.3.20) takes the form:

$$u_j \frac{\partial u_j}{\partial x} + v_j \frac{\partial u_j}{\partial y} = v_j \frac{\partial^2 u_j}{\partial y^2}, \quad \frac{\partial u_j}{\partial x} + \frac{\partial v_j}{\partial y} = 0,$$

$$u_j \frac{\partial c_j}{\partial x} + v_j \frac{\partial c_j}{\partial y} = D_j \frac{\partial^2 c_j}{\partial y^2}, \quad j = 1, 2, \tag{1.3.34}$$

with boundary conditions accounting for the continuity of the velocity distribution and the flows of momentum and mass at the face interphase:

$$x = 0, \quad u_j = u_{j0}, \quad c_j = c_{j0}, \quad j = 1, 2; \quad y = 0, \quad u_1 = u_2, \quad \mu_1 \frac{\partial u_1}{\partial y} = \mu_2 \frac{\partial u_2}{\partial y},$$

$$v_j = -\frac{MD_j}{\rho_{j0}^*} \frac{\partial c_j}{\partial y}, \quad j = 1, 2, \quad c_1 = \chi c_2, \quad \frac{D_1 \rho_1^*}{\rho_{10}} \frac{\partial c_1}{\partial y} = \frac{D_2 \rho_2^*}{\rho_{20}^*} \frac{\partial c_2}{\partial y};$$

$$y \to \infty, \quad u_1 = u_{10}, \quad c_1 = c_{10}; \quad y \to -\infty, \quad u_2 = u_{20}, \quad c_2 = c_{20}.$$

$$\tag{1.3.35}$$

The interphase mass transfer rate for a surface of length l is determined by averaging the local mass fluxes:

$$J = MK_1(c_{10} - \chi c_{20}) = -\frac{1}{l} \int_0^l I_1 dx = MK_2 \left(\frac{c_{10}}{\chi} - c_{20} \right) = -\frac{1}{l} \int_0^l I_2 dx, \tag{1.3.36}$$

where K_j ($j = 1, 2$) are the interphase mass transfer coefficients, while the local mass fluxes are obtained from (1.3.22):

$$I_j = -\frac{MD_j \rho_j^*}{\rho_{j0}^*} \left(\frac{\partial c_j}{\partial y} \right)_{y=0}, \quad j = 1, 2. \tag{1.3.37}$$

From (1.3.36) and (1.3.37), the Sherwood number is obtained:

$$\mathrm{Sh}_j = \frac{K_j l}{D_j} = \frac{\rho_j^*}{\rho_{j0}^*} \frac{\chi^{j-1}}{c_{10} - \chi c_{20}} \int_0^l \left(\frac{\partial c_j}{\partial y}\right)_{y=0} dx, \quad j = 1, 2. \tag{1.3.38}$$

Equations (1.3.34) and (1.3.35) can be solved introducing similarity variables:

$$u_j = 0.5 j u_{j0} \varepsilon_j \varphi_j', \quad v_j = (-1)^{j-1} 0.5 j \left(\frac{u_{j0} v_j}{x}\right)^{\frac{1}{2}} \left(\xi_j \varphi_j' - \varphi_j\right),$$

$$c_j = c_{j0} - (-\chi)^{1-j} (c_{10} - \chi c_{20}) \psi_j, \quad \varphi_j = \varphi_j(\xi_j), \quad \psi_j = \psi_j(\xi_j), \tag{1.3.39}$$

$$\xi_j = (-1)^{j-1} y \left(\frac{u_{j0}}{4 D_j x}\right)^{\frac{1}{2}}, \quad \varepsilon_j = \sqrt{Sc_j}, \quad Sc_j = \frac{v_j}{D_j}, \quad j = 1, 2.$$

Thus, the following result is obtained:

$$\varphi_j''' + j \varepsilon_j^{-1} \varphi_j \varphi_j'' = 0, \quad \psi'' + j \varepsilon_j \varphi_j \psi_j' = 0,$$

$$\varphi_j(0) = (-1)^j \theta_j + 2 \psi_j'(0), \quad \varphi_j'(\infty) = \frac{2}{j \varepsilon_j}, \quad \psi_j(\infty) = 0, \quad j = 1, 2;$$

$$\varphi_1'(0) = 2 \theta_1 \frac{\varepsilon_2}{\varepsilon_1} \varphi_2'(0), \quad \varphi_1''(0) = -0.5 \theta_2 \left(\frac{\varepsilon_1}{\varepsilon_2}\right)^2 \varphi_2''(0), \tag{1.3.40}$$

$$\psi_2'(0) = \frac{\chi}{\varepsilon_0} \psi_1'(0), \quad \psi_1(0) + \psi_2(0) = 1,$$

where

$$\theta_1 = \frac{u_{20}}{u_{10}}, \quad \theta_2 = \left(\frac{\mu_1}{\mu_2}\right) \left(\frac{v_1}{v_2}\right)^{-0.5} \left(\frac{u_{10}}{u_{20}}\right)^{1.5},$$

$$\theta_3 = \frac{M(c_{10} - \chi c_{20})}{\varepsilon_1 \rho_{10}^*}, \quad \theta_4 = \frac{M(c_{10} - \chi c_{20})}{2 \varepsilon_2 \rho_{20}^* \chi}, \tag{1.3.41}$$

It follows from (1.3.40) that the concentration of the absorbed material on the face interphase ($y = 0$) is constant. This allows a set of new boundary conditions to be used:

$$\psi_1(0) = A, \quad \psi_2(0) = 1 - A, \tag{1.3.42}$$

where A is determined from the conditions of the mass flow continuity on the phase interface. Thus, (1.3.42) permits to be solved of (1.3.40) as two independent problems.

The parameters θ_1 and θ_2 account for the kinetic and dynamic interactions between the phases, while θ_3 and θ_4—for the rate of the nonlinear effects in the gas and the liquid phases. For the cases of practical interest, $\theta_k < 1$ ($k = 1,\ldots, 4$) is valid and the problem could be solved making use of the perturbation method [6, 41], expressing the unknown functions by an expansion of the following types:

$$F = F^{(0)} + \theta_1 F^{(1)} + \theta_2 F^{(2)} + \theta_3 F^{(3)} + \theta_4 F^{(4)} + \cdots, \tag{1.3.43}$$

where F is a vector function

$$F = F(\varphi_1, \varphi_2, \psi_1, \psi_2, A). \tag{1.3.44}$$

The zero-order approximation is obtained from (1.3.40) when substituting $\theta_k = 0$, $k = 1, \dots, 4$.

The first-order approximations are obtainable from the equations:

$$\begin{aligned} \varphi_j'''^{(k)} + j\varepsilon_j^{-1}\left(\varphi_j''^{(k)}\varphi_j^{(0)} + \varphi_j''^{(0)}\varphi_j^{(k)}\right) &= 0, \\ \psi_j''^{(k)} + j\varepsilon_j\left(\varphi_j^{(k)}\psi_j'^{(0)} + \varphi_j^{(0)}\psi_j'^{(k)}\right) &= 0, \quad k = 1, \dots, 4, \quad j = 1, 2, \end{aligned} \tag{1.3.45}$$

with boundary conditions:

$$\varphi_j^{(0)}(0) = 0, \quad k = 1, 2, \quad j = 1, 2;$$
$$\varphi_1^{(3)}(0) = -\psi_1'^{(0)}(0), \quad \varphi_2^{(4)}(0) = -\psi_2'^{(2)}(0),$$
$$\varphi_1^{(4)}(0) = 0, \quad \varphi_2^{(3)}(0) = 0;$$
$$\varphi_1'^{(k)}(0) = 0, \quad k = 2, 3, 4; \quad \varphi_1'^{(1)}(0) = 2\frac{\varepsilon_2}{\varepsilon_1}\varphi_2'^{(0)}(0);$$
$$\varphi_j'^{(k)}(0) = 0, \quad k = 1, \dots 4, \quad j = 1, 2; \tag{1.3.46}$$
$$\psi_j'^{(k)}(0) = A^{(k)}, \quad \psi_j^{(k)}(\infty) = 0, \quad k = 1, \dots 4, \quad j = 1, 2;$$
$$\varphi_2'^{(k)}(0) = 0, \quad k = 1, 3, 4;$$
$$\varphi_2''^{(2)}(0) = -\frac{1}{2}\left(\frac{\varepsilon_1}{\varepsilon_2}\right)^2 \varphi_1''^{(0)}(0); \quad \psi_2^{(k)}(0) = -A^{(k)}, \quad k = 1, \dots, 4.$$

The values for $A^{(k)}$ ($k = 1, \dots, 4$) are calculated from the equation:

$$\psi_2'^{(k)}(0) = \frac{\chi}{\varepsilon_0}\psi_1'^{(k)}(0), \quad k = 1, \dots, 4. \tag{1.3.47}$$

The solutions of problems of the type (1.3.40) have been reported in a number of publications [78–86]. Using these solutions, the following can be written:

$$\varphi_1^{(0)}(\xi_1) = f(z), \quad z = \frac{2}{\varepsilon_1}\xi_1, \quad \psi_1^{(0)}(\xi_1) = A^{(0)}\left(1 - \frac{1}{\varphi_{10}}\right)\int\limits_0^z E(\varepsilon_1, p)\mathrm{d}p,$$

$$\psi_2^{(0)}(\xi_2) = \left(1 - A^{(0)}\right)\mathrm{erfc}\xi_2, \quad E(\varepsilon_1, p) = \exp\left[-\frac{\varepsilon_1^2}{2}\int\limits_0^p f(s)\mathrm{d}s\right],$$

$$\varphi_2^{(0)}(\xi_2) = \varepsilon_2^{-1}\xi_2, \quad \varphi_1^{(1)}(\xi_1) = \frac{1}{\alpha}f'(z), \quad \varphi_2^{(1)}(\xi_2) \equiv 0, \quad \alpha = f''(0),$$

$$\psi_1^{(1)}(\xi_2) = -A^{(1)}\mathrm{erfc}\xi_2, \quad A^{(1)} = -\frac{1}{\alpha\varphi_{10}}\frac{\alpha_0}{(1+a_0)^2}, \quad \psi_1^{(1)}(\xi_1) = A^{(1)}$$

$$+ \frac{A^{(0)}}{\alpha\varphi_{10}}[1 - E(\varepsilon_1, z)] - \left(\frac{A^{(1)}}{\varphi_{10}} + \frac{A^{(0)}}{\alpha\varphi_{10}^2}\right)\int\limits_0^z E(\varepsilon_1, p)\mathrm{d}p,$$

$$\varphi_1^{(2)}(\xi_1) \equiv 0, \quad \varphi_2^{(2)}(\xi_2) = \alpha\sqrt{\pi}\int\limits_0^{\frac{\varepsilon_2}{E_2}}\mathrm{erfc}p\,\mathrm{d}p,$$

$$\psi_1^{(2)}(\xi_1) = A^{(2)}\left[1 - \frac{1}{\varphi_{10}}\int\limits_0^z E(\varepsilon_1, p)\mathrm{d}p\right], \quad \psi_2^{(2)}(\xi_2) = -A^{(2)}$$

$$+ \left[A^{(2)} - 4\alpha\varepsilon_2\left(1 - A^{(0)}\right)\bar{\varphi}_2\right]\mathrm{erfc}\xi_2 + 4\alpha\varepsilon_2\left(1 - A^{(0)}\right)Q(\varepsilon_2, \xi_2),$$

(1.3.48)

$$Q(\varepsilon_2, \xi_2) = \int\limits_0^{\xi_2}\left[\exp(-q^2)\int\limits_0^q\left(\int\limits_0^{p/\varepsilon_2}\mathrm{erfc}s\,\mathrm{d}s\right)\mathrm{d}p\right]\mathrm{d}q$$

$$\approx \frac{1}{8}\sqrt{\frac{\pi}{Sc_2}}\mathrm{erf}\xi_2 - \frac{1}{4\sqrt{Sc_2}}\xi_2\exp(-\xi_2^2), \quad A^{(2)} = 4\alpha\varepsilon_2\bar{\varphi}_2\frac{a}{(1+a)^2},$$

$$\bar{\varphi}_2 = Q(\varepsilon_2, \infty) = \frac{1}{8}\sqrt{\frac{\pi}{Sc_2}}, \quad \varphi_1^{(3)}(\xi_1) = \frac{2A^{(0)}}{\varepsilon_1\varphi_{10}}\phi(z), \quad \varphi_2^{(3)}(\xi_2) \equiv 0,$$

$$\psi_1^{(3)}(\xi_1) = A^{(3)} - \left(\frac{A^{(3)}}{\varphi_{10}} + \frac{\varepsilon_1 A^{(0)}\varphi_{13}}{\varphi_{10}^3}\right)\int\limits_0^z E(\varepsilon_1, p)\mathrm{d}p$$

$$+ \frac{\varepsilon_1 A^{(0)}}{\varphi_{10}^2}\int\limits_0^z\left[\int\limits_0^p \varphi(s)\mathrm{d}s\right]E(\varepsilon_1, p)\mathrm{d}p, \quad \psi_2^{(3)}(\xi_2) = -A^{(3)}\mathrm{erfc}\xi_2,$$

$$A^{(3)} = -\frac{\varepsilon_1\varphi_{13}}{\varphi_{10}^2}\frac{a_0}{(1+a_0)^2}, \quad \psi_1^{(4)}(\xi_1) = A^{(4)}\left(1 - \frac{1}{\varphi_{10}}\int\limits_0^z E(\varepsilon_1, p)\mathrm{d}p\right),$$

$$\psi_2^{(4)}(\xi_2) = -A^{(4)} - \frac{4\varepsilon_2}{\pi}\left(1 - A^{(0)}\right)^2 + \frac{4\varepsilon_2}{\pi}\left(1 - A^{(0)}\right)^2\exp(-\xi_2^2)$$

$$+ \left[A^{(4)} + \frac{4\varepsilon_2}{\pi}\left(1 - A^{(0)}\right)^2\right]\mathrm{erf}\xi_2,$$

$$\psi_1^{(3)}(\xi_1) = A^{(3)} - \left(\frac{A^{(3)}}{\varphi_{10}} + \frac{\varepsilon_1 A^{(0)} \varphi_{13}}{\varphi_{10}^3}\right) \int_0^z E(\varepsilon_1, p)\mathrm{d}p$$

$$+ \frac{\varepsilon_1 A^{(0)}}{\varphi_{10}^2} \int_0^z \left[\int_0^p \varphi(s)\mathrm{d}s\right] E(\varepsilon_1, p)\mathrm{d}p, \quad \psi_2^{(3)}(\xi_2) = -A^{(3)}\mathrm{erfc}\,\xi_2,$$

$$A^{(3)} = -\frac{\varepsilon_1 \varphi_{13}}{\varphi_{10}^2}\frac{a_0}{(1+a_0)^2}, \quad \psi_1^{(4)}(\xi_1) = A^{(4)}\left(1 - \frac{1}{\varphi_{10}}\int_0^z E(\varepsilon_1, p)\mathrm{d}p\right),$$

$$\psi_2^{(4)}(\xi_2) = -A^{(4)} - \frac{4\varepsilon_2}{\pi}\left(1 - A^{(0)}\right)^2 + \frac{4\varepsilon_2}{\pi}\left(1 - A^{(0)}\right)^2\exp\left(-\xi_2^2\right)$$

$$+ \left[A^{(4)} + \frac{4\varepsilon_2}{\pi}\left(1 - A^{(0)}\right)^2\right]\mathrm{erf}\,\xi_2,$$

where f and φ are solutions of (1.1.19), φ_{10} and φ_{13} are expressed as:

$$\varphi_{10} = \int_0^\infty E(\varepsilon_1, p)\mathrm{d}p \approx 3.01 Sc_1^{-0.35},$$

$$\varphi_{13} = \int_0^\infty \left[\int_0^p \varphi(s)\mathrm{d}s\right] E(\varepsilon_1, p)\mathrm{d}p \approx 6.56 Sc_1^{-0.8}, \tag{1.3.49}$$

i.e., there values can be obtained from φ_0 and φ_3 in (3.30) with substituting $\varepsilon = \varepsilon_1$ (Sc = Sc_1).

The nonlinear interphase mass transfer rate (the Sherwood number) is obtainable from (1.3.38):

$$\mathrm{Sh}_j = -\frac{\rho_j^*}{\rho_{j0}^*}\sqrt{Pe_j}\,\psi_j'(0), \quad Pe_j = \frac{u_{j0}L}{D_j}, \quad j = 1, 2, \tag{1.3.50}$$

where $\psi'_1(0)$ and $\psi'_2(0)$ can be determined from (1.3.48):

$$\psi_1'(0) = -\frac{2}{\varepsilon_1 \varphi_{10}}\frac{1}{1+a} - \frac{2\theta_1}{\alpha\varphi_{10}^2\varepsilon_1}\frac{1}{(1+a)^2} - 8\theta_2\alpha\frac{\varepsilon_2}{\varepsilon_1}\frac{\bar{\varphi}_2}{\varphi_{10}}\frac{a}{(1+a)^2}$$

$$- 2\theta_3\frac{\varphi_{13}}{\varphi_{10}^3}\frac{1}{(1+a)^2} + 8\theta_4\frac{\varepsilon_2}{\pi\varphi_{10}\varepsilon_1}\frac{a^2}{(1+a)^3},$$

$$\psi_2'(0) = -\frac{2}{\sqrt{\pi}}\frac{a}{1+a} - \theta_1\frac{2}{\sqrt{\pi}\alpha\varphi_{10}}\frac{a}{(1+a)^2} - 8\theta_2\frac{\alpha\varepsilon_2\bar{\varphi}_2}{\sqrt{\pi}}\frac{a^2}{(1+a)^2}$$

$$- 2\theta_3\frac{\varepsilon_1\varphi_{13}}{\sqrt{\pi}\varphi_{10}^2}\frac{a}{(1+a)^3} + 8\theta_4\frac{\varepsilon_2}{\pi\sqrt{\pi}}\frac{a^3}{(1+a)^3}. \tag{1.3.51}$$

In the cases where the rate of the interphase mass transfer is limited by the diffusion resistance in the gas phase, from the last condition in (1.3.40) it follows that $\chi/\varepsilon_0 \to 0$, i.e., $a \to 0$. Thus, the Sherwood number can be expressed in the form:

$$\mathrm{Sh}_1 = \frac{\rho_1^*}{\rho_{10}^*}\mathrm{Pe}_1^{0.5}\left(\frac{2}{\varepsilon_1\varphi_{10}} + \frac{2\theta_1}{\varepsilon_1\alpha\varphi_{10}^2} + 2\theta_3\frac{\varphi_{13}}{\varphi_{10}^3}\right). \tag{1.3.52}$$

When the process is limited by the resistance in the liquid phase, $\chi/\varepsilon_0 \to \infty$, $a \to \infty$ i.e.

$$\mathrm{Sh}_2 = \frac{\rho_2^*}{\rho_{20}^*}\mathrm{Pe}_2^{0.5}\left(\frac{2}{\sqrt{\pi}} + 8\theta_2\frac{\alpha\varepsilon_2\bar{\varphi}_2}{\sqrt{\pi}} - 8\theta_4\frac{\varepsilon_2}{\pi\sqrt{\pi}}\right). \tag{1.3.53}$$

The comparison of the nonlinear effects in both the gas and the liquid [86] shows that the ratio of the parameters θ_3 and θ_4 takes the form

$$\frac{\theta_3}{\theta_4} = \frac{2\varepsilon_2\rho_{20}^*\chi}{\varepsilon_1\rho_{10}^*} \gg 1 \tag{1.3.54}$$

and is always greater than unity. The minimum value of this ratio occurs in cases of gases of high solubility, where θ_3 is greater than θ_4 by more than two orders of magnitude; i.e., for numerical calculation, it is always possible to assume $\theta_4 = 0$.

A numerical solution of Eqs. (1.3.45) and (1.3.46) has been performed as a check of the asymptotic theory [86, 87]. The analysis of the results demonstrates that the nonlinear effects are most significant in cases, where the nonlinear interphase mass transfer is limited by the mass transfer in the gas phase ($\chi/\varepsilon_0 = 0$). When the diffusion resistances are commensurable ($\chi/\varepsilon \approx 1$), the nonlinear effects are considerably smaller and their appearance in the liquid phase is a result from the hydrodynamic influence of the gas phase. However, these effects are totally absent when the process is limited by the mass transfer in the liquid phase. The influence of the direction of the interphase mass transfer on the kinetics of the mass transfer in the gas–liquid systems is similar, which has been observed in the systems gas–liquid–solid surface; i.e., the diffusion transfer in the case of absorption is greater than in the case of desorption.

The results of the asymptotic theory (1.3.52) show that in the cases of absorption and desorption, the deviation of the nonlinear mass transfer from linearity ($\theta = 0$) is symmetrical, while the numerical results show a non-symmetric deviation. This "contradicts" with the asymptotic theory and is possible to be explained by the absence of the quadratic terms (proportional to θ_3^2).

It is evident that the asymptotic theory has to be made more precise and to include all the quadratic terms. In the cases of a nonlinear interphase mass transfer limited by the mass transfer in the gas phase, Eq. (1.3.40) take the form:

$$\varphi''' + \varepsilon^{-1}\varphi_1\varphi_1'' = 0, \quad \varphi_2''' + 2\varepsilon_2^{-1}\varphi_2\varphi_2'' = 0, \quad \psi_1''' + \varepsilon_1\varphi_1\psi_1' = 0;$$

$$\varphi_1(0) = -\theta_3\psi_1'(0), \quad \varphi_2(0) = 0, \quad \varphi_1'(\infty) = \frac{2}{\varepsilon_1}, \quad \varphi_2'(\infty) = \frac{1}{\varepsilon_2},$$

$$\varphi_1'(0) = 2\theta_1\frac{\varepsilon_2}{\varepsilon_1}\varphi_2'(0), \quad \varphi_2''(0) = -0.5\theta_2\left(\frac{\varepsilon_1}{\varepsilon_2}\right)^2\varphi_1''(0),$$

$$\psi_1(0) = 1, \quad \psi_1(\infty) = 0.$$

(1.3.55)

In order to solve the problem (1.3.55), the expansion (1.3.43) is used, where the terms $\theta_1^2 F^{(11)} + \theta_3^2 F^{(33)} + \theta_1\theta_3 F^{(13)}$ should be added and $a_0 = 0$ substituted in the relationships (1.3.48).

Approximations proportional to θ_1^2 have been obtained in [80, 81]:

$$\varphi_1^{(11)}(\xi_1) = F(z), \quad \varphi_2^{(11)}(\xi_2) \equiv 0,$$

$$\psi_1^{(11)}(\xi_1) = \left(\frac{\varepsilon_1^4\varphi_{11}}{8\alpha^2\varphi_{10}^3} - \frac{\varepsilon_1^2\varphi_{12}}{2\varphi_{10}^2} - \frac{\varepsilon_1}{2\alpha^2\varphi_{10}^3}\right)\int_0^z E(\varepsilon_1, p)dp$$

$$+ \frac{\varepsilon_1^2}{2\varphi_{10}}\int_0^z\left[\int_0^p F(s)ds\right]E(\varepsilon_1, p)dp + \frac{\varepsilon_1}{2\alpha^2\varphi_{10}^2}[1 - E(\varepsilon_1, z)]$$

$$- \frac{\varepsilon_1^4}{8\alpha^2\varphi_{10}}\cdot\int_0^z f^2(p)E(\varepsilon_1, p)dp,$$

(1.3.56)

where the function F is the solution of the problem:

$$2F''' + fF'' + f''F = -\frac{1}{\alpha^2}f'f''', \quad F(0) = F'(0) = F'(\infty) = 0,$$

(1.3.57)

and has been tabulated in [38], while φ_{11} and φ_{12} have been obtained in [80]:

$$\varphi_{11} = \int_0^\infty f^2(p)E(\varepsilon_1, p)dp \approx 3.01Sc_1^{-1.608},$$

$$\varphi_{12} = \int_0^\infty\left[\int_0^p F(s)ds\right]E(\varepsilon_1, p)dp \approx 3.05Sc_1^{-1.285}.$$

(1.3.58)

Approximations proportional to θ_3^2 have been obtained in [88]:

$$\varphi_1^{(33)}(\xi_1) = \frac{2\varphi_3}{\varphi_{10}^3}\varphi(z) - \frac{4}{\varepsilon_1^2\varphi_{10}^2}\bar{\varphi}(z), \quad \varphi_2^{(33)}(\xi_2) \equiv 0,$$

$$\psi_1^{(33)}(\xi_1) = \left(-\frac{\varepsilon_1^2\varphi_{13}^2}{\varphi_{10}^5} + \frac{\varepsilon_1^2\varphi_{133}}{2\varphi_{10}^4} + \frac{2\bar{\varphi}_{133}}{\varphi_{10}^4}\right)\int_0^z E(\varepsilon_1,p)dp$$

$$+ \frac{\varepsilon_1^2\varphi_{13}}{\varphi_{10}^4}\int_o^z\left[\int_0^p\varphi(s)ds\right]E(\varepsilon_1,p)dp - \frac{\varepsilon_1^2}{2\varphi_{10}^3}\int_0^z\left[\int_0^p f(s)ds\right]^2 E(\varepsilon_1,p)dp$$

$$- \frac{2}{\varphi_{10}^3}\int_0^z\left[\int_0^p\bar{\varphi}(s)ds\right]E(\varepsilon_1,p)dp,$$

$$(1.3.59)$$

where $\bar{\varphi}$ is the solution of (1.1.29). Thus, φ_{133} and $\bar{\varphi}_{133}$ take the forms:

$$\varphi_{133} = \int_0^\infty\left[\int_0^p\varphi(s)ds\right]^2 E(\varepsilon_1,p)dp \approx 24Sc_1^{-1.3},$$

$$\bar{\varphi}_{133} = \int_0^\infty\left[\int_0^p\bar{\varphi}(s)ds\right]E(\varepsilon_1,p)dp \approx 0.326Sc_1^{-1.63},$$

$$(1.3.60)$$

i.e., they may be obtained from φ_{33} and $\bar{\varphi}_{33}$ in (1.3.30) via the substitution $\varepsilon = \varepsilon_1$ (Sc = Sc$_1$). From (1.3.28) and (1.3.59), it is obvious that $\psi_1^{(33)}(\xi_1) \equiv \psi_2(\eta)$, if $\varepsilon_1 = \varepsilon$. The approximations proportional to $\theta_1\theta_3$ have been obtained in [88]:

$$\varphi_1^{(13)} = \frac{1}{\alpha\varphi_{10}^2}\varphi(z) - \frac{2}{\varepsilon_1\alpha\varphi_{10}}\bar{\varphi}(z), \quad \varphi_2^{(13)}(\xi_2) \equiv 0,$$

$$\psi_1^{(13)}(\xi_1) = \left(-\frac{\varepsilon_1\varphi_{13}}{2\alpha\varphi_{10}^4} + \frac{\varepsilon_1\varphi_{113}}{\alpha\varphi_{10}^3} + \frac{\varepsilon_1\bar{\varphi}_{113}}{\alpha\varphi_{10}^3} - \frac{2\varepsilon_1\varphi_{13}}{\alpha\varphi_{10}^4}\right)\int_0^z E(\varepsilon_1,p)dp$$

$$+ \left(\frac{\varepsilon_1}{\alpha\varphi_{10}^3} + \frac{\varepsilon_1^2}{2\alpha\varphi_{10}^3}\right)\int_0^z\left[\int_0^p\varphi(s)ds\right]E(\varepsilon_1,p)dp$$

$$(1.3.61)$$

$$- \frac{\varepsilon_1}{\alpha\varphi_{10}^2}\int_0^z\left[\int_0^p\bar{\varphi}(s)ds\right]E(\varepsilon_1,p)dp + \frac{\varepsilon_1}{\alpha\varphi_{10}^2}E(\varepsilon_1,z)\int_0^z\varphi(p)dp$$

$$- \frac{\varepsilon_1}{\alpha\varphi_{10}^2}\int_0^z\varphi(p)E(\varepsilon_1,p)dp + \frac{\varepsilon_1\varphi_{13}}{\alpha\varphi_{10}^3}[1 - E(\varepsilon_1,z)],$$

where $\bar{\bar{\varphi}}$ is the solution of the problem:

$$2\bar{\bar{\varphi}}''' + f\bar{\bar{\varphi}}'' + f''\bar{\bar{\varphi}} = f'\varphi'' + f'''\varphi, \quad \bar{\bar{\varphi}}(0) = \bar{\bar{\varphi}}'(0) = \bar{\bar{\varphi}}(\infty) = 0, \qquad (1.3.62)$$

while φ_{133} and $\bar{\varphi}_{133}$ have been obtained in [89]:

$$\varphi_{113} = \int\limits_0^\infty \left[\int\limits_0^p \bar{\bar{\varphi}}(s)ds \right] E(\varepsilon_1, p)dp \approx Sc_1^{-1.3},$$

$$\bar{\varphi}_{113} = \int\limits_0^\infty \varphi(p)E(\varepsilon_1, p)dp \approx 4.18 Sc_1^{-0.46}. \qquad (1.3.63)$$

The expressions derived allow to determinate the rate of the nonlinear interphase mass transfer in the gas–liquid system when the process is limited by the mass transfer in the gas phase. From (1.3.50), the following is found:

$$\mathrm{Sh}_1 = \frac{K_1 L}{D_1} = \frac{\rho_1^*}{\rho_{10}^*}\sqrt{\mathrm{Pe}_1}\psi_1'(0), \qquad (1.3.64)$$

where $\psi'(0)$ is calculated taking all the quadratic approximations into account:

$$-\psi_1'(0) = \frac{2}{\varepsilon_1\varphi_{10}} + \theta_1 \frac{2}{\varepsilon_1\alpha\varphi_{10}^2} + \theta_3 \frac{2\varphi_{13}}{\varphi_{10}^3} + \theta_1\theta_3\left(\frac{\varepsilon_1\varphi_{13}}{\alpha\varphi_{10}^4} - \frac{2\varphi_{113}}{\alpha\varphi_{10}^3} - \frac{2\bar{\varphi}_{113}}{\alpha\varphi_{10}^3} + \frac{4\varphi_{13}}{\alpha\varphi_{10}^4}\right)$$

$$+ \theta_1^2\left(-\frac{\varepsilon_1^3\varphi_{11}}{4\alpha^2\varphi_{10}^2} + \frac{\varepsilon_1\varphi_{12}}{\varphi_{10}^2} + \frac{2}{\varepsilon_1\alpha^2\varphi_{10}^3}\right) + \theta_3^2\left(\frac{2\varepsilon_1\varphi_{13}^2}{\varphi_{10}^5} - \frac{\varepsilon_1\varphi_{133}}{\varphi_{10}^4} - \frac{4\bar{\varphi}_{133}}{\varepsilon_1\varphi_{10}^4}\right). $$

$$(1.3.65)$$

The expression (1.3.65) is the main result from the asymptotic theory of the nonlinear interphase mass transfer in the gas–liquid systems and is in good agreement with the results from the numerical solution of the problem (1.3.40), obtained in [87].

1.3.5 Nonlinear Mass Transfer and Marangoni Effect

Intensification of the mass transfer in the industrial gas–liquid systems is obtained quite often by the creation of large concentration gradients. This can be reached in a number of cases as a result of a chemical reaction of the transferred substance in the liquid phase. The thermal effect of the chemical reactions creates temperature gradients. The temperature and concentration gradients can affect considerably the mass transfer kinetics in gas–liquid systems. Hence, the experimentally obtained mass transfer coefficients differ significantly from those predicted by the linear mass transfer theory.

As it was shown in a number of papers [90–102], the temperature and concentration gradients on the gas–liquid or liquid–liquid interphase surface can create a surface tension gradient. As a result of this, a secondary flow is induced. The velocity of the induced flow is directed tangentially to the interface. It leads to a change in the velocity distribution in the boundary layer and therefore to a change in the mass transfer kinetics. These effects are thought to be of the Marangoni type and propose an explanation to all experimental deviations from the prediction of the linear the mass transfer theory.

Obviously, the Marangoni effect is possible to exist together with the effect of the large concentration gradients. These two effects can manifest themselves separately as well as in combination. That is why their influence on the mass transfer kinetics has to be assessed.

Co-current gas and liquid flows in the laminar boundary layer along the flat phase surface will be considered. One of the gas components is absorbed by the liquid and reacts with a component in the liquid phase. The chemical reaction rate is of a first order. The thermal effect of the chemical reaction creates a temperature gradient; i.e., the mass transfer together with a heat transfer can be observed. Under these conditions, the mathematical model takes the following form:

$$u_j \frac{\partial u_j}{\partial x} + v_j \frac{\partial u_j}{\partial y} = v_j \frac{\partial^2 u_j}{\partial y^2}, \quad \frac{\partial u_j}{\partial x} + \frac{\partial v_j}{\partial y} = 0,$$

$$u_j \frac{\partial c_j}{\partial x} + v_j \frac{\partial c_j}{\partial y} = D_j \frac{\partial^2 c_j}{\partial y^2} - (j-1)kc_j, \qquad (1.3.66)$$

$$u_j \frac{\partial t_j}{\partial x} + v_j \frac{\partial t_j}{\partial y} = a_j \frac{\partial^2 t_j}{\partial y^2} + (j-1)\frac{q}{\rho_j c_{pj}} kc_j, \quad j = 1, 2,$$

where the indexes 1 and 2 refer to the gas and the liquid, respectively. The influence of the temperature on the chemical reaction rate is not included in (1.3.66) because it has no considerable affect in the comparative analysis of these two effects.

The boundary conditions of (1.3.66) determine the potential two-phase flows far from the phase boundary. Thermodynamic equilibrium and continuity of velocity and stress tensor, mass and heat fluxes can be detected on the phase boundary. It has been shown in [103] that in the gas–liquid systems the effect of nonlinear mass transfer is confined into the gas phase. Taking into account these considerations, the boundary conditions assume the following form:

$$x = 0, \quad u_j = u_{j0}, \quad c_1 = c_{10}, \quad c_2 = 0, \quad t_j = t_0;$$

$$y \to \infty, \quad u_1 = u_{10}, \quad c_1 = c_{10}, \quad t_1 = t_0;$$

$$y \to -\infty, \quad u_2 = u_{20}, \quad c_2 = 0, \quad t_2 = t_0;$$

$$y = 0, \quad u_1 = u_2, \quad \mu_1 \frac{\partial u_1}{\partial y} = \mu_2 \frac{\partial u_2}{\partial y} - \frac{\partial \sigma}{\partial x},$$

$$v_1 = -\frac{MD_1}{\rho_{10}^*} \frac{\partial c_1}{\partial y}, \quad v_2 = 0, \quad \rho_1^* = \rho_{10}^* + Mc_1^*, \qquad (1.3.67)$$

$$c_1 = \chi c_2, \quad D_1 \frac{\rho_1^*}{\rho_{10}^*} \frac{\partial c_1}{\partial y} = D_2 \frac{\partial c_2}{\partial y}, \quad t_1 = t_2,$$

$$\lambda_1 \frac{\partial t_1}{\partial y} + \rho_1 c_{p1} v_1 t_1 = \lambda_2 \frac{\partial t_2}{\partial y}, \quad j = 1, 2.$$

At high enough values of c_0, a large concentration gradient directed normally to the interface $(\partial c_1/\partial y)_{y=0}$ can be observed, which induces a secondary flow with the rate v_1. The tangential concentration and temperature gradients along the phase boundary create surface tension gradient

$$\frac{\partial \sigma}{\partial x} = \frac{\partial \sigma}{\partial c_2} \frac{\partial c_2}{\partial x} + \frac{\partial \sigma}{\partial t_2} \frac{\partial t_2}{\partial x}, \qquad (1.3.68)$$

inducing a tangential secondary flow, which velocity is proportional to $\partial \sigma / \partial x$. Later, the use of substance, which is not surface active, i.e., $\partial \sigma / \partial c_2 \approx 0$, will be examined.

The mass transfer rate (J_c) and the heat transfer rate (J_t) can be determined from the local mass (I_c) and heat (I_t) fluxes after taking the average of these fluxes along the length (l) of the interface:

$$J_c = k_c c_0 = \frac{1}{l} \int_0^l I_c dx, \quad I_c = \frac{MD_1 \rho_1^*}{\rho_{10}^*} \left(\frac{\partial c_1}{\partial y}\right)_{y=0},$$

$$(1.3.69)$$

$$J_t = k_t t_0 = \frac{1}{l} \int_0^l I_t dx, \quad I_t = -\lambda_1 \left(\frac{\partial t_1}{\partial y}\right)_{y=0} + \rho_1 c_{p1} (v_1 t_1)_{y=0},$$

where c_1 and t_1 are determined upon solving the problems (1.3.66) and (1.3.68). In order to do this, the following dimensionless (generalized) variables are introduced:

$$x = lX, \quad y = (-1)^{j+1}\delta_j Y_j, \quad \delta_j = \sqrt{\frac{v_j l}{u_{j0}}},$$

$$u_j = u_{j0}U_j(X, Y_j), \quad v_j = (-1)^{j+1}u_{j0}\frac{\delta_j}{l}V_j(X, Y_j),$$

$$c_j = (-\chi)^{1-j}c_0 C_j(X, Y_j), \quad t_j = t_0 + (-1)^{j+1}t_0 T_j(X, Y_i), \quad j = 1, 2.$$

(1.3.70)

The introduction of (1.3.70) into (1.3.66) and (1.3.67) leads to the following equations:

$$U_j\frac{\partial U_j}{\partial X} + V_j\frac{\partial U_j}{\partial Y_j} = \frac{\partial^2 U_j}{\partial Y_j^2}, \quad \frac{\partial U_j}{\partial X} + \frac{\partial V_j}{\partial Y_j} = 0,$$

$$U_j\frac{\partial C_j}{\partial X} + V_j\frac{\partial C_j}{\partial Y_j} = \frac{1}{Sc_j}\frac{\partial^2 C_j}{\partial Y_j^2} - (j-1)DaC_j,$$

$$U_j\frac{\partial T_j}{\partial X} + V_j\frac{\partial T_j}{\partial Y_j} = \frac{1}{Pr_j}\frac{\partial^2 T_j}{\partial Y_j^2} + (j-1)QDaC_j, \quad j = 1, 2;$$

$$X = 0, \quad U_j = 1, \quad C_1 = 1, \quad C_2 = 0, \quad T_j = 0, \quad j = 1, 2;$$

$$Y_1 \rightarrow \infty, \quad U_1 = 1, \quad C_1 = 1, \quad T_1 = 0;$$

$$Y_2 \rightarrow \infty, \quad U_2 = 1, \quad C_2 = 0, \quad T_2 = 0;$$

$$Y_1 = Y_2 = 0, \quad U_1 = \theta_1 U_2, \quad \theta_2\frac{\partial U_1}{\partial Y_1} = -\frac{\partial U_2}{\partial Y_2} + \theta_4\frac{\partial T_2}{\partial X}, \quad \theta_5\frac{\partial C_1}{\partial Y_1} = \frac{\partial C_2}{\partial Y_2},$$

$$\theta_6\frac{\partial T_1}{\partial Y_1} = \frac{\partial T_2}{\partial Y_2}, \quad V_1 = -\theta_3\frac{\partial C_1}{\partial Y_1}, \quad V_2 = 0, \quad C_1 + C_2 = 0, \quad T_1 + T_2 = 0,$$

(1.3.71)

where

$$Da = \frac{kl}{u_{20}}, \quad Q = \frac{qc_0}{\chi\rho_2 c_{p2}t_0}, \quad Sc_j = \frac{v_j}{D_j}, \quad Pr_j = \frac{v_j}{a_j}, \quad j = 1, 2.$$

$$\theta_1 = \frac{u_{20}}{u_{10}}, \quad \theta_2 = \frac{\mu_1}{\mu_2}\sqrt{\frac{v_2}{v_1}}\left(\frac{u_{20}}{u_{10}}\right)^{3/2}, \quad \theta_3 = \frac{Mc_0}{\rho_{10}^* Sc_1},$$

(1.3.72)

$$\theta_4 = \frac{\partial\sigma}{\partial t_2}\frac{t_0}{u_{20}\mu_2}\sqrt{\frac{v_2}{u_{20}l}}, \quad \theta_5 = \chi\frac{D_1}{D_2}\frac{\rho_1^*}{\rho_{10}^*}\sqrt{\frac{u_{10}v_2}{u_{20}v_1}}, \quad \theta_6 = \frac{\lambda_1}{\lambda_2}\sqrt{\frac{u_{10}v_2}{u_{20}v_1}}.$$

From (1.3.69) and (1.3.70), the expressions for the Sherwood and Nusselt numbers are directly obtained:

$$Sh = \frac{k_c l}{D_1} = M \sqrt{Re_1} \int_0^1 \left(1 + \theta_3 Sc_1 C_1^*\right) \left(\frac{\partial C_1}{\partial Y_1}\right)_{Y_1=0} dX,$$

$$Nu = \frac{k_t l}{\lambda_1} = -Re_1 \left[\int_0^1 \left(\frac{\partial T_1}{\partial Y_1}\right)_{Y_1=0} dX + \theta_3 Pr_1 \int_0^1 \left(1 + T_1^*\right) \left(\frac{\partial C_1}{\partial Y_1}\right)_{Y_1=0} dX\right],$$

$$C_1^* = C_1(X,0), \quad T_1^* = T_1(X,0), \quad Re_1 = \frac{u_1 0 l}{v_1}.$$

$$(1.3.73)$$

The solution of (1.3.71) allows the determination of:

$$J_1 = \int_0^1 \left(\frac{\partial C_1}{\partial Y_1}\right)_{Y_1=0} dX, \quad J_2 = \int_0^1 C_1(X,0) \left(\frac{\partial C_1}{\partial Y_1}\right)_{Y_1=0} dX,$$

$$(1.3.74)$$

$$J_3 = \int_0^1 \left(\frac{\partial T_1}{\partial Y_1}\right)_{Y_1=0} dX, \quad J_4 = \int_0^1 T_1(X,0) \left(\frac{\partial C_1}{\partial Y_1}\right)_{Y_1=0} dX.$$

The introduction of (1.3.74) into (1.3.73) allows determining the Sherwood and Nusselt numbers:

$$Sh = M \sqrt{Re_1}(J_1 + \theta_3 Sc_1 J_2), \quad Nu = -\sqrt{Re_1}[J_3 + \theta_3 Pr_1(J_1 + J_4)]. \quad (1.3.75)$$

The problem (1.3.71) can be solved conveniently using the iterative algorithm, where 6 problems are solved consecutively, until a convergence with respect to the integral J_1 in (1.3.74):

$$U_1^{(k)} \frac{\partial U_1^{(k)}}{\partial X} + V_1^{(k)} \frac{\partial U_1^{(k)}}{\partial Y_1} = \frac{\partial^2 U_1^{(k)}}{\partial Y_1^2}, \quad \frac{\partial U_1^{(k)}}{\partial X} + \frac{\partial V_1^{(k)}}{\partial Y_1} = 0;$$

$$X = 0, \quad U_1^{(k)} = 1; \quad Y_1 = 0, \quad U_1^{(k)} = \theta_1 U_2^{(k-1)}, \quad V_1^{(k)} = -\theta_3 \frac{\partial C_1^{(k-1)}}{\partial Y_1};$$

$$Y_1 \to \infty \quad (Y_1 \geq Y_{1\infty}), \quad U_1^{(k)} = 1; \quad 0 \leq X \leq 1, \quad 0 \leq Y_1 \leq Y_{1\infty};$$

$$\theta_1 = 0.1, \quad Y_{1\infty} = 6, \quad \text{(at the first iteration } \theta_1 = \theta_3 = 0).$$

$$(1.3.76)$$

$$U_2^{(k)}\frac{\partial U_2^{(k)}}{\partial X}+V_2^{(k)}\frac{\partial U_2^{(k)}}{\partial Y_2}=\frac{\partial^2 U_2^{(k)}}{\partial Y_2^2},\quad \frac{\partial U_2^{(k)}}{\partial X}+\frac{\partial V_2^{(k)}}{\partial Y_2}=0;$$

$$X=0,\quad U_2^{(k)}=1;$$

$$Y_2=0,\quad \frac{\partial U_2^{(k)}}{\partial Y_2}=-\theta_2\left(\frac{\partial U_1^{(k)}}{\partial Y_1}\right)_{Y_1=0}+\theta_4\left(\frac{\partial T_2^{(k-1)}}{\partial X}\right)_{Y_2=0},\quad V_2^{(k-1)}=0;$$

$$Y_2\to\infty\quad(Y_2\geq Y_{2\infty}),\quad U_2^{(k)}=1;\quad 0\leq X\leq 1,\quad 0\leq Y_2\leq Y_{2\infty};$$

$$\theta_2=0.145,\quad Y_{2\infty}=6,\quad\text{(at the first iteration }\theta_4=0\text{)}.$$

$$(1.3.77)$$

$$U_1^{(k)}\frac{\partial C_1^{(k)}}{\partial X}+V_1\frac{\partial C_1^{(k)}}{\partial Y_1}=\frac{1}{Sc_1}\frac{\partial^2 C_1^{(k)}}{\partial Y_1^2};$$

$$X=0,\quad C_1^{(k)}=1;\quad Y_1=0,\quad C_1^{(k)}=-C_2^{(k-1)}(X,0);\qquad(1.3.78)$$

$$Y_1\to\infty\quad\left(Y_1\geq\overline{Y}_1\right),\quad C_1^{(k)}=1;\quad 0\leq X\leq 1,\quad 0\leq Y_1\leq\overline{Y}_1;$$

$$Sc_1=0.735,\quad \overline{Y}_1=7;\quad\text{(at the first iteration }C_2^{(k)}(X,0)=0\text{)}.$$

$$U_2^{(k)}\frac{\partial C_2^{(k)}}{\partial X}+V_2^{(k)}\frac{\partial C_2^{(k)}}{\partial Y_2}=\frac{1}{Sc_2}\frac{\partial^2 C_2^{(k)}}{\partial Y_2^2}-DaC_2^{(k)};$$

$$X=0,\quad C_2^{(k)}=0;\quad Y_2=0,\quad \frac{\partial C_2^{(k)}}{\partial Y_2}=\theta_5\left(\frac{\partial C_1^{(k)}}{\partial Y_1}\right)_{Y_1=0};\qquad(1.3.79)$$

$$Y_2\to\infty\quad(Y_2\geq\overline{Y}_2),\quad C_2^{(k)}=0;\quad 0\leq X\leq 1,\quad 0\leq Y_2\leq\overline{Y}_2;$$

$$Sc_2=564,\quad \theta_5=18.3,\quad \overline{Y}_2=0.26,\quad Da=10.$$

$$U_1^{(k)}\frac{\partial T_1^{(k)}}{\partial X}+V_1^k\frac{\partial T_1^{(k)}}{\partial Y_1}=\frac{1}{Pr_1}\frac{\partial^2 T_1^{(k)}}{\partial Y_1^2};$$

$$X=0,\quad T_1^{(k)}=0;\quad Y_1=0,\quad T_1^{(k)}=-T_2^{(k-1)}(X,0);\qquad(1.3.80)$$

$$Y_1\to\infty\quad(Y_1\geq\overline{\overline{Y}}_1),\quad T_1^{(k)}=0;\quad 0\leq X\leq 1,\quad 0\leq Y_1\leq\overline{\overline{Y}}_1;$$

$$Pr_1=0.666,\quad \overline{\overline{Y}}_1=7.4;\quad\text{(at the first iteration }T_2^{(k)}(X,0)=0\text{)}.$$

$$U_2^{(k)}\frac{\partial T_2^{(k)}}{\partial X}+V_2^{(k)}\frac{\partial T_2^{(k)}}{\partial Y_2}=\frac{1}{Pr_2}\frac{\partial^2 T_2^{(k)}}{\partial Y_2^2}+QDaC_2^{(k)};$$

$$X=0,\quad T_2^{(k)}=0;\quad Y_2=0,\quad \frac{\partial T_2^{(k)}}{\partial Y_2}=\theta_6\left(\frac{\partial T_1^{(k)}}{\partial Y_1}\right)_{Y_1=0};\qquad(1.3.81)$$

$$Y_2\to\infty\quad(Y_2\geq\overline{\overline{Y}}_2),\quad T_2^{(k)}=0;\quad 0\leq X\leq 1,\quad 0\leq Y_2\leq\overline{\overline{Y}}_2;$$

$$Pr_2=6.54,\quad \theta_6=0.034,\quad \overline{\overline{Y}}_2=2.4,\quad QDa=8.6.$$

Table 1.3 Influence of the nonlinear mass transfer effect and Marangoni effect on the heat and mass transfer kinetics in gas–liquid systems

Gas–liquid $\theta_1 = 0.1$ $\theta_2 = 0.145$

No.	θ_3	θ_4	J_1	J_2	J_3	J_4
1	0	0	0.5671	0.09721	0.01855	−0.01337
2	0.2	0	0.6129	0.01155	0.02143	−0.01554
3	−0.2	0	0.5274	0.08542	0.01623	−0.01162
4	0	10^{-4}	0.5671	0.09721	0.01855	−0.01338
5	0	10^{-3}	0.5671	0.09721	0.01855	−0.01337
6	0	10^{-2}	0.5670	0.09718	0.01857	−0.01339
7	0	10^{-1}	0.5658	0.09696	0.01879	−0.01364
8	0	1	0.5658	0.09696	0.01879	−0.01364
9	0	5	0.5660	0.09696	0.01854	0.01345

Table 1.4 Influence of the nonlinear mass transfer effect and Marangoni effect on the heat and mass transfer kinetics in liquid–liquid systems

Liquid–liquid $\theta_1 = 0.9$ $\theta_2 = 3$ $(u_2(X, Y_2) = 1)$

No.	θ_{31}	θ_{32}	θ_4	J_1	J_2	J_3	J_4
1	0	0	0	21.1000	4.8778	0.3320	−0.0524
2	4.10^{-4}	-8.10^{-4}	0	22.5419	5.7854	0.4288	−0.0628
3	0	0	2.10^{-4}	21.1000	4.8778	0.3320	−0.0524
4	0	0	1.10^{-3}	21.0999	4.8778	0.3320	−0.0524
5	0	0	1.10^{-2}	21.0990	4.8774	0.3320	−0.0524
6	0	0	1.10^{-1}	21.0899	4.8736	0.3319	−0.0524
7	0	0	5	20.5698	4.6527	0.3291	−0.0513

The values of the parameters in (1.3.76)–(1.3.81) are calculated for the process of absorption of NH_3 in water or water solutions of strong acids. The results obtained by solving these problems are shown in Tables 1.3, 1.4, and 1.5.

The comparative analysis of the nonlinear mass transfer effect and the Marangoni effect in gas–liquid and liquid–liquid systems [103, 104] shows (Tables 1.3 and 1.4) that the Marangoni effect does not affect the heat and mass transfer kinetics, because in real systems the parameter θ_4 is very small.

However, in cases where the velocity of the second phase is very low the occurrence of the Marangoni effect is to be expected because of its velocity dependence as $\left(u_{20}^{-3/2}\right)$. In order to evaluate the above case, systems with the velocity in the volume of the second phase equals to zero $(u_{20} = 0)$ have been investigated. The numerical results (Table 1.5) show that under these conditions the Marangoni effect is negligible, too.

Table 1.5 Influence of the non- linear mass transfer effect and Marangoni effect on the heat and mass transfer kinetics in liquid–liquid systems, when the second liquid is immobile

Liquid–liquid $\theta_1 = 1$ $\theta_2 = 1$ $(u_2(X, Y_2) = 10^{-4})$							
No.	θ_{31}	θ_{32}	θ_4	J_1	J_2	J_3	J_4
1	0	0	0	16.9333	3.3960	0.3041	−0.0460
2	4.10^{-4}	-8.10^{-4}	0	18.3164	4.0715	0.3967	−0.0551
3	0	0	2.10^{-4}	16.9333	3.3960	0.3041	−0.0460
4	0	0	1.10^{-3}	16.9331	3.3959	0.3042	−0.0460
5	0	0	1.10^{-2}	16.9314	3.3952	0.3041	−0.0596
6	0	0	1.10^{-1}	16.9145	3.3885	0.3040	−0.0592
7	0	0	1	16.7421	3.3201	0.3026	−0.0456
8	0	0	5	15.8955	2.9669	0.2968	−0.0437

The results obtained show that the Marangoni effect is negligible in two-phase systems with movable phase boundary and absence of surface active agents. The deviations from the linear mass transfer theory have to be explained by the non-linear mass transfer effect in conditions of the large concentration gradients.

1.4 Interphase Mass Transfer Resistances

In the thermodynamic approximations (1.0.4) and (1.0.5) from (1.1.23), it is seen that the interphase mass transfer rate J in two phases (gas–liquid and liquid–liquid) systems

$$J = K_1(c_{10} - \chi c_{20}) = K_2\left(\frac{c_{10}}{\chi} - c_{20}\right), \quad K_2 = \chi K_1, \qquad (1.4.1)$$

is equal to the mass transfer rate in the phases

$$J = k_1\left(c_{10} - c_1^*\right) = k_2\left(c_2^* - c_{20}\right), \quad c_1^* = \chi c_2^*, \qquad (1.4.2)$$

where the interphase mass transfer coefficients K_j $(j = 1, 2)$ and mass transfer coefficients k_j $(j = 1, 2)$ are processes rates constants; i.e., their reciprocal values K_j^{-1}, k_j^{-1} $(j = 1, 2)$ may be regarded as resistance constants of the processes. The interphase mass transfer rate J is equal to the mass transfer rate in the phase, where the mass transfer resistance constant is bigger and the intensification of the interphase mass transfer is related with a decrease of this bigger resistance.

At constant interphase mass transfer rate J, from (1.4.2) follows:

$$k_1^{-1} \to 0, \quad k_1 \to \infty, \quad \left(c_{10} - c_1^*\right) \to 0, \quad c_1^* \to c_{10}, \quad c_1 \equiv c_{10}, \quad c_2^* = \frac{c_{10}}{\chi};$$

$$k_2^{-1} \to 0, \quad k_2 \to \infty, \quad \left(c_{20} - c_2^*\right) \to 0, \quad c_2^* \to c_{20}, \quad c_2 \equiv c_{20}, \quad c_1^* = \chi c_{20},$$

$$(1.4.3)$$

i.e., in the case $k_1^{-1} \to 0 \left(k_2^{-1} \to 0\right)$, the interphase mass transfer is limited by the mass transfer in the phase 2 (1) and the concentration in the phase 1 (2) is a constant.

From (1.1.26), it is possible to obtain the mass transfer resistances distribution in the phases:

$$\rho_1 + \rho_2 = 1, \quad \rho_1 = \frac{K_1}{k_1} = \frac{K_2}{\chi k_1}, \quad \rho_2 = \frac{\chi K_1}{k_2} = \frac{K_2}{k_2} \qquad (1.4.4)$$

where ρ_j ($j = 1, 2$) are the portions (parts) of the interphase mass transfer resistance in the phases, and from (1.1.23), the following applies:

$$\rho_1 = \frac{c_{10} - c_1^*}{c_{10} - \chi c_{20}}, \quad \rho_2 = \chi \frac{c_2^* - c_{20}}{c_{10} - \chi c_{20}}. \qquad (1.4.5)$$

1.4.1 Film Theory Approximation

According to the film theory approximation (1.1.9), from (1.1.23) follows:

$$1 = \frac{K_1}{k_1} + \frac{K_1 \chi}{k_2} = \frac{K_1}{k_1} + \frac{K_2}{k_2} = \frac{K_2}{k_1 \chi} + \frac{K_2}{k_2} = \rho_1 + \rho_2,$$

$$\rho_1 = \frac{K_1}{k_1} = \frac{K_2}{k_1 \chi} = \frac{c_{10} - c_1^*}{c_{10} - \chi c_{20}}, \quad \rho_2 = \frac{K_1 \chi}{k_2} = \frac{K_2}{k_2} = \chi \frac{c_2^* - c_{20}}{c_{10} - \chi c_{20}}.$$

$$(1.4.6)$$

i.e., the expression (1.4.5) is film theory approximation, because in both cases (1.4.5, 1.4.6) the distribution of the interphase mass transfer resistance in the phases is determined without taking into account the effect of the movement of the phases on the mass transfer rate (like the thermodynamic approximation).

1.4.2 Boundary Layer Theory Approximation

The interphase mass transfer is a result of the convective and diffusive mass transfer, where convection dominates, but this is not accounted for by the film theory. A more precise analysis of the distribution of the interphase mass transfer resistance in two-phase systems requires the use of the estimate of the boundary layer theory [105]:

$$u_1 \frac{\partial c_1}{\partial x} + v_1 \frac{\partial c_1}{\partial y} = D_1 \frac{\partial^2 c_1}{\partial y^2}; \quad u_2 \frac{\partial c_2}{\partial x} + v_2 \frac{\partial c_2}{\partial y} = D_2 \frac{\partial^2 c_2}{\partial y^2};$$

$$x = 0, \quad c_1 = c_{10}, \quad c_2 = 0; \quad y = 0, \quad c_1 = \chi c_2, \quad D_1 \frac{\partial c_1}{\partial y} = D_2 \frac{\partial c_2}{\partial y}; \quad (1.4.7)$$

$$y \rightarrow \infty, \quad c_1 = c_{10}; \quad y \rightarrow -\infty, \quad c_2 = 0,$$

where $u_G = u_G(x, y)$, $v_G = v_G(x, y)$, $u_L = u_L(x, y)$, $v_L = v_L(x, y)$ are the velocities components in the gas and liquid phases.

Equation (1.4.7) permit to obtain the distribution of the interphase mass transfer resistance. For this purpose, a qualitative process analysis must be made using generalized (dimensionless) variables [6]:

$$X = \frac{x}{l}, \quad Y_i = \frac{y}{\delta_i}, \quad U_i = \frac{u_i}{u_{i0}}, \quad V_i = \frac{v_i}{\varepsilon_i u_{i0}},$$

$$C_i = \frac{c_i}{c_{i0}}, \quad c_{20} = \frac{c_{10}}{\chi}, \quad \varepsilon_i = \frac{\delta_i}{l}, \quad i = 1, 2, \qquad (1.4.8)$$

where u_{10} and u_{20} are characteristic velocities of the phases, and δ_1 and δ_2 are the thicknesses of the diffusion boundary layers in the phases:

$$\delta_i \sim \sqrt{\frac{D_i l}{u_{i0}}} = l P e_i^{-0.5}, \quad P e_i = \frac{u_{i0} l}{D_i}, \quad i = 1, 2, \qquad (1.4.9)$$

In this way, (1.4.7) yields:

$$U_1 \frac{\partial C_1}{\partial X} + V_1 \frac{\partial C_1}{\partial Y_1} = Fo_1 \frac{\partial^2 C_1}{\partial Y_1^2}, \quad U_2 \frac{\partial C_2}{\partial X} + V_2 \frac{\partial C_2}{\partial Y_2} = Fo_2 \frac{\partial^2 C_2}{\partial Y_2^2};$$

$$X = 0, \quad C_1 = 1, \quad C_2 = 0; \quad Y_1 = Y_2 = 0, \quad C_1 = C_2, \quad \frac{\chi}{\varepsilon_0} \frac{\partial C_1}{\partial Y_1} = \frac{\partial C_2}{\partial Y_2};$$

$$Y_1 \rightarrow \infty, \quad C_1 = 1; \quad Y_2 \rightarrow -\infty, \quad C_2 = 0;$$

$$(1.4.10)$$

where

$$Fo_i = \frac{D_i l}{u_{i0} \delta_i^2}, \quad i = 1, 2, \quad \varepsilon_0 = \frac{D_2 \delta_1}{D_1 \delta_2} = \sqrt{\frac{D_2 u_{20}}{D_1 u_{10}}}. \qquad (1.4.11)$$

From (1.4.10), it can be seen that in the cases of $\varepsilon_0/\chi = 0$, the solution of the first equation is $C_1 \equiv 1$, i.e., the interphase mass transfer is limited by the mass transfer in the phase 2, while in the opposite case $\chi/\varepsilon_0 = 0$, the solution of the second equation is $C_2 \equiv 0$ and the interphase mass transfer is limited by the mass transfer in the phase 1. The mass transfer resistances are comparable, when $\chi/\varepsilon_0 \sim 1$. As a result from (1.4.10), it follows that as dimensionless mass transfer resistances in the first (ρ_1) and second (ρ_2) phases it is possible to use:

$$\rho_1 = \left(\frac{\partial C_1}{\partial Y_1}\right)_{Y_1=0}, \quad \rho_2 = \left(\frac{\partial C_2}{\partial Y_2}\right)_{Y_2=0}, \quad \frac{\chi}{\varepsilon_0}\rho_1 = \rho_2, \quad \rho_1 + \rho_2 = 1,$$

$$\rho_1 = \frac{1}{1 + \frac{\chi}{\varepsilon_0}}, \quad \rho_2 = \frac{\frac{\chi}{\varepsilon_0}}{1 + \frac{\chi}{\varepsilon_0}}.$$

$$(1.4.12)$$

This theoretical result is obviously more precise from the film theory conclusions because the influence of the characteristic velocities (see 1.4.11) is taken into account.

From (1.4.11), it follows that if u_{10} (u_{20}) increase, ε_0 decreases (increases), and as a result from (1.4.12), it follows, that ρ_1 (ρ_2) decrease.

1.4.3 Interphase Mass Transfer Intensification

The interphase mass transfer intensification is possible to be realized by the decrease of the bigger (limited) mass transfer resistance in the phases, i.e., by the increase of the mass transfer rate in the limiting phase.

The mass transfer process has convective and diffusive components. An increase of the characteristic velocity in the phase leads to the increase of the convective mass transfer directly, while the increase of the diffusive mass transfer is a result of the decrease of the diffusion boundary layer thickness (see 1.4.9) and increase of the concentration gradients in the diffusion boundary layer.

The increase of the characteristic velocity in liquid drops and gas bubbles is limited practically; i.e., they must be used for the phases with small mass transfer resistance (non-limited phases). As a result, the absorption of highly soluble gases $(\chi \to \infty)$ must be realized in gas–liquid drops systems, while for absorption of low soluble gases $(\chi \to 0)$, the liquid–gas bubbles systems are suitable.

1.5 Three-Phase Mass Transfer Processes

The absorption of gas components (in gas mixtures) with high (HCl), middle (SO_2), and low (CO_2) solubility in water, with two-phase absorbents, where the active components ($CaCO_3$, $Ca(OH)_2$) are water suspensions, as used in power plants for gas cleaning (absorption of SO_2 from waste gases), or neutralization of lime solutions by CO_2 absorption (calcined soda production). Here, SO_2 is possible to be regarded as a high solubility gas.

The presence of the active component in the absorbent as solution and solid phase leads to an increase of the absorption capacity of the absorbent, but the introduction of a new process (the solution of the solid phase) creates conditions for variation of the absorption mechanism (interphase mass transfer through two-face interphases—gas/liquid and liquid/solid).

1.5.1 Physical Model

As a physical model, a laminar co-current gas–liquid flow in vertical canal with flat wools (Fig. 1.1) will be used [106], where the liquid is flowing down as a film.

The gas flow is a mixture, and an active component dissolves in the liquid film, where it reacts with the active component of the absorbent, dissolved in the liquid. The solid walls are made from the same active component of the absorbent and dissolve in the liquid film. The chemical reaction between active components (reagents) of the gas and solid phases is equimolecular.

For physical model definiteness, a frame of reference will be used (Fig. 1.1), where the liquid films with length l flow on the solid walls $y = \pm r$. The film thickness is h, and the gas–liquid face interphase is $y = \pm(r - h)$, where the interphase velocity is u_S (m s^{-1}).

The inlet $(x = 0)$ concentration of the active gas component in the gas phase is c_0 (kg mol m^{-3}). The inlet $(x = 0)$ concentration of the active sold component in liquid phase is c_{20} (kg mol m^{-3}), which is equal to the equilibrium solubility of the solid phase on the walls $y = \pm r$. The inlet $(x = 0)$ concentration of the active gas component in the liquid phase is $c_1 = 0$ (kg mol m^{-3}). On the face interphase

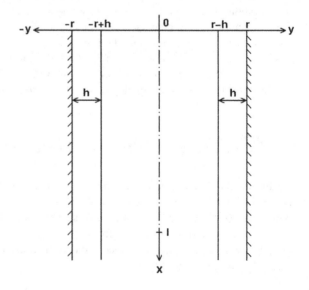

Fig. 1.1 Physical model of laminar co-current gas–liquid flow in vertical canal with flat wools

$y = \pm(r - h)$, a thermodynamic equilibrium exists $c^* = \chi c_1^*$, where c^* and c_1^* are the equilibrium concentrations of the active gas component in the gas and liquid phases, χ is Henry's number. For the chemical reaction rate $k c_1 c_2$ (kgmol m^{-3} s^{-1}) will be used, where c_1 and c_2 (kg mol m^{-3}) are the volumetric concentrations of the reagents and k—the chemical reaction rate constant.

1.5.2 Hydrodynamic Model

Let us consider a co-current flow of a laminar gas flow and a laminar liquid film flow with $\bar{u}_G$ and $\bar{u}_L$ being the average velocities (m s^{-1}):

$$\bar{u}_G = \frac{Q_G}{2r}, \quad \bar{u}_L = \frac{Q_L}{2h}, \tag{1.5.1}$$

where Q_G and Q_L (m^3 m^{-1} s^{-1}) are the gas and liquid flow rates in a canal with one meter width.

The velocity distributions in the gas and liquid phases will be obtained in the stratified flow approximations [9, 10]:

$$u_G = u_G(y), \quad u_L = u_L(y). \tag{1.5.2}$$

The big difference between gas and liquid viscosity coefficient values permits to neglect the friction force between gas and liquid phases. As a result, the hydrodynamic model assumes the form:

$$\frac{\partial^2 u_G}{\partial y^2} = \frac{1}{\mu} \frac{\partial p}{\partial x} = -A, \quad \frac{\partial p}{\partial x} < 0, \quad A > 0,$$

$$0 \le x \le l, \quad -r + h \le y \le r - h;$$

$$\frac{\partial^2 u_L}{\partial y^2} = -\frac{g}{v_L}, \quad 0 \le x \le l, \quad \pm(r - h) \le y \le \pm r;$$

$$y = 0, \quad \frac{\partial u_G}{\partial y} = 0; \quad y = \pm r, \quad u_L = 0; \tag{1.5.3}$$

$$y = \pm(r - h), \quad u_G = u_S, \quad \frac{\partial u_L}{\partial y} = 0,$$

where μ_G (kg m^{-1} s^{-1}) and v_L (m^2 s^{-1}) are the dynamic and kinetic coefficients of the viscosity of the gas and liquid phases.

The solutions of the equations in (1.5.3) for $y \ge 0$ are possible to be obtained immediately:

$$u_G = \bar{u}_G \left(\frac{u_S + u_0}{\bar{u}_G} - \frac{u_0}{\bar{u}_G} \frac{y^2}{(r-h)^2} \right), \quad u_L = u_S \left[2\frac{r-y}{h} - \frac{(r-y)^2}{h^2} \right],$$

$$u_S = \frac{gh^2}{2\nu_L}, \quad u_0 = \frac{3}{2}(\bar{u}_G - u_S). \tag{1.5.4}$$

The introduction of the characteristic scales (maximal or average values of the variables) permits to obtain generalized variables [6] (dimensionless variables with order of magnitude one). In the present case, they have the form:

$$u_G(y) = \bar{u}_G U_G(Y_G), \quad Y_G = \frac{y}{r-h};$$

$$u_L(y) = u_S U_L(Y), \quad Y = \frac{r-y}{h}. \tag{1.5.5}$$

The introduction of (1.5.5) into (1.4.4) leads to:

$$U_G = \alpha - \beta Y_G^2, \quad U_L = 2Y - Y^2, \quad \alpha = \frac{u_S + u_0}{\bar{u}_G}, \quad \beta = \frac{u_0}{\bar{u}_G}. \tag{1.5.6}$$

1.5.3   Interphase Mass Transfer Model

The obtained solutions (1.5.6) of the hydrodynamic problem permit the use of the convection–diffusion equations for the formulation of the interphase mass transfer model in thin layer approximations $\left(0 = \left(\frac{r}{l}\right)^2 \ll 1, \quad 0 = \left(\frac{h}{l}\right)^2 \ll 1 \right)$:

$$u_G \frac{\partial c}{\partial x} = D \frac{\partial^2 c}{\partial y^2}, \quad 0 \le x \le l, \quad 0 \le y \le r - h;$$

$$x = 0, \quad c = c_0; \quad y = 0, \quad \frac{\partial c}{\partial y} = 0; \quad y = r - h, \quad D_1 \frac{\partial c_1}{\partial y} = D \frac{\partial c}{\partial y}. \tag{1.5.7}$$

$$u_L \frac{\partial c_1}{\partial x} = D_1 \frac{\partial^2 c_1}{\partial y^2} - k c_1 c_2, \quad 0 \le x \le l, \quad r - h \le y \le r;$$

$$x = 0, \quad c_1 = 0; \quad y = r, \quad \frac{\partial c_1}{\partial y} = 0; \quad y = r - h, \quad c = \chi c_1. \tag{1.5.8}$$

$$u_L \frac{\partial c_2}{\partial x} = D_2 \frac{\partial^2 c_2}{\partial y^2} - k c_1 c_2, \quad 0 \le x \le l, \quad r - h \le y \le r;$$

$$x = 0, \quad c_2 = c_{20}; \quad y = r, \quad c_2 = c_{20}; \quad y = r - h, \quad \frac{\partial c_2}{\partial y} = 0. \tag{1.5.9}$$

In (1.5.7), (1.5.8), and (1.5.9), D, D_1, D_2 (m^2 s^{-1}) are the diffusivities of the reagents in the gas and liquid.

The theoretical analysis of the mechanism of gas absorption with two-phase absorbents will use (1.5.5) and the next generalized variables:

$$c(x, y) = c_0 C(X, Y_G), \quad c_1(x, y) = c_{10} C_1(X, Y),$$
$$c_2(x, y) = c_{20} C_2(X, Y), \quad x = lX, \quad c_{10} = \frac{c_0}{\chi}. \tag{1.5.10}$$

As a result, the interphase mass transfer model has the form:

$$\left(\alpha - \beta Y_G^2\right)\frac{\partial C}{\partial X} = Fo\frac{\partial^2 C}{\partial Y_G^2}; \quad X = 0, \quad C = 1;$$
$$Y_G = 0, \quad \frac{\partial C}{\partial Y_G} = 0; \quad Y = 1, \quad Y_G = 1, \quad \frac{\partial C}{\partial Y_G} = -\gamma_1\frac{\partial C_1}{\partial Y}. \tag{1.5.11}$$

$$\left(2Y - Y^2\right)\frac{\partial C_1}{\partial X} = Fo_1\frac{\partial^2 C_1}{\partial Y^2} - K_1 C_1 C_2;$$
$$X = 0, \quad C_1 = 0; \quad Y = 0, \quad \frac{\partial C_1}{\partial Y} = 0; \quad Y = 1, \quad Y_G = 1, \quad C = C_1. \tag{1.5.12}$$

$$\left(2Y - Y^2\right)\frac{\partial C_2}{\partial X} = Fo_2\frac{\partial^2 C_2}{\partial Y^2} - K_2 C_1 C_2;$$
$$X = 0, \quad C_2 = 1; \quad Y = 0, \quad C_2 = 1; \quad Y = 1, \quad \frac{\partial C_2}{\partial Y} = 0. \tag{1.5.13}$$

In (1.5.11), (1.5.12), and (1.5.13), the following expressions are used:

$$Fo = \frac{Dl}{\bar{u}_G(r - h)^2}, \quad Fo_1 = \frac{D_1 l}{u_S h^2}, \quad Fo_2 = \frac{D_2 l}{u_S h^2},$$
$$K_1 = \frac{k l c_{20}}{u_S}, \quad K_2 = \frac{k l c_0}{u_S}, \quad \gamma_1 = \frac{D_1(r - h)}{Dh\chi}. \tag{1.5.14}$$

1.5.4 Absorption Kinetics

The presented theoretical analysis of the gas absorption with two-phase absorbents shows that the absorption rate depends on the rates of four processes:

- mass transfer from the gas volume to the gas–liquid interphase;
- mass transfer from the gas–liquid interphase to the liquid volume;

- mass transfer from the solid–liquid interphase to the liquid volume;
- chemical reaction in the liquid volume.

In the general case, the absorption rate is possible to be obtained after solution of the problem (1.5.11), (1.5.12), and (1.5.13), but in special cases, the process is possible to be limited by one of the four processes and the model (1.5.11), (1.5.12), and (1.5.13) will be simplified greatly.

The absorption rate J (kg mol m^{-2} s^{-1}) is possible to be presented using an interphase mass transfer coefficient and the maximal concentration difference between the phases or as the average mass flow across the gas–liquid surface of the liquid film with l (m) length and 1 (m) width [6]:

$$J = k_G c_0 = \frac{D}{l} \int_0^l \left(\frac{\partial c}{\partial y}\right)_{y=r-h} dx = k_L \frac{c_0}{\chi} = \frac{D_1}{l} \int_0^l \left(\frac{\partial c_1}{\partial y}\right)_{y=r-h} dx, \quad (1.5.15)$$

where k_G and k_L (m s^{-1}) are interphase mass transfer coefficients [6] presented by the concentration gradients in the gas and liquid phases.

In generalized variables (1.5.5), (1.5.10) from (1.5.15), the expressions for Sherwood numbers in gas and liquid phases follow:

$$Sh_G = \frac{k_G(r-h)}{D} = \int_0^1 \left(\frac{\partial C}{\partial Y_G}\right)_{Y_G=1} dX, \quad Sh_L = \frac{k_L h}{D_1} = -\int_0^1 \left(\frac{\partial C_1}{\partial Y}\right)_{Y=1} dX,$$

$$(1.5.16)$$

where $C(X, Y_G)$ and $C_1(X, Y)$ are the solutions of the problem (1.5.11), (1.5.12), and (1.5.13).

1.5.5   Absorption Mechanism

The identification of the absorption mechanism is a result of the reagents mass balance in the film flow volume $\Delta v = h \Delta x$ (m^3) with width of 1 (m), thickness of h (m), and length of Δx (m).

The diffusion mass flux of the active gas component J_D (kg mol m^{-1} s^{-1}), which enters in the volume Δv across the gas–liquid surface $\Delta s_1 = \Delta x$ (m^2) (with width of 1 (m)), is:

$$J_D = -D_1 \left(\frac{\partial c_1}{\partial y}\right)_{y=r-h} \Delta x. \quad (1.5.17)$$

The convective mass flux of the active gas component J_C (kgmol m^{-1} s^{-1}), which enters in the volume Δv across the liquid surface $\Delta s_2 = h$ (m^2) (with of 1 (m) width), is:

$$J_C = \int_{r-h}^{r} u_L \Delta c_1 dy, \tag{1.5.18}$$

where

$$\Delta c_1 = c_1(x + \Delta x, y) - c_1(x, y). \tag{1.5.19}$$

The local mass flux of the active gas component (for each x) J_1) (kgmol m^{-2} s^{-1}) is:

$$J_1 = \lim_{\Delta s_1 \to 0} \frac{J_D + J_C}{\Delta s_1} = \lim_{\Delta x \to 0} \frac{J_D + J_C}{\Delta x}$$
$$= -D_1 \left(\frac{\partial c_1}{\partial y} \right)_{y=r-h} + \int_{r-h}^{r} u_L \frac{\partial c_1}{\partial x} dy. \tag{1.5.20}$$

The local mass flux of the active solid component J_2 (kg mol m^{-2} s^{-1}) is possible to be obtained by analogy:

$$J_2 = D_2 \left(\frac{\partial c_2}{\partial y} \right)_{y=r} + \int_{r-h}^{r} u_L \frac{\partial c_2}{\partial x} dy. \tag{1.5.21}$$

The chemical reaction is equimolecular, i.e., $J_1 = J_2$ and from (1.5.20) and (1.5.21), it follows:

$$D_1 \left(\frac{\partial c_1}{\partial y} \right)_{y=r-h} + D_2 \left(\frac{\partial c_2}{\partial y} \right)_{y=r} = \int_{r-h}^{r} u_L \left(\frac{\partial c_1}{\partial x} - \frac{\partial c_2}{\partial x} \right) dy. \tag{1.5.22}$$

In generalized variables (1.5.5), (1.5.10) from (1.5.22), it is possible to obtain:

$$\left(\frac{\partial C_1}{\partial Y} \right)_{Y=1} + \frac{D_2 \delta}{D_1} \left(\frac{\partial C_2}{\partial Y} \right)_{Y=0} = F,$$
$$F = \left(\frac{h}{l} \right)^2 \int_{0}^{1} U_L \left(\delta \frac{\partial C_2}{\partial X} - \frac{\partial C_1}{\partial X} \right) dY \ll 1, \quad Pe_1 = \frac{u_S l}{D_1}, \quad \delta = \frac{c_{20}}{c_0}, \tag{1.5.23}$$

where the parameter δ is the ratio of the maximal (equilibrium) solubilities of the gas and solid reagents in the liquid and the value of this parameter is possible to be in a very large interval. In the next analysis, the approximations $0 = F \ll 1$ and $(h/l)^2 \ll \delta^{-1}$ will be used.

The boundary conditions in (1.5.12) and (1.5.23) permit to obtain the relations between the mass fluxes at phase boundaries ($Y = 0, Y = 1, Y_G = 1$):

$$\left(\frac{\partial C}{\partial Y_G}\right)_{Y_G=1} = -\gamma_1 \left(\frac{\partial C_1}{\partial Y}\right)_{Y=1} \tag{1.5.24}$$

$$\left(\frac{\partial C_1}{\partial Y}\right)_{Y=1} = -\frac{D_2\delta}{D_1} \left(\frac{\partial C_2}{\partial Y}\right)_{Y=0} \tag{1.5.25}$$

$$\left(\frac{\partial C}{\partial Y_G}\right)_{Y_G=1} = \delta\gamma_2 \left(\frac{\partial C_2}{\partial Y}\right)_{Y=0}, \quad \gamma_2 = \frac{D_2(r-h)}{Dh\chi}, \tag{1.5.26}$$

where $\gamma_2 \sim \gamma_1$.

1.5.6   Absorption of Highly Soluble Gases

The relations (1.5.24), (1.5.25), and (1.5.26) permit to obtain the influence of the mass fluxes at the phase boundaries on the process mechanism.

In the cases of gases with high solubility (SO_2, HCl) $\chi \ll 1$ and $\gamma_1 \gg 1$ (e.g., $\chi_{SO_2} \sim 10^{-2}$, $\chi_{HCl} \sim 10^{-3}$) and the problem (1.5.12) is possible to be solved in approximation $0 = \gamma_1^{-1} \ll 1$. As a result, $\gamma_1 \gg 1$, $\left(\frac{\partial C_1}{\partial Y}\right)_{Y=1} \ll 1, C_1 \equiv 0$ and the model (1.5.11), (1.5.12), and (1.5.13) must be simplified:

$$(\alpha - \beta Y_G^2)\frac{\partial C}{\partial X} = Fo\frac{\partial^2 C}{\partial Y_G^2}; \quad X = 0, \quad C = 1;$$

$$Y_G = 0, \quad \frac{\partial C}{\partial Y_G} = 0; \quad Y = 1, \quad Y_G = 1, \quad \left(\frac{\partial C}{\partial Y_G}\right)_{Y_G=1} = \delta\gamma_2 \left(\frac{\partial C_2}{\partial Y}\right)_{Y=0}. \tag{1.5.27}$$

$$(2Y - Y^2)\frac{\partial C_2}{\partial X} = Fo_2\frac{\partial^2 C_2}{\partial Y^2};$$

$$X = 0, \quad C_2 = 1; \quad Y = 0, \quad C_2 = 1; \quad Y = 1, \quad \frac{\partial C_2}{\partial Y} = 0, \tag{1.5.28}$$

i.e., (1.5.12) is replaced by $C_1 \equiv 0$ and $-\gamma_1 \left(\frac{\partial C_1}{\partial Y}\right)_{Y=1}$ in (1.5.11) is replaced by $\delta\gamma_2 \left(\frac{\partial C_2}{\partial Y}\right)_{Y=0}$, using (0.5.26).

The solution of (1.5.27), (1.5.28) permits to obtain $\left(\frac{\partial C}{\partial Y_G}\right)_{Y_G=1}$ and k_G using (1.5.16).

From (1.5.27), it follows that if $(\delta\gamma_2)^{-1} \ll 1$, i.e., the problems (1.5.27), (1.5.28) are possible to be solved in approximation $0 = (\delta\gamma_2)^{-1} \ll 1$, i.e., $C_2 \equiv 1$ and the last boundary condition in (0.5.27) is replaced by $Y_G = 1$, $C = 1$:

$$(\alpha - \beta Y_G^2)\frac{\partial C}{\partial X} = Fo\frac{\partial^2 C}{\partial Y_G^2};$$

$$X = 0, \quad C = 1; \quad Y_G = 0, \quad \frac{\partial C}{\partial Y_G} = 0; \quad Y_G = 1, \quad C = 1.$$

$$(1.5.29)$$

In these cases, the interphase mass transfer is limited by the mass transfer in the gas phase, i.e., the diffusion resistance in the liquid phase is negligible. The solution of (0.5.29) permits to obtain $\left(\frac{\partial C}{\partial Y_G}\right)_{Y_G=1}$ and k_G using (1.5.16).

In the cases of $\delta\gamma_2 \ll 1$, the problems (1.5.27), (1.5.28) are possible to be solved in approximation $0 = \delta\gamma_2 \ll 1$, i.e. $C \equiv 1$ and the last boundary condition in (1.5.28) is replaced by $Y = 1, C_2 = 0$:

$$(2Y - Y^2)\frac{\partial C_2}{\partial X} = Fo_2\frac{\partial^2 C_2}{\partial Y^2};$$

$$X = 0, \quad C_2 = 1; \quad Y = 0, \quad C_2 = 1; \quad Y = 1, \quad C_2 = 0.$$

$$(1.5.30)$$

In these cases, the interphase mass transfer is limited by the mass transfer in the liquid phase (solid-phase dissolution); i.e., the diffusion resistance in the gas phase is negligible. The solution of (1.5.30) permits to obtain $\left(\frac{\partial C_2}{\partial Y}\right)_{Y=0}$ and k_L using (1.5.16), where $\left(\frac{\partial C_1}{\partial Y}\right)_{Y=1}$ is replaced by $-\frac{D_2\delta}{D_1}\left(\frac{\partial C_2}{\partial Y}\right)_{Y=0}$ (see 1.5.25):

$$Sh_L = \frac{k_L h}{D_1} = \frac{\delta D_2}{D_1}\int\limits_0^1 \left(\frac{\partial C_2}{\partial Y}\right)_{Y=0} dX,$$

$$(1.5.31)$$

The value of $\delta\gamma_2$ is related to the concentration of the active gas component in gas phase c_0 and increases for big concentration values of c_0; i.e., the diffusion resistance in gas phase increases. The similar result is possible to be obtained if the equilibrium solubility of the solid phase c_{20} decreases (replacement of $Ca(OH)_2$ by $CaCO_3$). An influence of the diffusion resistance in the liquid phase at gas–liquid interphase is possible too, because for SO_2 $\gamma_1 \sim 10$. As a result, a commensurability of the diffusion resistances in the gas and liquid phases is possible.

1.5.7 Absorption of Low Soluble Gases

In the cases of gases with low solubility (CO_2), $\chi \gg 1$ and $\gamma_1 \ll 1$ (e.g., $\chi_{CO_2} \sim 1$, $\gamma_1 \sim 10^{-2}$) and the problem (1.5.11) is possible to be solved in approximation $0 = \gamma_1 \ll 1$. As a result, the problem (1.5.11) must be replaced by $C \equiv 1$ and the last boundary condition in (1.5.12) by $Y = 1, C_1 = 1$:

$$\left(2Y - Y^2\right)\frac{\partial C_1}{\partial X} = Fo_1 \frac{\partial^2 C_1}{\partial Y^2} - K_1 C_1 C_2;$$

$$X = 0, \quad C_1 = 0; \quad Y = 0, \quad \frac{\partial C_1}{\partial Y} = 0; \quad Y = 1, \quad C_1 = 1. \tag{1.5.32}$$

$$\left(2Y - Y^2\right)\frac{\partial C_2}{\partial X} = Fo_2 \frac{\partial^2 C_2}{\partial Y^2} - K_2 C_1 C_2;$$

$$X = 0, \quad C_2 = 1; \quad Y = 0, \quad C_2 = 1; \quad Y = 1, \quad \frac{\partial C_2}{\partial Y} = 0. \tag{1.5.33}$$

The solution of (1.5.32), (1.5.33) permits to obtain $\left(\frac{\partial C_1}{\partial Y}\right)_{Y=1}$ and k_L using (1.5.16).

From (1.5.32), it follows that $K_1 \ll 1$ for small values of c_{20} and the problem (1.5.32) is possible to be solved in approximation $0 = K_1 \ll 1$:

$$\left(2Y - Y^2\right)\frac{\partial C_1}{\partial X} = Fo_1 \frac{\partial^2 C_1}{\partial Y^2};$$

$$X = 0, \quad C_1 = 0; \quad Y = 0, \quad \frac{\partial C_1}{\partial Y} = 0; \quad Y = 1, \quad C_1 = 1. \tag{1.5.34}$$

In these cases, the processes are limited by the diffusion resistance in the liquid phase at the gas–liquid interphase and the solution of (1.5.34) permits to obtain $\left(\frac{\partial C_1}{\partial Y}\right)_{Y=1}$ and k_L using (1.5.16).

From (1.5.33), it follows that $K_2 \ll 1$ for small values of c_0 and the problem (1.5.33) is possible to be solved in the approximation $0 = K_2 \ll 1$; i.e., $C_2 \equiv 1$ must be replaced in (1.5.32).

In the cases of neutralization of lime solutions by CO_2, the values of c_0 and c_{20} are possible to be very big and the problem (1.5.32), (1.5.33) is possible to be used as a model of the absorption process.

From (1.5.32), (1.5.33), it is seen that if K_1 and K_2 increase, the influence of the hydrodynamics (velocity distributions) decreases.

The two-phase absorbents (water suspensions of $CaCO_3$ or $Ca(OH)_2$) are used in power plants for waste gas cleaning (absorption of SO_2) and the neutralization of lime solutions by CO_2 in the calcinated soda production plants.

The presented theoretical analysis shows that the using two-phase absorbents leads to modification of the absorption mechanism as a result of the new mass transfer resistance at the liquid–solid interphase. The absorption mechanisms for the practically interesting systems $SO_2/CaCO_3$, $SO_2/Ca(OH)_2$, and $CO_2/Ca(OH)_2$ are related with the values of the parameter δ (the ratio of the maximal (equilibrium) solubilities of the gas and solid reagents in liquid).

For the $SO_2/CaCO_3$ system, $\delta \sim 1$, is possible, i.e., the interphase mass transfer is limited by the mass transfer in the gas phase and the liquid phase, i.e., the mass transfer resistances in the gas and liquid phases are commensurable.

In the system, $SO_2/Ca(OH)_2$ $\delta \gg 1$ and the interphase mass transfer is limited by the mass transfer in the gas phase; i.e., the solid phase does not influence the process mechanism.

The determination of the phase, where the mass transfer controls the interphase mass transfer rate, permits to intensify the interphase mass transfer by an increase of the convective transfer in this phase. In the practical cases, the gas absorption is realized in moving gas–liquid dispersion systems and an intensification of the convective transfer in the gas (liquid) phase is possible if the dispersed phase is the liquid (gas).

1.6   Conclusions

The modeling and simulation are the main approach for the quantitative description of the processes in the chemical, power, biotechnological, and other industries. The modeling in the chemical engineering is possible to be realized on the basis of the physical approximations of the mechanics of continua, where the mathematical point is equivalent to an elementary physical volume, which is sufficiently small with respect to the apparatus volume, but at the same time sufficiently large with respect to the intermolecular volumes in the medium.

The big part of the industrial processes is realized in one-, two-, or three-phase systems as a result of volume (homogeneous) and surface (heterogeneous) reactions, i.e., appearance (disappearance) of the phase components (reagents) in the elementary volumes in the phases or on the interphase surfaces, as a result of volume chemical reactions and surfaces reactions (catalytic reactions, interphase mass transfer, adsorption).

The volume reactions lead to different concentrations of the reagents in the phase volumes, and as a result, two mass transfer processes are realized—convective transfer (caused by the movement of the phases) and diffusion transfer (caused by the concentration gradients in the phases). The mass transfer models are a mass balance in the phases, where components are convective transfer, diffusion transfer, and volume reactions (volume mass sources or sinks). The surface reactions participate as mass sources or sinks in the boundary conditions of the model equations.

The presented theoretical analysis of the mass transfer theories shows that the predictions of the diffusion boundary layer theories are more accurate than the model theories conclusions.

A fundamental prerequisite for the use of the interphase mass transfer theory is the existence of a theoretical possibility to determine the velocity distribution in the phases and the interphase boundaries (surfaces). In the cases of modeling of the interphase mass transfer processes (absorption, adsorption, and catalytic reactions) in column apparatuses, the velocity and the interphase boundaries are unknown, and as a result, this theory is useful.

The use of the physical approximations of the mechanics of continua for the interphase mass transfer process modeling in industrial column apparatuses is possible if the mass appearance (disappearance) of the reagents on the interphase surfaces of the elementary physical volumes (as a result of the heterogeneous reactions) are replaced by the mass appearance (disappearance) of the reagents in the same elementary physical volumes (as a result of the equivalent homogenous reactions); i.e., the surface mass sources (sinks), caused by absorption, adsorption, or catalytic reactions must be replaced with equivalent volume mass sources (sinks).

A new approach to modeling the mass transfer processes in industrial column apparatuses is the creation of the convection–diffusion and average-concentration models.

The convection–diffusion models permit the qualitative analysis of the processes only, because the velocity distribution in the column is unknown. On this base, it is possible to be obtained the role of the different physical effects in the process and to reject those processes, whose relative influence is less than 1%, i.e., to be made process mechanism identification.

The average-concentration models permit the quantitative analysis of the processes. They are obtained from the convection–diffusion models, where average velocities and concentrations are introduced. The velocity distributions are introduced by the parameters in the model, which must to be determined experimentally.

References

1. Keizer J (1987) Statistical thermodynamics of nonequilibrium processes. Springer, New York
2. Rousseau RW (1987) Handbook of separation process technology. Wiley, New York
3. Perry R, Green DW (1999) Perry's chemical engineers' handbook. McGraw Hill, New York
4. Towler G, Sinnott R (2008) Chemical engineering design: principles, practice and economics of plant and process design. Elsevier, Amsterdam
5. Amundson WR, Pontinen AJ (1958) Ind Eng Chem 50:730
6. Boyadjiev C (2010) Theoretical chemical engineering. Modeling and simulation. Springer, Berlin
7. Schlichting H, Gerstein K (2000) Boundary layer theory, 8th revised and enlarged edition. Springer, Berlin
8. Nernst W (1904) Z Phys Chem 47:52
9. Boyadjiev C, Beshkov V (1988) Mass transfer in following liquid films. Mir, Moscow (in Russian)
10. Chr Boyadjiev, Beshkov V (1984) Mass transfer in liquid film flows. Bulgarian Academy of Sciences, Sofia
11. Langmuir I (1912) Phys Rev 34:321
12. Lewis WK, Whitman WG (1924) Ind Eng Chem 16:1215

13. Higbie R (1935) Trans Am Inst Chem Eng 31:365
14. Boyadjiev C, Levich VG, Krylov VS (1968) The effect of surface active materials on mass transfer in laminar film flow. I. Improvement of the theory of the convective diffusion. Int Chem Eng 8(3):393–396
15. Boyadjiev C (1971) Mass transfer during the simultaneous motion of a laminar liquid film and a laminar gas stream. Int Chem Eng 11(3):459–464
16. Prandtl L (1910) Z Phys 2:1072
17. Taylor GJ (1916) British advisory communications for aeronautics. R and M No 272
18. Kishinevsky MK, Pamfilov AV (1949) J Appl Chem (Russia) 22:1173
19. Kishinevsky MK (1951) J Appl Chem (Russia) 24:542
20. Danckwerts PV (1951) Ind Eng Chem 43:1960
21. Toor HL, Marchello JM (1958) AIChEJ 4:97
22. Ruckenstein E (1958) Chem Eng Sci 7:265
23. Ruckenstein E (1963) Chem Eng Sci 18:233
24. Ruckenstein E (1967) Chem Eng Sci 22:474
25. Kh Kishinevsky M (1965) Int Heat Mass Transfer 8:1181
26. Reichardt H, Angew Z (1951) Math Mech 7:31
27. Elrod HG (1957) J Aeronaut Sci 24:468
28. Wasan DT, Tien CL, Wilke CR (1963) AIChEJ 4:4
29. Dilman VV (1967) Theor Fundam Chem Technol (Russia) 1:438
30. Deissler RG (1955) NACA, Report No. 1210
31. Levich VG (1962) Physicochemical hydrodynamics. Prentice-Hall, New York
32. Son JS, Hanratty TJ (1967) AIChEJ 13:689
33. Van Driest ER (1956) J Aeronaut Sci 23:1007
34. Loytsiansky LG (1960) Apll Math Mech (Russia) 4:24
35. Reichardt H (1957) National Advisory Committee for Aeronautics Technical Note 1408
36. Boyadjiev C, Toshev E (1989) Asymptotic theory of nonlinear transport phenomena in boundary layers. 1. Mass transfer. Hung J Ind Chem 17:457–463
37. Pohlhausen E (1921) ZAMM 1:115
38. Boyadjiev C, Piperova M (1971) The hydrodynamics of certain two-phase flow. 4. Evaluation of some special functions. Int Chem Eng 11(3):479–487
39. Chr Boyadjiev, Pl Mitev (1977) On the concentration boundary layer theory at a moving interface. Chem Eng J 14:225–228
40. Chr Boyadjiev (1971) The hydrodynamics of certain two-phase flows. 1. The laminar boundary layer at a flat gas-liquid interface. Int Chem Eng 11(3):465–469
41. Chr Boyadjiev, Vulchanov N (1988) Non linear mass transfer in boundary layers—1. Asymptotic theory. Int J Heat Mass Transfer 31(4):795–800
42. Chr Boyadjiev, Pl Mitev, Tsv Sapundjiev (1976) Laminar boundary layers of co-current gas-liquid stratified flows-1. Theory. Int J Multiphase Flow 3(1):51–55
43. Pl Mitev, Chr Boyadjiev (1976) Laminar boundary layers of co-current gas-liquid stratified flows—2. Velocity measurements. Int J Multiphase Flow 3:57–60
44. Pl Mitev, Boyadjiev C (1978) Mass transfer by co-current gas-liquid stratified flow. Letters Heat Mass Transfer 5:349–354
45. Boyadjiev C, Mitev P, Beschkov V (1976) Laminar boundary layers at a moving interface generated by counter-current gas-liquid stratified flow. Int J Multiphase Flow 3:61–66
46. Chr Boyadjiev, Doichinova M (2000) Opposite-current flows in gas-liquid boundary layers —I. velocity distribution. Int J Heat Mass Transfer 43:2701–2706
47. Chr Boyadjiev, Vabishchevich P (1992) Numerical simulation of opposite currents. J Theor Appl Mech (Bulgaria) 23:114
48. Tersenov SA (1985) Parabolic equations with changing direction of time. Science, Novosibirsk (in Russian)
49. Larkin IA, Novikov VA, Ianenko NN (1983) Nonlinear equations from changed type. Science, Novosibirsk (in Russian)

50. Doichinova M, Chr Boyadjiev (2000) Opposite-current flows in gas-liquid boundary layers
    —II. Mass transfer kinetics. Int J Heat Mass Transfer 43:2707–2710
51. Boyadjiev CB, Babak VN (2000) Non-linear mass transfer and hydrodynamic stability.
    Elsevier, Amsterdam
52. Krylov VS, Bogoslovsky VE, Mihnev NN (1976) J Appl Chem (Russia) 49:1769
53. Bird RB, Stewart WE, Lightfoot EN (1960) Transport phenomena. Willey, New York
54. Brounstain BI, Fishbain GA (1977) Hydrodynamics mass and heat transfer in disperse
    systems. Himia, Leningrad (in Russian)
55. Chang WS (1973) Int J Heat Mass Transfer 16:811
56. Duda JL, Vrentas JS (1971) Int J Heat Mass Transfer 14:395
57. Parlange JY (1973) Acta Mech 18:157
58. Ranz WE, Dickinson PF (1965) Ind Eng Chem Funds 4:345
59. Golovin AM, Rubinina NM, Hohrin VM (1971) Theor Fundam Chem Technol (Russia)
    5:651
60. Emanuel AS, Olander DR (1964) Int J Heat Mass Transfer 7:539
61. Nienow AW, Unahabhokha R, Mullin JW (1968) Chem Eng Sci 24:1655
62. Olander DR (1962) Int J Heat Mass Transfer 5:765
63. Unahabhokha R, Nienow AW, Mullin JW (1972) Chem Eng Sci 26:357
64. Uan SU, Lic Czja-Czjao (eds) (1963) Cooling by means of liquid films—turbulence flows
    and heat transfer. Inostrannaja Literatura, Moscow (in Russian)
65. Olander DR (1962) J Heat Transfer Trans ASME Ser C 84:185
66. Ross SM (1974) J Fluid Mech 63:157
67. Sparrow EW, Gregg JL (1960) J Heat Transfer Trans ASME Ser C 82:294
68. Yuan SW, Finkelstein AB (1956) J Heat Transfer Trans ASMESer C 78:719
69. Hirshfelder J, Kertis E, Berd R (1961) Molecular theory of gases and liquids. Inostrannaja
    Literatura, Moscow (in Russian)
70. Franc-Kamenetskii VA (1969) Diffusion and heat transfer in chemical kinetics. Plenum
    Press, New York
71. Krylov VS, Davidov AD (1972) Proceedings of advanced methods in electrochemical
    machining. Shtiinca, Kishinev, p. 13 (in Russian)
72. Krylov VS, Malienko VN (1972) Thermodynamics of irreversible processes and its
    applications. In: Proceedings of 1st USSR conference, Chernovcy, p. 69 (in Russian)
73. Krylov VS, Malienko VN (1973) Electrochemistry (Russia) 9:3
74. Boyadjiev C, Vulchanov N (1987) Effect of the direction of the interphase mass transfer on
    the rate of mass transfer. C R Acad Bulg Sci 40(11):35–38
75. Boyadjiev C, Vulchanov N (1990) Influence of the interphase mass transfer on the rate of
    mass transfer—1. The system 'solid-fluid (gas)'. Int J Heat Mass Transf 33(9):2039–2044
76. Vulchanov N, Boyadjiev C (1988) The influence of nonlinear mass transfer on the laminar
    boundary layer. Theor Appl Mech (Bulgaria) 19(4):74–78
77. Boyadjiev C (1991) Asymptotic theory of the non-linear mass transfer. J Eng Physics
    (Russia) 60(5):845–862
78. Boyadjiev C (1972) On the absorption theory. Theor Fundam Chem Technol (Russia) 6
    (1):118–121
79. Chr Boyadjiev, Velchev L (1971) Gas absorption in horizontal channel. Theor Fundam
    Chem Technol (Russia) 5(6):912–915
80. Boyadjiev C (1971) Mass transfer during the simultaneous motion of a laminar liquid film
    and a laminar gas stream. Int Chem Eng 11(3):459–464
81. Boyadjiev C (1971) The hydrodynamics of certain two-phase flows. 1. The laminar
    boundary layer at a flat gas-liquid interface. Int Chem Eng 11(3):465–469
82. Boyadjiev C (1971) The hydrodynamics of certain two-phase flows. 2. Simultaneous motion
    of a gas and a liquid flow. Int Chem Eng 11(3):470–474
83. Chr Boyadjiev, Elenkov D (1971) The hydrodynamics of certain two-phase flows. 3. The
    effect of surface-active materials. Int Chem Eng 11(3):474–476

84. Krylov VS, Chr Boyadjiev, Levich VG (1967) On the convection-diffusion theory in liquid film flow. C R USSR Acad Sci (Russia) 175(1):156–159

85. Chr Boyadjiev (1992) On the kinetics of the intensive interphase mass transfer. Russ J Eng Thermophys 2(4):289–297

86. Vulchanov N, Chr Boyadjiev (1988) Non-linear mass transfer in boundary layers—2. Numerical investigation. Int J Heat Mass Transfer 31(4):801–805

87. Vulchanov N, Boyadjiev C (1990) Influence of the interphase mass transfer on the rate of mass transfer—2. The system 'gas-liquid'. Int J Heat Mass Transfer 33(9):2045–2049

88. Boyadjiev C (1998) Non-linear mass transfer in gas-liquid systems. Hung J Ind Chem 26:181–187

89. Chr Boyadjiev (1998) Non-linear interphase mass transfer in multi-component gas-liquid systems. Hung J Ind Chem 26:245–249

90. Sternling CV, Scriven LE (1959) AIChEJ 5:514

91. Linde H, Schwarz E, Groger K (1967) Chem Eng Sci 22:823

92. Ruckenstein E, Berbente C (1964) Chem Eng Sci 19:329

93. Thomas WJ, Nicholl E Mc K (1969) Trans Inst Chem Eng 47(10):325

94. Dilman VV, Kulov NN, Lothov VA, Kaminski VA, Najdenov VI (1998) Theor Fundam Chem Technol (Russia) 32:377

95. Porter KE, Cantwell ADC, Dermott CM (1971) AIChEJ 17:536

96. Hennenberg M, Bisch PM, Vignes-Adler M, Sanfeld A (1979) Interfacial instability and longitudinal waves in liquid-liquid systems. Lecture note in physics, vol 105. Springer, Berlin, p 229

97. Linde H, Schwartz P, Wilke H (1979) Dissipative structures and nonlinear kinetics of the Marangoni—instability. Lecture note in physics, vol 105. Springer, Berlin 75

98. Sanfeld A, Steinchen A, Hennenberg M, Bisch PM, Van Lamswerde-Galle D, Dall-Vedove W (1979) Mechanical and electrical constraints and hydrodynamic interfacial instability. Lecture note in physics, vol 105. Springer, Berlin, p 168

99. Savistowski H (1981) Interfacial Convection Ber Bunsenges Phys Chem 85:905

100. Sorensen TS, Hennenberg M (1979) Instability of a spherical drop with surface chemical reaction and transfer of surfactants. Lecture note in physics, vol 105. Springer, Berlin, p 276

101. Scriven LE, Sterling CV (1960) Nature (London) 127(4733):186

102. Velarde J, Gastillo L, Zierep J, Oertel H (eds) (1981) Transport and reactive phenomena leading to interfacial instability, convective transport and instability phenomena. Verlag G Braun, Kalsruhe

103. Chr Boyadjiev, Halatchev I (1998) Non-linear mass transfer and Marangoni effect in gas-liquid systems. Int J Heat Mass Transfer 41(1):197–202

104. Chr Boyadjiev, Doichinova M (1999) Non-linear mass transfer and Marangoni effect. Hungarian J Ind Chem 27:215–219

105. Boyadjiev C, Doichinova M, Boyadjiev B (2013) Some problems in the column apparatuses modeling. Private communication

106. Boyadjiev C (2011) Mechanism of gas absorption with two-phase absorbents. Int J Heat Mass Transfer 54:3004–3008

Part I
Qualitative Analysis of Column Apparatuses Processes

Convection–Diffusion-Type Models

The complex processes in the column apparatuses are a combination of hydrodynamic processes, convective and diffusive mass (heat) transfer processes, and chemical reactions between the reagents (components of the phases).

The fundamental problem in the column apparatuses modeling stems from the complicated hydrodynamic behavior of the flows in the columns, and as a result, the velocity distributions in the columns are unknown.

In the general case, a multicomponent $(i = 1, 2, ..., i_0)$ and multiphase $(j = 1, 2, 3$ for gas, liquid, and solid phases) flow in a cylindrical column with radius r_0 [m] and active zone height l [m] will be considered. If F_0 is the fluid flow rate in the column and F_j, $j = 1, 2, 3$ are the phase flow rates [m^3 s^{-1}], the parts of the column volume occupied by the gas, liquid, and solid phases, respectively, i.e., the phase volumes [m^3] in 1 m^3 of the column volume (holdup coefficients of the phases), are:

$$\varepsilon_j = \frac{F_j}{F_0}, \quad j = 1, 2, 3, \quad \sum_j^3 \varepsilon_j = 1. \tag{I.1}$$

The input velocities of the phases in the column u_j^0 [m s^{-1}], $j = 1, 2, 3$ are possible to be defined as:

$$u_j^0 = \frac{F_j}{\varepsilon_j \pi r_0^2}, \quad j = 1, 2, 3; \quad F_0 = \sum_{j=1}^3 F_j. \tag{I.2}$$

The column apparatuses are possible to be modeled using a new approach [1–4] on the basis of the physical approximations of the mechanics of continua, where the mathematical point (in the phase volume or on the surface between the phases) is equivalent to a small (elementary) physical volume, which is sufficiently small with respect to the apparatus volume, but at the same time sufficiently large with respect to the intermolecular volumes in the medium. All models in this part will be created on this basis.

The physical elementary column volumes contain the elementary phase volumes ε_j, $j = 1, 2, 3$ and will be presented as mathematical points M in a cylindrical coordinate system (r, z), where r and z [m] are radial and axial coordinates. As a result, the mathematical point $M(r, z)$ is equivalent to the elementary phase volumes, too.

The concentrations [kg-mol m^{-3}] of the reagents (components of the phases) are c_{ij}, $i = 1, 2, ..., i_0$, $j = 1, 2, 3$, i.e., the quantities of the reagents (kg-mol) in 1 m^3 of the phase volumes in the column.

In the cases of a stationary motion of fluids in cylindrical column apparatus, $u_j(r, z)$, $v_j(r, z)$, $j = 1, 2, 3$ [m s^{-1}] are the axial and radial velocity components of the phases in the elementary phase volumes.

In the column apparatuses, the phase boundaries are unknown, and therefore, the heterogeneous reactions (absorption, adsorption, catalytic reactions) are introduced as a volume sources (sinks) in the elementary phase volumes.

The volume reactions [kg-mol m^{-3} s^{-1}] in the phases (homogeneous chemical reaction and interphase mass transfer, as a volume source or sink in the phase volume in the column) are $Q_{ij}(c_{ij})$, $j = 1, 2, 3$, $i = 1, 2, ..., i_0$. The reagent (substance) concentrations in the elementary phase volumes can be created ($Q_{ij} > 0$) or disappear ($Q_{ij} < 0$), and the reaction rates Q_{ij} are determined by these concentrations $c_{ij}(t, r, z)$ [kg-mol m^{-3}], where $t(s)$ is the time.

The volume reactions lead to different values of the reagent (substance) concentrations in the elementary phase volumes, and as a result, two mass transfer effects exist—convective transfer (caused by the fluid motion) and diffusion transfer (caused by the concentration gradient).

The convective transfer in column apparatus is caused by a laminar flow. In a small (elementary) phase volume around the point $M(r, z)$ in the column, the mass transfer in this volume, as a result of the convection, is $u_j \frac{\partial c_{ij}}{\partial z} + v_j \frac{\partial c_{ij}}{\partial r}$ [kg-mol m^3 s^{-1}], $j = 1, 2, 3$, $i = 1, 2, ..., i_0$, i.e., convective transfer rate (kg-mol s^{-1}) in 1 m^3 of the phase volume.

The molecular diffusive transfer is $D_{ij} \left(\frac{\partial^2 c_{ij}}{\partial z^2} + \frac{1}{r} \frac{\partial c_{ij}}{\partial r} + \frac{\partial^2 c_{ij}}{\partial r^2} \right)$ [kg-mol m^{-3} s^{-1}], i.e., diffusive transfer rate (kg-mol s^{-1}) in 1 m^3 of the phase volume, and D_{ij} [m^2 s^{-1}] are the diffusivities of the reagents ($i = 1, 2, ..., i_0$) in the phases ($j = 1, 2, 3$).

The mathematical model of the processes in the column apparatuses, in the physical approximations of the mechanics of continua, represents the mass balances in the phase volumes (phase parts in the elementary column volume) between the convective transfer, the diffusive transfer, and the volume mass sources (sinks) (as a result of the chemical reactions, interphase mass transfer, adsorption or catalytic reaction). The sum total of these three effects is equal to $\partial c_{ij} / \partial t$,

$j = 1, 2, 3, \ i = 1, 2, \ldots, i_0$. In the case of balance between these three effects, the mass transfer process is stationary $\left(\partial c_{ij} / \partial t = 0 \right)$.

In the stationary case, the convection–diffusion equations (as a mathematical structure of the mass transfer process models in the column apparatuses) are:

$$u_j \frac{\partial c_{ij}}{\partial z} + v_j \frac{\partial c_{ij}}{\partial r} = D_{ij} \left(\frac{\partial^2 c_{ij}}{\partial z^2} + \frac{1}{r} \frac{\partial c_{ij}}{\partial r} + \frac{\partial^2 c_{ij}}{\partial r^2} \right) + Q_{ij}(c_{ij}),$$

$$j = 1, 2, 3, \ i = 1, 2, \ldots, i_0.$$

(I.3)

The axial and radial velocity components $u_j(r, z)$ and $v_j(r, z)$, $j = 1, 2, 3$ satisfy the continuity equations

$$\frac{\partial u_j}{\partial z} + \frac{\partial v_j}{\partial r} + \frac{v_j}{r} = 0;$$

$$z = 0, \ u_j \equiv u_j(r, 0); \quad r = r_0, \ v_j(r_0, z) \equiv 0; \quad j = 1, 2, 3.$$

(I.4)

The model of the mass transfer processes in the column apparatuses (I.3) includes boundary conditions, which express a symmetric concentration distributions $(r = 0)$, impenetrability of the column wall $(r = r_0)$, constant input concentrations c_{ij}^0, and mass balances at the column input $(z = 0)$:

$$r = 0, \quad \frac{\partial c_{ij}}{\partial r} \equiv 0; \quad r = r_0, \quad \frac{\partial c_{ij}}{\partial r} \equiv 0;$$

$$z = 0, \quad c_{ij} \equiv c_{ij}^0, \quad u_j^0 c_{ij}^0 \equiv u_j c_{ij}^0 - D_{ij} \left(\frac{\partial c_{ij}}{\partial z} \right)_{z_j = 0},$$

$$j = 1, 2, 3, \quad i = 1, 2, \ldots, i_0.$$

(I.5)

In (I.5), it is supposed that a symmetric radial velocity distribution at the column cross-sectional area will lead to a symmetric concentration distribution, too.

This convection–diffusion-type model (I.3), (I.4), and (I.5) permits a qualitative analysis of the process (model) to be made in order to obtain the main, small, and slight physical effects (mathematical operators) and to discard the slight effect (operators). As a result, the process mechanism identification becomes possible. This model permits to determine the mass transfer resistances in the gas and liquid phases and to find the optimal dispersion system in gas absorption (gas–liquid drops or liquid–gas bubbles). The convection–diffusion model is a base of the average-concentration models, which allow a quantitative analysis of the processes in column apparatuses.

References

1. Boyadjiev C.B. (2006) Diffusion models and scale-up. Int J Heat Mass Transfer 49:796–799
2. Boyadjiev C.B. (2010) Theoretical chemical engineering. Modeling and simulation. Springer, Berlin, Heidelberg
3. Doichinova M, Boyadjiev Chr (2012) On the column apparatuses modeling. Int J Heat Mass Transfer 55:6705–6715
4. Boyadjiev C.B. (2013) A new approach for the column apparatuses modeling in chemical engineering. J Pure Appl Maths: Adv Appl 10(2):131–150

Chapter 2
One-Phase Processes

The fundamental problem of the one-phase processes modeling in the column apparatuses comes from the complicated hydrodynamic behavior of the flow, and as a result, the velocity distribution in the column is unknown. This problem can be avoided using a new approach on the basis of the physical approximations of the mechanics of continua [1–4].

One-phase fluid motion in cylindrical column apparatus [4] with radius r_0 (m) and active zone height l (m) will be considered. The convection–diffusion model is possible to be obtained from (I.3) to (I.5), where (in the case of one-phase fluid motion) the phase index $j = 1, 2, 3$ is possible to be ignored. As a result, $\varepsilon_1 = 1$ ($\varepsilon_2 = \varepsilon_3 = 0$) or $\varepsilon_2 = 1$ ($\varepsilon_1 = \varepsilon_3 = 0$), $u_j(r, z) = u(r, z)$, $v_j(r, z) = v(r, z)$, $c_{ij}(t, r, z) = c_i(t, r, z)$, $Q_{ij}(c_{ij}) = Q_i(c_1, c_2, \ldots, c_{i_0})$, $i = 1, 2, \ldots, i_0$:

$$u\frac{\partial c_i}{\partial z} + v\frac{\partial c_i}{\partial r} = D_i\left(\frac{\partial^2 c_i}{\partial z^2} + \frac{1}{r}\frac{\partial c_i}{\partial r} + \frac{\partial^2 c_i}{\partial r^2}\right) + Q_i(c_1, c_2, \ldots, c_{i_0});$$

$$r = 0, \quad \frac{\partial c_i}{\partial r} \equiv 0; \quad r = r_0, \quad \frac{\partial c_i}{\partial r} \equiv 0; \quad (2.0.1)$$

$$z = 0, \quad c_i \equiv c_i^0, \quad u^0 c_i^0 \equiv uc_i^0 - D_i\frac{\partial c_i}{\partial z}; \quad i = 1, 2, \ldots, i_0.$$

The axial and radial velocity components $u(r, z)$ and $v(r, z)$ satisfy the continuity equation (I.4).

2.1 Column Chemical Reactor

The main process in one-phase column apparatuses is mass transfer of a component of the moving fluid complicated with volume chemical reaction. The quantitative description of this process in column chemical reactors is possible if the axial

© Springer International Publishing AG, part of Springer Nature 2018
C. Boyadjiev et al., *Modeling of Column Apparatus Processes*,
Heat and Mass Transfer, https://doi.org/10.1007/978-3-319-89966-4_2

distribution of the average concentration $\bar{c}(z)$ over the cross-sectional area of the column is known:

$$\bar{c} = \bar{c}(z), \quad 0 \leq z \leq l, \quad \bar{c}(0) = c^0, \quad \bar{c}(l) = c^l,$$

$$G = \frac{c^0 - c^l}{c^0}, \quad c^0 > c^l, \tag{2.1.1}$$

where $z = 0$ $(z = l)$ is the column inlet (outlet), and G is the conversion degree. Two main problems are possible to be solved on this basis:

- modeling (design) problem, i.e., to obtain l if G and c_0 are given;
- simulation (control) problem, i.e., to obtain G if l and c_0 are given.

The axial distribution of the average concentration $\bar{c}(z)$ is to be obtained as a solution of the mass transfer model equations. The modeling problems of the column chemical reactors are possible to be solved using a convection–diffusion-type model.

2.1.1   Convection–Diffusion-Type Model

In the stationary case, the convection–diffusion model of a two-component chemical reaction in the column apparatuses [3] has the form:

$$u\frac{\partial c_i}{\partial z} + v\frac{\partial c_i}{\partial r} = D_i\left(\frac{\partial^2 c_i}{\partial z^2} + \frac{1}{r}\frac{\partial c_i}{\partial r} + \frac{\partial^2 c_i}{\partial r^2}\right) + Q_i(c_1, c_2), \quad i = 1, 2, \tag{2.1.2}$$

where D_i, $i = 1, 2$, are the diffusivities of the reagents in the fluid $(m^2\ s^{-1})$.

The axial and radial velocity components $u(r, z)$ and $v(r, z)$ satisfy the continuity equation:

$$\frac{\partial u}{\partial z} + \frac{\partial v}{\partial r} + \frac{v}{r} = 0; \quad r = r_0, \quad v(r_0, z) \equiv 0, \quad z = 0, \quad u \equiv u(r, 0). \tag{2.1.3}$$

The model of the mass transfer processes in the column apparatuses (2.1.2) includes boundary conditions, which express a symmetric concentration distribution $(r = 0)$, impenetrability of the column wall $(r = r_0)$, a constant inlet concentrations c_i^0, $i = 1, 2$, (kg mol m^{-3}) and mass balance at the column input $(z = 0)$; i.e., the inlet mass flow $(u^0 c_i^0)$ is divided into a convective mass flow $(u c_i^0)$ and a diffusion mass flow $(-D_i\partial c_i/\partial z)$:

$$r = 0, \quad \frac{\partial c_i}{\partial r} \equiv 0; \quad r = r_0, \quad \frac{\partial c_i}{\partial r} \equiv 0;$$

$$z = 0, \quad c_i \equiv c_i^0, \quad u^0 c_i^0 \equiv u c_i^0 - D_i \frac{\partial c_i}{\partial z}, \quad i = 1, 2, \tag{2.1.4}$$

where u^0 (m s^{-1}) is the velocity at the column input. In (2.1.4), it is supposed that a symmetric radial velocity distribution will lead to a symmetric concentration distribution, too. The terms $Q_i(c_1, c_2)$, $i = 1, 2$ in (2.1.2) represent the volume chemical reaction rate (chemical kinetics model).

The mass transfer efficiency (g_i) in the column and conversion degree (G_i) is possible to be obtained using the inlet and outlet average convective mass flux at the cross-sectional area surface in the column:

$$g_i = u^0 c_i^0 - \frac{2}{r_0^2} \int\limits_0^{r_0} r u(r, l) c_i(r, l) dr, \quad G_i = \frac{g_i}{u^0 c_i^0}, \quad i = 1, 2. \tag{2.1.5}$$

The average values of the velocity at the column cross-sectional area can be presented as

$$\bar{u}(z) = \frac{2}{r_0^2} \int\limits_0^{r_0} r u(r, z) \, dr, \quad \bar{v}(z) = \frac{2}{r_0^2} \int\limits_0^{r_0} r v(r, z) \, dr, \tag{2.1.6}$$

The velocity distributions assume to be presented by the average functions (2.1.6):

$$u(r, z) = \bar{u}(z) \tilde{u}(r, z), \quad v(r, z) = \bar{v}(z) \tilde{v}(r, z), \tag{2.1.7}$$

where $\tilde{u}(r, z)$, $\tilde{v}(r, z)$ represent the radial non-uniformity of the velocity distributions, satisfying the conditions:

$$\frac{2}{r_0^2} \int\limits_0^{r_0} r \tilde{u}(r, z) \, dr = 1, \quad \frac{2}{r_0^2} \int\limits_0^{r_0} r \tilde{v}(r, z) \, dr = 1. \tag{2.1.8}$$

A differentiation of $u(r, z)$ in (2.1.7) with respect to z leads to:

$$\frac{\partial u}{\partial z} = \frac{d\bar{u}}{dz} \tilde{u} + \bar{u} \frac{\partial \tilde{u}}{\partial z}. \tag{2.1.9}$$

Practically, the cross-sectional area surface in the columns is a constant and the average velocity is a constant too $(d\bar{u}/dz = 0, \quad \bar{u} = u^0)$, i.e., $\partial u/\partial z \equiv 0$ if $\partial \tilde{u}/\partial z \equiv 0$ $(u = u(r), \quad \tilde{u} = \tilde{u}(r))$. In this case (practically $\partial \tilde{u}/\partial z \equiv 0$ in column apparatuses with big radius values, where the laminar boundary layer thickness at the column wall is negligible with respect to the column radius value) from (2.1.3) follows:

$$\frac{dv}{dr} + \frac{v}{r} = 0; \quad r = r_0, \quad v = 0 \tag{2.1.10}$$

and the solution is $v(r) \equiv 0$. This leads to a new form of the convection–diffusion-type model [4]:

$$u\frac{\partial c_i}{\partial z} = D_i\left(\frac{\partial^2 c_i}{\partial z^2} + \frac{1}{r}\frac{\partial c_i}{\partial r} + \frac{\partial^2 c_i}{\partial r^2}\right) + Q_i(c_1, c_2);$$

$$r = 0, \quad \frac{\partial c_i}{\partial r} \equiv 0; \quad r = r_0, \quad \frac{\partial c_i}{\partial r} \equiv 0; \tag{2.1.11}$$

$$z = 0, \quad c_i \equiv c_i^0, \quad u^0 c_i^0 \equiv u c_i^0 - D_i\frac{\partial c_i}{\partial z}; \quad i = 1, 2.$$

The presented convection–diffusion-type model (2.1.11) is possible to be used for the qualitative analysis of different chemical processes in the column apparatuses.

2.1.2   Complex Chemical Reaction Kinetics

The complex chemical reaction rate is a function of the reagent concentrations. When the reaction rate is denoted by y and the reagent concentrations by $x_1, \ldots, x_m$ the next model equation will be used:

$$y = f(x_1, \ldots, x_m). \tag{2.1.12}$$

The function f (like models of all physical processes) is invariant regarding the dimension transformations of the reagent concentration; i.e., this mathematical structure is invariant regarding similarity transformations [3]:

$$\bar{x}_i = k_i x_i, \quad i = 1, \ldots, m, \tag{2.1.13}$$

i.e.,

$$ky = f(k_1 x_1, \ldots, k_m x_m) = \varphi(k_1, \ldots, k_m) \cdot f(x_1, \ldots, x_m),$$
$$k = \varphi(k_1, \ldots, k_m). \tag{2.1.14}$$

From (2.1.14), it follows that f is a homogenous function; i.e., the relation between the dependent and independent variables in the models is possible to be presented (approximated) by a homogenous function, when the model equations are invariant regarding similarity transformations.

A short recording of (2.1.14) is:

$$f[\bar{x}_i] = \phi[k_i]f[x_i].\tag{2.1.15}$$

The problem consists in finding a function f that satisfies Eq. (2.1.15). A differentiation of Eq. (2.1.15) concerning k_1 leads to:

$$\frac{\partial f[\bar{x}_i]}{\partial k_1} = \frac{\partial \phi}{\partial k_1}f(x_i).\tag{2.1.16}$$

On the other hand,

$$\frac{\partial f[\bar{x}_i]}{\partial k_1} = \frac{\partial f[\bar{x}_i]}{\partial \bar{x}_1}\frac{\partial \bar{x}_1}{\partial k_1} = \frac{\partial f[\bar{x}_i]}{\partial \bar{x}_1}x_1.\tag{2.1.17}$$

From (2.1.16) and (2.1.17) follows

$$\frac{\partial f[\bar{x}_i]}{\partial \bar{x}_1}x_1 = b_1 f[x_i],\tag{2.1.18}$$

where

$$b_1 = \left(\frac{\partial \phi}{\partial k_1}\right)_{k_i=1}.\tag{2.1.19}$$

Equation (2.1.18) is valid for different values of k_i including $k_i = 1$ $(i = 1,\ldots,m)$. As a result, $\bar{x}_i = x_i$, $i = 1,\ldots m$ and from (2.1.18) follows

$$\frac{1}{f}\frac{\partial f}{\partial x_1} = \frac{b_1}{x_1},\tag{2.1.20}$$

i.e.,

$$f = c_1 x_1^{b_1}.\tag{2.1.21}$$

When the above operations are repeated for $x_2,\ldots,x_m$, the homogenous function f assumes the form:

$$f = c x_1^{b_1},\ldots,x_m^{b_m},\tag{2.1.22}$$

i.e., the function f is homogenous if it represents a power functions complex and as a result is invariant with respect to similarity (metric) transformations.

The result obtained shows that the chemical reaction rate in (2.1.11) is possible to be presented as

$$\frac{\partial c_i}{\partial t} = Q_i(c_1, c_2) = k_i c_1^m c_2^n, \quad i = 1, 2. \tag{2.1.23}$$

2.1.3 Two Components Chemical Reaction

Let us consider a complex chemical reaction in the column and $c_i(r, z)$ $i = 1, 2$ are the concentrations (kg mol m^{-3}) of the reagents. In this case, the model (2.1.11) has the form:

$$u\frac{\partial c_i}{\partial z} = D_i\left(\frac{\partial^2 c_i}{\partial z^2} + \frac{1}{r}\frac{\partial c_i}{\partial r} + \frac{\partial^2 c_i}{\partial r^2}\right) - k_i c_1^m c_2^n;$$

$$r = 0, \quad \frac{\partial c_i}{\partial r} \equiv 0; \quad r = r_0, \quad \frac{\partial c_i}{\partial r} \equiv 0; \tag{2.1.24}$$

$$z = 0, \quad c \equiv c_i^0, \quad u^0 c_i^0 \equiv u c_i^0 - D_i\frac{\partial c_i}{\partial z}, \quad i = 1, 2.$$

The qualitative analysis of the model (2.1.24) will be made using generalized variables [3]:

$$r = r_0 R, \quad z = lZ, \quad u(r) = u(r_0 R) = u^0 U(R), \quad \tilde{u}(r) = \tilde{u}(r_0 R) = U(R),$$

$$c_i(r, z) = c_i(r_0 R, lZ) = c_i^0 C_i(R, Z) \quad (i = 1, 2), \quad \varepsilon = \left(\frac{r_0}{l}\right)^2, \tag{2.1.25}$$

where r_0, l, u^0, c_i^0 $(i = 1, 2)$ are the characteristic (inherent) scales (maximal or average values) of the variables. The introduction of the generalized variables (2.1.25) in (2.1.24) leads to:

$$U\frac{\partial C_i}{\partial Z} = Fo_i\left(\varepsilon\frac{\partial^2 C_i}{\partial Z^2} + \frac{1}{R}\frac{\partial C_i}{\partial R} + \frac{\partial^2 C_i}{\partial R^2}\right) - Da_i\,C_1^m C_2^n;$$

$$R = 0, \quad \frac{\partial C_i}{\partial R} \equiv 0; \quad R = 1, \quad \frac{\partial C_i}{\partial R} \equiv 0;$$

$$Z = 0, \quad C_i \equiv 1, \quad 1 \equiv U - Pe_i^{-1}\frac{\partial C_i}{\partial Z}; \tag{2.1.26}$$

$$Fo_i = \frac{D_i l}{u_0 r_0^2}, \quad Da_i = \theta^{i-1} Da_i^0, \quad Pe_i = \frac{u_0 l}{D_i},$$

$$Da_i^0 = \frac{k_i l}{u_0}\left(c_1^0\right)^{m-1}\left(c_2^0\right)^n, \quad \theta = \frac{c_1^0}{c_2^0}; \quad i = 1, 2,$$

where Fo, Da, and Pe are the Fourier, Damkohler, and Peclet numbers, respectively.

2.1.4 Comparison Qualitative Analysis

As already noted [3, 4] when variable scales in (2.1.25), the maximal or average variable values are used. As a result, the unity is the order of magnitude of all functions and their derivatives in (2.1.26), i.e., the effects of the physical and chemical phenomena [the contribution of the terms in (2.1.26)], are determined by the orders of magnitude of the dimensionless parameters in (2.1.26). If all equations in (2.1.26) are divided by the dimensionless parameter, which has the maximal order of magnitude, all terms in the model equations will be classified in three parts:

1. The parameter is unity or its order of magnitude is unity; i.e., this mathematical operator represents a main physical effect;
2. The parameter's order of magnitude is 10^{-1}; i.e., this mathematical operator represents a small physical effect;
3. The parameter's order of magnitude is $\leq 10^{-2}$; i.e., this mathematical operator represents a very small (negligible) physical effect and has to be neglected, because it is not possible to be measured experimentally.

Here and throughout the book, it has to be borne in mind that the process (model) is composed of individual effects (mathematical operators), and if their relative role (influence) in the overall process (model) is less than 10^{-2}, they have to be ignored, because the inaccuracy of the experimental measurements is above 1%.

2.1.5 Pseudo-First-Order Reactions

In the cases of big difference between inlet concentrations of the reagents $\left(c_1^0 \ll c_2^0\right)$ in (2.1.24), the problem described by (2.1.26) is possible to be solved in zero approximation with respect to the very small parameter $\theta\ (0 = \theta \leq 10^{-2})$, and as a result, $Da_2 = 0$, $C_2 \equiv 1$. Very often $m = 1$ and from (2.1.26) follows:

$$U\frac{\partial C}{\partial Z} = Fo\left(\varepsilon\frac{\partial^2 C}{\partial Z^2} + \frac{1}{R}\frac{\partial C}{\partial R} + \frac{\partial^2 C}{\partial R^2}\right) - Da\,C; \quad \varepsilon = Fo^{-1}Pe^{-1};$$

$$R = 0, \quad \frac{\partial C}{\partial R} \equiv 0; \quad R = 1, \quad \frac{\partial C}{\partial R} \equiv 0; \quad Z = 0, \quad C \equiv 1, \quad 1 \equiv U - Pe^{-1}\frac{\partial C}{\partial Z};$$

$$(2.1.27)$$

where $C = C_1$, $Da = Da_1^0$ and model (2.1.27) of column apparatuses with pseudo-first-order chemical reaction is obtained. The parameters ε and Fo are related with the column radius r_0, and as a result, the convection–diffusion type of model (2.1.27) is possible to be used for solving the scale-up problem.

2.1.6   Similarity Conditions

From (2.1.27), it follows that two mass transfer processes in column apparatuses are similar if the parameter values of Fo, Da, Pe, and ε are identical; i.e., these parameters are similarity criteria. In the real cases when the difference between two similar processes is in the parameter values r_0^s, l^s, u_0^s, $s = 1, 2$, from the similarity conditions, it follows:

$$Fo = \frac{Dl^s}{u^{0s}r_0^{s2}}, \quad Da = \frac{kl^s}{u^{0s}}, \quad Pe = \frac{u^{0s}l^s}{D}, \quad \varepsilon = \left(\frac{r_0^s}{l^s}\right)^2, \quad s = 1, 2. \quad (2.1.28)$$

From (2.1.28), follow three expressions for the characteristic velocity:

$$u^{0s} = \frac{D}{\sqrt{\varepsilon r_0^s}\, Fo}, \quad u^{0s} = \frac{kr_0^s}{\sqrt{\varepsilon}\, Da}, \quad u^{0s} = \frac{D\, Pe\sqrt{\varepsilon}}{r_0^s}, \quad s = 1, 2, \quad (2.1.29)$$

i.e., the similarity criteria Fo, Da are incompatible, because from (2.1.29) follows, that (at constant values of Fo, Da, and ε) the increase of the radius r_0 (from laboratory model to industrial apparatus) leads to decrease and increase of the velocity u^{0s} simultaneously. The increase of the radius r_0 is not possible to be compensated by the changes of velocity u^{0s} (practically, the change of r_0^s is not possible to be compensated by the changes of D and k). These results show that the physical modeling is not possible to be used for a quantitative description of the mass transfer processes in column chemical reactors; i.e., the convection–diffusion model with radius r_0^1 is not physical model of the real process with radius r_0^2 if $r_0^1 \neq r_0^2$. The similar situation exists in two-phase processes with chemical reaction.

2.2   Model Approximations

The presentation of the models in generalized variables [3] permits to obtain different approximations of the models, i.e., the approximations of small ($\sim 10^{-1}$) and very small ($\leq 10^{-2}$, negligible) parameters.

2.2.1   Short Columns Model

For short columns $\varepsilon = (r_0/l)^2 = Fo^{-1}Pe^{-1}$ is a small parameter ($\varepsilon \sim 10^{-1}$), i.e., $Pe^{-1} \leq 10^{-1}Fo$, and for $Fo \leq 1$, the next small parameter is Pe^{-1} ($Pe^{-1} \leq 10^{-1}$). In these cases, the problem (2.1.27) is possible to be solved using the perturbation method (see Chap. 7 and [6]):

$$C(R,Z) = C^{(0)}(R,Z) + \varepsilon C^{(1)}(R,Z) + \varepsilon^2 C^{(2)}(R,Z) + \cdots, \tag{2.2.1}$$

where $C^{(0)}$, $C^{(1)}$, $C^{(2)}, \ldots$ are solutions of the next problems:

$$U\frac{\partial C^{(0)}}{\partial Z} = Fo\left(\frac{1}{R}\frac{\partial C^{(0)}}{\partial R} + \frac{\partial^2 C^{(0)}}{\partial R^2}\right) - Da\, C^{(0)};$$

$$R = 0, \quad \frac{\partial C^{(0)}}{\partial R} \equiv 0; \quad R = 1, \quad \frac{\partial C^{(0)}}{\partial R} \equiv 0; \tag{2.2.2}$$

$$Z = 0, \quad C^{(0)} \equiv 1.$$

$$U\frac{\partial C^{(s)}}{\partial Z} = Fo\left(\frac{1}{R}\frac{\partial C^{(s)}}{\partial R} + \frac{\partial^2 C^{(s)}}{\partial R^2}\right) - Da\, C^{(s)} + Fo\frac{\partial^2 C^{(s-1)}}{\partial Z^2};$$

$$R = 0, \quad \frac{\partial C^{(s)}}{\partial R} \equiv 0; \quad R = 1, \quad \frac{\partial C^{(s)}}{\partial R} \equiv 0; \tag{2.2.3}$$

$$Z = 0, \quad C^{(s)} \equiv 0; \quad s = 1, 2, \ldots..$$

A multistep procedure has to be used for solving (2.2.2) and (2.2.3):

1. Solving (2.2.2) and calculating $\frac{\partial^2 C^{(0)}}{\partial Z^2}$;
2. Solving of (2.2.3) and calculating $\frac{\partial^2 C^{(s-1)}}{\partial Z^2}$, $s = 1, 2, \ldots..$

2.2.2 High-Column Model

For high columns, ε is a very small parameter and the problem (2.1.27) is possible to be solved in zero approximation with respect to ε $(0 = \varepsilon \leq 10^{-2})$, i.e., $Pe^{-1} \leq 10^{-2} Fo$, and for $Fo \leq 1$, the next very small parameter is Pe^{-1} $(0 = Pe^{-1} \leq 10^{-2})$, i.e., $C = C^{(0)}$:

$$U\frac{\partial C}{\partial Z} = Fo\left(\frac{1}{R}\frac{\partial C}{\partial R} + \frac{\partial^2 C}{\partial R^2}\right) - Da\, C;$$

$$R = 0, \quad \frac{\partial C}{\partial R} \equiv 0; \quad R = 1, \quad \frac{\partial C}{\partial R} \equiv 0; \quad Z = 0, \quad C \equiv 1. \tag{2.2.4}$$

2.2.3 Effect of the Chemical Reaction Rate

The effect of the chemical reaction rate is negligible if $0 = Da \leq 10^{-2}$ and from (2.2.4) follows $C \equiv 1$.

When fast chemical reactions take place $(Da \geq 10^2)$, the terms in (2.2.4) must be divided by Da and the approximation $0 = Da^{-1} \leq 10^{-2}$ has to be applied. The result is:

$$0 = \frac{Fo}{Da}\left(\frac{1}{R}\frac{dC}{dR} + \frac{d^2C}{dR^2}\right) - C; \quad R = 0, \quad \frac{dC}{dR} \equiv 0; \quad R = 1, \quad \frac{dC}{dR} \equiv 0, \quad (2.2.5)$$

i.e., the model (2.2.5) is diffusion type.

2.2.4   Convection Types Models

In the cases of big values of the average velocity $(0 = Fo \leq 10^{-2})$, from the convection–diffusion-type model (2.1.27) is possible to obtain a convection-type model when putting $Fo = 0$:

$$U(R)\frac{dC}{dZ} = -DaC; \quad Z = 0, \quad C \equiv 1. \quad (2.2.6)$$

2.3   Effect of the Radial Non-uniformity of the Velocity Distribution

The radial non-uniformity of the axial velocity distribution influences the conversion degree, concentration distribution, and scale effect.

2.3.1   Conversion Degree

As an example will be used the case [4] of parabolic velocity distribution (Poiseuille flow):

$$u(r) = \bar{u}\left(2 - 2\frac{r^2}{r_o^2}\right). \quad (2.3.1)$$

From (2.1.25) and (2.3.1) follows

$$U(R) = 2 - 2R^2. \quad (2.3.2)$$

Table 2.1 Conversion degree

	G	G_0
Da = 1, Fo = 0	0.5568	0.6734
Da = 1, Fo = 0.1	0.5938	0.6452
Da = 1, Fo = 1	0.6211	0.6281
Da = 2, Fo = 0	0.7806	0.8516
Da = 2, Fo = 0.1	0.8115	0.8502
Da = 2, Fo = 1	0.8481	0.8538

The solutions of the problem (2.2.4) for $Da = 1, 2$ and $Fo = 0, 0.1, 1.0$ permit to obtain $C(R, Z)$ and $\overline{C}(Z) = \bar{c}(z)/c_0$:

$$\overline{C}(Z) = 2 \int_0^1 RC(Z, R)\,\mathrm{d}R. \tag{2.3.3}$$

As a result, it is possible to obtain (Table 2.1) the conversion degree (2.1.5) in the cases of presence (G) and absence (G_0) of a radial non-uniformity of the axial velocity in the column:

$$G = 1 - 2 \int_0^1 RU(R)C(R, 1)\,\mathrm{d}R, \quad G_0 = 1 - \overline{C}(1). \tag{2.3.4}$$

Table 1.1 shows that the radial non-uniformity of the axial velocity component leads to substantial decrease of the conversion degree, but an increase of the diffusion transfer (Fo) leads to decrease of the convective transfer (all hydrodynamic effects), and as a result, the effect of the radial non-uniformity of the axial velocity decreases.

2.3.2   Concentration Distribution

Different expressions for the velocity distribution in the column apparatuses permit to analyze [4] the influence of the velocity distribution radial non-uniformities on the concentration distribution:

$$u^0 = \bar{u}, \quad u^1(r) = \bar{u}\left(2 - 2\frac{r^2}{r_o^2}\right);$$

$$u^s(r) = \bar{u}\left(1 + a_s\frac{r^2}{r_0^2} + b_s\frac{r^4}{r_0^4}\right), \quad s = 2, 3, \quad a_2 = 2, \quad a_3 = -2, \quad b_2 = -3, \quad b_3 = 3;$$

$$u^s(r) = \bar{u}\left(\frac{n+1}{n} - \frac{2r^2}{nr_0^2}\right), \quad n = 2, \quad s = 4,$$

$$\tag{2.3.5}$$

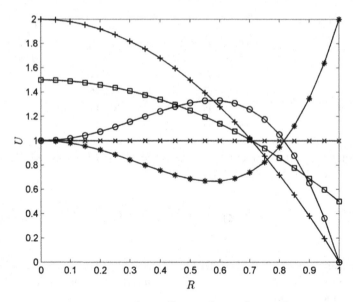

Fig. 2.1 Velocity distributions: $\times$—U^0; $+$—U^1; $\bigcirc$—U^2; $*$—U^3; $\square$—U^4

where $n = 1$ is the Poiseuille flow.

From (2.3.5), it is possible to obtain the following dimensionless velocity distributions $U^s(R) = u^s(r)/\bar{u}$:

$$U^0(R) = 1, \quad U^1(R) = 2 - 2R^2, \quad U^2(R) = 1 + 2R^2 - 3R^4,$$
$$U^3(R) = 1 - 2R^2 + 3R^4, \quad U^4(R) = \frac{3}{2} - R^2. \tag{2.3.6}$$

The differences between maximal and minimal velocity values $\Delta U_s = U_s^{\max} - U_s^{\min}$ $(s = 1, \ldots, 4)$ are the velocity distribution radial non-uniformity parameters $\left(\Delta U_1 = 2, \ \Delta U_2 = \Delta U_3 = \frac{4}{3}, \ \Delta U_4 = 1\right)$. The velocity distributions $U^0, \ldots, U^4$ are presented on Fig. 2.1.

The numerical solutions of (2.2.4) using different velocity distributions (2.3.6) present the effect of the velocity radial non-uniformity on the conversion degree (G) and column height (H) in comparison with the plug flow.

The concentration distributions obtained with the solution of (2.2.4) for $Fo = 0.1$ and $Da = 2$ are shown in Fig. 2.2.

Table 2.2 presents the values of the conversion degree $G_0, \ldots, G_3$ at $Da = 2$ and $Fo = 0.01, 0.1$. The column heights $Z = H_1, \ldots, H_3$, for which the maximum conversion degree of the plug flow $G_0 = 0.8643 \ (0.8645)$ is reached, were calculated.

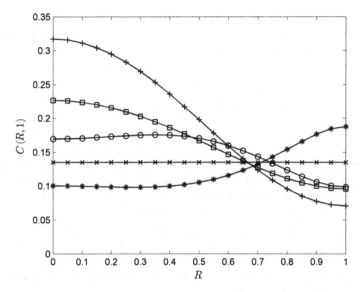

Fig. 2.2 Concentration distributions using the five velocity profiles: $\times$—U^0; $+$—U^1; $\bigcirc$—U^2; $*$—U^3; $\square$—U^4

Table 2.2 Process efficiency G at $Z = 1$ and column height $H = Z$ at $G_0 = 0.8643$

Fo	U^0	U^1	U^2	U^3
0.1 Laboratory	$G_0 = 0.8643$ $H_0 = 1$	$G_1 = 0.8143$ $H_1 = 1.2$	$G_2 = 0.8516$ $H_2 = 1.05$	$G_3 = 0.8513$ $H_3 = 1.05$
0.01 Industrial	$G_0 = 0.8645$ $H_0 = 1$	$G_1 = 0.7870$ $H_1 = 1.34$	$G_2 = 0.8349$ $H_2 = 1.12$	$G_3 = 0.8371$ $H_3 = 1.12$

Table 2.3 Effect of the velocity radial non-uniformity on the process efficiency and column height

Fo	U^1	U^2	U^3
0.1 Laboratory	$\Delta G_1 = 6\%$ $\Delta H_1 = 20\%$	$\Delta G_2 = 1.4\%$ $\Delta H_2 = 5\%$	$\Delta G_3 = 1.5\%$ $\Delta H_3 = 5\%$
0.01 Industrial	$\Delta G_1 = 9.8\%$ $\Delta H_1 = 34\%$	$\Delta G_2 = 3.5\%$ $\Delta H_2 = 12\%$	$\Delta G_3 = 3.3\%$ $\Delta H_3 = 12\%$

Table 2.3 presents the effect of the velocity radial non-uniformity on the relative conversion degree and column height at $G_0 = 0.8643$:

$$\Delta G_s = \frac{G_0 - G_s}{G_s} \cdot 100, \quad \Delta H_s = \frac{H_s - H_0}{H_0} \cdot 100, \quad s = 1, 2, 3, \quad (2.3.7)$$

where G_0 is the conversion degree in the case of plug flow.

The numerical results (Table 2.3) show the necessity of an essential augmentation of the column height in order to compensate the velocity distribution radial non-uniformity effect. The comparison of the results in Tables 2.2 and 2.3 shows that the effects of ΔU_2

and ΔU_3 are similar; i.e., the velocity distribution radial non-uniformity effects are caused by the velocity non-uniformity $\Delta U_s = U_s^{\max} - U_s^{\min}$ $(s = 1, \ldots, 4)$, but not by the velocity distribution U_s, $(s = 1, \ldots, 4)$.

2.3.3   Influence of the Velocity Radial Non-uniformity Shape

The influence of the shape of the velocity profile and the average velocity value in a column chemical reactor on the conversion degree has been presented in [8]. The effect of a simple velocity distribution (Poiseuille type)

$$0 \le r \le R_0, \quad u = \bar{u}\left(2 - 2\frac{r^2}{r_0^2}\right), \tag{2.3.8}$$

is compared with three complicated velocity distributions, which shapes change at different values of $b = b_0, b_1, b_2, b_3$:

$$0 \le r_1 \le r_0, \quad u_1(r_1) = \bar{u}\left(2 - 2\frac{r_1^2}{r_0^2}\right), \quad r_0 = \frac{R_0}{1+b};$$

$$r_0 \le r_2 \le R_0, \quad u_2(r_2) = \bar{u} \cdot F(b) \cdot \Phi(r_2);$$

$$\Phi(r_2) = \frac{r_2^2}{r_0^2} - 1 - \frac{2b+b^2}{\ln(1+b)}\ln(r_2/r_0),$$

$$\overline{\Phi}(b) = \frac{2}{R_0^2 - r_0^2}\int\limits_{r_0}^{R_0} r_2\Phi(r_2)dr_2, \quad F(b) = \left[\overline{\Phi}(b)\right]^{-1}, \tag{2.3.9}$$

$$F(b) = \frac{2\ln(1+b)}{b^2 + 2b - (b^2 + 2b + 2)\ln(1+b)}, \quad b = \frac{R_0 - r_0}{r_0},$$

where $b_0 = 0$ (Poiseuille-type flow), $b_1 = 1$, $b_2 = 0.42$, $b_3 = 0.11$. As a result, two convection–diffusion equations are considered:

$$u_1\frac{\partial c_1}{\partial z} = D\left(\frac{1}{r_1}\frac{\partial c_1}{\partial r_1} + \frac{\partial^2 c_1}{\partial r_1^2}\right) - kc_1,$$

$$u_2\frac{\partial c_2}{\partial z} = D\left(\frac{1}{r_2}\frac{\partial c_2}{\partial r_2} + \frac{\partial^2 c_2}{\partial r_2^2}\right) - kc_2. \tag{2.3.10}$$

The boundary conditions of (2.3.10) are:

$$r_1 = 0, \quad \frac{\partial c_1}{\partial r_1} = 0; \quad r_2 = R_0, \quad \frac{\partial c_2}{\partial r_2} = 0;$$

$$r_1 = r_2 = r_0, \quad c_1 = c_2, \quad \frac{\partial c_1}{\partial r_1} = \frac{\partial c_2}{\partial r_2}; \tag{2.3.11}$$

$$z = 0, \quad c_1 = c_2 = c_0.$$

The introduction of the dimensionless variables

$$Z = \frac{z}{L}, \quad R_1 = \frac{r_1}{R_0}, \quad R_2 = \frac{r_2}{R_0},$$

$$U_1(R_1) = \frac{u_1(r_1)}{\bar{u}}, \quad U_2(R_2) = \frac{u_2(r_2)}{\bar{u}}, \tag{2.3.12}$$

$$C_1(R_1, Z) = \frac{c_1}{c_0}, \quad C_2(R_2, Z) = \frac{c_2}{c_0},$$

in (2.3.8)–(2.3.11) leads to

$$U_1 \frac{\partial C_1}{\partial Z} = Fo\left(\frac{1}{R_1} \frac{\partial C_1}{\partial R_1} + \frac{\partial^2 C_1}{\partial R_1^2}\right) - DaC_1,$$

$$U_1(R_1) = 2 - 2(1 + b_i)^2 R_1^2, \quad 0 \le R_1 \le \frac{1}{1 + b_i};$$

$$U_2 \frac{\partial C_2}{\partial Z} = Fo\left(\frac{1}{R_2} \frac{\partial C_2}{\partial R_2} + \frac{\partial^2 C_2}{\partial R_2^2}\right) - DaC_2, \tag{2.3.13}$$

$$U_2(R_2) = F(b_i)\left[(1 + b_i)^2 R_2^2 - 1 - \frac{2b_i + b_i^2}{\ln(1 + b_i)} \ln(1 + b_i) R_2\right],$$

$$\frac{1}{1 + b_i} \le R_2 \le 1.$$

$$R_1 = 0, \quad \frac{\partial C_1}{\partial R_1} = 0; \quad R_2 = 1, \quad \frac{\partial C_2}{\partial R_2} = 0;$$

$$R_1 = R_2 = \frac{1}{1 + b_i}, \quad C_1 = C_2, \quad \frac{\partial C_1}{\partial R_1} = \frac{\partial C_2}{\partial R_2}, \quad i = 0, 1, 2, 3; \tag{2.3.14}$$

$$Z = 0, \quad C_1 = C_2 = 1.$$

The dimensionless velocity profiles in (2.3.13) are shown in Figs. 2.3, 2.4, 2.5, and 2.6.

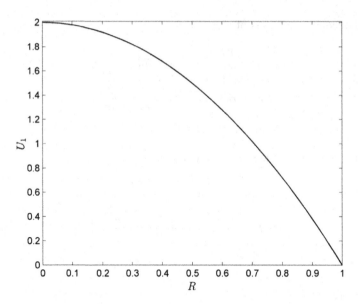

Fig. 2.3 Dimensionless velocity profiles (2.3.13) at $b = b_0 = 0$

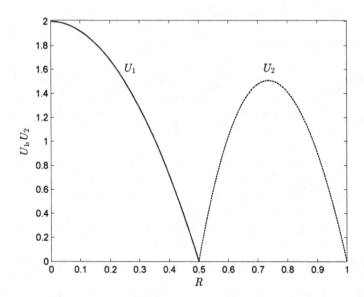

Fig. 2.4 Dimensionless velocity profiles (2.3.13) at $b = b_1 = 1$

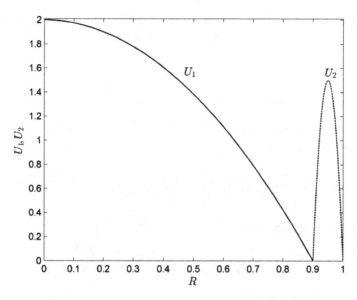

Fig. 2.5 Dimensionless velocity profiles (2.3.13) at $b = b_2 = 0.11$

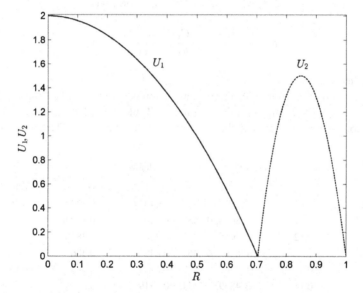

Fig. 2.6 Dimensionless velocity profiles (2.3.13) at $b = b_3 = 0.42$

At the boundary condition $R_1 = R_2$ given by (2.3.14), the concentrations have to be presented as a polynomial by three parameters:

$$R_1 = R_2 = \frac{1}{1+b_i}, \quad i = 1, 2, 3,$$

$$C_1(R_1, Z) = C_2(R_2, Z) = 1 + a_1^i Z + a_2^i Z^2 + a_3^i Z^3, \tag{2.3.15}$$

where the parameters $a_1^i, a_2^i, a_3^i, \ i = 1, 2, 3$, have to be obtained by the minimization of the function:

$$F\left(a_1^i, a_2^i, a_3^i\right) = \int\limits_0^1 f\left(a_1^i, a_2^i, a_3^i, Z\right) dZ,$$

$$f\left(a_1^i, a_2^i, a_3^i, Z\right) = \left[\alpha_1\left(a_1^i, a_2^i, a_3^i, Z\right) - \alpha_2\left(a_1^i, a_2^i, a_3^i, Z\right)\right]^2, \tag{2.3.16}$$

$$\alpha_1\left(a_1^i, a_2^i, a_3^i, Z\right) = \left(\frac{\partial C_1}{\partial R_1}\right)_{R_1=R_2}, \quad \alpha_2\left(a_1^i, a_2^i, a_3^i, Z\right) = \left(\frac{\partial C_2}{\partial R_2}\right)_{R_2=R_1}.$$

The obtained parameter values $a_1^i, a_2^i, a_3^i, \ i = 1, 2, 3$, are presented in Table 2.4.

Figure 2.7 presents three cases of the concentration gradient difference $f\left(a_1^i, a_2^i, a_3^i, Z\right), \ i = 1, 2, 3$, which show that the conditions

$$R_1 = R_2 = \frac{1}{1+b_i}, \quad \frac{\partial C_1}{\partial R_1} = \frac{\partial C_2}{\partial R_2}, \quad i = 1, 2, 3, \tag{2.3.17}$$

are satisfied.

These solutions permit to obtain the conversion degree (2.3.4), and the results for different values of Da and Fo are presented in Table 2.4, where it is seen that the conversion degree increases if the average velocity increases. The impact of the

Table 2.4   Parameter values and values of conversion degree

Fo, Da	b	a_1, a_2, a_3	G	G_0
Da = 0.5 Fo = 0.05	0	−0.2807, −0.1735, 0.04094	0.3722	0.3934
	0.11	−0.8000, 0.4553, −0.1502	0.3794	
	0.42	−0.8035, 0.6453, −0.2888	0.3846	0.3866
	1	−0.8169, 0.7366, −0.3487	0.3867	
Fo = 0.1 Da = 1	0	−0.9902, 0.4558, −0.09779	0.5938	0.6321
	0.11	−1.4869, 1.2044, −0.4319	0.6072	
	0.42	−1.3512, 1.1535, −0.4741	0.6186	0.6320
	1	−1.3204, 1.1857, −0.5171	0.6229	
Fo = 0.2 Da = 2	0	−1.9348, 1.6008, −0.5326	0.8241	0.8650
	0.11	−2.6316, 3.0105, −1.2783	0.8386	
	0.42	−2.3088, 2.4350, −1.0111	0.8518	0.8645
	1	−2.2008, 2.2693, −0.9443	0.8572	

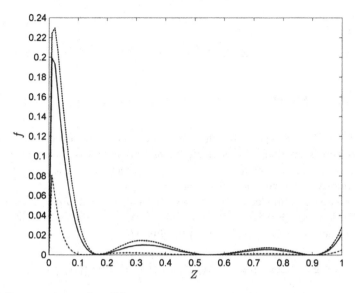

Fig. 2.7 Concentration gradient differences

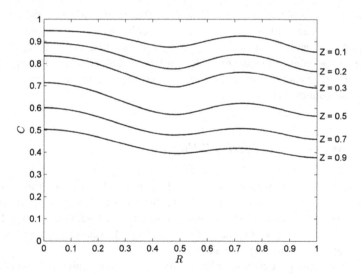

Fig. 2.8 Concentration distribution

different shapes of velocity profile ($b_0 = 0$, $b_1 = 1$, $b_2 = 0.42$, $b_3 = 0.11$) on the column apparatus efficiency is negligible compared with that of Damkohler number.

The concentration profiles $C_1(R_1, Z)$, $C_2(R_2, Z)$, which are solution of (2.3.13) for $Fo = 0.1$, $Da = 1$, $b_1 = 1$ and column height $Z = 0.1$; 0.3; 0.5; 0.7; 0.9, are shown in Fig. 2.8.

Table 2.5 Comparison of the scaling effect between different velocity profiles

	U^1 (%)	U^2 (%)	U^3 (%)
ΔG_{scale}	3.5	1.9	1.7
ΔH_{scale}	11.6	6.6	6.6

2.3.4 Scale Effect

The analyses [1, 3, 8, 9] of the influence of the column size on the mass transfer efficiency shows that the process efficiency in column apparatuses decreases with the column diameter increase. This scale-up effect is a result of the radial non-uniformity of the velocity distribution.

Let us consider "model" column ($r_0 = 0.2\,\text{(m)}$, $Da = 2$, $Fo = 0.1$) and "industrial" column ($r_0 = 0.5\,\text{(m)}$, $Da = 2$, $Fo = 0.01$) [7]. The scaling effects on the conversion degrees $\Delta G^s_{\text{scale}}$ and column heights $\Delta H^s_{\text{scale}}$:

$$\Delta G^s_{\text{scale}} = \frac{G^s_{\text{mod}} - G^s_{\text{ind}}}{G^s_{\text{ind}}} \cdot 100\%, \quad \Delta H^j_{\text{scal}} = \frac{H^s_{\text{ind}} - H^s_{\text{mod}}}{H^s_{\text{mod}}} \cdot 100\%, \Delta\, s = 1, \ldots, 3,$$

$$(2.3.18)$$

are possible to be obtained using Table 2.2. The results obtained are shown in Table 2.5. The comparison between the two columns on the basis of (2.3.7) (ΔQ_{mod}, ΔQ_{ind}) and (2.3.18) (ΔH_{mod}, ΔH_{ind}) shows that the scale-up leads to decrease of the conversion degree (for constant column height). If consider the columns with constant conversion degree, it leads to the column height increase as result of the column radius increase.

2.3.5 On the "Back Mixing" Effect

The reduction of the conversion degree in the column chemical reactors, resulting from the radial non-uniformity in the velocity distribution at the cross-sectional area of the column, is explained [9–12] by the mechanism of a back mass transfer ("back mixing" effect). The new approach for modeling of column apparatuses [1–5] permits a new explanation of this effect [27].

Let us consider a pseudo-first-order chemical reaction in a high column (2.1.27). The concentration distributions in the column as solutions of (2.1.27) were obtained in the case of $\varepsilon = 0.05$ using the perturbation method [6]. As a radial non-uniformity at the velocity distribution at the cross sectional area of the column will be used

$$U = 2 - 2R^2. \qquad (2.3.19)$$

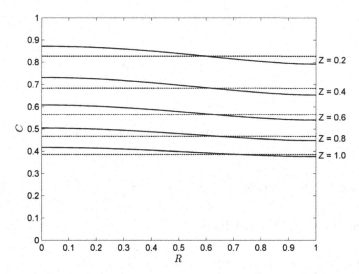

Fig. 2.9 Concentration distributions $C(R, Z)$ for $\varepsilon = 0.05$, $Fo = 1$, $Da = 1$ and different Z: $U = 2 - 2R^2$ (solid lines); $U = 1$ (dotted lines)

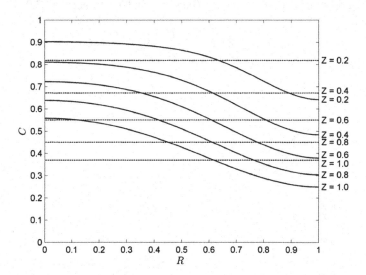

Fig. 2.10 Concentration distributions $C(R, Z)$ for $\varepsilon = 0.05$, $Fo = 0.1$, $Da = 1$ and different Z: $U = 2 - 2R^2$ (solid lines); $U = 1$ (dotted lines)

Figures 2.9 and 2.10 present comparison of the results obtained in the cases $Fo = 1$, $Da = 1$ and $Fo = 0.1$, $Da = 1$ for different values of Z (solid lines) with the case of absence of the radial non-uniformity in the velocity distribution $U = 1$ (dotted lines).

From Figs. 2.9 and 2.10, it is possible to obtain the average concentrations $\overline{C}(Z)$:

$$\overline{C}(Z) = 2 \int\limits_{0}^{1} RC(R, Z)\, dR. \qquad (2.3.20)$$

The results are presented in Fig. 2.11.

The convection–diffusion mass flux in the column $\mathbf{j}$ (kg mol m^{-2} s^{-1}) is possible to be presented as

$$\mathbf{j}(r, z) = \mathbf{u}c - D\,\mathbf{grad}\,c = \left[u(r)c(r, z) - D\frac{\partial c}{\partial z} \right]\hat{\mathbf{z}} - D\frac{\partial c}{\partial r}\hat{\mathbf{r}}, \qquad (2.3.21)$$

or in generalized variables (2.1.25) as:

$$\mathbf{J}(R, Z) = \frac{\mathbf{j}(r, z)}{u^0 c^0} = \left[U(R)C(R, Z) - Pe^{-1}\frac{\partial C}{\partial Z} \right]\hat{\mathbf{z}} - \varepsilon^{-0.5}Pe^{-1}\frac{\partial C}{\partial R}\hat{\mathbf{r}}, \qquad (2.3.22)$$

where $\hat{\mathbf{r}}$ and $\hat{\mathbf{z}}$ are the unit vectors, $U = 2 - 2R^2$, C—the solution of the problem (2.1.27). From the solution, it is seen (Figs. 1.9 and 1.10) that in (2.3.22)

$$U(R)C(R, Z) \geq 0, \quad \frac{\partial C}{\partial Z} \leq 0, \quad \frac{\partial C}{\partial R} \leq 0, \qquad (2.3.23)$$

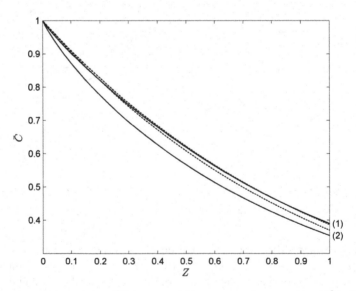

Fig. 2.11 Average concentration $\overline{C}(Z)$: (1) $\varepsilon = 0.05$, $Fo = 1$, $Da = 1$; $U = 2 - 2R^2$ (solid lines); $U = 1$ (dotted lines), (2) $\varepsilon = 0.05$, $Fo = 0.1$, $Da = 1$; $U = 2 - 2R^2$ (solid lines); $U = 1$ (dotted lines)

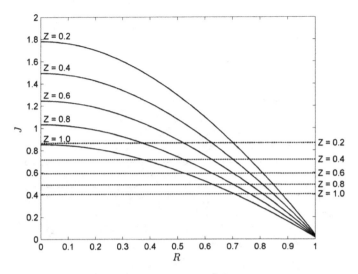

Fig. 2.12 Mass flux $J(R, Z)$ for different Z: $U = 2 - 2R^2$, $\varepsilon = 0.05$, $Fo = 1$, $Da = 1$ (solid lines); $U = 1$, $\varepsilon = 0.05$, $Fo = 1$, $Da =$ (dotted lines)

i.e., the vector components of $\mathbf{J}(R, Z)$ are positive, and there are no conditions for a backward mass transfer ("back mixing" effect).

The mass flux in every point (r, z) in the column (see the lines on Figs. 2.12 and 2.13) is possible to be obtained from (2.3.21):

$$j(r, z) = \left\{ \left[u(r)c(r, z) - D\frac{\partial c}{\partial z} \right]^2 + \left[D\frac{\partial c}{\partial r} \right]^2 \right\}^{0.5} \tag{2.3.24}$$

or in generalized variables (2.1.25):

$$J(R, Z) = \frac{j(r, z)}{u^0 c^0} = \left\{ \left[U(R)C(R, Z) - Pe^{-1}\frac{\partial C}{\partial Z} \right]^2 + \left[\varepsilon^{-0.5} Pe^{-1}\frac{\partial C}{\partial R} \right]^2 \right\}^{0.5}, \tag{2.3.25}$$

where $Pe^{-1} = \varepsilon Fo$.

The average mass flux in the cross-sectional area of the column in generalized variables (2.1.25)

$$\bar{J}(Z) = 2 \int_0^1 RJ(R, Z)\, dR \tag{2.3.26}$$

is presented in Fig. 2.14.

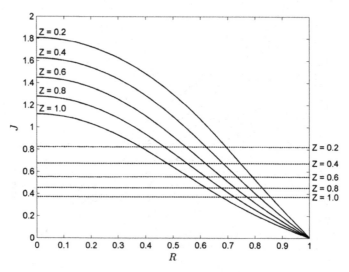

Fig. 2.13 Mass flux $J(R, Z)$ for different Z: $U = 2 - 2R^2$, $\varepsilon = 0.05$, $Fo = 0.1$, $Da = 1$ (solid lines); $U = 1$, $\varepsilon = 0.05$, $Fo = 0.1$, $Da =$ (dotted lines)

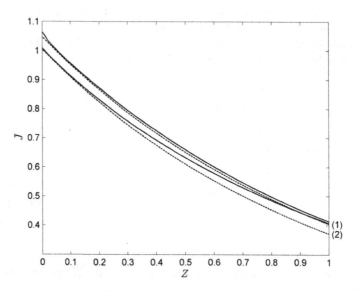

Fig. 2.14 Average mass flux $\overline{J}(Z)$: (1) $\varepsilon = 0.05$, $Fo = 1$, $Da = 1$; $U = 2 - 2R^2$ (solid lines); $U = 1$ (dotted lines), (2) $\varepsilon = 0.05$, $Fo = 0.1$, $Da = 1$; $U = 2 - 2R^2$ (solid lines); $U = 1$ (dotted lines)

Table 2.6 Conversion degree

	Fo = 1, $U = 2 - 2R^2$	Fo = 1, $U = 1$	Fo = 0.1, $U = 2 - 2R^2$	Fo = 0.1, $U = 1$
$\bar{J}(0)$	1.0634	1.0473	1.0085	1.0049
$\bar{J}(1)$	0.4137	0.4048	0.4080	0.3716
G	0.6110	0.6135	0.5954	0.6302

The conversion degree is possible to be obtained using the difference between the average mass fluxes in the cross-sectional area at the column's ends:

$$G = \frac{\bar{J}(0) - \bar{J}(1)}{\bar{J}(0)} \qquad (2.3.27)$$

and the results are presented in Table 2.6.

It is seen from Table 2.6 that the conversion degree decreases as a result of the radial non-uniformity in the velocity distribution in the cross-sectional area of the column. As was shown, this effect cannot be explained by "back mixing" effect, but may be explained by the residence times of the flows in the column.

The radial non-uniformity in the velocity distribution in the cross-sectional area of the column leads to flows with different axial velocities, different residence times and chemical reaction times of these flows, which results in non-uniformity of the concentration distribution in the cross-sectional area of the column. The conversion degree is related to the average residence time and the average reaction times in these flows in the column.

Let us consider the cases of presence ($u = u(r)$) and absence ($u = u^0$) of radial non-uniformity in the velocity distribution in the cross-sectional area of the column. The residence times of the flows in the column in these cases are:

$$\theta(r) = \frac{l}{u(r)}, \quad \theta_0 = \frac{l}{u^0}. \qquad (2.3.28)$$

The average residence times at the cross-sectional area of the column are

$$\bar{\theta} = \frac{2}{r_0^2} \int_0^{r_0} r \frac{l}{u(r)} dr, \quad \bar{\theta}_0 = \frac{l}{u^0}. \qquad (2.3.29)$$

The using of generalized variables (2.1.25) and θ_0 as a scale leads to

$$\theta(r) = \theta_0 \Theta(R), \quad \bar{\theta}_0 = \theta_0 \Theta_0, \quad \Theta = \frac{1}{U(R)}, \quad \Theta_0 = 1 \qquad (2.3.30)$$

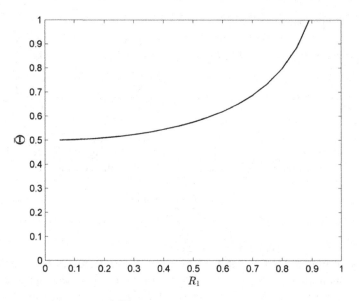

Fig. 2.15 Average residence time $\overline{\Theta}(R_1)$ in the interval $(0, R_1)$

and the dimensionless average residence times are:

$$\overline{\Theta} = 2 \int_0^1 R \frac{1}{U(R)} dR, \quad \overline{\Theta}_0 = 1. \qquad (2.3.31)$$

Figure 2.15 presents the average residence time for different values of R_1 in the interval $(0, 1)$:

$$\overline{\Theta}(R_1) = \frac{1}{R_1^2} \int_0^{R_1} \frac{R}{1 - R^2} dR. \qquad (2.3.32)$$

It can be seen, that in the interval $0 \leq R_1 < 0.9$ $\overline{\Theta}(R_1) < \overline{\Theta}_0 = 1$, which explains the low conversion degree in this interval.

A comparison of the average mass fluxes in the intervals $(0, R_1)$ and $(R_1, 1)$:

$$\overline{J}_1(R_1, Z) = \frac{2}{R_1^2} \int_0^{R_1} R J(R, Z) dR, \quad \overline{J}_2(R_1, Z) = \frac{2}{1 - R_1^2} \int_{R_1}^1 R J(R, Z) dR \qquad (2.3.33)$$

and the average mass flux (2.3.26) in the interval $(0, 1)$ for Fo $= 1$, 0.1; $Z = 1$, $R_1 = 0.9$ (see Table 2.7, where $\overline{J}_0(1)$ is the average mass flux (2.3.26) in the

Table 2.7

$Fo = 1$	$\bar{J}(1) = 0.4137$	$\bar{J}_0(1) = 0.4048$	$\bar{J}_1(0.9, 1) = 0.4902$	$\bar{J}_2(0.9, 1) = 0.0875$
$Fo = 0.1$	$\bar{J}(1) = 0.4080$	$\bar{J}_0(1) = 0.3716$	$\bar{J}_1(0.9, 1) = 0.4920$	$\bar{J}_2(0.9, 1) = 0.0496$

interval $(0, 1)$ in the case $U = 1$) reveals that in the interval $(0, R_1)$: the residence time $\overline{\Theta}(R_1)$ is less, the average mass flux $\bar{J}_1(R_1, Z)$ is larger and the conversion degree is less than in the case $U = 1$. The average mass flux $\bar{J}_2(R_1, Z)$ is much smaller than the average mass flux $\bar{J}_1(R_1, Z)$ and, as a result, the average mass flux (2.3.26) in the interval $(0, 1)$ is larger than the average mass flux $\bar{J}_0(1)$, i.e. the conversion degree is less than in the case $U = 1$.

The presented theoretical analysis shows that the reduction of the conversion degree in the column chemical reactors, which results from the radial non-uniformity in the velocity distribution in the cross-sectional area of the column, is not possible to be explained by the mechanism of a back mass transfer ("back mixing" effect). The new approach for modeling of column apparatuses permits to provide a new explanation of this effect. The radial non-uniformity in the velocity distribution in the cross-sectional area of the column leads to decrease of the average residence time of the flow in the column (chemical reaction time), increase of the average mass flux at the column outlet, and thus to decrease of the conversion degree in the column. This effect increases if the convection part of the convection–diffusion flow in the column increases due to the average velocity increase or the flow viscosity reduction.

2.4 Examples

The presented new approach for modeling of chemical processes in one-phase column apparatuses is used for quantitative theoretical analysis of the effect of the tangential input flow in the columns on the radial non-uniformity of the axial velocity, the simultaneous mass and heat transfer processes, the effect of circulation zones in column apparatuses, and the mass transfer in countercurrent flows in column apparatuses.

2.4.1 Effect of the Tangential Flow

Let us consider a cylindrical column [14] in cylindrical coordinate system (r, z, φ), where the velocity has axial $u_z = u_z(r, z, \varphi)$, radial $u_r = u_r(r, z, \varphi)$, and angular $u_\varphi(r, z, \varphi)$ components. In the case of axial input of the gas (liquid) flow (Fig. 2.16a) $u_\varphi \equiv 0$ and the axial and radial velocity components u_z, u_r satisfy the continuity equation

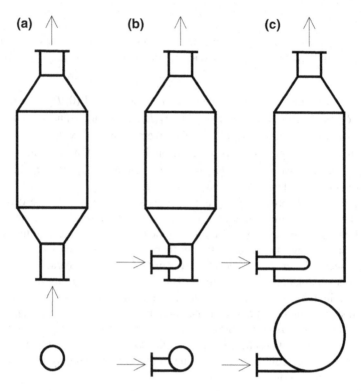

Fig. 2.16 Cylindrical column with: **a** axial gas (liquid) flow; **b** tangential gas (liquid) flow in column input; **c** tangential gas (liquid) flow in column working area

$$\frac{\partial u_z}{\partial z} + \frac{\partial u_r}{\partial r} + \frac{u_r}{r} = 0 \qquad (2.4.1)$$

with boundary conditions

$$r = r_0, \quad u_r(r_0, z) \equiv 0; \quad z = 0, \quad u_z(r, 0) \equiv u_z^0(r), \qquad (2.4.2)$$

where $u_z^0(r)$ is the input distribution of the axial velocity component and is a result of the geometric conditions at the axial input of the column. The velocity components u_z, u_r are possible to be obtained as a solution of the Navier–Stokes equations in boundary layer approximation, i.e., to solve the problem of the gas (liquid) jet in immobile gas (liquid) phase. As a result, a radial non-uniformity of the axial velocity component is obtained. In these conditions, the conversion degree is possible to increase if the radial non-uniformity of the axial velocity component decreases as a result of special geometric conditions at the axial input of the column.

A possibility for partial reduction of the radial non-uniformity of the axial velocity component is the use of a column with tangential entering [15] of the gas (liquid) flow (Fig. 2.16b) in the column input.

Maximal reduction of the radial non-uniformity of the axial velocity component is achieved by using of a column with tangential entering [16] of the gas (liquid) flow (Fig. 2.16c) in the column working area. In this case, the velocity components $u_z = u_z(r, z, \varphi)$, $u_r = u_r(r, z, \varphi)$, $u_\varphi(r, z, \varphi)$ satisfy the continuity equation:

$$\frac{\partial u_z}{\partial z} + \frac{1}{r}\frac{\partial u_\varphi}{\partial \varphi} + \frac{\partial u_r}{\partial r} + \frac{u_r}{r} = 0; \tag{2.4.3}$$

with boundary conditions

$$r = r_0, \quad 0 < z \le l, \quad 0 \le \varphi \le 2\pi, \quad u_r(z, r_0, \varphi) \equiv 0;$$

$$z = 0, \quad 0 \le r < r_0, \quad 0 \le \varphi \le 2\pi, \quad u_z(0, r, \varphi) \equiv \bar{u} = \frac{F}{\pi r_0^2};$$

$$z = 0, \quad \varphi = 0, \quad u_\varphi(0, r_0, 0) \equiv u_\varphi^0 = \frac{F}{\pi r_{00}^2}, \tag{2.4.4}$$

where F (m^3 s^{-1}) is the gas (liquid) flow rate in the column, r_0—the column radius, r_{00}—the column feed-pipe radius.

The use of the generalized variables:

$$z = lZ, \quad r = r_0 R, \quad \varphi = 2\pi\Phi, \quad u_z = \bar{u}U_z, \quad u_r = \bar{u}\frac{r_0}{l}U_r, \quad u_\varphi = u_\varphi^0 U_\varphi, \tag{2.4.5}$$

leads to

$$\frac{1}{R}\frac{\partial U_\varphi}{\partial \varphi} + 2\pi\frac{\bar{u}r_0}{u_\varphi^0 l}\left(\frac{\partial U_z}{\partial Z} + \frac{\partial U_r}{\partial R} + \frac{U_r}{R}\right) = 0;$$

$$R = 1, \quad 0 < Z \le 1, \quad 0 \le \Phi \le 1, \quad U_r(Z, 1, \Phi) \equiv 0; \tag{2.4.6}$$

$$Z = 0, \quad 0 \le R < 1, \quad 0 \le \Phi \le 1, \quad U_z(0, R, \Phi) \equiv 1;$$

$$Z = 0, \quad \Phi = 0, \quad U_\varphi(0, 1, 0) \equiv 1.$$

Practically, $\bar{u} \ll u_\varphi^0$ and the approximation $0 = 2\pi\frac{\bar{u} r_0}{u_\varphi^0 l} \le 10^{-2}$ is possible to be used:

$$\frac{\partial U_\varphi}{\partial \varphi} = 0 \tag{2.4.7}$$

and from (2.4.6) it follows:

$$\frac{\partial U_z}{\partial Z} + \frac{\partial U_r}{\partial R} + \frac{U_r}{R} = 0;$$

$$R = 1, \quad 0 < Z \leq 1, \quad U_r(1, Z) \equiv 0; \tag{2.4.8}$$

$$Z = 0, \quad 0 \leq R < 1, \quad U_z(R, 0) \equiv 1.$$

From (2.4.8) follows that practically $U_z(R, Z) \equiv 1$, $U_r(R, Z) \equiv 0$ (except for a thin boundary layer on the wall).

The presented theoretical analysis shows that the use of tangential input of the flows in the columns area lead to a significant decrease of the velocity radial non-uniformity and as a result an increase of the conversion degree in the columns.

2.4.2   Simultaneous Mass and Heat Transfer Processes

The heat and mass transfer kinetics theory shows [3] that the process rate depends on the characteristic velocity in the boundary layer. The big difference between the velocities in the cases of axial and tangential input of the flows $\left(\bar{u} \ll u_{\varphi}^0 \right)$ leads to a substantial increase of the heat transfer rate through the column wall [14].

Let us consider a simultaneous mass and heat transfer processes in a column chemical reactor, where the velocity, concentration, and temperature θ (deg) distributions in the column are denoted as $u = u(r)$, $c = c(r, z)$, $\theta = \theta(r, z)$. The mass and heat transfer models in the physical approximations of the mechanics of continua [1–5] can be expressed as:

$$u\frac{\partial c}{\partial z} = D\left(\frac{\partial^2 c}{\partial z^2} + \frac{1}{r}\frac{\partial c}{\partial r} + \frac{\partial^2 c}{\partial r^2}\right) - kc;$$

$$r = 0, \quad \frac{\partial c}{\partial r} \equiv 0; \quad r = r_0, \quad \frac{\partial c}{\partial r} \equiv 0; \tag{2.4.9}$$

$$z = 0, \quad c \equiv c^0, \quad \bar{u}c^0 \equiv uc^0 - D\frac{\partial c}{\partial z};$$

$$u\frac{\partial \theta}{\partial z} = \frac{\lambda}{\rho c_p}\left(\frac{\partial^2 \theta}{\partial z^2} + \frac{1}{r}\frac{\partial \theta}{\partial r} + \frac{\partial^2 \theta}{\partial r^2}\right) + \frac{q}{\rho c_p}kc;$$

$$r = 0, \quad \frac{\partial \theta}{\partial r} \equiv 0; \quad r = r_0, \quad -\lambda\frac{\partial \theta}{\partial r} \equiv k_0(\theta - \theta^*); \tag{2.4.10}$$

$$z = 0, \quad \theta \equiv \theta_0, \quad \bar{u}\theta_0 \equiv u\theta_0 - \frac{\lambda}{\rho c_p}\frac{\partial \theta}{\partial z},$$

where ρ (kg m^{-3}) is the density, c_p (J kg^{-1} deg^{-1})—the specific heat at constant pressure, λ (J m^{-1} s^{-1} deg^{-1})—the thermal conductivity, q (J kg^{-1})—the heat effect of the chemical reaction, k_0 (J m^{-2} s^{-1} deg^{-1})—the interphase heat transfer coefficient, θ^* (deg)—the temperature outside the column. In the model (2.4.9) and (2.4.10), D, k, λ, ρ, c_p, q, k_0 are temperature functions, where $\theta_0 \leq \theta(r, z) \leq \theta$ (r, l) or $\theta(r, l) \leq \theta \leq \theta_0$, ($\theta_0$—inlet temperature, $\theta(r, l)$—temperature at the column outlet) in the case of endothermic ($q < 0$) or exothermic ($q > 0$) chemical reaction. Practically, the difference $|\theta_0 - \theta(r, l)|$ is not so big, and in (2.4.9) and (2.4.10), it is possible to use constant values of $D, k, \lambda, \rho, c_p, q, k_0$ at $\theta = \theta_0$; i.e., the temperature effect is related the heat generation rate.

In the isothermal case, the volume heat generation in the column is equal to the interface heat transfer (heat flux) through the column wall:

$$2\pi \int_0^{r_0} rqkc \, dr \equiv -2\pi r_0 \lambda \left(\frac{\partial \theta}{\partial r}\right)_{r=r_0}, \quad -\lambda \left(\frac{\partial \theta}{\partial r}\right)_{r=r_0} \equiv \frac{r_0 qk}{2} \bar{c}(z) \equiv k_0(\theta - \theta^*).$$

$$(2.4.11)$$

A qualitative analysis of the models (2.4.9) and (2.4.10) will be made using generalized variables:

$$r = r_0 R, \quad z = lZ, \quad u(r) = u(r_0 R) = \bar{u} U(R),$$
$$c(r, z) = c(r_0 R, lZ) = c^0 C(R, Z),$$
$$\theta(r, z) = \theta(r_0 R, lZ) = \theta_0 \Theta(R, Z), \quad \theta^* = \theta_0 \Theta^*,$$

$$(2.4.12)$$

where r_0, l, $\bar{u}$, c^0, θ_0 are the characteristic (inherent) scales (maximal or average values) of the variables. The introduction of the generalized variables (2.4.12) in (2.4.9) and (2.4.10) leads to:

$$U(R)\frac{\partial C}{\partial Z} = \frac{Dl}{\bar{u}r_0^2}\left(\frac{r_0^2}{l^2}\frac{\partial^2 C}{\partial Z^2} + \frac{1}{R}\frac{\partial C}{\partial R} + \frac{\partial^2 C}{\partial R^2}\right) - \frac{kl}{\bar{u}}C;$$

$$R = 0, \quad \frac{\partial C}{\partial R} \equiv 0; \quad R = 1, \quad \frac{\partial C}{\partial R} \equiv 0; \qquad (2.4.13)$$

$$Z = 0, \quad C \equiv 1, \quad 1 \equiv U(R) - \frac{D}{\bar{u}l}\frac{\partial C}{\partial Z}.$$

$$U(R)\frac{\partial \Theta}{\partial Z} = \frac{\lambda l}{\bar{u}\rho c_p r_0^2}\left(\frac{r_0^2}{l^2}\frac{\partial^2 \Theta}{\partial Z^2} + \frac{1}{R}\frac{\partial \Theta}{\partial R} + \frac{\partial^2 \Theta}{\partial R^2}\right) + \frac{qlkc^0}{\bar{u}\rho c_p \theta_0}C;$$

$$R = 0, \quad \frac{\partial \Theta}{\partial R} \equiv 0; \quad R = 1, \quad \frac{\partial \Theta}{\partial R} \equiv -\frac{kr_0}{\lambda}(\Theta - \Theta^*); \qquad (2.4.14)$$

$$Z = 0, \quad \Theta \equiv 1, \quad 1 \equiv U(R) - \frac{\lambda}{\bar{u}\rho c_p l}\frac{\partial \Theta}{\partial Z}.$$

In the cases of very high columns, it is possible to use the approximation $0 = r_0^2/l^2 \leq 10^{-2}$ and the model equations (2.4.13) and (2.4.14) are of a parabolic type:

$$U(R)\frac{\partial C}{\partial Z} = \frac{Dl}{\bar{u}r_0^2}\left(\frac{1}{R}\frac{\partial C}{\partial R} + \frac{\partial^2 C}{\partial R^2}\right) - \frac{kl}{\bar{u}}C;$$

$$R = 0, \quad \frac{\partial C}{\partial R} \equiv 0; \quad R = 1, \quad \frac{\partial C}{\partial R} \equiv 0; \quad Z = 0, \quad C \equiv 1. \tag{2.4.15}$$

$$U(R)\frac{\partial \Theta}{\partial Z} = \frac{\lambda l}{\bar{u}\rho c_p r_0^2}\left(\frac{1}{R}\frac{\partial \Theta}{\partial R} + \frac{\partial^2 \Theta}{\partial R^2}\right) + \frac{qlkc^0}{\bar{u}\rho c_p \theta_0}C;$$

$$R = 0, \quad \frac{\partial \Theta}{\partial R} \equiv 0; \quad R = 1, \quad \frac{\partial \Theta}{\partial R} \equiv -\frac{kr_0}{\lambda}(\Theta - \Theta^*); \quad Z = 0, \quad \Theta \equiv 1.$$

$$\tag{2.4.16}$$

If the average velocity $\bar{u}$ is very big, it is possible to use the approximations $0 = \frac{Dl}{\bar{u}r_0^2} \leq 10^{-2}$ and $0 = \frac{\lambda l}{\bar{u}\rho c_p r_0^2} \leq 10^{-2}$; i.e., the models (2.4.13) and (2.4.14) are of a convective type:

$$U(R)\frac{dC}{dZ} = -\frac{kl}{\bar{u}}C; \quad Z = 0, \quad C \equiv 1. \tag{2.4.17}$$

$$U(R)\frac{d\Theta}{dZ} = \frac{qlkc^0}{\bar{u}\rho c_p \theta_0}C; \quad Z = 0, \quad \Theta \equiv 1. \tag{2.4.18}$$

The presented models are the basis for qualitative analysis of simultaneous mass and heat transfer processes and construction of models for the average concentrations and temperatures.

2.4.3   Circulation Zones in Column Apparatuses

In some cases, the radial non-uniformity of the axial velocity component is a result of circulation flows in the stagnation zone and the conversion degree in the column decreases. The application of the convection–diffusion type models [13, 17] permits to obtain the concentration distribution in these cases and evaluate the zones' breadths influence on the mass transfer rate.

Let us consider fluid motion in a column with radius R_0, where the concentration of the fluid components decreases as a result of a first-order chemical reaction. The presence of solid barriers (see Fig. 2.17) leads to a circulation zone. As a mathematical model of the concentration distribution, $c(r, z)$ in the column will be used the convection–diffusion equation

Fig. 2.17 Circulation zones
in column apparatuses

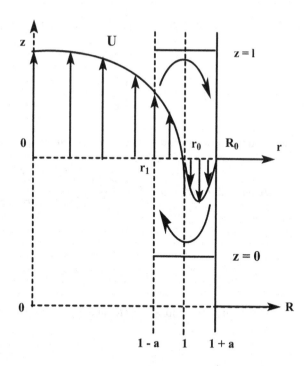

$$u\frac{\partial c}{\partial z} = D\left(\frac{\partial^2 c}{\partial z^2} + \frac{1}{r}\frac{\partial c}{\partial r} + \frac{\partial^2 c}{\partial r^2}\right) - kc. \tag{2.4.19}$$

The solution of this equation is possible if the velocity distribution $u(r,z)$ is positive. The change of the velocity direction (see Fig. 2.17) leads to solution of (2.4.19) in different zones in the column: $0 \leq r \leq r_0$ $(u_1(r,z) \geq 0)$ and $r_0 \leq r \leq R_0$ $(u_2(r,z) \leq 0)$, where:

$$u = u_1(r) = \bar{u}\left(2 - 2\frac{r^2}{r_0^2}\right), \quad u = u_2(r) = A\left[r_0^2 - r^2 + \frac{R_0^2 - r_0^2}{\ln(R_0/r_0)}\ln(r/r_0)\right].$$
$$\tag{2.4.20}$$

The parameter A has to be obtained from the physical condition in the circulation zone, where the absolute values of the average velocities in the zones $r_1 \leq r \leq r_0$ and $r_0 \leq r \leq R_0$ are equal:

$$A = \frac{2\bar{u}(2a - a^2)}{\left[R_0^2 + r_0^2 - \frac{R_0^2 - r_0^2}{\ln(R_0/r_0)}\right]}, \quad \frac{R_0 - r_0}{r_0} = a, \tag{2.4.21}$$

where $2ar_0$ is the breadth of the stagnation zone.

From (2.4.20), it is possible to obtain the average velocities $\bar{u}_1$ and $\bar{u}_2$:

$$\bar{u}_1 = \bar{u},$$

$$\bar{u}_2 = \frac{2}{R_0^2 - r_0^2} \int_{r_0}^{R_0} ru_2(r)\,\mathrm{d}r = \frac{A}{2}\left[R_0^2 + r_0^2 - \frac{R_0^2 - r_0^2}{\ln(R_0/r_0)}\right] = \bar{u}(2a - a^2), \qquad (2.4.22)$$

and the dimensionless forms of the velocity distributions are:

$$U_1(R) = 2 - 2R^2, \quad U_2(R) = \frac{2(2a - a^2)\left[1 - R^2 + \frac{2a + a^2}{\ln(1+a)}\ln R\right]}{2 + 2a + a^2 - \frac{2a + a^2}{\ln(1+a)}}, \qquad (2.4.23)$$

where

$$R = \frac{r}{r_0}, \quad \frac{R_0}{r_0} = 1 + a, \quad U_1(R) = \frac{u_1(r)}{\bar{u}}, \quad U_2(R) = \frac{u_2(r)}{\bar{u}}. \qquad (2.4.24)$$

The velocity distributions $U_1(R)$ and $U_2(R)$ are shown in Fig. 2.18.

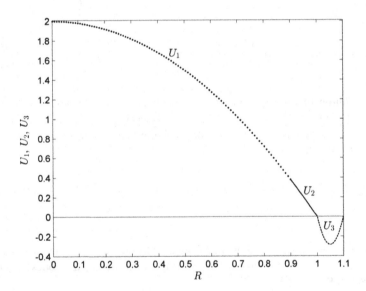

Fig. 2.18 Velocity distribution at $a = 0.1$

The sign change of the velocity at $r = r_0$ leads to necessity the problem (2.4.19) to be presented as three problems in two-coordinate systems (r, z_1) and (r, z_2), where $z_1 + z_2 = l$ with l being the circulation zone height:

$$u_1 \frac{\partial c_1}{\partial z_1} = D\left(\frac{1}{r}\frac{\partial c_1}{\partial r} + \frac{\partial^2}{\partial r^2}\right) - kc_1;$$

$$(0 \leq r \leq r_1, \quad 0 \leq z_1 \leq l, \quad u_1 = u_1(r), \quad c_1 = c_1(r, z_1))$$

$$r = 0, \quad \frac{\partial c_1}{\partial r} \equiv 0; \quad r = r_1, \quad c_1 \equiv c_2, \quad \frac{\partial c_1}{\partial r} \equiv \frac{\partial c_2}{\partial r}; \quad z_1 = 0, \quad c_1 \equiv \bar{c}_0.$$

$$(2.4.25)$$

$$u_1 \frac{\partial c_2}{\partial z_1} = D\left(\frac{1}{r}\frac{\partial c_2}{\partial r} + \frac{\partial^2}{\partial r^2}\right) - kc_2;$$

$$(r_1 \leq r \leq r_0, \quad 0 \leq z_1 \leq l, \quad u_1 = u_1(r), \quad c_2 = c_2(r, z_1))$$

$$r = r_1, \, c_1 \equiv c_2, \, \frac{\partial c_1}{\partial r} \equiv \frac{\partial c_2}{\partial r}; \, r = r_0, \, c_2 \equiv c_3, \, \frac{\partial c_2}{\partial r} \equiv \frac{\partial c_3}{\partial r};$$

$$z_1 = 0, \quad c_2 \equiv (\bar{c}_3)_{z_2=l}.$$

$$(2.4.26)$$

$$u_2 \frac{\partial c_3}{\partial z_2} = D\left(\frac{1}{r}\frac{\partial c_3}{\partial r} + \frac{\partial^2 c_3}{\partial r^2}\right) - kc_3;$$

$$(r_0 \leq r \leq R_0, \quad 0 \leq z_2 \leq l, \quad u_2 = u_2(r), \quad c_3 = c_3(r, z_2))$$

$$r = r_0, \quad c_2 \equiv c_3, \quad \frac{\partial c_2}{\partial r} \equiv \frac{\partial c_3}{\partial r}, \quad r = R_0, \quad \frac{\partial c_3}{\partial z} \equiv 0;$$

$$z_2 = 0, \quad c_3 \equiv (\bar{c}_2)_{z_1=l}.$$

$$(2.4.27)$$

In (2.4.26) and (2.4.27), $(\bar{c}_2)_{z_1=l}$ and $(\bar{c}_3)_{z_2=l}$ are the average concentrations:

$$(\bar{c}_2)_{z_1=l} = \frac{2}{r_0^2 - r_1^2}\int_{r_1}^{r_0} rc_2(r, l)\, dr, \quad (\bar{c}_3)_{z_2=l} = \frac{2}{R_0^2 - r_0^2}\int_{r_0}^{R_0} rc_3(r, l)\, dr. \quad (2.4.28)$$

The solution of the problem (2.4.25)–(2.4.27) permits to obtain the column mass transfer efficiency q in the case of circulation zone:

$$q = u_1^0 c^0 - \frac{2}{r_1^2}\int_0^{r_1} ru_1 c_1(r, l)\, dr, \quad u_1^0 = \frac{2}{r_1^2}\int_0^{r_1} ru_1(r)\, dr = \bar{u}(1 + 2a - a^2)$$

$$(2.4.29)$$

where u_1^0 is the average velocity of the convective mass flux in a column with solid barriers.

For the solution of the problem (2.4.25)–(2.4.27), dimensionless variables have to be used:

$$
\begin{aligned}
r = r_0 R, \quad z = l Z_1 = l Z_2, \quad Z_1 + Z_2 = 1, \\
c_1 = c^0 C_1(R, Z_1), \quad c_2 = c^0 C_2(R, Z_1), \quad c_3 = c^0 C_3(R, Z_2).
\end{aligned}
\tag{2.4.30}
$$

Introducing (2.4.30) into the problem (2.4.25)–(2.4.27) leads to the following set of equations:

$$
\begin{aligned}
&U_1 \frac{\partial C_1}{\partial Z_1} = Fo \left(\frac{1}{R} \frac{\partial C_1}{\partial R} + \frac{\partial^2 C_1}{\partial R^2} \right) - Da\, C_1, \quad (0 \le R \le 1 - a); \\
&R = 0, \quad \frac{\partial C_1}{\partial R} \equiv 0; \quad R = 1 - a, \quad \frac{\partial C_1}{\partial R} \equiv \delta_1(Z_1), \quad \delta_1(Z_1) = \frac{\partial C_2}{\partial R}; \\
&Z_1 = 0, \quad C_1 \equiv 1.
\end{aligned}
\tag{2.4.31}
$$

$$
\begin{aligned}
&U_1 \frac{\partial C_2}{\partial Z_1} = Fo \left(\frac{1}{R} \frac{\partial C_2}{\partial R} + \frac{\partial^2 C_2}{\partial R^2} \right) - Da\, C_2, \quad (1 - a \le R \le 1); \\
&R = 1 - a, \quad C_2(R, Z_1) \equiv C_1(R, Z_1); \\
&R = 1, \quad \frac{\partial C_2}{\partial Z_1} \equiv \delta_2(Z_1), \quad \delta_2(Z_1) = \left(\frac{\partial C_3}{\partial Z_2} \right)_{Z_2 = 1 - Z_1}; \\
&Z_1 = 0, \quad C_2 \equiv \alpha_2, \quad \alpha_2 = \frac{2}{2a + a^2} \int\limits_{1}^{1+a} R C_3(R, 1)\, dR.
\end{aligned}
\tag{2.4.32}
$$

$$
\begin{aligned}
&U_2 \frac{\partial C_3}{\partial Z_2} = Fo \left(\frac{1}{R} \frac{\partial C_3}{\partial R} + \frac{\partial^2 C_3}{\partial R^2} \right) - Da\, C_3, \quad (1 \le R \le 1 + a); \\
&R = 1, \quad C_3 \equiv \delta_3, \quad \delta_3(Z_2) = (C_2)_{Z_1 = 1 - Z_2}; \\
&R = 1 + a, \quad \frac{\partial C_3}{\partial R} \equiv 0; \\
&Z_2 = 0, \quad C_3 \equiv \alpha_3, \quad \alpha_3 = \frac{2}{a} \int\limits_{1-a}^{1} R C_2(R, 1)\, dR;
\end{aligned}
\tag{2.4.33}
$$

where Fo and Da are similar to the Fourier and Damkohler numbers:

$$
Fo = \frac{Dl}{\bar{u} r_0^2}, \quad Da = \frac{kl}{\bar{u}}.
\tag{2.4.34}
$$

The column mass transfer efficiency (2.4.28) in dimensionless variables is:

$$Q = \frac{q}{u_1^0 c^0} = 1 - \frac{4}{(1-a)^2(1+2a-a^2)} \int\limits_{0}^{1-a} R(1-R^2)C_1(R,1)\,dR. \qquad (2.4.35)$$

The solution of the problem (2.4.31)–(2.4.33) is possible to be obtained consecutively using the next algorithm:

1. The problem (2.4.31) is possible to be solved independently if putting $\delta_1(Z_1) \equiv 0$. As a result, the zero approximation of the concentration $C_1^0(R, Z_1)$ is obtained.

2. The problem (2.4.32) is possible to be solved independently if putting $\alpha_2 = 0$ and $\delta_2(Z_1) \equiv 0$. As a result, the zero approximation of the concentration is obtained $C_2^0(R, Z_2)$, $\delta_1^0(Z_1) = \frac{\partial C_2^0}{\partial R}$ $(Z_2 = 1 - Z_1)$, $\delta_3^0(Z_2) = C_2^0(1, Z_1)$, $Z_1 = 1 - Z_2$ and $\alpha_3^0 = \frac{2}{2a-a^2} \int\limits_{1-a}^{1} RC_2^0(R,1)\,dR$.

3. The problem (2.4.33) is possible to be solved independently if putting $\alpha_3 = \alpha_3^0$ and $\delta_3(Z_2) = \delta_3^0(Z_2)$. As a result, the zero approximation of the concentration is obtained $\qquad C_3^0(R, Z_2)$, $\delta_2^0(Z_1) = \left(\frac{\partial C_3^0}{\partial Z_2}\right)_{R=1, Z_2=1-Z_1}$, and $\alpha_2^0 = \frac{2}{2a+a^2} \int\limits_{1}^{1+a} RC_3^0(R,1)\,dR$. The obtained zero approximations permits to make the ith step of the algorithm.

4. Solution of (2.4.31), where $\delta_1(Z_1) = \delta_1^{i-1}(Z_1)$. As a result, the ith approximation of the concentration $C_1^i(R, Z_1)$ is obtained.

5. Solution of (2.4.32), where $\alpha_2 = \alpha_2^{i-1}$ and $\delta_2(Z_1) = \delta_2^{i-1}(Z_1)$. The result is obtaining the ith approximation of the concentration $C_2^i(R, Z_2)$, $\delta_1^i(Z_1) = \frac{\partial C_2^i}{\partial R}$ $(Z_2 = 1 - Z_1)$, $\delta_3^i(Z_2) = C_2^i(1, Z_1)$ $(Z_1 = 1 - Z_2)$, $\alpha_3^i = \frac{2}{2a-a^2} \int\limits_{1-a}^{1} RC_2^i(R,1)\,dR$.

6. Solution of (2.4.33), where $\alpha_3 = \alpha_3^i$ and $\delta_3(Z_2) = \delta_3^i(Z_2)$. The result is obtaining the zero approximation of the concentration $C_3^i(R, Z_2)$, $\delta_2^i(Z_1) = \left(\frac{\partial C_3^i}{\partial Z_2}\right)_{R=1, Z_2=1-Z_1}$, $\alpha_2^i = \frac{2}{2a+a^2} \int\limits_{1}^{1+a} RC_3^i(R,1)\,dR$.

7. Return to step 4.

The criterion of the iterative procedure stopping is the minimization of the difference (P) between the concentration distributions of two consecutive iterations:

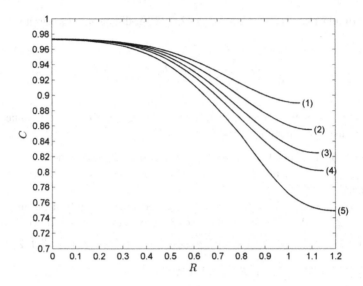

Fig. 2.19 Comparison of the concentration distributions for different stagnation zone breadths a and $Z_1 = Z_2 = 0.5$; (1) $a = 0.05$, (2) $a = 0.1$, (3) $a = 0.13$, (4) $a = 0.15$, (5) $a = 0.2$

$$P = \int_{0}^{1-a} \left(C_1^i - C_1^{i-1}\right)^2 dR + \int_{1-a}^{1} \left(C_2^i - C_2^{i-1}\right)^2 dR + \int_{1}^{1+a} \left(C_3^i - C_3^{i-1}\right)^2 dR.$$

(2.4.36)

The comparison of the results in [9] shows that the concentration distributions after the fourth iteration satisfy the differential equations and the boundary conditions (2.4.31)–(2.4.33), and the criterion of the iterative procedure stopping is $P \leq 10^{-5}$.

The influence of the stagnation zone breadth on the concentration distributions is shown in Fig. 2.19.

The fluid flow rate in a column with barriers is:

$$F = u_1^0 \pi r_1^2 = \bar{u}_0 \pi R_0^2,$$

(2.4.37)

where $\bar{u}_0$ is the average velocity in the column in the absence of the barriers and the flow is of Poiseuille type:

$$\bar{u}_0 = \bar{u}\left(1 + 2a - a^2\right) \frac{(1-a)^2}{(1+a)^2}.$$

(2.4.38)

The velocity distribution in a column without barriers is:

$$u_0 = 2\bar{u}_0\left(1 - \frac{r^2}{R_0^2}\right) = 2\bar{u}\left(1 + 2a - a^2\right)\frac{(1-a)^2}{(1+a)^2}\left(1 - \frac{r^2}{R_0^2}\right). \tag{2.4.39}$$

The convection–diffusion equation in the case of absence of barriers is possible to be obtained from (2.4.19) with the replacement $u = u_0$ and $c = c_0$:

$$u_0\frac{\partial c_0}{\partial z} = D\left(\frac{\partial^2 c_0}{\partial z^2} + \frac{1}{r}\frac{\partial c_0}{\partial r} + \frac{\partial^2 c_0}{\partial r^2}\right) - kc_0 \tag{2.4.40}$$

with boundary conditions:

$$r = 0, \quad \frac{\partial c_0}{\partial r} = 0; \quad r = R_0, \quad \frac{\partial c_0}{\partial r} = 0; \quad z = 0, \quad c_0 = c^0. \tag{2.4.41}$$

The introduction of the dimensionless variable

$$r = r_0 R, \quad z = lZ, \quad c_0 = c^0 C_0(R, Z), \quad u_0(r) = \bar{u}U_0(R), \tag{2.4.42}$$

leads to:

$$U_0\frac{\partial C_0}{\partial Z_1} = Fo\left(\frac{1}{R}\frac{\partial C_0}{\partial R} + \frac{\partial^2 C_0}{\partial R^2}\right) - Da\,C_0;$$

$$R = 0, \quad \frac{\partial C_0}{\partial R} \equiv 0; \quad R = 1 + a, \quad \frac{\partial C_0}{\partial R} \equiv 0; \quad Z = 0, \quad C_0 \equiv 1; \tag{2.4.43}$$

where

$$U_0(R) = \left(1 + 2a - a^2\right)\left(\frac{1-a}{1+a}\right)^2\left[2 - \frac{2}{(1+a)^2}R^2\right]. \tag{2.4.44}$$

The mass transfer efficiency in this case is possible to be presented as

$$g_0 = \bar{u}_0 c^0 - \frac{2}{R_0^2}\int_0^{R_0} ru_0 c_0\,dr \tag{2.4.45}$$

and in the dimensionless form (conversion degree)

$$G_0 = \frac{g_0}{\bar{u}_0 c^0} = 1 - \frac{4}{(1+a)^2}\int_0^{1+a} R\left(1 - \frac{1}{(1+a)^2}R^2\right)C_0(R, 1)\,dR. \tag{2.4.46}$$

Table 2.8 Conversion
degree in columns with
(G) and without (G₀)
stagnation zones, relative
column efficiency (ΔG), and
relative stagnation zone
volume (ΔW)

a	G_0	G	ΔG	ΔW	$\Delta G/\Delta W$
0.05	0.1042	0.0865	0.1699	0.009	18.67
0.10	0.1165	0.0803	0.3107	0.033	9.39
0.13	0.1253	0.0770	0.3855	0.053	7.29
0.15	0.1317	0.0751	0.4298	0.068	6.31
0.20	0.1502	0.0707	0.5293	0.111	4.76

Let us consider the effect of the parameter a (which is related with the breadth of the stagnation zone) on the conversion degree in a column and comparing it with the process efficiency in a column without stagnation zones and a Poiseuille-type flow. Table 2.8 shows comparison results of the conversion degree in columns with (G) and without (G₀) stagnation zones (circulation flows as a result of the barriers); the relative conversion degree decreases (ΔG) as a result of the stagnation zones:

$$\Delta G = \frac{G_0 - G}{G_0}. \tag{2.4.47}$$

The relative stagnation zone volume with respect to the column volume

$$\Delta F = \frac{(R_0 - r_1)^2}{R_0^2} = \frac{4a^2}{(1+a)^2} \tag{2.4.48}$$

and the relative influence of the stagnation zone volume on the conversion degree $\Delta G/\Delta F$ are demonstrated in Table 2.8, too.

The proposed method [17] for modeling of the mass transfer in column chemical reactors with change of the velocity sign permits to analyze the influence of the stagnation zones on the conversion degree. The results obtained show that the increase of the stagnation zone breadth leads to decrease in the conversion degree G, while G_0 increases as a result of the average velocity $\bar{u}_0$ decrease. The increase of the relative stagnation zone volume ΔW leads to a relative column efficiency ΔG increase, while the relative influence of the stagnation zone volume on the column efficiency $\Delta G/\Delta W$ decreases.

The elimination of the negative effect of the stagnation zone is possible in the cases of absence of solid barriers, i.e., empty columns.

2.4.4   Mass Transfer in One-Phase Countercurrent Flow

The increase of the efficiency of apparatuses at lowest achievable costs is a main tendency in the industrial chemistry. In this respect, one of the main approaches to achieve complete processing of raw materials and reduce waste is the introduction of countercurrent organization of fluid flows aiming to increase the driving force of the mass transfer between the phases and to decrease the final concentrations of harmful substances. This mode of fluid flow has received industrial application in condensers, evaporators, absorption columns, chemical reactors, etc.

The mathematical modeling of interphase mass transfer in countercurrent flows is associated with the appearance of areas in the velocity field, where it becomes negative [19–21] which leads to negative Laplacian in the convection–diffusion equations. The problem is solved in the cases of gas–liquid [22–24] and liquid–liquid [25, 26] countercurrent boundary layer flows using similarity variables. In column apparatuses, this problem is solved in the cases of stagnant zones, where the circulation flows change the velocity sign and two-coordinate systems are used [18].

The main problem of the mass transfer modeling in column apparatuses is the change of the velocity sign and as a result the necessity to use two-coordinate systems, because the velocity must be positive only. For this purpose, a polynomial approximation for the axial concentration distribution at the zero velocity cylindrical surface, when the velocity changes its sign, is proposed [18].

Let us consider one-phase countercurrent flow organization of the fluid in the column, where the velocity distributions are positive when presented in two-coordinate systems:

$$u_1(r, z_1) \geq 0, \quad 0 \leq r \leq r_0, \quad 0 \leq z_1 \leq l; \quad 0 = \left(\frac{r_0}{l}\right)^2 \leq 10^{-2};$$

$$u_2(r, z_2) \geq 0, \quad r_0 \leq r \leq r_1, \quad 0 \leq z_2 \leq l; \quad z_1 + z_2 = l,$$

(2.4.49)

where

$$u_1(r, z_1) = u_1(r) = \bar{u}_1 \left(\frac{n_1 + 1}{n_1} - \frac{2r^2}{n_1 r_0^2}\right),$$

$$u_2(r, z_2) = u_2(r) = \bar{u}_2 \left(\frac{n_2 + 1}{n_2} - \frac{2}{n_2} \frac{r^2}{r_0^2}\right),$$

(2.4.50)

$$n_1 = 1, \quad n_2 = 1.75, \quad \bar{u}_1 = 1, \quad \bar{u}_2 = 0.5.$$

In these coordinate systems, the convection–diffusion equations in one-phase countercurrent flows have the form:

$$u_1 \frac{\partial c_1}{\partial z_1} = D\left(\frac{1}{r}\frac{\partial c_1}{\partial r} + \frac{\partial^2 c_1}{\partial r^2}\right) - kc_1;$$

$$u_2 \frac{\partial c_2}{\partial z_2} = D\left(\frac{1}{r}\frac{\partial c_2}{\partial r} + \frac{\partial^2 c_2}{\partial r^2}\right) - kc_2;$$

$$r = 0, \quad \frac{\partial c_1}{\partial r} \equiv 0; \quad r = r_1, \quad \frac{\partial c_2}{\partial r} \equiv 0;$$

(2.4.51)

$$r = r_0, \quad c_1 \equiv c_2, \quad \frac{\partial c_1}{\partial r} \equiv \frac{\partial c_2}{\partial r};$$

$$z_1 = 0, \quad c_1 \equiv c_1^0; \quad z_2 = l, \quad c_2 \equiv c_2^0;$$

where $c_1 = c_1(r, z_1)$, $0 \leq r \leq r_0$, $0 \leq z_1 \leq l$ and $c_2 = c_2(r, z_2)$, $r_0 \leq r \leq r_1$, $0 \leq z_2 \leq l$ are the concentrations distributions in the countercurrent flows and a cylindrical surface with radius r_0 and height l is the zero velocity surface between the two countercurrent flows.

In order to solve the problem (2.4.51), dimensionless variables have to be used:

$$z_1 = lZ_1, \quad z_2 = lZ_2, \quad Z_1 + Z_2 = 1, \quad r = r_1 R, \quad r_0 = r_1 R_0, \quad R_0 = 0.922,$$

$$u_1 = u_1^0 U_1(R), \quad u_2 = u_2^0 U_2(R), \quad u_1^0 = \frac{2}{r_0^2} \int_0^{r_0} r u_1(r) \, dr,$$

$$u_2^0 = \frac{2}{r_1^2 - r_0^2} \int_{r_0}^{r_1} r u_2(r) \, dr, \quad c_1 = c_1^0 C_1(R, Z_1), \quad c_2 = c_2^0 C_2(R, Z_2),$$

$$Fo_1 = \frac{Dl}{u_1^0 r_1^2}, \quad Fo_2 = \frac{Dl}{u_2^0 r_1^2}, \quad Da_1 = \frac{kl}{u_1^0}, \quad Da_2 = \frac{kl}{u_2^0}.$$

$$(2.4.52)$$

The introduction of the dimensionless variables (2.4.52) into (2.4.51) leads to the next set of dimensionless equations:

$$U_1 \frac{\partial C_1}{\partial Z_1} = Fo_1 \left(\frac{1}{R} \frac{\partial C_1}{\partial R} + \frac{\partial^2 C_1}{\partial R^2} \right) - Da_1 C_1;$$

$$U_2 \frac{\partial C_2}{\partial Z_2} = Fo_2 \left(\frac{1}{R} \frac{\partial C_2}{\partial R} + \frac{\partial^2 C_2}{\partial R^2} \right) - Da_2 C_2;$$

$$R = 0, \quad \frac{\partial C_1}{\partial R} \equiv 0; \quad R = R_0, \quad C_1 \equiv C_2, \quad \frac{\partial C_1}{\partial R} \equiv \frac{\partial C_2}{\partial R}; \quad R = 1, \quad \frac{\partial C_2}{\partial R} \equiv 0;$$

$$Z_1 = 0, \quad C_1 \equiv 1; \quad Z_2 = 0, \quad C_2 \equiv 1;$$

$$(2.4.53)$$

where

$$U_1 = 1.2143 - 1.4286 R^2, \quad 0 \leq R \leq R_0 = 0.922;$$
$$U_2 = 1.4286 R^2 - 1.2143, \quad R_0 \leq R \leq 1.$$

$$(2.4.54)$$

The problem (2.4.53) is possible to be solved, if the polynomial approximations of the concentrations at the zero velocity surface between two phases ($R = R_0$) are used:

$$C_1 = 1 + a_1 Z_1 + a_2 Z_1^2, \quad C_2 = 1 + b_1 Z_2 + b_2 Z_2^2, \qquad (2.4.55)$$

where the functions (2.4.55) satisfy the boundary condition $C_1(R_0, Z_1) = C_2(R_0, Z_2)$, $(Z_1 + Z_2 = 1)$ if $a_1 = a$, $a_2 = -a$, $b_1 = a$, $b_2 = -a$, i.e.,

$$C_1 = 1 + aZ_1 - aZ_1^2, \quad C_2 = 1 + aZ_2 - aZ_2^2. \qquad (2.4.56)$$

It is possible to use higher degrees of the polynomial approximations:

$$\begin{aligned} C_1 &= 1 + (a_2 - a_1)Z_1 + a_1 Z_1^2 - a_2 Z_1^3, \\ C_2 &= 1 + (2a_2 - a_1)Z_2 + (a_1 - 3a_2)Z_2^2 + a_2 Z_2^3. \end{aligned} \qquad (2.4.57)$$

$$\begin{aligned} C_1 &= 1 + (-a_1 - a_2 - a_3)Z_1 + a_1 Z_1^2 + a_2 Z_1^3 + a_3 Z_1^4, \\ C_2 &= 1 + (-a_1 - 2a_2 - 3a_3)Z_2 + (a_1 + 3a_2 + 6a_3)Z_2^2 + (-a_2 - 4a_3)Z_1^3 + a_3 Z_1^4. \end{aligned} \qquad (2.4.58)$$

This approach permits to solve the problem (2.4.53) as two problems:

$$U_1 \frac{\partial C_1}{\partial Z_1} = Fo_1 \left(\frac{1}{R} \frac{\partial C_1}{\partial R} + \frac{\partial^2 C_1}{\partial R^2} \right) - Da_1 C_1;$$

$$R = R_0, \quad C_1 = 1 + aZ_1 - aZ_1^2; \quad R = 0, \quad \frac{\partial C_1}{\partial R} \equiv 0; \quad Z_1 = 0, \quad C_1 \equiv 1; \qquad (2.4.59)$$

$$U_2 \frac{\partial C_2}{\partial Z_2} = Fo_2 \left(\frac{1}{R} \frac{\partial C_2}{\partial R} + \frac{\partial^2 C_2}{\partial R^2} \right) - Da_2 C_2;$$

$$R = 1, \quad \frac{\partial C_2}{\partial R} \equiv 0; \quad R = R_0, \quad C_2 = 1 + aZ_2 - aZ_2^2; \quad Z_2 = 0, \quad C_2 \equiv 1. \qquad (2.4.60)$$

The solution of the set of Eqs. (2.4.59) and (2.4.60) is possible after the minimization of the least-square function $P(a)$:

$$P(a) = \int_0^1 [\alpha_1(a, Z_1) - \bar{\alpha}_1(a, Z_1)]^2 dZ_1, \quad P(a) \to \min, \quad a \to a_0, \qquad (2.4.61)$$

where:

$$\begin{aligned} \alpha_1(a, Z_1) &= \left(\frac{\partial C_1}{\partial R} \right)_{R=R_0}, \quad \alpha_2(a, Z_2) = \left(\frac{\partial C_2}{\partial R} \right)_{R=R_0}, \\ \bar{\alpha}_1(a, Z_1) &= \alpha_2(a, 1 - Z_1), \end{aligned} \qquad (2.4.62)$$

i.e., after the minimization of (2.4.61), the polynomial approximations of the concentrations (2.4.56) satisfy the boundary conditions in (2.4.53):

$$R = R_0, \quad C_1 \equiv C_2, \quad \frac{\partial C_1}{\partial R} \equiv \frac{\partial C_2}{\partial R}. \tag{2.4.63}$$

The problems (2.4.59), (2.4.60) were solved [18] using polynomial approximations for concentration with three parameters (2.4.58) in two cases:

$$\begin{aligned} a &= (a_1, a_2, a_3), \quad 0 \leq Z_1 \leq 1; \\ a &= (a_1, a_2, a_3), \quad 0 \leq Z_1 \leq 0.85; \quad a = (a_{11}, a_{21}, a_{31}), \quad 0.85 \leq Z_1 \leq 1 \end{aligned} \tag{2.4.64}$$

and the results are presented in Table 2.9. The functions are shown in Figs. 2.20 and 2.21.

Figures 2.22 and 2.23 present the concentration profiles obtained at different axial coordinates.

Table 2.9 Parameter values of the polynomial approximations (2.4.58)

Parameter	Value	Z_1
a_1	27.128	$0 \leq Z_1 \leq 1$
a_2	−43.368	
a_3	23.252	
a_{11}	−1.782	$0.85 \leq Z_1 \leq 1$
a_{21}	−76.245	
a_{31}	58.279	

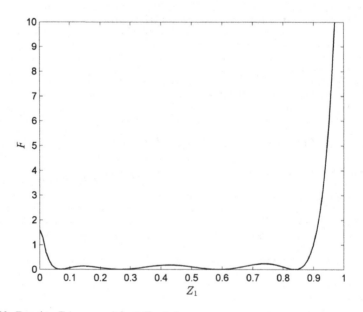

Fig. 2.20 Function $F(a_1, a_2, a_3)$ $0 \leq Z_1 \leq 1$

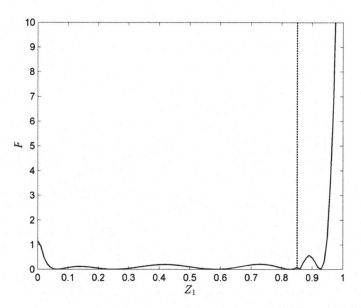

Fig. 2.21 Functions $F(a_1, a_2, a_3)$ $0 \leq Z_1 \leq 0.85$ and $F(a_{11}, a_{21}, a_{31})$ $0.85 \leq Z_1 \leq 1$

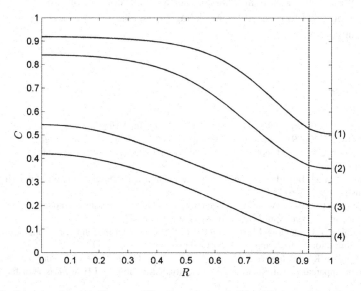

Fig. 2.22 Concentration profiles at different column height (1) $Z_1 = 0.1$, (2) $Z_1 = 0.2$, (3) $Z_1 = 0.6$, (4) $Z_1 = 0.8$

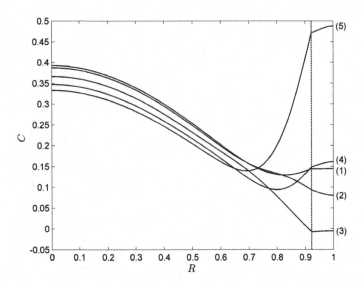

Fig. 2.23 Concentration profiles at different column height (1) $Z_1 = 0.85$, (2) $Z_1 = 0.86$, (3) $Z_1 = 0.9$, (4) $Z_1 = 0.94$, (5) $Z_1 = 0.97$

The presented theoretical analysis shows that convection–diffusion type of models is possible to be used for modeling of the chemical processes in one-phase column apparatuses.

References

1. Boyadjiev C (2006) Diffusion models and scale-up. Int J Heat Mass Transfer 49:796–799
2. Boyadjiev C (2009) Modeling of column apparatuses. Trans Academenergo 3:7–22
3. Boyadjiev C (2010) Theoretical chemical engineering. modeling and simulation. Springer, Berlin
4. Doichinova M, Boyadjiev C (2012) On the column apparatuses modeling. Int J Heat Mass Transfer 55:6705–6715
5. Boyadjiev C (2013) A new approach for the column apparatuses modeling in chemical engineering. J Pure Appl Math Adv Appl 10(2):131–150
6. Boyadjiev B, Doichinova M, Boyadjiev C (2015) Computer modeling of column apparatuses. 3. Perturbation method approach. J Eng Thermophys. 24(4):371–380
7. Panayotova K, Doichinova M, Boyadjiev C (2009) On the scale effect and scale-up in the column apparatuses 1. Influence of the velocity distribution. Int J Heat Mass Transfer 52:543–547
8. Doichinova M (2015) Influence of the velocity non-uniformity structure on the column apparatuses efficiency. Private communication
9. Rozen AM et al (eds) (1980) Scale-up in chemical technology. Chemistry, Moscow
10. Mc Mullen AK, Miyauchi T, Vermenlen T (1958) UCRI—3911. U.S. Atomic Energy Comm
11. Sleicher CAJr (1959). AIChEJ, 5: 145
12. Miyauchi T, Vermenlen T (1963) Longitudinal dispersion in two-phase continuous-flow operations. Ind Eng Chem (Fund) 2:113

13. Panayotova K, Doichinova M, Boyadjiev C (2009) On the scale effect and scale-up in the column apparatuses 2. Scale effect modeling. Int J Heat Mass Transfer 52:2358–2361
14. Boyadjiev C, Doichinova M, Popova-Krumova P, Boyadjiev B (2014) Intensive column apparatus for chemical reactions. Open Access Libr J 1(3):1–9
15. Lietuvos Respublika Patent 3884, 1994
16. Boyadjiev C, Boyadjiev B, Doichinova M, Popova-Krumova P (2013) Column reactor for chemical processes. Utility model, BG 1776 U1, 17 June 2013
17. Panayotova K, Doichinova M, Boyadjiev C (2010) On the scale effect and scale-up in the column apparatuses 3. Circulation zones. Int J Heat Mass Transfer 53:2128–2132
18. Doichinova M, Boyadjiev C (2015) Mass transfer in counter-current flow column apparatuses. Private communication
19. Tersenov SA (1985) Parabolic equations with changing direction of time. 1st edn. Science, Novosibirsk
20. Boyadjiev C, Mitev P, Beshkov V (1976) Laminar boundary layers at a moving interface generated by counter-current gas–liquid stratified flow. Int J Multiphase Flow 3:61–66
21. Boyadjiev C, Vabishchevich P (1992) Numerical simulation of opposite currents. J Theor Appl Mech (Bulgaria) 23:114–119
22. Boyadjiev C, Doichinova M (2000) Opposite-current flows in gas–liquid boundary layers-I. Velocity distribution. Int J Heat Mass Transfer 43:2701–2706
23. Doichinova M, Boyadjiev C (2000) Opposite-current flows in gas-liquid boundary layers-II. Mass transfer kinetics. Int J Heat Mass Transfer 43:2707–2710
24. Doichinova M, Boyadjiev C (2001) Opposite-current flows in gas-liquid boundary layers-III. Non-linear mass transfer. Int J Heat Mass Transfer 44:2121–2125
25. Horvath E, Nagy E, Boyadjiev C, Gyenis J (2007) Interphase mass transfer between liquid–liquid counter-current flows. I. Velocity distribution. J Eng Phys Thermophys 80(4):721–727
26. Horvath E, Nagy E, Boyadjiev C, Gyenis J (2007) Interphase mass transfer between liquid–liquid counter-current flows. II. Mass Transfer kinetics. J Eng Phys Thermophys 80(4):728–733
27. Boyadjiev B, Doichinova M, Boyadjiev C (2015) On the "back mixing" effect in column chemical reactors. Int J Modern Trends Eng Res 2(8):168–175

Chapter 3
Two-Phase Processes

The modeling of two-phase gas–liquid ($j = 1, 2$), gas–solid ($j = 1, 3$), and liquid–solid ($j = 2, 3$) interphase mass transfer processes in column apparatuses is possible to be used in the case of absorption, adsorption, and heterogeneous (catalytic) chemical reactions. For the modeling of two-phase processes, [1–5] the model equations (I.3–I.5) have to be used, i.e., component mass balances ($i = 1, 2, \ldots, i_0$) in the phases, where according (2.1.10) the radial velocity components are equal to zero ($v_j \equiv 0, j = 1, 2, 3$):

$$u_j \frac{\partial c_{ij}}{\partial z_j} = D_{ij} \left(\frac{\partial^2 c_{ij}}{\partial z_j^2} + \frac{1}{r} \frac{\partial c_{ij}}{\partial r} + \frac{\partial^2 c_{ij}}{\partial r^2} \right) + Q_{ij}(c_{ij});$$

$$r = 0, \quad \frac{\partial c_{ij}}{\partial r} \equiv 0; \quad r = r_0, \quad \frac{\partial c_{ij}}{\partial r} \equiv 0; \tag{3.0.1}$$

$$z_j = 0, \quad c_{ij} \equiv c_{ij}^0, \quad u_j^0 c_{ij}^0 \equiv u_j c_{ij}^0 - D_{ij} \left(\frac{\partial c_{ij}}{\partial z_j} \right)_{z_j = 0};$$

$$i = 1, 2, \ldots, i_0, \quad j = 1, 2 = 1, 3 = 2, 3.$$

In (3.0.1), $u_j = u_j(r)$ (m s^{-1}) and $c_{ij} = c_{ij}(r, z_j)$ (kg mol m^{-3}) are the axial velocity components and transferred substance concentrations in the phases, D_{ij} (m^2 s^{-1}) are the diffusivities in the phases, and u_j^0 and c_{ij}^0 are the inlet velocities and the concentrations in the phases. The concentrations of the transferred substance in the phases are presented as kg mol of the transferred substance in 1 m^3 of the phase volume. The holdup coefficients (m^3 of the phase volume in 1 m^3 of the column volume) and the inlet velocities in the column are obtained from the ratios $\varepsilon_j = F_j/F_0$ and $u_j^0 = F_j/\varepsilon_j \pi r_0^2$, where r_0 is the column radius (m), F_j are the phase flow rates (m^3 s^{-1}) in the column, $j = 1, 2, 3$, and $F_0 = \sum_{j=1}^{3} F_j$ (m^3 s^{-1}) is the total flow rate of the fluids in the column. The volume reactions terms $Q_{ij}, j = 1, 2, 3$ (kg mol m^{-3} s^{-1}) are the rates of the chemical reactions and interphase mass

© Springer International Publishing AG, part of Springer Nature 2018
C. Boyadjiev et al., *Modeling of Column Apparatus Processes*,
Heat and Mass Transfer, https://doi.org/10.1007/978-3-319-89966-4_3

transfer, as volume sources ($Q_{ij} > 0$) or sinks ($Q_{ij} < 0$), in the phase parts of the elementary column volume and participate in the mass balance in the elementary phase volumes.

The model (3.0.1) is possible to be used for co-current two-phase flows ($z_1 = z_2 = z$) or for countercurrent ones ($z_1 + z_2 = l$, where l is the active zone height (m) of the column). In the countercurrent flows, the mass transfer process models have to be presented in two-coordinate systems [1] because in a one-coordinate system, one of the equations has no solution due to the negative Laplacian value. The solution method of the equation set in two-coordinate systems will be presented in Chap. 8.

3.1 Absorption Processes

The convection–diffusion-type models of the absorption processes [5, 6] in the gas–liquid systems are possible to be obtained from (3.0.1) if $j = 1, 2$ ($1 = \varepsilon_1 + \varepsilon_2$), $i = 1, 2$. The kinetic terms $Q_j, j = 1, 2$ are the interphase mass transfer rates $(-1)^j k_0 (c_{11} - \chi c_{12}), j = 1, 2$ in the gas and liquid phases [see (1.0.4)] and the chemical reaction rate $(-k c_{12} c_{22})$ in the liquid phase, as volume sources or sinks of the substances in the phase parts of the elementary (column) volume (kg mol m^{-3} s^{-1}), where k_0 (s^{-1}) is the interphase mass transfer coefficient, χ—Henry's number, k—the chemical reaction rate constant. The same models are possible to be used for modeling of the extraction processes if χ is the redistribution factor.

The concentration of the transferred substance ($i = 1$) in the gas (liquid) phase is $c_{11}(c_{12})$ (kg mol m^{-3}), i.e. kg mol of the transferred substance in the gas (liquid) phase in 1 m^3 of the phase (elementary) volume, while the concentration of the reagent ($i = 2$) in the liquid phase is c_{22} (kg mol m^{-3}) (in 1 m^3 of the phase elementary volume).

The inlet concentration of the transferred substance in the gas (liquid) phase is $c_{11}^0(c_{12}^0)$. In the cases of absorption (desorption), $c_{12}^0 = 0$ ($c_{11}^0 = 0$) practically. The input velocities u_j^0 ($j = 1, 2$) (m s^{-1}) of the gas and liquid phases are equal to the average velocities $\bar{u}_j$ ($j = 1, 2$) of the phases in the column, which are defined as

$$u_j^0 = \frac{F_j}{\varepsilon_j \pi r_0^2} = \bar{u}_j = \frac{2}{r_0^2} \int_0^{r_0} r u_j(r) dr, \quad j = 1, 2, \tag{3.1.1}$$

where $F_j, j = 1, 2$ are the gas- and liquid-phase flow rates (m^3 s^{-1}) in the column volume.

3.1.1 Physical Absorption

The physical absorption is an interphase mass transfer of one substance from the gas to the liquid phase. The opposite is desorption. In these cases, $i_0 = 1$ and the substance index i is possible to be ignored; i.e., the concentrations will be designated as $c_j, j = 1, 2$. As a result, the convection–diffusion-type model for the steady-state physical absorption in the column apparatuses has the form:

$$u_j \frac{\partial c_j}{\partial z_j} = D_j \left(\frac{\partial^2 c_j}{\partial z_j^2} + \frac{1}{r} \frac{\partial c_j}{\partial r} + \frac{\partial^2 c_j}{\partial r^2} \right) + (-1)^j k_0 (c_1 - \chi c_2), \quad j = 1, 2, \quad (3.1.2)$$

where u_j (m s^{-1}), D_j (m^2 s^{-1}), and ε_j ($j = 1, 2$) are the velocities, the diffusivities, and the holdup coefficients in the gas and liquid phases. The boundary conditions of (3.1.2) are different in the cases of co-current and countercurrent gas–liquid flows.

Let us consider a countercurrent gas–liquid bubble column with an active zone height l, where $c_1(r, z_1)$ and $c_2(r, z_2)$ are the concentrations of the absorbed substance in the gas and the liquid phase $(z_1 + z_2 = l)$. The boundary conditions of (3.1.2) have the form:

$$r = 0, \quad \frac{\partial c_1}{\partial r} = \frac{\partial c_2}{\partial r} \equiv 0; \quad r = r_0, \quad \frac{\partial c_1}{\partial r} = \frac{\partial c_2}{\partial r} \equiv 0;$$

$$z_1 = 0, \, c_1(r, 0) \equiv c_1^0, \quad u_1^0 c_1^0 \equiv u_1(r) c_1^0 - D_1 \left(\frac{\partial c_1}{\partial z_1} \right)_{z_1 = 0}; \qquad (3.1.3)$$

$$z_2 = 0, c_2(r, 0) \equiv c_2^0, \quad u_2^0 c_2^0 \equiv u_2(r) c_2^0 - D_2 \left(\frac{\partial c_2}{\partial z_2} \right)_{z_2 = 0},$$

where $u_j^0, j = 1, 2$ are the inlet (average) velocities in the gas and the liquid phase. In the case of gas absorption, $c_2^0 = 0$ is practically valid. In the cases of co-current flows, $z_1 = z_2 = z$.

The presented convection–diffusion-type models (3.1.2), (3.1.3) permit a qualitative analysis of the physical absorption processes [5, 6] to be made using dimensionless (generalized) variables:

$$R = \frac{r}{r_0}, \quad Z_1 = \frac{z_1}{l}, \quad Z_2 = \frac{z_2}{l},$$

$$U_1 = \frac{u_1}{u_1^0}, \quad U_2 = \frac{u_2}{u_2^0}, \qquad (3.1.4)$$

$$C_1 = \frac{c_1}{c_1^0}, \quad C_2 = \frac{c_2 \chi}{c_1^0}.$$

If (3.1.4) is put into (3.1.2), (3.1.3), the model in generalized variables assumes the form:

$$U_1(R)\frac{\partial C_1}{\partial Z_1} = Fo_{11}\left(\varepsilon\frac{\partial^2 C_1}{\partial Z_1^2} + \frac{1}{R}\frac{\partial C_1}{\partial R} + \frac{\partial^2 C_1}{\partial R^2}\right) - K_0(C_1 - C_2);$$

$$U_2(R)\frac{\partial C_2}{\partial Z_2} = Fo_{12}\left(\varepsilon\frac{\partial^2 C_2}{\partial Z_2^2} + \frac{1}{R}\frac{\partial C_2}{\partial R} + \frac{\partial^2 C_2}{\partial R^2}\right) + K_0\frac{u_1^0\chi}{u_2^0}(C_1 - C_2);$$

$$R = 0, \quad \frac{\partial C_j}{\partial R} \equiv 0; \quad R = 1, \quad \frac{\partial C_j}{\partial R} \equiv 0; \quad j = 1, 2; \tag{3.1.5}$$

$$Z_1 = 0, \quad C_1 \equiv 1, \quad 1 \equiv U_1(R) - Pe_{11}^{-1}\left(\frac{\partial C_1}{\partial Z_1}\right)_{Z_1=0};$$

$$Z_2 = 0, \quad C_2 \equiv 0, \quad \left(\frac{\partial C_2}{\partial Z_2}\right)_{Z_2=0} \equiv 0,$$

where

$$\varepsilon = \frac{r_0^2}{l^2}, \quad K_0 = \frac{k_0 l}{u_1^0}, \quad Fo_{1j} = \frac{D_{1j}l}{u_j^0 r_0^2}, \quad Pe_{11} = \frac{u_1^0 l}{D_{11}}, \quad j = 1, 2. \tag{3.1.6}$$

From (3.1.5), it is possible to obtain directly the models of the physical absorption in the cases of highly ($\chi \to 0, C_2 \equiv 0$) and slightly ($\chi \to \infty, C_1 \equiv 1$) soluble gases.

The approximations of the film theory and the boundary layer theory of the mass transfer are not valid for the interphase mass transfer in the column apparatuses, and the expressions for the distribution of the interphase mass transfer resistance between the gas and liquid phases (0.4.5, 0.4.12) are not possible to be used.

From (3.1.5) it follows that $0 = K_0 = \frac{k_0 l}{u_1^0} \leq 10^{-2}, C_1 \equiv 1$ in the cases of a big average gas velocity $\bar{u}_1 = u_1^0$; i.e., the solution of the first equation in (3.1.5) is equal to unity. The concentration gradient in the gas phase is equal to zero as a result of the very big convective mass transfer rate in the gas phase; i.e., the mass transfer resistance in the gas phase is very small, and the process is limited by the mass transfer in the liquid phase.

In the cases $0 = K_0\frac{u_1^0\chi}{u_2^0} \leq 10^{-2}, C_2 \equiv 0$; i.e., the solution of the second equation in (3.1.5) is equal to zero. The concentration gradient in the liquid phase is equal to zero as a result of the very big convective mass transfer rate (big average liquid velocity $\bar{u}_2 = u_2^0$); i.e., the mass transfer resistance in the liquid phase is equal to zero, and the process is limited by the mass transfer in the gas phase.

These results show that the convection–diffusion-type model permits to be obtained the dimensionless mass transfer resistances in the gas (ρ_1) and liquid (ρ_2) phases:

$$\rho_1 = K_0, \quad \rho_2 = \rho_0 \rho_1, \quad \rho_0 = \frac{u_1^0 \chi}{u_2^0},$$

$$\rho_1 + \rho_2 = 1, \quad \rho_1 = \frac{1}{1 + \rho_0}, \quad \rho_2 = \frac{\rho_0}{1 + \rho_0}. \tag{3.1.7}$$

From (3.1.7), it is possible to obtain directly models of the physical absorption in the cases of highly $(\chi \to 0, \rho_0 \to 0, \rho_2 \to 0, C_2 \equiv 0)$ and slightly $(\chi \to \infty, \rho_0 \to \infty, \rho_1 \to 0, C_1 \equiv 1)$ soluble gases.

The intensification of the absorption processes is possible to be realized by intensification of the mass transfer in the limiting phase (practically by increasing of the convective mass transfer), i.e. phase with the higher mass transfer resistance. The increasing of the convective transfer in the liquid drops and gas bubbles has a limit; i.e., the optimal organization of the absorption process is the absorption in gas–liquid drops systems, when the resistance is in the gas phase $(\rho_2 \leq 10^{-2}, \rho_0 \leq 10^{-2})$, or absorption in liquid–gas bubbles systems, when the resistance is in the liquid phase $(\rho_1 \leq 10^{-2}, \rho_0 \geq 10^2)$.

For high columns, the parameter ε is very small $(0 = \varepsilon \leq 10^{-2})$ and the problem (3.1.5) is possible to be solved in zero approximation with respect to ε:

$$U_1(R) \frac{\partial C_1}{\partial Z_1} = Fo_1 \left(\frac{1}{R} \frac{\partial C_1}{\partial R} + \frac{\partial^2 C_1}{\partial R^2} \right) - K_0(C_1 - C_2);$$

$$U_2(R) \frac{\partial C_2}{\partial Z_2} = Fo_2 \left(\frac{1}{R} \frac{\partial C_2}{\partial R} + \frac{\partial^2 C_2}{\partial R^2} \right) + K_0 \frac{u_1^0 \chi}{u_2^0} (C_1 - C_2); \tag{3.1.8}$$

$$R = 0, \quad \frac{\partial C_i}{\partial R} \equiv 0; \quad R = 1, \quad \frac{\partial C_i}{\partial R} \equiv 0; \quad i = 1, 2;$$

$$Z_1 = 0, \quad C_1 \equiv 1; \quad Z_2 = 0, \quad C_2 \equiv 0.$$

For big values of the average velocities $0 = Fo_1 \leq 10^{-2}, 0 = Fo_2 \leq 10^{-2}$ and from (3.1.5) follows the convective-type model

$$U_1(R) \frac{dC_1}{dZ_1} = -K_0(C_1 - C_2);$$

$$U_2(R) \frac{dC_2}{dZ_2} = K_0 \frac{u_1^0 \chi}{u_2^0} (C_1 - C_2); \tag{3.1.9}$$

$$Z_1 = 0, \quad C_1 \equiv 1; \quad Z_2 = 0, \quad C_2 \equiv 0.$$

For small values of the average velocities $0 = K_0^{-1} \leq 10^{-2}$, $0 = \left(K_0 \frac{u_1^0 \chi}{u_2^0} \right)^{-1} \leq 10^{-2}$ from (3.1.5) follows the diffusion-type model:

$$0 = K_0^{-1} Fo_1 \left(\varepsilon \frac{\partial^2 C_1}{\partial Z_1^2} + \frac{1}{R} \frac{\partial C_1}{\partial R} + \frac{\partial^2 C_1}{\partial R^2} \right) - (C_1 - C_2);$$

$$0 = \left(K_0 \frac{u_1^0 \chi}{u_2^0} \right)^{-1} Fo_2 \left(\varepsilon \frac{\partial^2 C_2}{\partial Z_2^2} + \frac{1}{R} \frac{\partial C_2}{\partial R} + \frac{\partial^2 C_2}{\partial R^2} \right) + (C_1 - C_2);$$

$$R = 0, \quad \frac{\partial C_i}{\partial R} \equiv 0; \quad R = 1, \quad \frac{\partial C_i}{\partial R} \equiv 0; \quad i = 1, 2; \qquad (3.1.10)$$

$$Z_1 = 0, \quad C_1 \equiv 1, \quad 1 \equiv U_1(R) - Pe_1^{-1} \left(\frac{\partial C_1}{\partial Z_1} \right)_{Z_1=0};$$

$$Z_2 = 0, \quad C_2 \equiv 0, \quad \left(\frac{\partial C_2}{\partial Z_2} \right)_{Z_2=0} \equiv 0.$$

The solution of the model equations of a countercurrent physical absorption in two-coordinate systems is presented in Chap. 8 and [6].

3.1.2 Chemical Absorption

Two reagents $(i_0 = 2)$ participate in the chemical absorption. The first is in the gas phase $(i = 1, j = 1)$, and the second is in the liquid phase $(i = 2, j = 2)$. The chemical absorption will be presented in a co-current column $(z_1 = z_2 = z)$. Considering that $c_{11}(c_{12})$ is the concentration of the first reagent in the gas (liquid) phase and c_{22} is the concentration of the second reagent in the absorbent, the mass sources (sinks) in the medium elementary volume (in the physical approximations of the mechanics of continua) are equal to the chemical reaction rate—$kc_{12}c_{22}$ and the interphase mass transfer rate—$k_0(c_{11} - \chi c_{12})$. As a result, the convection–diffusion model in a column has the form:

$$u_1 \frac{\partial c_{11}}{\partial z} = D_{11} \left(\frac{\partial^2 c_{11}}{\partial z^2} + \frac{1}{r} \frac{\partial c_{11}}{\partial r} + \frac{\partial^2 c_{11}}{\partial r^2} \right) - k_0(c_{11} - \chi c_{12}),$$

$$u_2 \frac{\partial c_{12}}{\partial z} = D_{12} \left(\frac{\partial^2 c_{12}}{\partial z^2} + \frac{1}{r} \frac{\partial c_{12}}{\partial r} + \frac{\partial^2 c_{12}}{\partial r^2} \right) + k_0(c_{11} - \chi c_{12}) - kc_{12}c_{22}, \quad (3.1.11)$$

$$u_2 \frac{\partial c_{22}}{\partial z} = D_{22} \left(\frac{\partial^2 c_{22}}{\partial z^2} + \frac{1}{r} \frac{\partial c_{22}}{\partial r} + \frac{\partial^2 c_{22}}{\partial r^2} \right) - kc_{12}c_{22},$$

where $u_1(r), u_2(r)$ are the velocity distributions in the gas and liquid phases, $c_{ij}(r, z)$ and D_{ij} $(i = 1, 2; j = 1, 2)$ are the concentration distributions and the diffusivities of the first reagent in the gas and liquid phases and of the second reagent in the liquid phase.

Let us consider a co-current liquid–gas bubble column with a radius r_0 and working zone height l. The boundary conditions of the model equation (3.1.11) have the form:

$$r = 0, \quad \frac{\partial c_{11}}{\partial r} = \frac{\partial c_{12}}{\partial r} = \frac{\partial c_{22}}{\partial r} \equiv 0; \quad r = r_0, \quad \frac{\partial c_{11}}{\partial r} = \frac{\partial c_{12}}{\partial r} = \frac{\partial c_{22}}{\partial r} \equiv 0;$$

$$z = 0, \quad c_{11} \equiv c_{11}^0, \quad u_1^0 c_{11}^0 \equiv u_1(r)c_{11}^0 - D_{11}\left(\frac{\partial c_{11}}{\partial z}\right)_{z=0};$$

$$z = 0, \quad c_{12} \equiv c_{12}^0, \quad u_2^0 c_{12}^0 \equiv u_2(r)c_{12}^0 - D_{12}\left(\frac{\partial c_{12}}{\partial z}\right)_{z=0};$$

$$z = 0, \quad c_{22} \equiv c_{22}^0, \quad u_2^0 c_{22}^0 \equiv u_2(r)c_{22}^0 - D_{22}\left(\frac{\partial c_{22}}{\partial z}\right)_{z=0},$$

$$(3.1.12)$$

where $u_j^0, c_{ij}^0, i = 12, j = 1, 2$ are the inlet velocities and concentrations in the gas and liquid phases. In the cases of gas absorption, $c_{12}^0 = 0$ is practically valid.

A qualitative analysis of the model is possible to be made using dimensionless (generalized) variables:

$$R = \frac{r}{r_0}, \quad Z = \frac{z}{l}, \quad U_1 = \frac{u_1}{u_1^0}, \quad U_2 = \frac{u_2}{u_2^0},$$

$$C_{11} = \frac{c_{11}}{c_{11}^0}, \quad C_{12} = \frac{c_{12}\chi}{c_{11}^0}, \quad C_{22} = \frac{c_{22}}{c_{22}^0}.$$

$$(3.1.13)$$

The models (3.1.11), (3.1.12) in generalized variables (3.1.13) have the form:

$$U_1(R)\frac{\partial C_{11}}{\partial Z} = Fo_{11}\left(\varepsilon\frac{\partial^2 C_{11}}{\partial Z^2} + \frac{1}{R}\frac{\partial C_{11}}{\partial R} + \frac{\partial^2 C_{11}}{\partial R^2}\right) - K_0(C_{11} - C_{12});$$

$$U_2(R)\frac{\partial C_{12}}{\partial Z} = Fo_{12}\left(\varepsilon\frac{\partial^2 C_{12}}{\partial Z^2} + \frac{1}{R}\frac{\partial C_{12}}{\partial R} + \frac{\partial^2 C_{12}}{\partial R^2}\right)$$

$$+ K_0\frac{u_1^0\chi}{u_2^0}(C_{11} - C_{12}) - Da\frac{c_{22}^0\chi}{c_{11}^0}C_{12}C_{22};$$

$$U_2(R)\frac{\partial C_{22}}{\partial Z} = Fo_{22}\left(\varepsilon\frac{\partial^2 C_{22}}{\partial Z^2} + \frac{1}{R}\frac{\partial C_{22}}{\partial R} + \frac{\partial^2 C_{22}}{\partial R^2}\right) - DaC_{12}C_{22};$$

$$(3.1.14)$$

$$R = 0, \quad \frac{\partial C_s}{\partial R} \equiv 0; \quad R = 1, \quad \frac{\partial C_s}{\partial R} \equiv 0; \quad s = 11, 12, 22;$$

$$Z = 0, \quad C_{11} \equiv 1, \quad 1 \equiv U_1(R) - Pe_{11}^{-1}\left(\frac{\partial C_{11}}{\partial Z}\right)_{Z=0},$$

$$C_{12} \equiv 0, \quad \left(\frac{\partial C_{12}}{\partial Z}\right)_{Z=0} \equiv 0;$$

$$Z = 0, \quad C_{22} \equiv 1, \quad 1 \equiv U_2(R) - Pe_{22}^{-1}\left(\frac{\partial C_{22}}{\partial Z}\right)_{Z=0},$$

where

$$K_0 = \frac{k_0 l}{u_1^0}, \quad Fo_{11} = \frac{D_{11} l}{u_1^0 r_0^2}, \quad Fo_{12} = \frac{D_{12} l}{u_2^0 r_0^2}, \quad Fo_{22} = \frac{D_{22} l}{u_2^0 r_0^2},$$

$$Da = \frac{k l c_{11}^0}{u_2^0 \chi}, \quad Pe_{11} = \frac{u_1^0 l}{D_{11}}, \quad Pe_{22} = \frac{u_2^0 l}{D_{22}}. \tag{3.1.15}$$

From (3.1.14) follows that the absence of a chemical reaction in liquid phase—$k = 0$ (or $c_3^0 = 0$) leads to $Da = 0, C_{22} \equiv 1$, (or $C_{22} \equiv 0$), and as a result, the model of the physical absorption is obtained (3.1.5). The same result is possible to be obtained in the cases $0 = Da \frac{c_{22}^0 \chi}{c_{11}^0} \leq 10^{-2}$; i.e., the chemical reaction effect is negligible (it is not possible to be measured experimentally).

In the cases, when the interphase mass transfer is a result of the chemical reaction in the liquid phase $\left(Da \frac{c_{22}^0 \chi}{c_{11}^0} \geq 1 \right)$, the second equation in (3.1.14) should be divided by $Da \frac{c_{22}^0 \chi}{c_{11}^0} \geq 1$; i.e.,

$$U_1(R) \frac{\partial C_{11}}{\partial Z} = Fo_{11} \left(\varepsilon \frac{\partial^2 C_{11}}{\partial Z^2} + \frac{1}{R} \frac{\partial C_{11}}{\partial R} + \frac{\partial^2 C_{11}}{\partial R^2} \right) - K_0 (C_{11} - C_{12});$$

$$Da^{-1} \frac{c_{11}^0}{c_{22}^0 \chi} U_2(R) \frac{\partial C_{12}}{\partial Z} = \frac{Fo_{12} c_{11}^0}{Da c_{22}^0 \chi} \left(\varepsilon \frac{\partial^2 C_{12}}{\partial Z^2} + \frac{1}{R} \frac{\partial C_{12}}{\partial R} + \frac{\partial^2 C_{12}}{\partial R^2} \right)$$

$$+ K_0 \frac{u_1^0 c_{11}^0}{Da u_2^0 c_{22}^0} (C_{11} - C_{12}) - C_{12} C_{22};$$

$$Da^{-1} U_2(R) \frac{\partial C_{22}}{\partial Z} = Da^{-1} Fo_{22} \left(\varepsilon \frac{\partial^2 C_{22}}{\partial Z^2} + \frac{1}{R} \frac{\partial C_{22}}{\partial R} + \frac{\partial^2 C_{22}}{\partial R^2} \right) - C_{12} C_{22};$$

$$R = 0, \quad \frac{\partial C_s}{\partial R} \equiv 0; \quad R = 1, \quad \frac{\partial C_s}{\partial R} \equiv 0; \quad s = 11, 12, 22;$$

$$Z = 0, \quad C_{11} \equiv 1, \quad 1 \equiv U_1(R) - Pe_{11}^{-1} \left(\frac{\partial C_{11}}{\partial Z} \right)_{Z=0},$$

$$C_{12} \equiv 0, \quad \left(\frac{\partial C_{12}}{\partial Z} \right)_{Z=0} \equiv 0, \quad C_{22} \equiv 1, \quad 1 \equiv U_2(R) - Pe_{22}^{-1} \left(\frac{\partial C_{22}}{\partial Z} \right)_{Z=0}.$$

$$\tag{3.1.16}$$

In the cases of very fast chemical reactions $\left(Da\frac{c_{22}^0 \chi}{c_{11}^0} \geq 10^2\right)$ from (3.1.14) is possible to obtain:

$$U_1(R)\frac{\partial C_{11}}{\partial Z} = Fo_{11}\left(\varepsilon\frac{\partial^2 C_{11}}{\partial Z^2} + \frac{1}{R}\frac{\partial C_{11}}{\partial R} + \frac{\partial^2 C_{11}}{\partial R^2}\right) - K_0(C_{11} - C_{12});$$

$$0 = \frac{Fo_{12}c_{11}^0}{Dac_{22}^0\chi}\left(\varepsilon\frac{\partial^2 C_{12}}{\partial Z^2} + \frac{1}{R}\frac{\partial C_{12}}{\partial R} + \frac{\partial^2 C_{12}}{\partial R^2}\right) + K_0\frac{u_1^0 c_{11}^0}{Dau_2^0 c_{22}^0}(C_{11} - C_{12}) - C_{12}C_{22};$$

$$U_2(R)\frac{\partial C_{22}}{\partial Z} = Fo_{22}\left(\varepsilon\frac{\partial^2 C_{22}}{\partial Z^2} + \frac{1}{R}\frac{\partial C_{22}}{\partial R} + \frac{\partial^2 C_{22}}{\partial R^2}\right) - DaC_{12}C_{22};$$

$$R = 0, \quad \frac{\partial C_s}{\partial R} \equiv 0; \quad R = 1, \quad \frac{\partial C_s}{\partial R} \equiv 0; \quad s = 11, 12, 22;$$

$$Z = 0, \quad C_{11} \equiv 1, \quad 1 \equiv U_1(R) - Pe_{11}^{-1}\left(\frac{\partial C_{11}}{\partial Z}\right)_{Z=0}, \quad C_{12} \equiv 0, \quad \left(\frac{\partial C_{12}}{\partial Z}\right)_{Z=0} \equiv 0;$$

$$Z = 0, \quad C_{22} \equiv 1, \quad 1 \equiv U_2(R) - Pe_{22}^{-1}\left(\frac{\partial C_{22}}{\partial Z}\right)_{Z=0}.$$

$$(3.1.17)$$

For big values of the average velocities $0 = Fo_s \leq 10^{-2}, s = 1, 2, 3$ from (3.1.14) follows the convective-type model

$$U_1(R)\frac{dC_{11}}{dZ} = -K_0(C_{11} - C_{12});$$

$$U_2(R)\frac{dC_{12}}{dZ} = K_0\frac{u_1^0\chi}{u_2^0}(C_{11} - C_{12}) - Da\frac{c_{22}^0\chi}{c_{11}^0}C_{12}C_{22};$$

$$U_2(R)\frac{dC_{22}}{dZ} = -DaC_{12}C_{22};$$

$$Z = 0, \quad C_s \equiv 1, \quad s = 11, 22; \quad Z = 0, \quad C_{12} \equiv 0.$$

$$(3.1.18)$$

The concentration distribution in the chemical absorption case will be obtained for high columns, where the parameter ε in (3.1.14) is very small $(0 = \varepsilon \leq 10^{-2})$. The velocity distributions in the phases will be Poiseuille type, and the difference between the velocities of the phases will be in the average velocities only:

$$U_1 = U_2 = 2 - 2R^2. \tag{3.1.19}$$

As a result, the problem (3.1.14) takes the form:

$$\left(2 - 2R^2\right)\frac{\partial C_{11}}{\partial Z} = Fo_{11}\left(\frac{1}{R}\frac{\partial C_{11}}{\partial R} + \frac{\partial^2 C_{11}}{\partial R^2}\right) - K_0(C_{11} - C_{12});$$

$$\left(2 - 2R^2\right)\frac{\partial C_{12}}{\partial Z} = Fo_{12}\left(\frac{1}{R}\frac{\partial C_{12}}{\partial R} + \frac{\partial^2 C_{12}}{\partial R^2}\right) + K_0\frac{u_1^0\chi}{u_2^0}(C_{11} - C_{12}) - Da\frac{c_{22}^0\chi}{c_{11}^0}C_{12}C_{22};$$

$$\left(2 - 2R^2\right)\frac{\partial C_{22}}{\partial Z} = Fo_{22}\left(\frac{1}{R}\frac{\partial C_{22}}{\partial R} + \frac{\partial^2 C_{22}}{\partial R^2}\right) - DaC_{12}C_{22};$$

$$R = 0, \quad \frac{\partial C_s}{\partial R} \equiv 0; \quad R = 1, \quad \frac{\partial C_s}{\partial R} \equiv 0; \quad s = 11, 12, 22;$$

$$Z = 0, \quad C_{11} \equiv 1, \quad C_{12} \equiv 0, \quad C_{22} \equiv 1.$$

$$(3.1.20)$$

The solution of (3.1.20) is obtained in the case $Fo_{11} = Fo_{12} = Fo_{22} = 0.1$, $K_0 = Da = 1, \frac{u_1^0\chi}{u_2^0} = 1, \frac{c_{22}^0\chi}{c_{11}^0} = 2$, and the results are presented in Figs. 3.1 and 3.2.

In the chemical absorption case, the model [3.1.14] permits to obtain (similar to (3.1.17)] the interphase mass transfer resistance distribution between the gas and liquid phases:

$$\rho_1 = K, \quad \rho_2 = \rho_0\rho_1, \quad \rho_0 = \frac{u_1^0 c_{11}^0}{Da u_2^0 c_{22}^0},$$

$$\rho_1 + \rho_2 = 1, \quad \rho_1 = \frac{1}{1 + \rho_0}, \quad \rho_2 = \frac{\rho_0}{1 + \rho_0},$$

$$(3.1.21)$$

where the parameters ρ_1 and ρ_2 can be considered as mass transfer resistances in the gas and liquid phases. Very often, the big values of Da lead to small values of ρ_0 and as a result $\rho_1 \ll \rho_2$; i.e., the gas is the limiting phase, and the optimal organization of the absorption process is the absorption in gas–liquid drops systems.

Fig. 3.1 Concentration distribution $C_{11}(R, Z)$: (1) $Z = 0.2$; (2) $Z = 0.4$; (3) $Z = 0.8$; (4) $Z = 1$

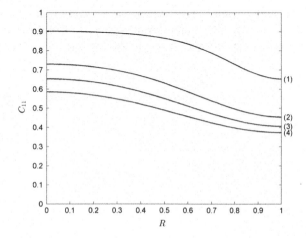

Fig. 3.2 Concentration distribution $C_{22}(R,Z)$: (1) $Z = 0.2$; (2) $Z = 0.4$; (3) $Z = 0.8$; (4) $Z = 1$

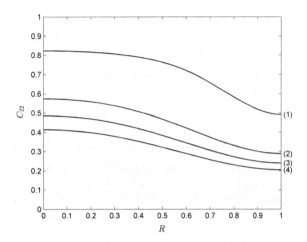

3.2 Adsorption Processes

In the adsorption process [7–10] participate two reagents ($i_0 = 2$), where the first is in the gas or liquid phase ($i = 1, j = 1, 2$) and the second is in the solid phase ($i = 2, j = 3$). The adsorption is the process of mass transfer of an active component (the substance, which is possible to be adsorbed) from the gas (liquid) volume to the solid interface due to a physical (van der Waals's) or chemical (valence) force [1]. The rate of the adsorption process determines analogically the chemical reaction rates, where law of mass action is changed by the law of surface action.

The convection–diffusion-type models of the adsorption processes in the gas (liquid)–solid systems are possible to be obtained from (3.0.1) if $j = 1, 3 = 2, 3$ ($1 = \varepsilon_1 + \varepsilon_3 = \varepsilon_2 + \varepsilon_3$), $i_0 = 2$, where $i = 1$ is the active component (AC) in the gas (liquid) phase, $i = 2$—the active sites (AS) in the adsorbent (solid phase). The volume adsorption rate in the case of a solid adsorbent is $Q_3 = b_0 Q_{03}$ (kg mol m^{-3} s^{-1}), where b_0 (m^2 m^{-3}) is m^2 of the inner surface in the solid phase (the surface of the capillaries in the solid phase) in 1 m^3 of the solid phase (adsorbent), Q_{03} (kg mol m^{-2} s^{-1})—the surface adsorption rate. A gas adsorption will be considered for convenience, where c_{11} (kg mol m^{-3}) is the volume concentration of the AC in the gas phase (elementary) volume, c_{13} (kg mol m^{-3})—the volume concentration of the AC in the void volume of the solid phase (adsorbent), c_{23} (kg eq m^{-3})—the volume concentration of the AS in the solid phase (elementary) volume (1 kg eq AS in the adsorbent combine 1 kg mol AC), $1 = \varepsilon_1 + \varepsilon_3$, $u_1 = u_1(r)$—velocity of the gas phase (m s^{-1}), $u_3 = 0$ (solid phase is immobile). All concentrations are in kg mol (kg eq) in 1 m^3 of the phase (elementary) volume. The inlet gas velocity in the column is $u_1^0 = F_1/\varepsilon_1 \pi r_0^2$, where r_0 is the column radius (m) and F_1—the gas-phase flow rates (m^3 s^{-1}). The average velocities $\bar{u}_1$ of the gas phases in the column are supposed to be equal to the inlet gas velocity in the column.

3.2.1 Physical Adsorption

In the cases of physical adsorption on a solid surface [7], the adsorption rate is proportional to the surface concentration of the free AS (which may be associated with the molecules of the AC) and the volume concentration of the AC:

$$Q_0^1 = k_1 c_{13}\left(1 - \frac{\Gamma}{\Gamma_\infty}\right), \tag{3.2.1}$$

where k_1 (m s^{-1}) is the adsorption rate constant, Γ (kg eq m^{-2})—the surface concentration of the AS, which is linked to the molecules of the AC (the surface concentration of the adsorbed AC), Γ_∞ (kg eq m^{-2})—the maximal surface concentration of the free AS. The surface concentration of the free AS is $(\Gamma_\infty - \Gamma)$.

The physical adsorption process is reversible, and the desorption rate could be obtained by analogical consideration, which is represented as:

$$Q_0^2 = k_2 \Gamma, \tag{3.2.2}$$

where k_2 (s^{-1}) is the desorption rate constant. The resultant adsorption rate is

$$Q_{03} = Q_0^1 - Q_0^2 = k_1 c_{13}\left(1 - \frac{\Gamma}{\Gamma_\infty}\right) - k_2 \Gamma. \tag{3.2.3}$$

The volume concentration of the free AS in the solid phase (adsorbent) c_{23} and its maximum value c_{23}^0 (kg eq m^{-3}) are possible to be obtained immediately:

$$c_{23} = b(\Gamma_\infty - \Gamma), \quad c_{23}^0 = b\Gamma_\infty \tag{3.2.4}$$

and from (3.2.3) and (3.2.4) follows the expression for the surface adsorption rate:

$$Q_{03} = k_1 c_{13} C_{23} - k_2 \frac{c_{23}^0}{b_0}(1 - C_{23}), \quad C_{23} = \frac{c_{23}}{c_{23}^0}. \tag{3.2.5}$$

Let us consider a non-stationary gas adsorption in a column apparatus, where the solid phase (adsorbent) is immobile. The convection–diffusion model of this process is possible to be obtained from (3.0.1), where the diffusivity of the free AS in the solid phase (adsorbent) volume is equal to zero. If the rate of the interphase mass transfer of the AC from the gas phase to the solid phase [see (1.0.3)] is $k_0(c_{11} - c_{13})$ and the process is non-stationary as a result of the free AS concentration decrease, the convection–diffusion model has the form:

$$\frac{\partial c_{11}}{\partial t} + u_1 \frac{\partial c_{11}}{\partial z} = D_{11}\left(\frac{\partial^2 c_{11}}{\partial z^2} + \frac{1}{r}\frac{\partial c_{11}}{\partial r} + \frac{\partial^2 c_{11}}{\partial r^2}\right) - k_0(c_{11} - c_{13}),$$

$$\frac{dc_{13}}{dt} = k_0(c_{11} - c_{13}) - b_0 k_1 c_{13}\frac{c_{23}}{c_{23}^0} + k_2 c_{23}^0\left(1 - \frac{c_{23}}{c_{23}^0}\right), \qquad (3.2.6)$$

$$\frac{dc_{23}}{dt} = -b_0 k_1 c_{13}\frac{c_{23}}{c_{23}^0} + k_2 c_{23}^0\left(1 - \frac{c_{23}}{c_{23}^0}\right),$$

where t is the time, D_{11} is the diffusivity of the AC in the gas phase, and k_0 is the interphase mass transfer coefficient (s^{-1}). In (3.2.6), $c_{11} = c_{11}(t, r, z)$ and (r, z) are parameters in $c_{13} = c_{13}(t, r, z)$ and $c_{23} = c_{23}(t, r, z)$. The concentration of the adsorbed AC is equal to the related AS concentration $(c_{23}^0 - c_{23})$.

The model (3.2.6) represents the decrease of the concentration of the AC (free AS) in the part of the elementary volume ε_1 (ε_3) due to the physical adsorption.

Let us consider an adsorption column with a radius r_0 and a height of the active volume l. The boundary conditions of (3.2.6) have the form:

$$t = 0, \quad c_{11} \equiv c_{11}^0, \quad c_{13} \equiv 0, \quad c_{23} \equiv c_{23}^0; \quad r = 0, \quad \frac{\partial c_{11}}{\partial r} = 0;$$

$$r = r_0, \quad \frac{\partial c_{11}}{\partial r} \equiv 0; \quad z = 0, \quad c_{11} \equiv c_{11}^0, \quad u_1^0 c_{11}^0 \equiv u_1(r)c_{11}^0 - D_{11}\left(\frac{\partial c_{11}}{\partial z}\right)_{z=0}; \qquad (3.2.7)$$

where u_1^0 is the inlet (average) velocity of the gas phase.

The use of dimensionless (generalized) variables [1] permits a qualitative analysis of the model (3.2.6), (3.2.7) to be made, where as characteristic scales are used the average velocity, the inlet concentrations, the characteristic time t_0 (s) and the column parameters (r_0, l):

$$T = \frac{t}{t^0}, \quad R = \frac{r}{r_0}, \quad Z = \frac{z}{h},$$

$$U = \frac{u_1}{u_1^0}, \quad C_{11} = \frac{c_{11}}{c_{11}^0}, \quad C_{13} = \frac{c_{13}}{c_{11}^0} \quad C_{23} = \frac{c_{23}}{c_{23}^0}. \qquad (3.2.8)$$

If (3.2.8) is put in (3.2.6), (3.2.7), the model in generalized variables takes the form:

$$\gamma\frac{\partial C_{11}}{\partial T} + U(R)\frac{\partial C_{11}}{\partial Z} = Fo\left(\varepsilon\frac{\partial^2 C_{11}}{\partial Z^2} + \frac{1}{R}\frac{\partial C_{11}}{\partial R} + \frac{\partial^2 C_{11}}{\partial R^2}\right) - K_0(C_{11} - C_{13});$$

$$\frac{dC_{13}}{dT} = K_3(C - C_1) - K_1 C_{13} C_{23} + K_2\frac{c_{23}^0}{c_{11}^0}(1 - C_{23});$$

$$\frac{dC_{23}}{dT} = -K_1\frac{c_{11}^0}{c_{23}^0}C_{13} C_{23} + K_2(1 - C_{23});$$

$$T = 0, \quad C_{11} \equiv 1, \quad C_{13} \equiv 0, \quad C_{23} \equiv 1; \quad R = 0, \quad \frac{\partial C_{11}}{\partial R} \equiv 0; R = 1,$$

$$\frac{\partial C_{11}}{\partial R} \equiv 0; \quad Z = 0, \quad C_{11} \equiv 1, \quad 1 \equiv U(R) - Pe^{-1}\left(\frac{\partial C_{11}}{\partial Z}\right)_{Z=0}, \tag{3.2.9}$$

where (R, Z) are parameters in $C_{13}(T, R, Z)$, $C_{23}(T, R, Z)$ and

$$Fo = \frac{D_{11}l}{u_1^0 r_0^2}, \quad Pe = \frac{u_1^0 l}{D_{11}}, \quad \gamma = \frac{l}{u_1^0 t^0}, \quad \varepsilon = \frac{r_0^2}{l^2},$$

$$K_0 = \frac{k_0 l}{u_1^0}, \quad K_1 = k_1 t^0 b_0, \quad K_2 = k_2 t^0 \quad K_3 = k_0 t^0. \tag{3.2.10}$$

Practically, for long duration processes $0 \le \gamma \le 10^{-2}$ and the problem (3.2.9) has the form:

$$U(R)\frac{\partial C_{11}}{\partial Z} = Fo\left(\varepsilon\frac{\partial^2 C_{11}}{\partial Z^2} + \frac{1}{R}\frac{\partial C_{11}}{\partial R} + \frac{\partial^2 C_{11}}{\partial R^2}\right) - K_0(C_{11} - C_{13});$$

$$\frac{dC_{13}}{dT} = K_3(C_{11} - C_{13}) - K_1 C_{13}C_{23} + K_2\frac{c_{23}^0}{c_{11}^0}(1 - C_{23});$$

$$\frac{dC_{23}}{dT} = -K_1\frac{c_{11}^0}{c_{23}^0}C_{13}C_{23} + K_2(1 - C_{23}); \tag{3.2.11}$$

$$T = 0, \quad C_{13} \equiv 0, \quad C_{23} \equiv 1;$$

$$R = 0, \quad \frac{\partial C_{11}}{\partial R} \equiv 0; \quad R = 1, \quad \frac{\partial C_{11}}{\partial R} \equiv 0;$$

$$Z = 0, \quad C_{11} \equiv 1, \quad 1 \equiv U(R) - Pe^{-1}\left(\frac{\partial C_{11}}{\partial Z}\right)_{Z=0},$$

where T is a parameter in $C_{11}(T, R, Z)$.

For big gas velocity $0 = Fo \le 10^{-2}$, $0 = \gamma \le 10^{-2}$ and from (3.2.9) follows the convection-type model

$$U(R)\frac{dC_{11}}{dZ} = -K_0(C_{11} - C_{13});$$

$$\frac{dC_{13}}{dT} = K_3(C_{11} - C_{13}) - K_1 C_{13}C_{23} + K_2\frac{c_{23}^0}{c_{11}^0}(1 - C_{23});$$

$$\frac{dC_{23}}{dT} = -K_1\frac{c_{11}^0}{c_{23}^0}C_{13}C_{23} + K_2(1 - C_{23}); \tag{3.2.12}$$

$$T = 0, \quad C_{13} \equiv 0, \quad C_{23} \equiv 1; \quad Z = 0, \quad C \equiv 1.$$

In the cases of high columns $(0 = \varepsilon \leq 10^{-2})$, the problem (3.2.11) has to be solved in zero approximation with respect to ε $(\varepsilon = 0, Pe^{-1} = \varepsilon Fo = 0)$:

$$U(R)\frac{\partial C_{11}}{\partial Z} = Fo\left(\frac{1}{R}\frac{\partial C_{11}}{\partial R} + \frac{\partial^2 C_{11}}{\partial R^2}\right) - K_0(C_{11} - C_{13});$$

$$R = 0, \quad \frac{\partial C_{11}}{\partial R} \equiv 0; \quad R = 1, \quad \frac{\partial C_{11}}{\partial R} \equiv 0; \quad Z = 0, \quad C_{11} \equiv 1. \tag{3.2.13}$$

$$\frac{dC_{13}}{dT} = K_3(C_{11} - C_{13}) - K_1 C_{13} C_{23} + K_2\frac{c_{23}^0}{c_{11}^0}(1 - C_{23}); \quad T = 0, \quad C_{13} \equiv 0. \tag{3.2.14}$$

$$\frac{dC_{23}}{dT} = -K_1\frac{c_{11}^0}{c_{23}^0}C_{13}C_{23} + K_2(1 - C_{23}); \quad T = 0, \quad C_{23} \equiv 1. \tag{3.2.15}$$

The solution of the model equations (3.2.13), (3.2.14), (3.2.15), using a multistep algorithm (see Chap. 9 and [10]), is obtained in the case of parabolic velocity distribution (Poiseuille flow) in the gas phase $U(R) = 2 - 2R^2$ and the parameters values $Fo = 10^{-1}, K_0 = K_1 = K_3 = c_{23}^0 = 1, \ K_2 = 10^{-3}, c_{11}^0 = 10^{-2}$. The concentration distributions $C_{11}(0.6, R, Z)$ and $C_{11}(T, 0.2, Z)$, for different T and Z, are presented in Figs. 3.3 and 3.4.

3.2.2 Chemical Adsorption

The presence of chemical bonds between the AC and AS at the solid surface leads to the next expression [1] for the adsorption rate:

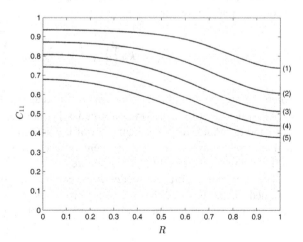

Fig. 3.3 Concentration distributions $C_{11}(0.6, R, Z)$:
(1) $Z = 0.2$; (2) $Z = 0.4$;
(3) $Z = 0.6$; (4) $Z = 0.8$;
(5) $Z = 1$

Fig. 3.4 Concentration
distributions $C_{11}(T, R, 0.2)$:
(1) $T = 0.2$; (2) $T = 0.4$;
(3) $T = 0.6$; (4) $T = 0.8$;
(5) $T = 1$

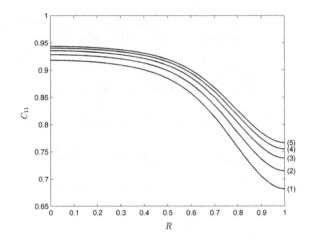

$$Q = k_0 \exp(-E/RT) \prod_{s=1}^{m} z_s^{\alpha_s} \prod_{i=1}^{n} p_i^{\beta_i} z_0^{\left(m_t - \sum_{s=1}^{m} \alpha_s \right)}, \qquad (3.2.16)$$

where z_s is the part of the face interphase occupied by the molecules of the substances A_s $(s = 1, \ldots, m)$, p_i—the partial pressures (volume concentrations) of the substances B_i $(i = 1, \ldots, n)$ in the gas (liquid), z_0—the part of the free surface, which is able to realize physical bonds with the molecules of the substances B_i $(j = 1, \ldots, n)$, m_t—the number of the AS at the interface, which realizes the physical bonds.

In (3.2.16), it is assumed that the molecules A_s $(s = 1, \ldots, m)$ from the solid surface react chemically with a part of the molecules B_i $(i = 1, \ldots, n)$, while the other part realizes physical bonds with the active places m_t. The heterogeneous reaction rate and the reactions orders are Q, α_s $(s = 1, \ldots, m)$ and β_i $(i = 1, \ldots, n)$.

In the cases of reversible heterogeneous chemical reactions, the equation of adsorption rate follows from and (3.2.16)

$$Q = k_{pi} \prod_{i=1}^{n} c_i / \left(1 + \sum_{i=1}^{n} k_{pi} c_i \right), \qquad (3.2.17)$$

where k_{pi} are the equilibrium constants of the reagents B_i $(i = 1, \ldots, n)$.

All equations of the adsorption kinetics [1] are based on the ideal adsorption layer model. Practically, the main part of the adsorption processes is related with real adsorption layers; i.e., the adsorbent surfaces are non-homogeneous as a result of the changes of the solid-phase structure.

The convection–diffusion model of the chemical adsorption [7] is possible to be obtained on the basis of the two-phase processes model, where ε_1 and ε_3 are the parts of the gas phase and sorbent particles (solid) phase $(\varepsilon_1 + \varepsilon_3 = 1)$ of the medium elementary volume in the column apparatus, where the solid phase is immobile ($u_3 = 0$). The volume concentrations of the AC in the gas phase and in the void volume of the solid phase (adsorbent) are c_{11} and c_{13} (kg mol m^{-3}). For the interphase (gas–solid) mass transfer rate, it is possible to use $k_0(c_{11} - c_{13})$, where k_0 is the interphase mass transfer coefficient. The adsorption rate in the solid phase (similar to two components chemical reaction) is presented as $kc_{13}c_{23}$, where c_{23} (kg eq m^{-3}) is the volume concentration of the (AS) in the solid phase (particles volume), k—the chemical reaction rate constant (1 kg eq AS in the adsorbent combine chemically 1 kg mol AC in gas phase). All concentrations are in kg mol (kg eq) in 1 m^3 of the phase (elementary) volume.

The diffusivity of the AC in the mobile gas phase is D_{11}. The convective transport of AC and AS in the solid phase is not possible. The diffusion mass transfer of AC in the solid phase (Knudsen diffusion) can be neglected due to the small value of the Knudsen diffusivity. The diffusivity of the AS in the adsorbent phase (particles volume) is equal to zero, too.

If the process is non-stationary as a result of the free AS concentration decrease, the convection–diffusion model of the chemical adsorption has the form:

$$\frac{\partial c_{11}}{\partial t} + u_1 \frac{\partial c_{11}}{\partial z} = D_{11}\left(\frac{\partial^2 c_{11}}{\partial z^2} + \frac{1}{r}\frac{\partial c_{11}}{\partial r} + \frac{\partial^2 c_{11}}{\partial r^2}\right) - k_0(c_{11} - c_{13}),$$

$$\frac{dc_{13}}{dt} = k_0(c_{11} - c_{13}) - kc_{13}c_{23} = 0, \quad \frac{dc_{23}}{dt} = -kc_{13}c_{23}. \tag{3.2.18}$$

The initial and boundary conditions of (3.2.18) are:

$$t = 0, \quad c_{11} \equiv c_{11}^0, \quad c_{13} \equiv 0, \quad c_{23} \equiv c_{23}^0;$$

$$r = 0, \quad \frac{\partial c_{11}}{\partial r} \equiv 0; \quad r = r_0, \quad \frac{\partial c_{11}}{\partial r} \equiv 0; \tag{3.2.19}$$

$$z = 0, \quad c_{11} \equiv c_{11}^0, \quad u_1^0 c_{11}^0 \equiv u_1(r)c_{11}^0 - D_{11}\left(\frac{\partial c_{11}}{\partial z}\right)_{z=0},$$

where u_1^0, c_{11}^0 are the inlet velocity and the concentration of the active component in the gas phase, c_{23}^0—the initial concentration of AS in the solid phase.

The using of dimensionless (generalized) variables

$$T = \frac{t}{t^0}, \quad R = \frac{r}{r_0}, \quad Z = \frac{z}{l}, \quad U = \frac{u_1}{u_1^0},$$

$$C_{11} = \frac{c_{11}}{c_{11}^0}, \quad C_{13} = \frac{c_{13}}{c_{11}^0}, \quad C_{23} = \frac{c_{23}}{c_{23}^0} \tag{3.2.20}$$

leads to:

$$\gamma \frac{\partial C_{11}}{\partial T} + U(R)\frac{\partial C_{11}}{\partial Z} = Fo\left(\varepsilon\frac{\partial^2 C_{11}}{\partial Z^2} + \frac{1}{R}\frac{\partial C_{11}}{\partial R} + \frac{\partial^2 C_{11}}{\partial R^2}\right) - K_0(C_{11} - C_{13});$$

$$\frac{dC_{13}}{dT} = K_3(C_{11} - C_{13}) - Kc^0_{23}C_{13}C_{23}; \qquad \frac{dC_{23}}{dT} = -Kc^0_{11}C_{13}C_{23};$$

$$T = 0, \quad C_{11} \equiv 1, \quad C_{13} \equiv 0, \quad C_{23} \equiv 1;$$

$$R = 0, \quad \frac{\partial C_{11}}{\partial R} \equiv 0; \quad R = 1, \quad \frac{\partial C_{11}}{\partial R} \equiv 0;$$

$$Z = 0, \quad C_{11} \equiv 1, \quad 1 \equiv U(R) - Pe^{-1}\left(\frac{\partial C_{11}}{\partial Z}\right)_{Z=0},$$

$$(3.2.21)$$

where (R, Z) are parameters in $C_{13}(T, R, Z)$, $C_{23}(T, R, Z)$ and

$$Fo = \frac{D_{11}l}{u^0_1 r^2_0}, \quad Pe = \frac{u^0_1 l}{D_{11}}, \quad \gamma = \frac{l}{u^0_1 t^0}, \quad \varepsilon = \left(\frac{r_0}{l}\right)^2,$$

$$K = kt^0, \quad K_0 = \frac{k_0 l}{u^0_1}, \quad K_3 = k_0 t^0.$$

$$(3.2.22)$$

For lengthy processes, it is possible to use the approximation $0 = \gamma \leq 10^{-2}$:

$$U(R)\frac{\partial C_{11}}{\partial Z} = Fo\left(\varepsilon\frac{\partial^2 C_{11}}{\partial Z^2} + \frac{1}{R}\frac{\partial C_{11}}{\partial R} + \frac{\partial^2 C_{11}}{\partial R^2}\right) - K_0(C_{11} - C_{13});$$

$$\frac{dC_{13}}{dT} = K_3(C_{11} - C_{13}) - Kc^0_{23}C_{13}C_{23}; \qquad \frac{dC_{23}}{dT} = -Kc^0_{11}C_{13}C_{23};$$

$$T = 0, \quad C_{13} \equiv 0, \quad C_{23} \equiv 1; \qquad\qquad (3.2.23)$$

$$R = 0, \quad \frac{\partial C_{11}}{\partial R} \equiv 0; \quad R = 1, \quad \frac{\partial C_{11}}{\partial R} \equiv 0;$$

$$Z = 0, \quad C_{11} \equiv 1, \quad 1 \equiv U(R) - Pe^{-1}\left(\frac{\partial C_{11}}{\partial Z}\right)_{Z=0},$$

where T is a parameter in $C_{11}(T, R, Z)$.

In the cases of high columns ($\varepsilon \leq 10^{-2}$), the problem (3.2.21) has to be solved in zero approximation with respect to ε ($\varepsilon = 0$):

$$\gamma\frac{\partial C_{11}}{\partial T} + U(R)\frac{\partial C_{11}}{\partial Z} = Fo\left(\frac{1}{R}\frac{\partial C_{11}}{\partial R} + \frac{\partial^2 C_{11}}{\partial R^2}\right) - K_0(C_{11} - C_{13});$$

$$\frac{dC_{13}}{dT} = K_3(C_{11} - C_{13}) - Kc^0_{23}C_{13}C_{23}; \qquad \frac{dC_{23}}{dT} = -Kc^0_{11}C_{13}C_{23};$$

$$T = 0, \quad C_{11} \equiv 1, \quad C_{13} \equiv 0, \quad C_{23} \equiv 1;$$

$$R = 0, \quad \frac{\partial C_{11}}{\partial R} \equiv 0; \quad R = 1, \quad \frac{\partial C_{11}}{\partial R} \equiv 0; \qquad (3.2.24)$$

$$Z = 0, \quad C_{11} \equiv 1.$$

For big gas velocity $0 = Fo \leq 10^{-2}, 0 = \gamma \leq 10^{-2}$ and from (3.2.23) follows the convection-type model

$$U(R)\frac{dC_{11}}{dZ} = -K_0(C_{11} - C_{13});$$

$$\frac{dC_{13}}{dT} = K_3(C_{11} - C_{13}) - Kc_{23}^0 C_{13}C_{23}; \quad \frac{dC_{23}}{dT} = -Kc_{11}^0 C_{13}C_{23}; \qquad (3.2.25)$$

$$T = 0, \quad C_{13} \equiv 0, \quad C_{23} \equiv 1; \quad Z = 0, \quad C_{11} \equiv 1,$$

where T is a parameter in $C_{11}(T, R, Z)$, while R and Z are parameters in $C_{13}(T, R, Z)$ and $C_{23}(T, Z, R)$.

3.3 Catalytic Processes

The catalytic process is a chemical reaction between three reagents $(i_0 = 3)$ in gas $(j = 1)$, liquid $(j = 2)$, or solid $(j = 3)$ phase [11]. For definiteness, catalytic processes in gas or gas-solid systems will be discussed.

The catalytic processes are of heterogeneous or homogeneous type. In the first case, the chemical reaction is implemented on a solid catalytic surface, where the first reagent is connected (adsorbed) physically or chemically with the third reagent (catalyst). The adsorption leads to a decrease of the activate energy E of the chemical reaction between the first and second reagents, and the chemical reaction rate increases. Analogous effects are possible in the cases of homogeneous chemical reactions, but they are result of the dissolved catalytic substances (third reagent), which change the chemical reaction route and as a result the general activate energy decreases, too.

The modeling of the homogeneous catalytic processes is possible to be realized using the model (2.1.12) for three-component chemical reaction $(i_0 = 3)$ and one-phase $(j = 1)$ column, where the concentration (c_{31}) of the third reagent (catalyst) is a constant and the catalytic effect is focused on the chemical kinetics term $kc_{11}^m c_{21}^n$, where the chemical reaction rate constant k is a function of the catalyst concentration (c_{31}).

The heterogeneous catalytic processes are a result of the chemical reaction between two reagents on the catalytic interface, wherein one of them is adsorbed physically or chemically on the free active sites (AS) of the solid catalytic surface. After the chemical reaction, the physical (van der Waals's) or chemical (valence) force between the obtained new substance and AS decreases and the new substance (reaction product) is desorbed from the solid surface. As a result, the process is stationary and the convection–diffusion models of the heterogeneous catalytic processes are possible to be created in the cases of physical adsorption mechanism (3.2.6) and chemical adsorption mechanism (3.2.18).

3.3.1 Physical Adsorption Mechanism

Let us consider a heterogeneous chemical reaction between two reagents (AC) in gas–solid system, where the first reagent is adsorbed physically on the free active sites (AS) of the solid catalytic surface. The reagents concentrations in the gas-phase elementary volume are c_{11}, c_{21} (kg mol m^{-3}), while in the void elementary volume of the solid phase (catalyst), the concentrations are c_{13}, c_{23}. The concentration of the free AS in the solid (catalytic)-phase elementary volume is c_{33} (kg eq m^{-3}). The maximal concentrations of AC and AS are c_{11}^0, c_{21}^0, c_{33}^0, where c_{11}^0, c_{21}^0 are input AC concentrations in the gas phase. The volume concentration of the adsorbed AC in the solid-phase elementary volume is equal to $c_{33}^0 - c_{33}$.

According to the physical adsorption mechanism at the gas–solid interphase, the interphase mass transfer rate of the first reagent (see [1.0.3]) is $k_{01}(c_{11} - c_{13})$, while that of the physical adsorption rate in the solid phase is $b_0 k_1 c_{13} \frac{c_{33}}{c_{33}^0} - k_2 c_{33}^0 \left(1 - \frac{c_{33}}{c_{33}^0}\right)$. The gas–solid interphase mass transfer rate of the second reagent is $k_{02}(c_{21} - c_{23})$, while the catalytic reaction rate is $k c_{23}(c_{33}^0 - c_{33})$. The difference between the interphase mass transfer coefficients k_{01}, k_{02} (s^{-1}) is a result of the difference between the diffusivities of the reagents in the gas phase. The concentration of AS decreases as a result of the physical adsorption and increases as a result of the catalytic reaction, because the reaction product does not have adsorption properties.

In the cases of a non-stationary catalytic process, the mass balance of AC and AS in the gas and solid phases leads to the convection–diffusion model of a heterogeneous catalytic chemical reaction in a column apparatus:

$$\frac{\partial c_{11}}{\partial t} + u_1 \frac{\partial c_{11}}{\partial z} = D_{11}\left(\frac{\partial^2 c_{11}}{\partial z^2} + \frac{1}{r}\frac{\partial c_{11}}{\partial r} + \frac{\partial^2 c_{11}}{\partial r^2}\right) - k_{01}(c_{11} - c_{13});$$

$$\frac{\partial c_{21}}{\partial t} + u_1 \frac{\partial c_{21}}{\partial z} = D_{21}\left(\frac{\partial^2 c_{21}}{\partial z^2} + \frac{1}{r}\frac{\partial c_{21}}{\partial r} + \frac{\partial^2 c_{21}}{\partial r^2}\right) - k_{02}(c_{21} - c_{23});$$

$$\frac{dc_{13}}{dt} = k_{01}(c_{11} - c_{13}) - b_0 k_1 c_{13}\frac{c_{33}}{c_{33}^0} + k_2 c_{33}^0\left(1 - \frac{c_{33}}{c_{33}^0}\right); \qquad (3.3.1)$$

$$\frac{dc_{23}}{dt} = k_{02}(c_{21} - c_{23}) - k c_{23}(c_{33}^0 - c_{33});$$

$$\frac{dc_{33}}{dt} = -b_0 k_1 c_{13}\frac{c_{33}}{c_{33}^0} + k_2 c_{33}^0\left(1 - \frac{c_{33}}{c_{33}^0}\right) + k c_{23}(c_{33}^0 - c_{33}),$$

where $u_1 = u_1(r)$ is the velocity distribution in the gas phase, and $\varepsilon_1, \varepsilon_3$ $(\varepsilon_1 + \varepsilon_3 = 1)$ are the parts of the gas and solid phases in the column volume.

The initial and boundary conditions of (3.3.1) are:

$$t = 0, \quad c_{11} \equiv c_{11}^0, \quad c_{21} \equiv c_{21}^0, \quad c_{13} \equiv 0, \quad c_{23} \equiv 0, \quad c_{33} \equiv c_{33}^0;$$

$$r = 0, \quad \frac{\partial c_{11}}{\partial r} = \frac{\partial c_{21}}{\partial r} \equiv 0; \quad r = r_0, \quad \frac{\partial c_{11}}{\partial r} = \frac{\partial c_{21}}{\partial r} \equiv 0;$$

$$z = 0, \quad c_{11} \equiv c_{11}^0, \quad u_1^0 c_{11}^0 \equiv u_1(r)c_{11}^0 - D_{11}\left(\frac{\partial c_{11}}{\partial z}\right)_{z=0}, \qquad (3.3.2)$$

$$c_{21} \equiv c_{21}^0, \quad u_1^0 c_{21}^0 \equiv u_1(r)c_{21}^0 - D_{21}\left(\frac{\partial c_{21}}{\partial z}\right)_{z=0},$$

where u_1^0 is the inlet velocity of the gas phase.

For a long duration process, the concentration of AS is a constant with respect to the time (as a result of the desorption of the reaction product) and the models (3.3.1) and (3.3.2) are stationary form:

$$u_1 \frac{\partial c_{11}}{\partial z} = D_{11}\left(\frac{\partial^2 c_{11}}{\partial z^2} + \frac{1}{r}\frac{\partial c_{11}}{\partial r} + \frac{\partial^2 c_{11}}{\partial r^2}\right) - k_{01}(c_{11} - c_{13});$$

$$u_1 \frac{\partial c_{21}}{\partial z} = D_{21}\left(\frac{\partial^2 c_{21}}{\partial z^2} + \frac{1}{r}\frac{\partial c_{21}}{\partial r} + \frac{\partial^2 c_{21}}{\partial r^2}\right) - k_{02}(c_{21} - c_{23});$$

$$k_{01}(c_{11} - c_{13}) - b_0 k_1 c_{13}\frac{c_{33}}{c_{33}^0} + k_2 c_{33}^0\left(1 - \frac{c_{33}}{c_{33}^0}\right) = 0;$$

$$k_{02}(c_{21} - c_{23}) - kc_{23}(c_{33}^0 - c_{33}) = 0;$$

$$- b_0 k_1 c_{13}\frac{c_{33}}{c_{33}^0} + k_2 c_{33}^0\left(1 - \frac{c_{33}}{c_{33}^0}\right) + kc_{23}(c_{33}^0 - c_{33}) = 0; \qquad (3.3.3)$$

$$r = 0, \quad \frac{\partial c_{11}}{\partial r} = \frac{\partial c_{21}}{\partial r} \equiv 0; \quad r = r_0, \quad \frac{\partial c_{11}}{\partial r} = \frac{\partial c_{21}}{\partial r} \equiv 0;$$

$$z = 0, \quad c_{11} \equiv c_{11}^0, \quad u_1^0 c_{11}^0 \equiv u_1(r)c_{11}^0 - D_{11}\left(\frac{\partial c_{11}}{\partial z}\right)_{z=0},$$

$$c_{21} \equiv c_{21}^0, \quad u_1^0 c_{21}^0 \equiv u_1(r)c_{21}^0 - D_{21}\left(\frac{\partial c_{21}}{\partial z}\right)_{z=0}.$$

The use of dimensionless (generalized) variables [1] permits to make a qualitative analysis of the model (3.3.3), where the inlet velocity and concentrations and the column parameters (r_0, l) are used as characteristic scales:

$$R = \frac{r}{r_0}, \quad Z = \frac{z}{l}, \quad U = \frac{u_1}{u_1^0}, \quad C_{11} = \frac{c_{11}}{c_{11}^0},$$

$$C_{21} = \frac{c_{21}}{c_{21}^0}, \quad C_{33} = \frac{c_{33}}{c_{33}^0}, \quad C_{13} = \frac{c_{13}}{c_{11}^0}, \quad C_{23} = \frac{c_{23}}{c_{21}^0}. \qquad (3.3.4)$$

If (3.3.4) is put in (3.3.3), the model in generalized variables takes the form:

$$U(R)\frac{\partial C_{11}}{\partial Z} = Fo_{11}\left(\varepsilon\frac{\partial^2 C_{11}}{\partial Z^2} + \frac{1}{R}\frac{\partial C_{11}}{\partial R} + \frac{\partial^2 C_{11}}{\partial R^2}\right) - K_{01}(C_{11} - C_{13});$$

$$U(R)\frac{\partial C_{21}}{\partial Z} = Fo_{21}\left(\varepsilon\frac{\partial^2 C_{21}}{\partial Z^2} + \frac{1}{R}\frac{\partial C_{21}}{\partial R} + \frac{\partial^2 C_{21}}{\partial R^2}\right) - K_{02}(C_{21} - C_{23});$$

$$R = 0, \quad \frac{\partial C_{i1}}{\partial R} \equiv 0; \quad R = 1, \quad \frac{\partial C_{i1}}{\partial R} \equiv 0;$$

$$Z = 0, \quad C_{i1} \equiv 1, \quad 1 \equiv U(R) - Pe_{i1}^{-1}\left(\frac{\partial C_{i1}}{\partial Z}\right)_{Z=0}; \quad i = 1, 2.$$

(3.3.5)

$$C_{13} = \frac{C_{11} + K_1(1 - C_{33})}{1 + K_2 C_{33}}, \quad C_{23} = \frac{C_{21}}{1 + K_3(1 - C_{33})}, \quad C_{33} = \frac{K_5 + C_{23}}{K_4 C_{13} + K_5 + C_{23}}.$$

(3.3.6)

In (3.3.5), (3.3.6), the following parameters are used:

$$K_{0i} = \frac{k_{0i}l}{u_1^0}, \quad Fo_{i1} = \frac{D_{i0}l}{u_1^0 r_0^2}, \quad Pe_{i1} = \frac{u_1^0 l}{D_{i0}}, \quad i = 1, 2, \quad \varepsilon = \frac{r_0^2}{l^2} = Fo_{i1}^{-1} Pe_{i1}^{-1},$$

$$K_1 = \frac{k_2}{k_{01}}\frac{c_{33}^0}{c_{11}^0}, \quad K_2 = \frac{b_0 k_1}{k_{01}}, \quad K_3 = \frac{k_{23} c_{33}^0}{k_{02}}, \quad K_4 = \frac{b_0 k_1}{k_{23} c_{21}^0}\frac{c_{11}^0}{c_{33}^0}, \quad K_5 = \frac{k_2}{k_{23} c_{21}^0}.$$

(3.3.7)

For high columns, the parameter ε is very small $(0 = \varepsilon \leq 10^{-2})$ and the problem (3.3.5) is possible to be solved in zero approximation with respect to ε:

$$U(R)\frac{\partial C_{11}}{\partial Z} = Fo_{11}\left(\frac{1}{R}\frac{\partial C_{11}}{\partial R} + \frac{\partial^2 C_{11}}{\partial R^2}\right) - K_{01}(C_{11} - C_{13});$$

$$U(R)\frac{\partial C_{21}}{\partial Z} = Fo_{21}\left(\frac{1}{R}\frac{\partial C_{21}}{\partial R} + \frac{\partial^2 C_{21}}{\partial R^2}\right) - K_{02}(C_{21} - C_{23});$$

$$R = 0, \quad \frac{\partial C_{i0}}{\partial R} \equiv 0; \quad R = 1, \quad \frac{\partial C_{i0}}{\partial R} \equiv 0; \quad Z = 0, \quad C_{i0} \equiv 1; \quad i = 1, 2.$$

(3.3.8)

For big values of the average velocities $0 = Fo_{11} \leq 10^{-2}, 0 = Fo_{21} \leq 10^{-2}$ and from (3.3.8) follows the convective type of model

$$U(R)\frac{dC_{11}}{dZ} = -K_{01}(C_{11} - C_{13});$$

$$U(R)\frac{dC_{21}}{dZ} = -K_{02}(C_{21} - C_{23}); \quad Z = 0, \quad C_{i0} \equiv 1; \quad i = 1, 2.$$

(3.3.9)

For small values of the average velocities $0 = K_{0i}^{-1} \leq 10^{-2}, i = 1, 2$, from (3.3.5) follows the diffusion type of model:

$$0 = K_{01}^{-1} \mathrm{Fo}_{11} \left(\varepsilon \frac{\partial^2 C_{11}}{\partial Z^2} + \frac{1}{R} \frac{\partial C_{11}}{\partial R} + \frac{\partial^2 C_{11}}{\partial R^2} \right) - (C_{11} - C_{13});$$

$$0 = K_{02}^{-1} \mathrm{Fo}_{21} \left(\varepsilon \frac{\partial^2 C_{21}}{\partial Z^2} + \frac{1}{R} \frac{\partial C_{21}}{\partial R} + \frac{\partial^2 C_{21}}{\partial R^2} \right) - (C_{21} - C_{23});$$

$$R = 0, \quad \frac{\partial C_{i0}}{\partial R} \equiv 0; \quad R = 1, \quad \frac{\partial C_{i0}}{\partial R} \equiv 0;$$

$$Z = 0, \quad C_{i0} \equiv 1, \quad 1 \equiv U(R) - Pe_{i0}^{-1} \left(\frac{\partial C_{i0}}{\partial Z} \right)_{Z=0}; \quad i = 1, 2.$$

(3.3.10)

The solution of the model equations (3.3.6), (3.3.8) requires a velocity distribution in the column. As an example, the case of parabolic velocity distribution (Poiseuille flow) in the gas phase will be presented [11]:

$$u_1 = \bar{u}_1 \left(2 - 2 \frac{r^2}{r_0^2} \right), \quad U(R) = 2 - 2R^2.$$

(3.3.11)

The solution of (3.3.8) depends on the two functions:

$$C_{13} = \frac{C_{11} + K_1(1 - C_{33})}{1 + K_2 C_{33}}, \quad C_{23} = \frac{C_{21}}{1 + K_3(1 - C_{33})},$$

(3.3.12)

where C_{33} is the solution of the cubic equation:

$$\begin{aligned}
&\omega_3 (C_{33})^3 + \omega_2 (C_{33})^2 + \omega_1 C_{33} + \omega_0 = 0, \\
&\omega_3 = K_3 (K_1 K_4 - K_2 K_5), \\
&\omega_2 = K_5 (K_2 + 2K_2 K_3 - K_3) - K_4 (K_1 + 2K_1 K_3 + K_3 C_{11}) + K_2 C_{21}, \\
&\omega_1 = K_4 (C_{11} + K_1)(1 + K_3) + K_5 (1 + 2K_3 - K_2 - K_2 K_3) + (1 - K_2) C_{21}, \\
&\omega_0 = -C_{21} - K_3 K_5 - K_5.
\end{aligned}$$

(3.3.13)

As a solution of (3.3.13), $0 \leq C_{33} \leq 1$ is to be used.

A solution of the problems (3.3.8), (3.3.12), (3.3.13) has been obtained for the case

$$\begin{aligned}
&K_{0i} = 1, \quad Fo_{i0} = 0.1, \quad i = 1, 2, \quad K_1 = 2.5, \\
&K_2 = 1, \quad K_3 = 1, \quad K_4 = 0.5, \quad K_5 = 1
\end{aligned}$$

(3.3.14)

as five-matrix forms:

$$C_{11}(R,Z) = \left\|C_{11(\rho\zeta)}\right\|, \quad C_{21}(R,Z) = \left\|C_{21(\rho\zeta)}\right\|, \quad C_{13}(R,Z) = \left\|C_{13(\rho\zeta)}\right\|,$$
$$C_{23}(R,Z) = \left\|C_{23(\rho\zeta)}\right\|, \quad C_{33}(R,Z) = \left\|C_{33(\rho\zeta)}\right\|;$$
$$R = \frac{\rho - 1}{\rho^0 - 1}, \quad \rho = 1,2,\dots,\rho^0; \quad Z = \frac{\zeta - 1}{\zeta^0 - 1}, \quad \zeta = 1,2,\dots,\zeta^0, \quad \rho^0 = \zeta^0.$$

$$(3.3.15)$$

The concentration distributions for different Z are presented in Figs. 3.5, 3.6, 3.7, 3.8, and 3.9.

Fig. 3.5 Radial distribution of the concentration $C_{11}(R,Z)$

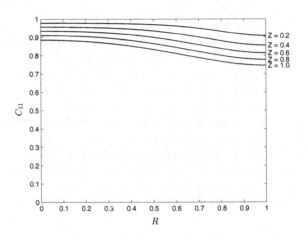

Fig. 3.6 Radial distribution of the concentration $C_{21}(R,Z)$

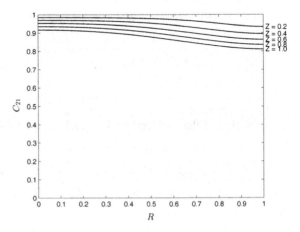

Fig. 3.7 Radial distribution of the concentration $C_{13}(R, Z)$

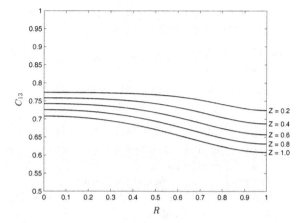

Fig. 3.8 Radial distribution of the concentration $C_{23}(R, Z)$

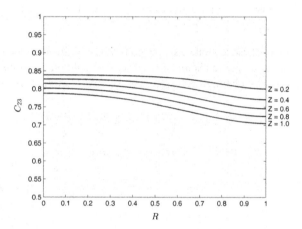

Fig. 3.9 Radial distribution of the concentration $C_{33}(R, Z)$

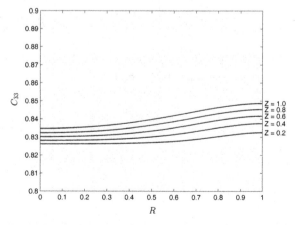

3.3.2   Chemical Adsorption Mechanism

The difference between the physical and chemical adsorption mechanisms (in the stationary case) is that in (3.3.3), the physical adsorption rate $-b_0 k_1 c_{13} \frac{c_{33}}{c_{33}^0} + k_2 c_{33}^0 \left(1 - \frac{c_{33}}{c_{33}^0}\right)$ has to be replaced by the chemical adsorption rate $-k_{13} c_{13} c_{33}$. As a result:

1. The gas–solid interphase mass transfer rate of the first reagent $k_{01}(c_{11} - c_{13})$ is equal to the chemical reaction between this reagent and AS in the solid-phase (catalyst) capillaries $k_{13} c_{13} c_{33}$.
2. The gas–solid interphase mass transfer rate of the second reagent $k_{02}(c_{21} - c_{23})$ is equal to the chemical reaction between this reagent and adsorbed reagent in the solid phase (catalyst) $k c_{23} \left(c_{33}^0 - c_{33}\right)$.
3. The adsorption rate of the first reagent $k_{13} c_{13} c_{33}$ must be equal to the desorption rate of the catalytic reaction product, i.e., to the catalytic reaction rate $k c_{23} \left(c_{33}^0 - c_{33}\right)$.

In these conditions, the convection–diffusion model of a stationary heterogeneous catalytic chemical reaction in a column apparatuses between two AC in the cases of chemical adsorption of one AC has the form:

$$u_1 \frac{\partial c_{11}}{\partial z} = D_{11} \left(\frac{\partial^2 c_{11}}{\partial z^2} + \frac{1}{r} \frac{\partial c_{11}}{\partial r} + \frac{\partial^2 c_{11}}{\partial r^2}\right) - k_{01}(c_{11} - c_{13});$$

$$u_1 \frac{\partial c_{21}}{\partial z} = D_{21} \left(\frac{\partial^2 c_{21}}{\partial z^2} + \frac{1}{r} \frac{\partial c_{21}}{\partial r} + \frac{\partial^2 c_{21}}{\partial r^2}\right) - k_{02}(c_{21} - c_{23});$$

$$r = 0, \quad \frac{\partial c_{11}}{\partial r} = \frac{\partial c_{21}}{\partial r} \equiv 0; \quad r = r_0, \quad \frac{\partial c_{11}}{\partial r} = \frac{\partial c_{21}}{\partial r} \equiv 0; \qquad (3.3.16)$$

$$z = 0, \quad c_{11} \equiv c_{11}^0, \quad u_1^0 c_{11}^0 \equiv u_1(r) c_{11}^0 - D_{11} \left(\frac{\partial c_{11}}{\partial z}\right)_{z=0},$$

$$c_{21} \equiv c_{21}^0, \quad u_1^0 c_{21}^0 \equiv u_1(r) c_{21}^0 - D_{21} \left(\frac{\partial c_{21}}{\partial z}\right)_{z=0}.$$

$$k_{01}(c_{11} - c_{13}) = k_{13} c_{13} c_{33}; \quad k_{02}(c_{21} - c_{23}) = k c_{23} \left(c_{33}^0 - c_{33}\right);$$
$$k_{13} c_{13} c_{33} = k c_{23} \left(c_{33}^0 - c_{33}\right). \qquad (3.3.17)$$

The introduction of the dimensionless variables (3.3.4) in (3.3.16), (3.3.17) leads to:

$$U(R)\frac{\partial C_{11}}{\partial Z} = Fo_{11}\left(\varepsilon\frac{\partial^2 C_{11}}{\partial Z^2} + \frac{1}{R}\frac{\partial C_{11}}{\partial R} + \frac{\partial^2 C_{11}}{\partial R^2}\right) - K_{01}(C_{11} - C_{13});$$

$$U(R)\frac{\partial C_{21}}{\partial Z} = Fo_{21}\left(\varepsilon\frac{\partial^2 C_{21}}{\partial Z^2} + \frac{1}{R}\frac{\partial C_{21}}{\partial R} + \frac{\partial^2 C_{21}}{\partial R^2}\right) - K_{02}(C_{21} - C_{23});$$

$$R = 0, \quad \frac{\partial C_{i1}}{\partial R} \equiv 0; \quad R = 1, \quad \frac{\partial C_{i1}}{\partial R} \equiv 0;$$

$$Z = 0, \quad C_{i1} \equiv 1, \quad 1 \equiv U(R) - Pe_{i1}^{-1}\left(\frac{\partial C_{i1}}{\partial Z}\right)_{Z=0}; \quad i = 1,2.$$

(3.3.18)

$$C_{13} = \frac{C_{11}}{1 + K_1 C_{33}}, \quad C_{23} = \frac{C_{21}}{1 + K_2(1 - C_{33})}, \quad C_{33} = \frac{C_{23}}{C_{23} + K_3 C_{13}}, \quad (3.3.19)$$

where

$$K_1 = \frac{k_{13}c_{33}^0}{k_{01}}, \quad K_2 = \frac{k c_{33}^0}{k_{02}}, \quad K_3 = \frac{k_{13}c_{11}^0}{k c_{21}^0}. \quad (3.3.20)$$

The models (3.3.5) and (3.3.18) are equivalent, and the theoretical analysis of the physical adsorption mechanism of the catalytic reactions in column chemical reactors (3.3.8), (3.3.9), (3.3.10) is valid in the chemical adsorption case. The difference is in the expressions (3.3.6), (3.3.19) of the volume concentrations in the solid phase (catalyst), only.

The solution of the model equations (3.3.18), (3.3.19) needs a velocity distribution in the column. The case of parabolic velocity distribution (3.3.11) will be presented [11] as an example.

The solution of (3.3.18) depends on the two functions (C_{13}, C_{23}) in (3.3.19), where C_{33} is the solution of the quadratic equation

$$(C_{21}K_1 - C_{11}K_2K_3)(C_{33})^2 + (C_{21} + C_{11}K_3 + C_{11}K_2K_3 - C_{21}K_1)C_{33} - C_{21} = 0. \quad (3.3.21)$$

As a solution of (3.3.21), $0 \leq C_{33} \leq 1$ is to be used.

A solution of the problem (3.3.18), (3.3.19), (3.3.21) is obtained for the case

$$K_{0i} = 1, \quad Fo_{i0} = 0.1, \quad \varepsilon = 0, \quad i = 1,2, \quad K_1 = 1, \quad K_2 = 0.5, \quad K_3 = 1, \quad (3.3.22)$$

as five-matrix forms (3.3.15). The concentration distributions for different Z are presented in Figs. 3.10, 3.11, 3.12, 3.13, and 3.14.

Fig. 3.10 Radial distribution
of the concentration $C_{11}(R, Z)$

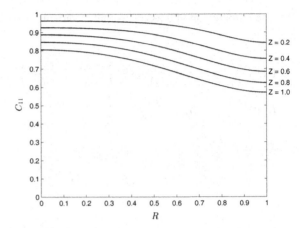

Fig. 3.11 Radial distribution
of the concentration $C_{21}(R, Z)$

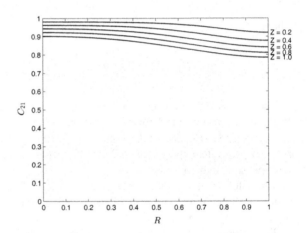

Fig. 3.12 Radial distribution
of the concentration $C_{13}(R, Z)$

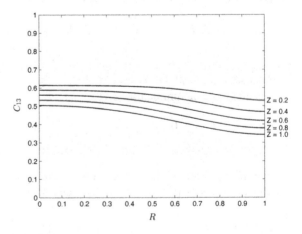

Fig. 3.13 Radial distribution
of the concentration $C_{23}(R, Z)$

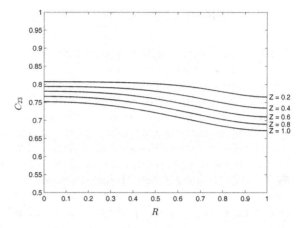

Fig. 3.14 Radial distribution
of the concentration $C_{33}(R, Z)$

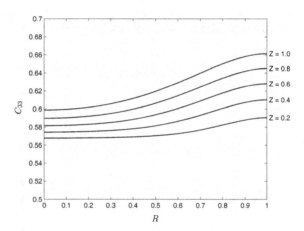

The presented new approach for modeling of two-phase processes in column apparatuses is a basis for qualitative analysis of particular processes and for the creation of the average-concentration models and quantitative analysis of the processes.

3.4   Examples

The presented convection–diffusion type of models is used for modeling of different practically interesting processes in airlift reactors, photo-bioreactors, etc.

3.4.1 Airlift Reactor

The airlift apparatuses for fluids transport are also used for interphase mass transfer and chemical reactions in two-phase gas–liquid systems [12], where the processes are continuous or periodic (batch).

Let us consider an airlift reactor (Fig. 3.15) with a cross-sectional area S_1 (m^2) for the riser zone (the inner tube, where the liquid–gas bubble mixture rises) and S_2 (m^2) for the downcomer zone (between the two cylinders, where the liquid moves down). The radii of the riser and the airlift are r_0, R_0 (m), while the length of the working zones is l (m). The gas flow rate is F_1 (m^3 s^{-1}), and the liquid flow rate is F_2 (m^3 s^{-1}). The gas and liquid holdups in the riser are ε_1 and ε_2 ($\varepsilon_1 + \varepsilon_2 = 1$), i.e., the gas and liquid parts in the small (elementary) volume in the riser zone volume (according to the physical approximations of the mechanics of continua [5]).

The concentration of the active gas component in the gas phase is $c_{11}(t, r, z_1)$, and in the liquid phase, it is $c_{12}(t, r, z_1)$ for the riser and $\hat{c}_{12}(t, r, z_2)$ for the downcomer, where $z_1 + z_2 = l$. The concentration of the active liquid component in

Fig. 3.15 Airlift reactor

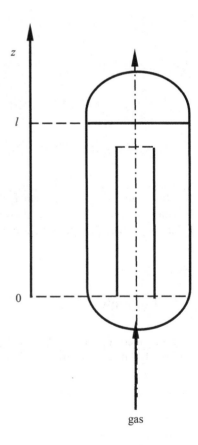

the downcomer is $\hat{c}_{22}(t, r, z_2)$, and in the riser, it is $c_{22}(t, r, z_1)$. All concentrations are in kg mol in 1 m³ of the phase (elementary) volumes.

The riser and downcomer inlet gas and liquid velocities are equal to the average velocities in gas and liquid phases in the riser and downcomer and are defined as:

$$u_1^0 = \frac{F_1}{\varepsilon_1 \pi r_0^2} = \bar{u}_1 = \frac{2}{r_0^2} \int_0^{r_0} r u_1(r) \mathrm{d}r,$$

$$u_2^0 = \frac{F_2}{\varepsilon_2 \pi r_0^2} = \bar{u}_2 = \frac{2}{r_0^2} \int_0^{r_0} r u_2(r) \mathrm{d}r, \tag{3.4.1}$$

$$\hat{u}_2^0 = \frac{F_2}{\pi (R_0^2 - r_0^2)} = \frac{2}{R_0^2 - r_0^2} \int_{r_0}^{R_0} r \hat{u}_2(r) \mathrm{d}r,$$

where $u_1 = u_1(r), u_2 = u_2(r), \hat{u}_2 = \hat{u}_2(r)$ are the gas and liquid velocity distributions in the riser and the liquid velocity distribution in the downcomer.

The interphase mass transfer rate in the riser is

$$Q_j = (-1)^j k_0 (c_{11} - \chi c_{22}), \quad j = 1, 2. \tag{3.4.2}$$

The chemical reaction rates in the riser and in the downcomer are:

$$Q_2 = k c_{12}^{\alpha_1} c_{22}^{\alpha_2}, \quad \hat{Q}_2 = k \hat{c}_{12}^{\alpha_1} \hat{c}_{22}^{\alpha_2}. \tag{3.4.3}$$

The convection–diffusion model of the chemical processes in an airlift reactor will be created on the basis of the differential mass balance in the reactor volume [1]; i.e., convection–diffusion equations with volume reactions (3.1.11) will be used.

The processes in airlift reactors are very often non-stationary, and the equations for the active gas component concentration distributions in the gas and liquid phases in the riser are:

$$\frac{\partial c_{11}}{\partial t} + u_1 \frac{\partial c_{11}}{\partial z_1} = D_{11} \left(\frac{\partial^2 c_{11}}{\partial z_1^2} + \frac{1}{r} \frac{\partial c_{11}}{\partial r} + \frac{\partial^2 c_{11}}{\partial r^2} \right) - k_0 (c_{11} - \chi c_{12}),$$

$$\frac{\partial c_{12}}{\partial t} + u_2 \frac{\partial c_{12}}{\partial z_1} = D_{12} \left(\frac{\partial^2 c_{12}}{\partial z_1^2} + \frac{1}{r} \frac{\partial c_{12}}{\partial r} + \frac{\partial^2 c_{12}}{\partial r^2} \right) + k_0 (c_{11} - \chi c_{12}) - k c_{12}^{\alpha_1} c_{22}^{\alpha_2},$$

$$\frac{\partial c_{22}}{\partial t} + u_2 \frac{\partial c_{22}}{\partial z_1} = D_{22} \left(\frac{\partial^2 c_{22}}{\partial z_1^2} + \frac{1}{r} \frac{\partial c_{22}}{\partial r} + \frac{\partial^2 c_{22}}{\partial r^2} \right) - k c_{12}^{\alpha_1} c_{22}^{\alpha_2}.$$

$$\tag{3.4.4}$$

The equations for the active liquid and gas concentration distributions in the liquid phase in the downcomer are:

$$\frac{\partial \hat{c}_{12}}{\partial t} + \hat{u}_2 \frac{\partial \hat{c}_{12}}{\partial z_2} = D_{12}\left(\frac{\partial^2 \hat{c}_{12}}{\partial z_2^2} + \frac{1}{r}\frac{\partial \hat{c}_{12}}{\partial r} + \frac{\partial^2 \hat{c}_{12}}{\partial r^2}\right) - k\hat{c}_{12}^{\alpha_1}\hat{c}_{22}^{\alpha_2},$$

$$\frac{\partial \hat{c}_{22}}{\partial t} + \hat{u}_2 \frac{\partial \hat{c}_{22}}{\partial z_2} = D_{22}\left(\frac{\partial^2 \hat{c}_{22}}{\partial z_2^2} + \frac{1}{r}\frac{\partial \hat{c}_{22}}{\partial r} + \frac{\partial^2 \hat{c}_{22}}{\partial r^2}\right) - k\hat{c}_{12}^{\alpha_1}\hat{c}_{22}^{\alpha_2}.$$

$$(3.4.5)$$

The initial conditions will be formulated for the case, when at $t = 0$, the process starts with the beginning of gas motion:

$$t = 0, \quad c_{11} \equiv c_{11}^0, \quad c_{12} \equiv 0, \quad c_{22} \equiv c_{22}^0, \quad \hat{c}_{12} \equiv 0, \quad \hat{c}_{22} \equiv c_{22}^0, \quad (3.4.6)$$

where c_{11}^0 and c_{22}^0 are the initial concentrations of the reagents in the two phases.

The boundary conditions are equalities of the concentrations and mass fluxes at the two ends of the working zones—$z_1 = 0$ $(z_2 = l)$ and $z_1 = l$ $(z_2 = 0)$.

The boundary conditions for c_{11}, c_{22}, c_{12} in the riser are:

$$r = 0, \quad \frac{\partial c_{11}}{\partial r} = \frac{\partial c_{22}}{\partial r} = \frac{\partial c_{12}}{\partial r} \equiv 0; \quad r = r_0, \quad \frac{\partial c_{11}}{\partial r} = \frac{\partial c_{22}}{\partial r} = \frac{\partial c_{12}}{\partial r} \equiv 0;$$

$$z_1 = 0, \quad c_{11} \equiv c_{11}^0, \quad u_1^0 c_{11}^0 \equiv u_1 c_{11}^0 - D_{11}\left(\frac{\partial c_{11}}{\partial z_1}\right)_{z_1=0},$$

$$c_{22} \equiv c_{22}^0, \quad u_2^0 c_{22}^0 \equiv u_2 c_{22}^0 - D_{22}\left(\frac{\partial c_{22}}{\partial z_1}\right)_{z_1=0},$$

$$c_{12} = c_{12}^0, \quad u_2^0 c_{12}^0 \equiv u_2 c_{12}^0 - D_{12}\left(\frac{\partial c_{12}}{\partial z_1}\right)_{z_1=0}.$$

$$(3.4.7)$$

The boundary conditions for $\hat{c}_{12}, \hat{c}_{22}$ are:

$$r = r_0, \quad \frac{\partial \hat{c}_{12}}{\partial r} = \frac{\partial \hat{c}_{22}}{\partial r} \equiv 0; \quad r = R_0, \quad \frac{\partial \hat{c}_{12}}{\partial r} = \frac{\partial \hat{c}_{22}}{\partial r} \equiv 0;$$

$$z_2 = 0, \quad \hat{c}_{12} \equiv \hat{c}_{12}^0, \quad \hat{u}_2^0 \hat{c}_{12}^0 \equiv \hat{u}_2 \hat{c}_{12}^0 - D_{12}\left(\frac{\partial \hat{c}_{12}}{\partial z_2}\right)_{z_2=0},$$

$$\hat{c}_{22} \equiv \hat{c}_{22}^0, \quad \hat{u}_2^0 \hat{c}_{22}^0 \equiv \hat{u}_2 \hat{c}_{22}^0 - D_{22}\left(\frac{\partial \hat{c}_{22}}{\partial z_2}\right)_{z_2=0}.$$

$$(3.4.8)$$

In (3.4.7) and (3.4.8), it is assumed that a stirring in the liquid transition between the riser and downcomer leads to averaging of the concentration distributions; i.e., the input (output) concentrations in the riser (downcomer) are equal to the average output (input) concentrations in the downcomer (riser), and as a result:

$$c_{12}^0 = \frac{2}{R_0^2 - r_0^2} \int_{r_0}^{R_0} r\hat{c}_{12}(t, r, l)\mathrm{d}r, \quad c_{22}^0 = \frac{2}{R_0^2 - r_0^2} \int_{r_0}^{R_0} r\hat{c}_{22}(t, r, l)\mathrm{d}r,$$

$$\hat{c}_{12}^0 = \frac{2}{r_0^2} \int_0^{r_0} rc_{12}(t, r, l)\mathrm{d}r, \quad \hat{c}_{22}^0 = \frac{2}{r_0^2} \int_0^{r_0} rc_{22}(t, r, l)\mathrm{d}r. \tag{3.4.9}$$

A qualitative analysis of the processes in the airlift apparatuses is possible to be made using dimensionless (generalized) variables:

$$T = \frac{t}{t_0}, \quad R_1 = \frac{r}{r_0}, \quad R_2 = \frac{r - r_0}{R_0 - r_0}, \quad Z_1 = \frac{z_1}{l}, \quad Z_2 = \frac{z_2}{l},$$

$$U_1 = \frac{u_1}{u_1^0}, \quad U_2 = \frac{u_2}{u_2^0}, \quad \hat{U}_2 = \frac{\hat{u}_2}{\hat{u}_2^0}, \quad C_{11} = \frac{c_{11}}{c_{11}^0},$$

$$C_{12} = \frac{c_{12}\chi}{c_{11}^0}, \quad C_{22} = \frac{c_{22}}{c_{22}^0}, \quad \hat{C}_{12} = \frac{\hat{c}_{12}\chi}{c_{12}^0}, \quad \hat{C}_{22} = \frac{\hat{c}_{22}}{c_{22}^0}. \tag{3.4.10}$$

As a result, the models (3.4.4)–(3.4.8) of the processes in the airlift apparatuses in dimensionless (generalized) variables have the form:

$$\gamma_1 \frac{\partial C_{11}}{\partial T} + U_1 \frac{\partial C_{11}}{\partial Z_1} = Fo_{11}\left(\varepsilon \frac{\partial^2 C_{11}}{\partial Z_1^2} + \frac{1}{R_1}\frac{\partial C_{11}}{\partial R_1} + \frac{\partial^2 C_{11}}{\partial R_1^2}\right) - K_{01}(C_{11} - C_{12});$$

$$\gamma_2 \frac{\partial C_{12}}{\partial T} + U_2 \frac{\partial C_{12}}{\partial Z_1} = Fo_{12}\left(\varepsilon \frac{\partial^2 C_{12}}{\partial Z_1^2} + \frac{1}{R_1}\frac{\partial C_{12}}{\partial R_1} + \frac{\partial^2 C_{12}}{\partial R_1^2}\right)$$

$$+ K_{02}(C_{11} - C_{12}) - K\frac{\delta\chi}{c_{11}^0} C_{12}^{\alpha_1} C_{22}^{\alpha_2};$$

$$\gamma_2 \frac{\partial C_{22}}{\partial T} + U_2 \frac{\partial C_{22}}{\partial Z_1} = Fo_{22}\left(\varepsilon \frac{\partial^2 C_{22}}{\partial Z_1^2} + \frac{1}{R_1}\frac{\partial C_{22}}{\partial R_1} + \frac{\partial^2 C_{22}}{\partial R_1^2}\right) - K\frac{\delta}{c_{22}^0} C_{12}^{\alpha_1} C_{22}^{\alpha_2};$$

$$\hat{\gamma}_2 \frac{\partial \hat{C}_{12}}{\partial T} + \hat{U}_2 \frac{\partial \hat{C}_{12}}{\partial Z_2} = \hat{Fo}_{12}\left(\varepsilon \frac{\partial^2 \hat{C}_{12}}{\partial Z_2^2} + \frac{1}{R_2}\frac{\partial \hat{C}_{12}}{\partial R_2} + \frac{\partial^2 \hat{C}_{12}}{\partial R_2^2}\right) - \hat{K}_2\frac{\delta\chi}{c_{11}^0} \hat{C}_{12}^{\alpha_1} \hat{C}_{22}^{\alpha_2};$$

$$\hat{\gamma}_2 \frac{\partial \hat{C}_{22}}{\partial T} + \hat{U}_2 \frac{\partial \hat{C}_{22}}{\partial Z_2} = \hat{Fo}_{22}\left(\varepsilon \frac{\partial^2 \hat{C}_{22}}{\partial Z_2^2} + \frac{1}{R_2}\frac{\partial \hat{C}_{22}}{\partial R_2} + \frac{\partial^2 \hat{C}_{22}}{\partial R_2^2}\right) - \hat{K}_2\frac{\delta}{c_{22}^0} \hat{C}_{12}^{\alpha_1} \hat{C}_{22}^{\alpha_2}. \tag{3.4.11}$$

The initial and boundary condition of (3.4.11) are:

$$T = 0, \quad C_{11} \equiv 1, \quad C_{12} \equiv 0, \quad C_{22} \equiv 1, \quad \hat{C}_{12} \equiv 0, \quad \hat{C}_{22} \equiv 1;$$

$$R_1 = R_2 = 0, \quad \frac{\partial C_{11}}{\partial R_1} = \frac{\partial C_{22}}{\partial R_1} = \frac{\partial C_{12}}{\partial R_1} = \frac{\partial \hat{C}_{12}}{\partial R_2} = \frac{\partial \hat{C}_{22}}{\partial R_2} \equiv 0;$$

$$R_1 = R_2 = 1, \quad \frac{\partial C_{11}}{\partial R_1} = \frac{\partial C_{22}}{\partial R_1} = \frac{\partial C_{12}}{\partial R_1} = \frac{\partial \hat{C}_{12}}{\partial R_2} = \frac{\partial \hat{C}_{22}}{\partial R_2} = 0;$$

$$Z_1 = 0, \quad C_{11} = 1, \quad 1 = U_1 - Pe_{11}^{-1}\left(\frac{\partial C_{11}}{\partial Z_1}\right),$$

$$C_{22} = C_{22}^0, \quad 1 = U_2 - Pe_{22}^{-1}\left(\frac{\partial C_{22}}{\partial Z_1}\right), \tag{3.4.12}$$

$$C_{12} = C_{12}^0, \quad 1 = U_2 - Pe_{12}^{-1}\left(\frac{\partial C_{12}}{\partial Z_1}\right);$$

$$Z_2 = 0, \quad \hat{C}_{12} = \hat{C}_{12}^0, \quad 1 = U_3 - \hat{Pe}_{12}^{-1}\left(\frac{\partial \hat{C}_{12}}{\partial Z_2}\right),$$

$$\hat{C}_{22} = \hat{C}_{22}^0, \quad 1 = U_3 - \hat{Pe}_{22}^{-1}\left(\frac{\partial \hat{C}_{22}}{\partial Z_2}\right).$$

The parameters in (3.4.11), (3.4.12) are

$$\gamma_j = \frac{l}{u_j^0 t_0}, \quad \varepsilon = \frac{r_0^2}{l^2}, \quad K_{0j} = \frac{kl}{u_j^0}, \quad j = 1, 2,$$

$$K = \frac{kl}{u_2^0}, \quad \hat{K}_2 = \frac{kl}{\hat{u}_2^0}, \quad Fo_{ij} = \frac{D_{ij} l}{u_j^0 r_0^2}, \tag{3.4.13}$$

$$Pe_{ij} = \frac{u_j^0 l}{D_{ij}}, \quad \hat{Fo}_{ij} = Fo_{ij} \frac{u_j^0}{\hat{u}_2^0},$$

$$\hat{Pe}_{ij} = Pe \frac{\hat{u}_2^0}{u_j^0}, \quad i = 1, 2, j = 1, 2, \quad \delta = \left(\frac{c_{11}^0}{\chi}\right)^{\alpha_1} (c_{22}^0)^{\alpha_2}.$$

For specific cases, different approximations of (3.4.11) are possible to be used. The high columns approximation is $\varepsilon = 0$. The long duration processes approximation is $0 = \hat{\gamma}_2 \sim \gamma_j \leq 10^{-2}, j = 1, 2$. For big values of the characteristic velocities, the model is convection type $(0 = Fo_{ij} \leq 10^{-2}, i = 1, 2, j = 1, 2)$, or the process is stationary $(0 = \hat{\gamma}_2 \sim \gamma_j \leq 10^{-2}, j = 1, 2)$. Other approximations are possible too.

3.4.2 Airlift Photo-bioreactor

The photo-bioprocesses include dissolution of an active gas component (CO_2, O_2) in liquid (H_2O) and its reaction with a photo-active material (cells). These two processes may take place in different systems, such as mixed bioreactors, bubble columns, or airlift photo-bioreactors [13–17]. The comparison of these systems shows apparent advantages in the use of airlift photo-bioreactors, because of the possibility of manipulation of the light-darkness history of the photosynthetic cells [18–20]

The hydrodynamic behavior of the gas and liquid in airlift reactors is very complicated, but in all cases, the process includes convective transport, diffusion transport, and volume reactions. That is why convection–diffusion equations with volume reaction may be used as a mathematical structure of the model. The models (2.4.4)–(2.4.9) will be a basis for mathematical description of a photosynthesis process in airlift photo-bioreactor, where the active components in the gas and liquid phases are CO_2 and plant cells, respectively [13].

In the riser, c_{11} (kg mol m^{-3}) is the volume concentration of CO_2 in the gas phase, c_{12} (kg mol m^{-3})—the volume concentration of CO_2 in the liquid phase (in kg mol in 1 m^3 phase elementary volume), c_{22} (kg eq m^{-3})—the volume concentration of the plant cells in the liquid phase (1 kg eq cells in the liquid phase combine 1 kg mol CO_2), while in the downcomer, $\hat{c}_{12}$ (kg mol m^{-3}) and $\hat{c}_{22}$ (kg eq m^{-3}) are the volume concentrations of CO_2 and plant cells in the liquid-phase elementary volume, respectively.

The photochemical reaction rates in the riser and downcomer are:

$$Q_2 = kc_{12}^{\alpha_1}c_{22}^{\alpha_2}P, \quad \hat{Q}_2 = k\hat{c}_{12}^{\alpha_1}\hat{c}_{22}^{\alpha_2}\hat{P}, \tag{3.4.14}$$

where $P, \hat{P}$ are the photon flux densities (E m^{-2} s^{-1}) in the riser and downcomer.

The photon flux densities $P, \hat{P}$ are functions of the cylindrical coordinates

$$P = P(t, r, z_1), \quad \hat{P} = \hat{P}(t, r, z_2) \tag{3.4.15}$$

and these relations is possible to be obtained using Fig. 3.16, where a cylindrical surface with radius R_0 and height 1 m is regularly illuminated with a photon flux density P_0.

Fig. 3.16 Cylindrical surface illuminated with a photon flux density

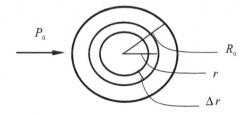

The photon flux densities over cylindrical surfaces with radiuses $r < R_0$ are represented by

$$p(r) = \frac{R_0 P_0}{r}. \tag{3.4.16}$$

The increasing of the photon flux density between r and $r - \Delta r$ is:

$$\Delta \hat{P}_1 = \frac{P_0 R_0}{r - \Delta r} - \frac{P_0 R_0}{r} = \frac{P_0 R_0 \Delta r}{r(r - \Delta r)}. \tag{3.4.17}$$

The volume between the cylindrical surfaces (m^3 liquid/m^2 surface) with height 1 m and radii r and $r - \Delta r$ is

$$V = \frac{\pi r^2 - \pi (r - \Delta r)^2}{2\pi r} = \Delta r \left(1 - \frac{\Delta r}{2r} \right). \tag{3.4.18}$$

The decrease of the photon flux density as a result of the light absorption (from the photoactive cells) in a volume V is

$$\Delta \hat{P}_2 = \hat{P}(t, r, z_2) \phi \hat{c}_{22} \Delta r \left(1 - \frac{\Delta r}{2r} \right), \tag{3.4.19}$$

where ϕ is the light absorption coefficient, and $\hat{c}_{22} = \hat{c}_{22}(t, r, z_2)$ is the concentration of the photoactive cells in the downcomer.

The difference between photon flux densities for r and $r - \Delta r$ is

$$\Delta \hat{P} = \Delta \hat{P}_1 - \Delta \hat{P}_2 = \frac{P_0 R_0 \Delta r}{r(r - \Delta r)} - \hat{P} \phi \hat{c}_{22} \Delta r \left(1 - \frac{\Delta r}{2r} \right). \tag{3.4.20}$$

As a result,

$$\lim_{\Delta r \to 0} \frac{\Delta \hat{P}}{\Delta r} = \frac{\partial \hat{P}}{\partial r} = \frac{R_0 P_0}{r^2} - \phi \hat{c}_{22} \hat{P}, \tag{3.4.21}$$

where $\hat{P}(R_0) = P_0$. The solution of (3.4.18) for $\hat{c}_{22} = \hat{c}_{22}(t, r, z_2)$ is

$$\hat{P}(t, r, z_2) = \exp \left(\phi \int_r^{R_0} \hat{c}_{22}(t, \rho, z_2) d\rho \right)$$
$$* \left\{ P_0 - R_0 P_0 \int_r^{R_0} \frac{1}{\rho^2} \exp \left[-\phi \int_\rho^{R_0} \hat{c}_{22}(t, \eta, z_2) d\eta \right] d\rho \right\}. \tag{3.4.22}$$

The photon flux density in the riser $P = P(t, r, z_1)$ is possible to be obtained analogously, as a solution of the problem

$$
\frac{\partial P}{\partial r} = \frac{r_0 P(t, r_0, z_1)}{r^2} - \phi c_{22} P,
$$

$$
r = r_0, \quad P(t, r_0, z_1) = \hat{P}(t, r_0, z_2), \quad z_2 = l - z_1. \tag{3.4.23}
$$

The mathematical model of the process in the airlift photo-bioreactor will be built on the basis of the differential mass balances in the reactor volume (3.4.4)–(3.4.8).

The equations for the concentration distribution in the riser of the active gas component (CO_2) in the gas and liquid phases and photo-active substance (cells) in the liquid phase are:

$$
\frac{\partial c_{11}}{\partial t} + u_1 \frac{\partial c_{11}}{\partial z_1} = D_{11} \left(\frac{\partial^2 c_{11}}{\partial z_1^2} + \frac{1}{r} \frac{\partial c_{11}}{\partial r} + \frac{\partial^2 c_{11}}{\partial r^2} \right) - k_0 (c_{11} - \chi c_{12});
$$

$$
\frac{\partial c_{12}}{\partial t} + u_2 \frac{\partial c_{12}}{\partial z_1} = D_{12} \left(\frac{\partial^2 c_{12}}{\partial z_1^2} + \frac{1}{r} \frac{\partial c_{12}}{\partial r} + \frac{\partial^2 c_{12}}{\partial r^2} \right) + k_0 (c_{11} - \chi c_{12}) - k c_{12}^{\alpha_1} c_{22}^{\alpha_2} P;
$$

$$
\frac{\partial c_{22}}{\partial t} + u_2 \frac{\partial c_{22}}{\partial z_1} = D_{22} \left(\frac{\partial^2 c_{22}}{\partial z_1^2} + \frac{1}{r} \frac{\partial c_{22}}{\partial r} + \frac{\partial^2 c_{22}}{\partial r^2} \right) - k c_{12}^{\alpha_1} c_{22}^{\alpha_2} P.
$$

$$
\tag{3.4.24}
$$

The equations for the concentration distribution of CO_2 and the cells in the downcomer are:

$$
\frac{\partial \hat{c}_{12}}{\partial t} + \hat{u}_2 \frac{\partial \hat{c}_{12}}{\partial z_2} = D_{12} \left(\frac{\partial^2 \hat{c}_{12}}{\partial z_2^2} + \frac{1}{r} \frac{\partial \hat{c}_{12}}{\partial r} + \frac{\partial^2 \hat{c}_{12}}{\partial r^2} \right) - k \hat{c}_{12}^{\alpha_1} \hat{c}_{22}^{\alpha_2} \hat{P};
$$

$$
\frac{\partial \hat{c}_{22}}{\partial t} + \hat{u}_2 \frac{\partial \hat{c}_{22}}{\partial z_2} = D_{22} \left(\frac{\partial^2 \hat{c}_{22}}{\partial z_2^2} + \frac{1}{r} \frac{\partial \hat{c}_{22}}{\partial r} + \frac{\partial^2 \hat{c}_{22}}{\partial r^2} \right) - k \hat{c}_{12}^{\alpha_1} \hat{c}_{22}^{\alpha_2} \hat{P}.
$$

$$
\tag{3.4.25}
$$

The initial conditions are formulated for the case of thermodynamic equilibrium between the gas and liquid phases; i.e., a full saturation of the liquid with the active gas component (CO_2) and the process starts with the starting of the illumination:

$$
t = 0, \quad c_{11} \equiv c_{11}^0, \quad c_{12} \equiv \frac{c_{11}^0}{\chi} 0, \quad c_{22} \equiv c_{22}^0, \quad \hat{c}_{12} \equiv \frac{c_{11}^0}{\chi}, \quad \hat{c}_{22} \equiv c_{22}^0, \tag{3.4.26}
$$

where c_{11}^0 and c_{22}^0 are the initial concentrations of the active gas component (CO_2) in the gas phase and the photo-active substance (sells) in the liquid phase.

The boundary conditions are equalities of the concentrations and the mass fluxes at the two ends of the working zones: $z_1 = 0$ ($z_2 = l$) and $z_1 = l$ ($z_2 = 0$).

The boundary conditions for the riser equations are

$$r = 0, \quad \frac{\partial c_{11}}{\partial r} = \frac{\partial c_{22}}{\partial r} = \frac{\partial c_{12}}{\partial r} \equiv 0;$$

$$r = r_0, \quad \frac{\partial c_{11}}{\partial r} = \frac{\partial c_{22}}{\partial r} = \frac{\partial c_{12}}{\partial r} \equiv 0;$$

$$z_1 = 0, \quad c_{11} \equiv c_{11}^0, \quad u_1^0 c_{11}^0 \equiv u_1 c_{11}^0 - D_{11} \left(\frac{\partial c_{11}}{\partial z_1} \right)_{z_1=0},$$

$$c_{22} \equiv c_{22}^0, \quad u_2^0 c_{22}^0 \equiv u_2 c_{22}^0 - D_{22} \left(\frac{\partial c_{22}}{\partial z_1} \right)_{z_1=0}, \qquad (3.4.27)$$

$$c_{12} \equiv c_{12}^0, \quad u_2^0 c_{12}^0 \equiv u_2 c_{12}^0 - D_{12} \left(\frac{\partial c_{12}}{\partial z_1} \right)_{z_1=0}.$$

The boundary conditions for the downcomer equations are

$$r = r_0, \quad \frac{\partial \hat{c}_{12}}{\partial r} = \frac{\partial \hat{c}_{22}}{\partial r} \equiv 0;$$

$$r = R_0, \quad \frac{\partial \hat{c}_{12}}{\partial r} = \frac{\partial \hat{c}_{22}}{\partial r} \equiv 0;$$

$$z_2 = 0, \quad \hat{c}_{12} \equiv \hat{c}_{12}^0, \quad \hat{u}_2^0 \hat{c}_{12}^0 \equiv \hat{u}_2 \hat{c}_{12}^0 - D_{12} \left(\frac{\partial \hat{c}_{12}}{\partial z_2} \right)_{z_2=0}, \qquad (3.4.28)$$

$$\hat{c}_{22} \equiv \hat{c}_{22}^0, \quad \hat{u}_2^0 \hat{c}_{22}^0 \equiv \hat{u}_2 \hat{c}_{22}^0 - D_{22} \left(\frac{\partial \hat{c}_{22}}{\partial z_2} \right)_{z_2=0}.$$

The presented airlift photo-bioreactor model (3.4.21)–(3.4.25) is possible to be analyzed qualitatively in a way similar to that for the airlift reactor models (3.4.4)–(3.4.8).

3.4.3   Moisture Adsorption

For countries with a cold climate, the large difference between indoor and outdoor temperatures leads to large heat loses in ventilation systems, moisture freezing at the systems exit and great reduction in the indoor humidity. A new method to regenerating moisture in ventilation systems has been proposed [21–24].

A new composite sorbent "CaCl$_2$/alumina" for moisture adsorption is used [22, 23], where the process results from the equilibrium process

$$CaCl_2 + 6H_2O \rightleftarrows CaCl_2.6H_2O. \qquad (3.4.29)$$

The covalent bond between $CaCl_2$ and H_2O allows to use the kinetic model of the physical adsorption [7].

Many models are used to describe the adsorption process in a fixed bed of solid adsorbent and a flux of humid air that is passing through the bed [24]. The convection–diffusion type of model [1, 7] permits to describe the process of physical adsorption of H_2O in the case of a fixed bed column filled with the new composite sorbent "CaCl$_2$/alumina."

In the cases of physical adsorption of H_2O in a solid adsorbent [7], the adsorption rate in the adsorbent volume is $Q_3 = b_0 Q_{03}$ (kg mol m^{-3} s^{-1}), where b_0 (m^2 m^{-3}) is the specific active interface in the solid-phase (adsorbent) volume; it is related with the interface physical adsorption rate (3.2.5) as:

$$Q_{03} = k_1 c_{13} C_{23} - k_2 \frac{c_{23}^0}{b_0} (1 - C_{23}), \quad C_{23} = \frac{c_{23}}{c_{23}^0}. \qquad (3.4.30)$$

In (3.4.30), c_{13} (kg mol m^{-3}) is the volume concentration of the active component (AC) in the sold phase (the free moisture in the capillaries of the adsorbent), c_{23} (kg eq m^{-3})—the volume concentration on the free active sites (AS) of the solid phase (1 kg eq AS in the adsorbent combine 1 kg mol AC in gas phase), c_{23}^0 (kg eq m^{-3})—the initial (maximal) concentration of free AS in the solid phase, $\left(c_{23}^0 - c_{23}\right)$—the concentration of the adsorbed moisture in the solid phase, where $c_{23}^0 = \omega c_0, 1 \le \omega \le 6$, c_0 (kg mol m^{-3}) is the concentration of $CaCl_2$ in the solid-phase volume, $\omega = c_{13}^{max}/c_0$, c_{13}^{max} (kg mol m^{-3})—the maximal concentration of the adsorbed moisture in the solid phase. All concentrations are in (kg mol) (kg eq) in 1 (m^3) of the column (elementary) volume. The parts of the gas and solid phases in the column (elementary) volume are $\varepsilon_1, \varepsilon_3$, $(\varepsilon_1 + \varepsilon_3 = 1)$.

The volume rate of the interphase mass transfer of the moisture between the gas and solid phases is $k_0(c_{11} - c_{13})$, where c_{11} (kg mol m^{-3}) is the moisture concentration in the gas phase and k_0 (s^{-1})—the interphase mass transfer coefficient (mass transfer of the moisture between the gas phase in the column volume and the gas phase in the absorbent capillaries volume).

Let us consider a non-stationary adsorption process of H_2O by sorbent particles in a gas–solid system, where the sorbent particle phase is immobile. The model will be presented in a cylindrical coordinate system (t, r, z), where in the cases of a constant axial distribution of the average gas velocity $\bar{u}_1$ (m s^{-1}) in the column cross-sectional areas [1], the axial velocity component is symmetric with respect to the longitudinal coordinate z; i.e., $u_1 = u_1(r)$, and the radial velocity component is equal to zero.

The mass balance of H_2O in the gas phase is a result of the convective transfer, diffusion transfer, and interphase mass transfer of the moisture between the gas and solid phases (as a volume mass sink).

The mass balance of AS in the solid phase is a result of the physical adsorption and the interphase mass transfer only, because the solid phase is immobile and diffusivity of the free AS in the solid-phase (adsorbent) volume is equal to zero.

The process is non-stationary as a result of the free AS concentration decreasing over time, and the convection–diffusion model has the form:

$$
\frac{\partial c_{11}}{\partial t} + u_1 \frac{\partial c_{11}}{\partial z} = D_{11}\left(\frac{\partial^2 c_{11}}{\partial z^2} + \frac{1}{r}\frac{\partial c_{11}}{\partial r} + \frac{\partial^2 c_{11}}{\partial r^2}\right) - k_0(c_{11} - c_{13});
$$

$$
\frac{dc_{13}}{dt} = k_0(c_{11} - c_{13}) - b_0 k_1 c_{13}\frac{c_{23}}{c_{23}^0} + k_2\left(c_{23}^0 - c_{23}\right); \qquad (3.4.31)
$$

$$
\frac{dc_{23}}{dt} = -b_0 k_1 c_{13}\frac{c_{23}}{c_{23}^0} + k_2\left(c_{23}^0 - c_{23}\right),
$$

where D_{11} ($m^2\ s^{-1}$) is the diffusivity of the AC (moisture) in the gas phase and (r, z) are parameters in $c_{13} = c_{13}(t, r, z)$ and $c_{23} = c_{23}(t, r, z)$.

Let us consider an adsorption column with a radius r_0 (m) and a height of the active volume l (m). The initial and boundary conditions [1, 2] of (3.4.31) have the form:

$$
t = 0, \quad c_{11} = c_{11}^0, \quad c_{13} = 0, \quad c_{23} = c_{23}^0;
$$

$$
r = 0, \quad \frac{\partial c_{11}}{\partial r} = 0; \quad r = r_0, \quad \frac{\partial c_{11}}{\partial r} = 0; \qquad (3.4.32)
$$

$$
z = 0, \quad c_{11} = c_{11}^0, \quad u_1^0 c_{11}^0 = u_1(r)c_{11}^0 - D_{11}\left(\frac{\partial c_{11}}{\partial z}\right)_{z=0},
$$

where $u_1^0 = \bar{u}_1$ is the inlet velocity of the gas phase.

The use of dimensionless (generalized) variables [1] permits to make a qualitative analysis of the model (3.4.31), (3.4.32), where as characteristic scales are used the average velocity, inlet concentrations, characteristic time (t_0) and column parameters (r_0, l):

$$
T = \frac{t}{t_0}, \quad R = \frac{r}{r_0}, \quad Z = \frac{z}{l}, \quad U = \frac{u_1}{\bar{u}_1},
$$

$$
C = \frac{c_{11}}{c_{11}^0}, \quad C_1 = \frac{c_{13}}{c_{11}^0}, \quad C_0 = \frac{c_{23}}{c_{23}^0}. \qquad (3.4.33)
$$

If (3.4.33) is put in (3.4.31), (3.4.32), the model in generalized variables takes the form:

$$\gamma \frac{\partial C}{\partial T} + U(R) \frac{\partial C}{\partial Z} = Fo \left(\varepsilon \frac{\partial^2 C}{\partial Z^2} + \frac{1}{R} \frac{\partial C}{\partial R} + \frac{\partial^2 C}{\partial R^2} \right) - K_{01}(C - C_1);$$

$$\frac{dC_1}{dT} = K_{03}(C - C_1) - K_{13} C_1 C_0 + K_{23} \frac{c_{23}^0}{c_{11}^0}(1 - C_0);$$

$$\frac{dC_0}{dT} = -K_{13} \frac{c_{11}^0}{c_{23}^0} C_1 C_0 + K_{23}(1 - C_0);$$

$$T = 0, \quad C = 1, \quad C_1 = 0, \quad C_0 = 1; \quad R = 0, \quad \frac{\partial C}{\partial R} = 0; \quad R = 1, \quad \frac{\partial C}{\partial R} = 0;$$

$$Z = 0, \quad C = 1, \quad 1 = U(R) - Pe^{-1} \left(\frac{\partial C}{\partial Z} \right)_{Z=0},$$

$$(3.4.34)$$

where (R, Z) are parameters in $C_0(T, R, Z)$ and

$$Fo = \frac{D_{11} l}{\bar{u}_1 r_0^2}, \quad Pe = \frac{\bar{u}_1 l}{D_{11}}, \quad \gamma = \frac{l}{\bar{u}_1 t_0}, \quad \varepsilon = \frac{r_0^2}{l^2},$$

$$K_{01} = \frac{k_0 l}{\bar{u}_1}, \quad K_{03} = k_0 t_0, \quad K_{13} = b_0 k_1 t_0, \quad K_{23} = k_2 t_0.$$

$$(3.4.35)$$

Experimental data for moisture adsorption in the new composite sorbent bed column [22, 23] are obtained in the following conditions:

$$r_0 = 0.2 \text{ (m)}, \quad l = 0.8 \text{ (m)}, \quad \bar{u}_1 = 2.25 \text{ (m s}^{-1}), \quad t_0 \sim 10^3 \text{ (s)},$$
$$D_{11} \sim 10^{-5} \text{ (m}^2 \text{ s}^{-1}), \quad c_{11}^0 = 0.02285 \text{ (kg mol m}^{-3}),$$
$$c_{23}^0 = 15.33 \text{ (kg eq m}^{-3}), \quad \varepsilon_1 = 0.346, \quad \varepsilon_3 = 0.654, \quad (3.4.36)$$
$$\gamma \sim 10^{-3}, \quad Fo \sim 10^{-4}, \quad Pe^{-1} \sim 10^{-5}, \quad \frac{c_{11}^0}{c_{23}^0} \sim 10^{-3}.$$

Under these conditions, it is possible to use in the model (3.4.34) the approximations

$$0 = \gamma \ll 1, \quad 0 = Fo = Pe^{-1} \ll 1. \qquad (3.4.37)$$

As a result, (3.4.34) assumes the convective form:

$$U(R)\frac{\partial C}{\partial Z} = -K_{01}(C - C_1);$$

$$\frac{dC_1}{dT} = K_{03}(C - C_1) - K_{13}C_1C_0 + K_{23}\frac{c_{23}^0}{c_{11}^0}(1 - C_0);$$

$$\frac{dC_0}{dT} = -K_{13}\frac{c_{11}^0}{c_{23}^0}C_1C_0 + K_{23}(1 - C_0); \tag{3.4.38}$$

$$T = 0, \quad C_1 = 0, \quad C_0 = 1; \quad R = 0, \quad \frac{\partial C}{\partial R} = 0; \quad R = 1, \quad \frac{\partial C}{\partial R} = 0;$$

$$Z = 0, \quad C = 1.$$

In Eq. (3.4.38), T is a parameter in the function $C(T, Z)$, because $C_1(T, Z)$ is function of the time, while Z is a parameter in $C_1(T, Z)$ and $C_0(T, Z)$.

References

1. Boyadjiev C (2010) Theoretical chemical engineering. Modeling and simulation. Springer, Berlin, Heidelberg
2. Boyadjiev C (2006) Diffusion models and scale-up. Int J Heat Mass Transfer 49:796–799
3. Doichinova M, Boyadjiev C (2012) On the column apparatuses modeling. Int J Heat Mass Transfer 55:6705–6715
4. Boyadjiev C (2013) A new approach for the column apparatuses modeling in chemical engineering. J Pure Appl Math: Adv Appl 10(2):131–150
5. Doichinova M, Boyadjiev C (2015) A new approach for the column apparatuses modeling in chemical and power engineering. Therm Sci 19(5):1747–1759
6. Boyadjiev B, Doichinova M, Boyadjiev C (2015) Computer modeling of column apparatuses. 1. Two-coordinate systems approach. J Eng Thermophys 24(3): 247–258
7. Boyadjiev C, Boyadjiev B, Popova-Krumova P, Doichinova M (2015) An innovative approach for adsorption column modeling. Chem Eng Tech 38(4):675–682
8. Close D, Banks B (1972) Chem Eng Sci 27:1155–1167
9. Ruthven DM, Farooq S, Knaebel KS (1994) Pressure swing adsorption. VCH, New York
10. Boyadjiev B, Doichinova M, Boyadjiev C (2014) Computer modeling of column apparatuses. 2. Multi-steps modeling approach. J Eng Thermophys 24(4): 362–370
11. Boyadjiev B, Boyadjiev C (2015) A new approach for the catalytic processes modeling in columns apparatuses. Int J Mod Trends Eng Res 2(8): 152–167
12. Boyadjiev C (2006) On the modeling of an airlift reactor. Int J Heat Mass Transfer 49:2053–2057
13. Boyadjiev C, Merchuk J (2008) On the modeling of an airlift photobioreactor. J Eng Thermophys 17(2):134–141
14. Lee JK, Low GS (1992) Productivity of out doors algae cultures in enclosed tubular photobioreactor. Biotechnol Bioeng 40:1119–1122
15. Frohlich BT, Webster IA, Ataai MM, Shuler ML (1983) Photobioreactors: models for interaction of light intensity, reactor design and algae physiology. Biotechnol Bioeng Symp 13:331–350

16. Ogbonna JC, Yada H, Masu H, Tanaka H (1966) A novel internally illuminated stirred tank photobioreactor for large scale cultivation of photosynthetic cells. J Fermentat Bioengug 82:61–67

17. Prokop A, Erickson LE (1994) Photobioreactors, in boireactor system design (Asenjo JA, Merchuk JC, Editors). Marcel Decker, New York

18. Merchuk JC, Ladwa JC, Bulmer M (1993) Improving the airlift reactor: the helical flow promoter. In: A Ninehow (eds) Bioprocesses and bioreactor fluid dynamics. BHRA, Elsevier, pp 61–68

19. Merchuk JC, Gluz M (1999) Airlift reactors. In: Flickinger MC, Drew SW (eds) Encyclopedia of bioprocess technology. Wiley, New York, pp 320–353

20. Schlotelburg C, Gluz M, Popovic M, Merchuk JC (1999) Characterization of an airlift reactor with helical flow promoters. Can J Chem Eng 77:804–810

21. Aristov Y, Mezentsev I, Mukhin V (2006) New approach to regenerate heat and moisture in a ventilation system: 1. Laboratory prototype. J Eng Thermophys 79:143–150

22. Aristov Y, Mezentsev I, Mukhin V (2006) New approach to regenerate heat and moisture in a ventilation system: 2. Prototype of real unit. J Eng Thermophys 79:151–157

23. Aristov Y, Mezentsev I, Mukhin V, Boyadjiev C, Doichinova M, Popova P (2006) New approach to regenerate heat and moisture in a ventilation system: experiment. In: Proceedings of 11th workshop on "transport phenomena in two-phase flow", Bulgaria, pp 77–85

24. Aristov Y, Gordeeva LG, Tokarev MM (2008) Composite sorbents "salt in porous matrix": synthesis, properties, application. Publishing house, Siberian Branch of the Russian Academy of Sciences, Novosibirsk

Chapter 4
Three-Phase Processes

The modeling of three-phase (gas–liquid–solid) interphase mass transfer processes in column apparatuses [1–4] is used in the case of absorption and adsorption in two-component ($i_0 = 2$), three-phase ($j = 1, 2, 3$) systems. For the modeling of three-phase processes, the three-equation model (I.3–I.5) has to be used, which features the mass balances in all the phases, having in mind that according to (2.1.10) the radial velocity components are equal to zero ($v_j \equiv 0, j = 1, 2, 3$):

$$u_j \frac{\partial c_{ij}}{\partial z_j} = D_{ij}\left(\frac{\partial^2 c_{ij}}{\partial z_j^2} + \frac{1}{r}\frac{\partial c_{ij}}{\partial r} + \frac{\partial^2 c_{ij}}{\partial r^2}\right) \pm Q_i(c_{ij});$$

$$r = 0, \quad \frac{\partial c_{ij}}{\partial r} = 0; \quad r = r_0, \quad \frac{\partial c_{ij}}{\partial r} = 0; \tag{4.0.1}$$

$$z_j = 0, \quad c_{ij} = c_{ij}^0, \quad u_j^0 c_{ij}^0 = u_j c_{ij}^0 - D_{ij}\frac{\partial c_{ij}}{\partial z_j}; \quad i = 1, 2, \; j = 1, 2, 3.$$

In (4.0.1), $u_j = u_j(r)$ (m s^{-1}) and $c_{ij} = c_{ij}(r, z_j)$ (kg mol m^{-3}) are the axial velocity components and the transferred substance concentrations in the phases, D_{ij} (m^2 s^{-1}) are the diffusivities in the phases, u_j^0 and c_{ij}^0 are the inlet velocities and concentrations in the column, where $i = 1, 2, j = 1, 2, 3$. The concentrations of the transferred substance in the phases are presented as kg mol of the transferred substance in the phase in 1 m^3 of the phase (elementary) volume. The inlet velocities in the column are defined as $u_j^0 = F_j/\varepsilon_j \pi r_0^2$, where r_0 is the column radius (m), $j = 1, 2, 3$, $F_0 = \sum_{j=1}^{3} F_j$, F_0 (m^3 s^{-1})—the flow rate in the column, $F_j, j = 1, 2, 3$—the phases flow rates (m^3 s^{-1}) in the column. The coefficients ε_j, $j = 1, 2, 3$ are in m^3 of the phase volume in 1 m^3 of the column volume.

The volume rates $Q_j, j = 1, 2, 3$ (kg mol m^{-3} s^{-1}) of the interphase mass transfer, in case of chemical reaction or adsorption, are volume sources or sinks in the phase part of the elementary column volumes and participate in the mass balance in the phase elementary volume.

© Springer International Publishing AG, part of Springer Nature 2018
C. Boyadjiev et al., *Modeling of Column Apparatus Processes*,
Heat and Mass Transfer, https://doi.org/10.1007/978-3-319-89966-4_4

4.1 Two-Phase Absorbent Processes

In many practical cases, two phases of absorbents [5–7] are used (e.g., water suspensions of $CaCO_3$ or $Ca(OH)_2$) because they are low priced and of big absorption capacity. The presence of the active component in the absorbent, as a solution and solid phase, leads to the introduction of a new process (the dissolution of the solid phase) and creates conditions for variations in the absorption mechanism (interphase mass transfer through two interphase surfaces—gas–liquid and liquid–solid) [5–7].

4.1.1 $CaCO_3/H_2O$ Absorbent

Many companies (e.g., Babcock & Wilcox Power Generation Group, Inc., Alstom Power Italy, Idreco-Insigma-Consortium) provide methods and apparatuses for waste gas purification from SO_2, using two-phase absorbent ($CaCO_3/H_2O$ suspension). The gas enters in the middle of a countercurrent column, contacts with the absorbent (suspension) drops and exits from the top. The collected absorbent in the bottom half of the column is returned to the top of the column. A theoretical analysis of this SO_2 absorption with two-phase absorbent is possible to be made using the convection–diffusion model approximation, where the $CaCO_3/H_2O$ suspension is considered as a liquid phase [5–7]. The process is presented as such involving two components and two phases ($i = 1, 2, j = 1, 2$).

Let the concentrations of SO_2 in the gas and the liquid phase are c_{11} and c_{12}, while c_{22} is the concentration of the dissolved $CaCO_3$ in the absorbent. The mass sources in the liquid elementary volume are equal to the rate of the interphase mass transfer across the gas–liquid boundary $+ k_0(c_{11} - \chi c_{12})$ and the rate of the interphase mass transfer across the liquid–solid boundary is $+ k_1 \left(c_{22}^0 - c_{22}\right)$, where c_{22}^0 is the maximal (equilibrium) solubility of $CaCO_3$ in water. The mass sinks in the gas and liquid elementary volumes are equal to the rate of the interphase mass transfer across the gas–liquid boundary $- k_0(c_{11} - \chi c_{12})$ and the rate of the chemical reaction $- k c_{12} c_{22}$. As a result, the convection–diffusion model of a column apparatus (presented for countercurrent absorption process in two cylindrical coordinate systems—$(r, z_1), (r, z_2), (z_1 + z_2 = l)$) has the form:

$$u_1 \frac{\partial c_{11}}{\partial z_1} = D_{11} \left(\frac{\partial^2 c_{11}}{\partial z_1^2} + \frac{1}{r} \frac{\partial c_{11}}{\partial r} + \frac{\partial^2 c_{11}}{\partial r^2} \right) - k_0(c_{11} - \chi c_{12});$$

$$u_2 \frac{\partial c_{12}}{\partial z_2} = D_{12} \left(\frac{\partial^2 c_{12}}{\partial z_2^2} + \frac{1}{r} \frac{\partial c_{12}}{\partial r} + \frac{\partial^2 c_{12}}{\partial r^2} \right) + k_0(c_{11} - \chi c_{12}) - k c_{12} c_{22}; \quad (4.1.1)$$

$$u_2 \frac{\partial c_{22}}{\partial z_2} = D_{22} \left(\frac{\partial^2 c_{22}}{\partial z_2^2} + \frac{1}{r} \frac{\partial c_{22}}{\partial r} + \frac{\partial^2 c_{22}}{\partial r^2} \right) - k c_{12} c_{22} + k_1 \left(c_{22}^0 - c_{22}\right),$$

where $u_1(r), u_2(r)$ are the velocity distributions in the gas and liquid phases, $c_{11}(r, z_1), c_{12}(r, z_2), c_{22}(r, z_2)$ and D_{11}, D_{12}, D_{22}—the concentration distributions and the diffusivities of SO_2 in the gas and liquid phases and of $CaCO_3$ in the liquid phase, k—the chemical reaction rate constant, k_0, k_1—interphase mass transfer coefficients.

The boundary conditions of (4.1.1) in a column with radius r_0 and working zone height l have the form:

$$r = 0, \quad \frac{\partial c_{11}}{\partial r} = \frac{\partial c_{12}}{\partial r} = \frac{\partial c_{22}}{\partial r} \equiv 0; \quad r = r_0, \quad \frac{\partial c_{11}}{\partial r} = \frac{\partial c_{12}}{\partial r} = \frac{\partial c_{22}}{\partial r} \equiv 0;$$

$$z_1 = 0, \quad c_{11}(r, 0) \equiv c_{11}^0, \quad u_1^0 c_{11}^0 \equiv u_1(r) c_{11}^0 - D_{11} \left(\frac{\partial c_{11}}{\partial z_1} \right)_{z_1 = 0};$$

$$z_2 = 0, \quad c_{12}(r, 0) \equiv 0, \quad \left(\frac{\partial c_{12}}{\partial z_2} \right)_{z_2 = 0} \equiv 0,$$

$$c_{22}(r, 0) \equiv c_{22}^0, \quad u_2^0 c_{22}^0 \equiv u_2(r) c_{22}^0 - D_{22} \left(\frac{\partial c_{22}}{\partial z_2} \right)_{z_2 = 0},$$

$$(4.1.2)$$

where $u_1^0, u_2^0, c_{11}^0, c_{12}^0$ are the average input velocities and the inlet concentrations of SO_2 in the gas and liquid phases, c_{22}^0 is the maximal (equilibrium) solubility of $CaCO_3$.

A qualitative analysis of the model (4.1.1) is possible to be made using dimensionless (generalized) variables:

$$R = \frac{r}{r_0}, \quad Z_1 = \frac{z_1}{l}, \quad Z_2 = \frac{z_2}{l_2}, \quad U_1 = \frac{u_1}{u_1^0}, \quad U_2 = \frac{u_2}{u_2^0},$$

$$C_1 = \frac{c_{11}}{c_{11}^0}, \quad C_2 = \frac{c_{12} \chi}{c_{11}^0}, \quad C_3 = \frac{c_{22}}{c_{22}^0}.$$

$$(4.1.3)$$

When (4.1.3) is put into (4.1.1), the model in generalized variables takes the form:

$$U_1 \frac{\partial C_1}{\partial Z_1} = \frac{D_{11} l}{u_1^0 r_0^2} \left(\frac{r_0^2}{l^2} \frac{\partial^2 C_1}{\partial Z_1^2} + \frac{1}{R} \frac{\partial C_1}{\partial R} + \frac{\partial^2 C_1}{\partial R^2} \right) - \frac{k_0 l}{u_1^0} (C_1 - C_2);$$

$$U_2 \frac{\partial C_2}{\partial Z_2} = \frac{D_{12} l}{u_2^0 r_0^2} \left(\frac{r_0^2}{l^2} \frac{\partial^2 C_2}{\partial Z_2^2} + \frac{1}{R} \frac{\partial C_2}{\partial R} + \frac{\partial^2 C_2}{\partial R^2} \right) + \frac{k_0 l \chi}{u_2^0} (C_1 - C_2) - \frac{k l c_{22}^0}{u_2^0} C_2 C_3;$$

$$U_2 \frac{\partial C_3}{\partial Z_2} = \frac{D_{22} l}{u_2^0 r_0^2} \left(\frac{r_0^2}{l^2} \frac{\partial^2 C_3}{\partial Z_2^2} + \frac{1}{R} \frac{\partial C_3}{\partial R} + \frac{\partial^2 C_3}{\partial R^2} \right) - \frac{k l c_{11}^0}{u_2^0 \chi} C_2 C_3 + \frac{k_1 l}{u_2^0} (1 - C_3),$$

$$(4.1.4)$$

where dimensionless chemical kinetic parameters are:

$$K_2 = \frac{klc_{22}^0}{u_2^0}, \quad K_3 = \frac{klc_{11}^0}{u_2^0\chi}. \tag{4.1.5}$$

The low SO_2 concentration ($c_{11}^0 \sim 10^{-4}$ kg mol m^{-3}) in the waste gases of the thermal power plants, the low concentration of the dissolved $CaCO_3$ in the absorbent ($c_{22}^0 \sim 10^{-4}$ kg mol m^{-3}) and Henry's number value of SO_2/H_2O ($\chi \sim 10^{-2}$) lead to:

$$\frac{K_2}{K_3} = \frac{\chi c_{22}^0}{c_{11}^0} \sim 10^{-2}, \tag{4.1.6}$$

i.e., if the mass transfer of the dissolved $CaCO_3$ in the absorbent is a result of the chemical reaction ($K_3 \sim 1$), the chemical reaction effect on the interphase mass transfer of SO_2 between the gas and liquid is possible to be neglected ($0 = K_2 \sim 10^{-2}$). In these conditions, the chemical reaction is very slow and as a result of the brief existence of the drops in the gas–liquid dispersion (~ 2 (s) in industrial conditions), the chemical reaction passes in the collected absorbent in the bottom half of the column.

From (4.1.6) follows that in the cases of waste gas purification from SO_2 using two-phase absorbent ($CaCO_3/H_2O$ suspension), the interphase mass transfer process in the gas–liquid drops system is practically physical absorption, as a result of the low concentration of the dissolved SO_2 and $CaCO_3$ in the water, and the mathematical model is possible to be obtained directly from the model (3.1.5).

In a real case of SO_2 physical absorption in a gas–liquid drops system $\chi = 2.86 \times 10^{-2}$, $\varepsilon_1 = 0.98$, $\varepsilon_2 = 0.02$, $u_1^0 = u_2^0 = 4$ (m s^{-1}) and from (3.1.7) follows $\rho_0 = 0.0286$, $\rho_1 = 0.972$, $\rho_2 = 0.0278$.

4.1.2   *Ca(OH)$_2$/H$_2$O* Absorbent

The chemical absorption in three-phase systems is used in the cases of SO_2 (CO_2) absorption by a $Ca(OH)_2/H_2O$ suspension, where the solubility of the solid $Ca(OH)_2$ is big.

The model (4.1.4) with boundary conditions (4.1.2) is possible to be presented [see (3.1.14)] as:

$$U_1(R)\frac{\partial C_1}{\partial Z} = Fo_1\left(\varepsilon\frac{\partial^2 C_1}{\partial Z^2} + \frac{1}{R}\frac{\partial C_1}{\partial R} + \frac{\partial^2 C_1}{\partial R^2}\right) - K_0(C_1 - C_2);$$

$$U_2(R)\frac{\partial C_2}{\partial Z} = Fo_2\left(\varepsilon\frac{\partial^2 C_2}{\partial Z^2} + \frac{1}{R}\frac{\partial C_2}{\partial R} + \frac{\partial^2 C_2}{\partial R^2}\right)$$
$$+ K_0\frac{u_1^0\chi}{u_2^0}(C_1 - C_2) - Da\frac{c_{22}^0}{c_{11}^0}C_2 C_3;$$

$$U_2(R)\frac{\partial C_3}{\partial Z} = Fo_3\left(\varepsilon\frac{\partial^2 C_3}{\partial Z^2} + \frac{1}{R}\frac{\partial C_3}{\partial R} + \frac{\partial^2 C_3}{\partial R^2}\right) - DaC_2 C_3 + K_1(1 - C_3);$$

$$R = 0, \quad \frac{\partial C_s}{\partial R} \equiv 0; \quad R = 1, \quad \frac{\partial C_s}{\partial R} \equiv 0; \quad s = 1,2,3;$$

$$Z = 0, \quad C_1 \equiv 1, \quad 1 \equiv U_1(R) - Pe_1^{-1}\left(\frac{\partial C_1}{\partial Z}\right)_{Z=0};$$

$$Z = 0, \quad C_3 \equiv 1, \quad 1 \equiv U_2(R) - Pe_3^{-1}\left(\frac{\partial C_3}{\partial Z}\right)_{Z=0};$$

$$Z = 0, \quad C_2 \equiv 0, \quad \left(\frac{\partial C_2}{\partial Z}\right)_{Z=0} \equiv 0,$$

$$(4.1.7)$$

where

$$K_0 = \frac{k_0 l}{u_1^0}, \quad K_1 = \frac{k_1 l}{u_2^0}, \quad Fo_1 = \frac{D_{11}l}{u_1^0 r_0^2}, \quad Fo_2 = \frac{D_{12}l}{u_2^0 r_0^2},$$
$$Fo_3 = \frac{D_{22}l}{u_2^0 r_0^2}, \quad Da = \frac{klc_{11}^0}{u_2^0\chi}, \quad Pe_1 = \frac{u_1^0 l}{D_{11}}, \quad Pe_3 = \frac{u_2^0 l}{D_{22}}.$$

$$(4.1.8)$$

In the cases, when the interphase mass transfer is a result of the chemical reaction in the liquid phase $\left(Dac_{22}^0/c_{11}^0 \geq 1\right)$, the second equation in (4.1.7) should be divided by $Dac_{22}^0/c_{11}^0 \geq 1$, i.e.,

$$U_1(R)\frac{\partial C_1}{\partial Z} = Fo_1\left(\varepsilon\frac{\partial^2 C_1}{\partial Z^2} + \frac{1}{R}\frac{\partial C_1}{\partial R} + \frac{\partial^2 C_1}{\partial R^2}\right) - K_0(C_1 - C_2);$$

$$Da^{-1}\frac{c_{11}^0}{c_{22}^0}U_2(R)\frac{\partial C_2}{\partial Z} = \frac{Fo_2 c_{11}^0}{Dac_{22}^0}\left(\varepsilon\frac{\partial^2 C_2}{\partial Z^2} + \frac{1}{R}\frac{\partial C_2}{\partial R} + \frac{\partial^2 C_2}{\partial R^2}\right)$$
$$+ K_0\frac{u_1^0\chi c_{11}^0}{Dau_2^0 c_{22}^0}(C_1 - C_2) - C_2 C_3;$$

$$U_2(R)\frac{\partial C_3}{\partial Z} = Fo_3\left(\varepsilon\frac{\partial^2 C_3}{\partial Z^2} + \frac{1}{R}\frac{\partial C_3}{\partial R} + \frac{\partial^2 C_3}{\partial R^2}\right) - DaC_2 C_3 + K_1(1 - C_3);$$

$$R = 0, \quad \frac{\partial C_s}{\partial R} \equiv 0; \quad R = 1, \quad \frac{\partial C_s}{\partial R} \equiv 0; \quad s = 1, 2, 3;$$

$$Z = 0, \quad C_1 \equiv 1, \quad 1 \equiv U_1(R) - Pe_1^{-1}\left(\frac{\partial C_1}{\partial Z}\right)_{Z=0};$$

$$Z = 0, \quad C_3 \equiv 1, \quad 1 \equiv U_2(R) - Pe_3^{-1}\left(\frac{\partial C_3}{\partial Z}\right)_{Z=0};$$

$$Z = 0, \quad C_2 \equiv 0, \quad \left(\frac{\partial C_2}{\partial Z}\right)_{Z=0} \equiv 0.$$

$$(4.1.9)$$

In a way similar to (2.1.19), the model (4.1.9) permits to obtain the interphase mass transfer resistance distribution between the gas and liquid phases:

$$\rho_1 = K_0, \quad \rho_2 = \rho_0\rho_1, \quad \rho_0 = \frac{u_1^0 \chi c_{11}^0}{Dau_2^0 c_{22}^0},$$

$$\rho_1 + \rho_2 = 1, \quad \rho_1 = \frac{1}{1 + \rho_0}, \quad \rho_2 = \frac{\rho_0}{1 + \rho_0},$$

$$(4.1.10)$$

where the parameters ρ_1 and ρ_2 can be considered as mass transfer resistances in the gas and liquid phases.

4.2 Absorption–Adsorption Processes

The adsorption process (gas–solid) can be used for the gases cleaning of low concentration impurities [8]. In many cases, however, the process is difficult as a result of the small values of the particles diameter and density of the adsorbent particles. This problem can be eliminated if the process is carried in a three-phase system (gas–liquid–solid), where the impurity is absorbed and adsorbed (physically or chemically) consecutively in the liquid and solid phases and simultaneously in the time [8].

4.2.1 Physical Adsorption Mechanism

Let us consider a physical adsorption [8] of gas impurity (GI) in a liquid–gas bubbles column, where the solid adsorbent particles are mixed with the liquid phase. The process is non-stationary because the concentration of the free active sites (AS) in the adsorbent decreases with the time as a result of the physical adsorption of the GI.

The intensive flow of gas bubbles in the column creates an ideal mixing regime in the liquid–solid phase, i.e., the concentrations in the gas phase are averaged over the cross section of the column, while the concentrations in the liquid and solid phases are constants in the column volume. As a result, the GI concentration in the gas phase $\bar{c}_{11}(t, z)$ (kg mol m^{-3}) is independent of the radial coordinate r (m), while the GI concentration (kg mol m^{-3}) in the liquid phase $\bar{c}_{12}(t)$ is a function of the time t (s) only. The GI concentration in the solid phase (in the capillary volume of the adsorbent) is $\bar{c}_{13}(t)$ (kg mol m^{-3}), while the concentration of the free active sites (AS) in the adsorbent (kg eq m^{-3}) is $\bar{c}_{23}(t)$. The inlet velocity is defined as $u_1^0 = F_1/\varepsilon_1 \pi r_0^2$ (m s^{-1}), where F_1 (m^3 s^{-1}) is the gas flow rate in the column, r_0— the column radius (m), and the inlet GI concentration in the gas phase is c_{11}^0 (kg mol m^{-3}).

The local absorption rate q and the average absorption rate Q_1 in the column (kg mol m^{-3} s^{-1}) are:

$$q(t, z) = k_{01}(\bar{c}_{11} - \chi \bar{c}_{12}),$$

$$Q_1(t) = \frac{1}{l} \int_0^l q(t, z) dz$$

$$= k_{01} \left[\frac{1}{l} \int_0^l \bar{c}_{11}(t, z) dz - \chi \bar{c}_{12}(t) \right],$$

(4.2.1)

where χ is Henry's number, k_{01}—an interphase mass transfer coefficient.

The average values (kg mol m^{-3} s^{-1}) of the interphase (liquid–solid) mass transfer rate in the column volume Q_2 and the physical adsorption rate [9] in the solid phase Q_3 [see (2.2.5)] are:

$$Q_2(t) = k_{02}(\bar{c}_{12} - \bar{c}_{13}), \quad Q_3(t) = bk_1\bar{c}_{13}\frac{\bar{c}_{23}}{c_{23}^0} - k_2(c_{23}^0 - \bar{c}_{23}),$$

(4.2.2)

where c_{23}^0 is the initial concentration (kg eq m^{-3}) of the AS.

The concentration of GI and AS in the phases is presented as kg mol (kg eq) of the substance in the phase in 1 m^3 of the phase elementary volumes and 1 kg eq AS in the adsorbent combine with 1 kg mol GI in the gas phase.

The presented process rates are possible to be introduced in (4.0.1), where $u_1 = u_1^0$, and the model of a physical absorption–adsorption process takes the form:

$$\frac{\partial \bar{c}_{11}}{\partial t} + u_1^0 \frac{\partial \bar{c}_{11}}{\partial z} = D_{11} \frac{\partial^2 \bar{c}_{11}}{\partial z^2} - k_{01}(\bar{c}_{11} - \chi \bar{c}_{12});$$

$$t = 0, \quad \bar{c}_{11}(0,z) \equiv c_{11}^0; \quad z = 0, \quad \bar{c}_{11}(t,0) \equiv c_{11}^0, \quad \left(\frac{\partial \bar{c}_{11}}{\partial z}\right)_{z=0} \equiv 0. \tag{4.2.3}$$

$$\frac{d\bar{c}_{12}}{dt} = k_{01}\left(\frac{1}{l}\int_0^l \bar{c}_{11} dz - \chi \bar{c}_{12}\right) - k_{02}(\bar{c}_{12} - \bar{c}_{13}); \quad t = 0, \quad \bar{c}_{12} \equiv \frac{c_{11}^0}{\chi}. \tag{4.2.4}$$

$$\frac{d\bar{c}_{13}}{dt} = k_{02}(\bar{c}_{12} - \bar{c}_{13}) - bk_1 \bar{c}_{13}\frac{\bar{c}_{23}}{c_{23}^0} + k_2\left(c_{23}^0 - \bar{c}_{23}\right); \quad t = 0, \quad \bar{c}_{13} \equiv \frac{c_{11}^0}{\chi}. \tag{4.2.5}$$

$$\frac{d\bar{c}_{23}}{dt} = -bk_1 \bar{c}_{13}\frac{\bar{c}_{23}}{c_{23}^0} + k_2\left(c_{23}^0 - \bar{c}_{23}\right); \quad t = 0, \quad \bar{c}_{23} \equiv c_{23}^0. \tag{4.2.6}$$

The model (4.2.3)–(4.2.6) presents a process that starts with the beginning of the adsorption process.

The qualitative and quantitative analyses of the physical absorption–adsorption processes are possible to be made using the following generalized variables:

$$T = \frac{t}{t_0}, \quad Z = \frac{z}{l}, \quad C_1 = \frac{\bar{c}_{11}}{c_{11}^0}, \quad C_2 = \frac{\bar{c}_{12}\chi}{c_{11}^0}, \quad C_3 = \frac{\bar{c}_{13}\chi}{c_{11}^0}, \quad C_{23} = \frac{\bar{c}_{23}}{c_{23}^0}. \tag{4.2.7}$$

The introducing of (4.2.7) in (4.2.3)–(4.2.6) leads to:

$$\gamma\frac{\partial C_1}{\partial T} + \frac{\partial C_1}{\partial Z} = Pe_1^{-1}\frac{\partial^2 C_1}{\partial Z^2} - K_{01}(C_1 - C_2);$$

$$T = 0, \quad C_1(0,Z) \equiv 1; \quad Z = 0, \quad C_1(T,0) \equiv 1, \quad \left(\frac{\partial C_1}{\partial Z}\right)_{Z=0} \equiv 0. \tag{4.2.8}$$

$$\frac{dC_2}{dT} = K_{12}\left(\int_0^1 C_1 dZ - C_2\right) - K_{02}(C_2 - C_3); \quad T = 0, \quad C_2 \equiv 1. \tag{4.2.9}$$

$$\frac{dC_3}{dT} = K_{02}\left(C_2 - C_3\right) - K_{13}C_3C_{23} + K_{23}\delta^{-1}(1 - C_{23}); \quad T = 0, \quad C_3 \equiv 1. \tag{4.2.10}$$

$$\frac{dC_{23}}{dT} = -K_{13}\delta C_3C_{23} + K_{23}(1 - C_{23}); \quad T = 0, \quad C_{23} \equiv 1. \tag{4.2.11}$$

In (4.2.8)–(4.2.11), the dimensionless parameters are:

$$\gamma = \frac{l}{t_0 u_1^0}, \quad Pe_1 = \frac{u_1^0 l}{D_{11}}, \quad K_{01} = \frac{k_{01} l}{u_1^0}, \quad K_{02} = k_{02} t_0,$$

$$K_{12} = k_{01} t_0 \chi, \quad K_{13} = b k_1 t_0, \quad K_{23} = k_2 t_0, \quad \delta = \frac{c_{11}^0}{c_{23}^0 \chi}. \tag{4.2.12}$$

In the practical cases, $0 = \gamma \le 10^{-2}, 0 = Pe_1^{-1} \le 10^{-2}$, and the problem (4.2.8) has the form:

$$\frac{dC_1}{dZ} = -K_{01}[C_1 - C_2(T)]; \quad Z = 0, \quad C_1(T,0) \equiv 1, \tag{4.2.13}$$

where T is a parameter. The solution of (4.2.13) is:

$$C_1(T,Z) = [1 - C_2(T)] \exp(-K_{01} Z) + C_2(T). \tag{4.2.14}$$

From (4.2.14), it is possible to obtain

$$\int_0^1 C_1(T,Z) dZ = \frac{1 - \exp(-K_{01})}{K_{01}} [1 - C_2(T)] + C_2(T) \tag{4.2.15}$$

and put it into (4.2.9).

The results (4.2.8)–(4.2.15) obtained show that the equations set (4.2.9)–(4.2.11), (4.2.14) is the mathematical model of the physical absorption–adsorption processes in column apparatuses. The algorithm for solving the equations set (4.2.9)–(4.2.11), (4.2.14) has two steps:

1. Solving equations set (4.2.9)–(4.2.11), (4.2.15);
2. Calculating (4.2.14).

The solutions of (4.2.9)–(4.2.11), (4.2.14) are presented on Figs. 4.1, 4.2, 4.3, 4.4, and 4.5 for the case

$$K_{01} = 1, \quad K_{02} = 1, \quad K_{12} = 1, \quad K_{13} = 1, \quad K_{23} = 1, \quad \delta = 1. \tag{4.2.16}$$

4.2.2 Chemical Adsorption Mechanism

An alternative for the gases cleaning of low concentration impurities is the chemical absorption–adsorption processes in column apparatuses. In the cases of the gas purification from SO_2, the synthetic anionites are suitable.

Fig. 4.1 Concentration distribution $C_1(T,Z)$ for different values of T

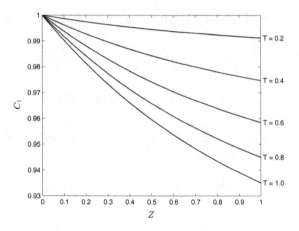

Fig. 4.2 Concentration distribution $C_1(T,Z)$ for different values of Z

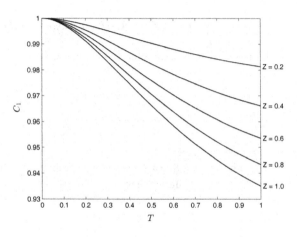

Fig. 4.3 Concentration distribution $C_2(T)$

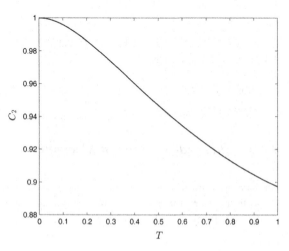

Fig. 4.4 Concentration distribution $C_3(T)$

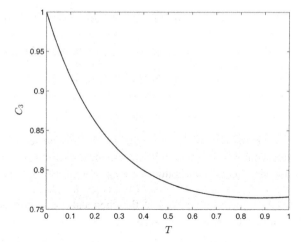

Fig. 4.5 Concentration distribution $C_{23}(T)$

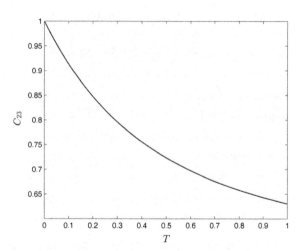

An use of synthetic anionites (basic anion-exchange resins—R–OH form of Amberlite, Duolite, Kastel, Varion, Wofatit) as adsorbent [10–12] for gas purification from SO_2 leads to possibilities for adsorbent regeneration. The chemical reaction of SO_2 with the synthetic anionites R-OH is possible to be presented by the stoichiometric equation

$$SO_2 + R{-}OH = R - HSO_3. \qquad (4.2.17)$$

After the synthetic anionite particles become saturated with sulfur dioxide, the regeneration of the adsorbent is possible to be carried out by water solution of NH_4OH:

$$R - HSO_3 + 2NH_4OH = R-OH + (NH_4)_2SO_3(\text{or } NH_4HSO_3). \quad (4.2.18)$$

The model of the chemical absorption–adsorption processes in column apparatuses is similar to the physical adsorption case (4.2.3)–(4.2.6), where the chemical adsorption rate [see (3.2.16)] $Q_3(t)$ has to be introduced:

$$Q_3(t) = k\bar{c}_{13}\bar{c}_{23}. \quad (4.2.19)$$

The adsorption rate in the solid phase is presented similar to two components chemical reaction, and k is the chemical reaction rate constant (1 kg eq AS in the adsorbent combine chemically 1 kg mol GI in gas phase). All concentrations are in (kg mol) (kg eq) in 1 (m^3) of the column volume.

As a result, the model of the chemical absorption–adsorption processes in column apparatuses has the form:

$$\frac{\partial \bar{c}_{11}}{\partial t} + u_1^0 \frac{\partial \bar{c}_{11}}{\partial z} = D_{11} \frac{\partial^2 \bar{c}_{11}}{\partial z^2} - k_{01}(\bar{c}_{11} - \chi\bar{c}_{12});$$

$$t = 0, \quad \bar{c}_{11}(0,z) \equiv c_{11}^0; \quad z = 0, \quad \bar{c}_{11}(t,0) \equiv c_{11}^0, \quad \left(\frac{\partial \bar{c}_{11}}{\partial z}\right)_{z=0} \equiv 0. \quad (4.2.20)$$

$$\frac{d\bar{c}_{12}}{dt} = k_{01}\left(\frac{1}{l}\int_0^l \bar{c}_{11}dz - \chi\bar{c}_{12}\right) - k_{02}(\bar{c}_{12} - \bar{c}_{13}); \quad t = 0, \quad \bar{c}_{12} \equiv \frac{c_{11}^0}{\chi}. \quad (4.2.21)$$

$$\frac{d\bar{c}_{13}}{dt} = k_{02}(\bar{c}_{12} - \bar{c}_{13}) - k\bar{c}_{13}\bar{c}_{23}; \quad t = 0, \quad \bar{c}_{13} \equiv \frac{c_{11}^0}{\chi}. \quad (4.2.22)$$

$$\frac{d\bar{c}_{23}}{dt} = -k\bar{c}_{13}\bar{c}_{23}; \quad t = 0, \quad \bar{c}_{23} \equiv c_{23}^0. \quad (4.2.23)$$

The qualitative and quantitative analyses of the physical absorption–adsorption processes are possible to be made using the generalized variables (4.2.7), i.e.,

$$\gamma \frac{\partial C_1}{\partial T} + \frac{\partial C_1}{\partial Z} = Pe_1^{-1} \frac{\partial^2 C_1}{\partial Z^2} - K_{01}(C_1 - C_2);$$

$$T = 0, \quad C_1(0,Z) \equiv 1; \quad Z = 0, \quad C_1(T,0) \equiv 1, \quad \left(\frac{\partial C_1}{\partial Z}\right)_{Z=0} \equiv 0. \quad (4.2.24)$$

$$\frac{dC_2}{dT} = K_{12}\left(\int_0^1 C_1 dZ - C_2\right) - K_{02}(C_2 - C_3); \quad T = 0, \quad C_2 \equiv 1. \quad (4.2.25)$$

$$\frac{dC_3}{dT} = K_{02}\frac{\varepsilon_2}{\varepsilon_3}(C_2 - C_3) - KC_3C_{23}; \quad T = 0, \quad C_3 \equiv 1. \quad (4.2.26)$$

$$\frac{dC_{23}}{dT} = -K\delta C_3 C_{23}; \quad T = 0, \quad C_{23} \equiv 1; \quad K = \frac{kt_0 c_{23}^0}{\varepsilon_3}. \tag{4.2.27}$$

In the practical cases, $0 = \gamma \leq 10^{-2}, 0 = Pe_1^{-1} \leq 10^{-2}$, and (4.2.24) is possible to be replaced by (4.2.14), (4.2.15).

The solution of the model equations (4.2.14), (4.2.25)–(4.2.27) is similar to the solution of (4.2.9)–(4.2.11), (4.2.14), and the results are presented on Figs. 4.6, 4.7, 4.8, 4.9, and 4.10 for the case:

$$K_{01} = 1, \quad K_{02} = 1, \quad K_{12} = 1, \quad K = 1, \quad \delta = 1. \tag{4.2.28}$$

Fig. 4.6 Concentration distribution $C_1(T,Z)$ for different values of T

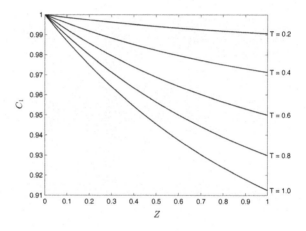

Fig. 4.7 Concentration distribution $C_1(T,Z)$ for different values of Z

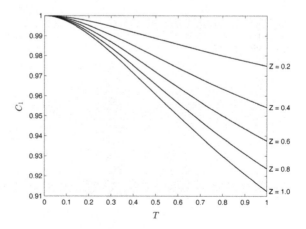

Fig. 4.8 Concentration
distribution $C_2(T)$

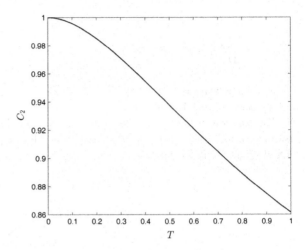

Fig. 4.9 Concentration
distribution $C_3(T)$

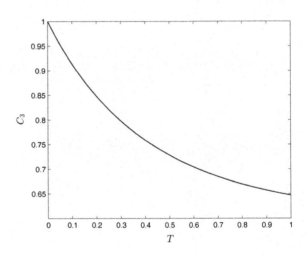

Fig. 4.10 Concentration
distribution $C_{23}(T)$

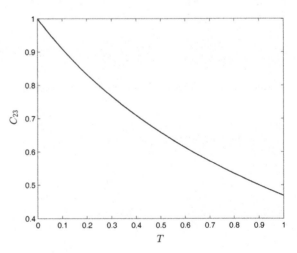

4.3   Three-Phase Catalytic Process

Let us consider a three-phase catalytic process in the case of co-current gas–liquid flow through immobile solid phase (catalyst). The concentrations of the first reagent in the gas, liquid, and solid phases are c_{11}, c_{12}, c_{13}. The concentrations of the second reagent in the liquid and solid phases are c_{22}, c_{23}. The concentration of the active sites in the solid phase is c_{33}. All concentrations are in kg mol (kg eq) in 1 m^3 of the phase (elementary) volume. The interphase mass transfer rates of the first reagent through the gas–liquid and the liquid–solid boundaries are $-k_{00}(c_{11} - \chi c_{12})$ and $k_{01}(c_{12} - c_{13})$, while the interphase mass transfer rates of the second reagent through the liquid–solid boundary is $-k_{02}(c_{22} - c_{23})$. The rates of the physical or chemical adsorption [see (3.3.1), (3.3.16)] of the first reagent in the solid (catalyst) phase are $-bk_1 c_{13} \frac{c_{33}}{c_{33}^0} + k_2 c_{33}^0 \left(1 - \frac{c_{33}}{c_{33}^0}\right)$ or $-k_{13} c_{13} c_{33}$. The rate of the chemical reaction between the two reagents in the solid phase is $kc_{23}\left(c_{33}^0 - c_{33}\right)$.

In a co-current gas–liquid flow and stationary case, the model of the three-phase catalytic process is possible to be obtained from (3.3.3):

$$u_1 \frac{\partial c_{11}}{\partial z} = D_{11}\left(\frac{\partial^2 c_{11}}{\partial z^2} + \frac{1}{r}\frac{\partial c_{11}}{\partial r} + \frac{\partial^2 c_{11}}{\partial r^2}\right) - k_{00}(c_{11} - \chi c_{12});$$

$$u_2 \frac{\partial c_{12}}{\partial z} = D_{12}\left(\frac{\partial^2 c_{12}}{\partial z^2} + \frac{1}{r}\frac{\partial c_{12}}{\partial r} + \frac{\partial^2 c_{12}}{\partial r^2}\right) + k_{00}(c_{11} - \chi c_{12}) - k_{01}(c_{12} - c_{13});$$

$$u_2 \frac{\partial c_{22}}{\partial z} = D_{21}\left(\frac{\partial^2 c_{22}}{\partial z^2} + \frac{1}{r}\frac{\partial c_{22}}{\partial r} + \frac{\partial^2 c_{22}}{\partial r^2}\right) - k_{02}(c_{22} - c_{23});$$

$$r = 0, \quad \frac{\partial c_{11}}{\partial r} = \frac{\partial c_{12}}{\partial r} = \frac{\partial c_{22}}{\partial r} \equiv 0; \quad r = r_0, \quad \frac{\partial c_{11}}{\partial r} = \frac{\partial c_{12}}{\partial r} = \frac{\partial c_{22}}{\partial r} \equiv 0;$$

$$z = 0, \quad c_{11} \equiv c_{11}^0, \quad u_1^0 c_{11}^0 \equiv u_1(r)c_{11}^0 - D_{11}\left(\frac{\partial c_{11}}{\partial z}\right)_{z=0},$$

$$c_{12} \equiv c_{12}^0, \quad u_2^0 c_{12}^0 \equiv u_2(r)c_{12}^0 - D_{12}\left(\frac{\partial c_{12}}{\partial z}\right)_{z=0},$$

$$c_{22} \equiv c_{22}^0, \quad u_2^0 c_{22}^0 \equiv u_2(r)c_{22}^0 - D_{22}\left(\frac{\partial c_{22}}{\partial z}\right)_{z=0}.$$

$$(4.3.1)$$

The concentrations of the reagents in the solid phase in the cases of physical and chemical adsorption mechanisms [see (3.3.3), (3.3.17)] are:

$$k_{01}(c_{12} - c_{13}) - bk_1 c_{13}\frac{c_{33}}{c_{33}^0} + k_2 c_{33}^0\left(1 - \frac{c_{33}}{c_{33}^0}\right) = 0;$$

$$k_{02}(c_{22} - c_{23}) - kc_{23}(c_{33}^0 - c_{33}) = 0; \qquad\qquad (4.3.2)$$

$$-bk_1 c_{13}\frac{c_{33}}{c_{33}^0} + k_2 c_{33}^0\left(1 - \frac{c_{33}}{c_{33}^0}\right) + kc_{23}(c_{33}^0 - c_{33}) = 0.$$

$$k_{01}(c_{12} - c_{13}) = k_{13} c_{13} c_{33};$$

$$k_{02}(c_{22} - c_{23}) = kc_{23}(c_{33}^0 - c_{33}); \qquad\qquad (4.3.3)$$

$$k_{13} c_{13} c_{33} = kc_{23}(c_{33}^0 - c_{33}).$$

The theoretical analysis of the model (4.3.1)–(4.3.3) is similar to that of the model (3.3.3).

References

1. Boyadjiev C (2010) Theoretical chemical engineering. Modeling and simulation. Springer, Berlin, Heidelberg
2. Boyadjiev C (2006) Diffusion models and scale-up. Int J Heat Mass Transfer 49:796–799
3. Doichinova M, Boyadjiev C (2012) On the column apparatuses modeling. Int J Heat Mass Transfer 55:6705–6715
4. Chr Boyadjiev (2013) A new approach for the column apparatuses modeling in chemical engineering. J Pure Appl Math: Adv Appl 10(2):131–150
5. Boyadjiev C (2011) Mechanism of gas absorption with two-phase absorbents. Int J Heat Mass Transfer 54:3004–3008
6. Boyadjiev C, Doichinova M, Popova P (2011) On the SO₂ problem in power engineering. 1. Gas absorption. In: Proceedings, 15th workshop on transport phenomena in two-phase flow, Bulgaria, pp 94–103
7. Boyadjiev C, Popova P, Doichinova M (2011) On the SO₂ problem in power engineering. 2. Two-phase absorbents. In: Proceedings 15th workshop on transport phenomena in two-phase flow, Bulgaria, pp 104–115
8. Boyadjiev B, Boyadjiev C (2017) An absorption-adsorption apparatus for gases purification from SO₂ in power plants. OALib J 4:e3546
9. Boyadjiev C, Boyadjiev B, Doichinova M, Popova-Krumova P (2015) An innovative approach for adsorption column modeling. Chem Eng Technol 38(4):675–682
10. Pantofchieva L, Boyadjiev C (1995) Adsorption of sulphur dioxide by synthetic anion exchangers. Bulg Chem Comm 28:780–794
11. Boyadjiev C, Pantofchieva L, Hristov J (2000) Sulphur dioxide adsorption in a fixed bed of a synthetic anionite. Theor Found Chem Eng 34(2):141–144
12. Hristov J, Boyadjiev C, Pantofchieva L (2000) Sulphur dioxide adsorption in a magnetically stabilized bed of a synthetic anionite. Theor Found Chem Eng 34(5):439–443

Part II
Quantitative Analysis of Column Apparatuses Processes

Average-Concentration-Type Models

In the Part I, it was shown that the column apparatuses are possible to be modeled using a new approach [1–4] on the basis of the physical approximations of the mechanics of continua, where the mathematical point is equivalent to a small (elementary) physical volume, which is sufficiently small with respect to the apparatus volume, but at the same time sufficiently large with respect to the intermolecular volumes in the medium. These convection–diffusion models are possible to be used for qualitative analysis only, because the velocity distribution functions are unknown and cannot be obtained. The problem can be solved by using average values of the velocity and concentration over the cross-sectional area of the column; i.e., the medium elementary volume (in the physical approximations of the mechanics of continua) will be equivalent to a small cylinder with column radius r_0 and a height, which is sufficiently small with respect to the column height and at the same time sufficiently large with respect to the intermolecular distances in the medium. All models in this part will be created on this basis.

Let us consider a cylinder with radius $R = R(\phi)$ in a cylindrical coordinate system (r, z, ϕ), where r, z, ϕ are the radial, axial, and angular coordinates, respectively. The average value of a function $f(r, z, \phi)$ at the cross-sectional area of the cylinder is:

$$\bar{f}(z) = \frac{\int_{(S)} f(r, z, \phi) \, dS}{S} \tag{II.1}$$

where

$$S = \int_0^{2\pi} \frac{[R(\phi)]^2}{2} \, d\phi, \quad \iint_{(S)} f(r, z, \phi) \, dS = \int_0^{2\pi} \left[\int_0^{R(\phi)} rf(r, z, \phi) \, dr \right] d\phi. \tag{II.2}$$

In the practical cases $\frac{\partial f}{\partial \phi} = 0$ and the cylinder is circular ($R = $ const); i.e., from (II.1) and (II.2) follows:

$$S = \pi R^2, \quad \iint_{(S)} f(r,z)\mathrm{d}S = 2\pi \int_0^R rf(r,z)\mathrm{d}r, \quad \bar{f}(z) = \frac{2}{R^2} \int_0^R rf(r,z)\mathrm{d}r. \quad \text{(II.3)}$$

Let us consider a column reactor with radius r_0 and height of the active volume l. The average-concentration model will be presented on the basis of a convection–diffusion model in the case of pseudo-first-order chemical reaction. Further, if the fluid circulation takes place, the process is non-stationary and the velocity and concentration distributions in the column must be defined as:

$$u = u(r,z), \quad v = v(r,z), \quad c = c(t,r,z), \quad \text{(II.4)}$$

i.e., the convection–diffusion model can be expressed as

$$\frac{\partial c}{\partial t} + u\frac{\partial c}{\partial z} + v\frac{\partial c}{\partial r} = D\left(\frac{\partial^2 c}{\partial z^2} + \frac{1}{r}\frac{\partial c}{\partial r} + \frac{\partial^2 c}{\partial r^2}\right) - kc; \quad \frac{\partial u}{\partial z} + \frac{\partial v}{\partial r} + \frac{v}{r} = 0;$$

$$t = 0 \quad c \equiv c^0 \quad r = 0, \quad \frac{\partial c}{\partial r} \equiv 0; \quad r = r_0, \quad \frac{\partial c}{\partial r} \equiv 0, \quad v \equiv 0; \quad \text{(II.5)}$$

$$z = 0 \quad c(t,r,0) \equiv \bar{c}(t,l), \quad u \equiv u^0, \quad u^0\bar{c}(t,l) \equiv u\bar{c}(t,l) - D\frac{\partial c}{\partial z}.$$

In (II.5) c^0 is the initial concentration, $\bar{c}(t,l)$ is the average concentration at the column outlet ($z = l$) and inlet ($z = 0$) (as a result of the fluid circulation in the column), and u^0 is the average velocity at the column inlet.

From (II.3) follow the average values of the velocity and concentration at the column cross-sectional area:

$$\bar{u}(z) = \frac{2}{r_0^2} \int_0^{r_0} ru(r,z)\mathrm{d}r, \quad \bar{v}(z) = \frac{2}{r_0^2} \int_0^{r_0} rv(r,z)\mathrm{d}r, \quad \bar{c}(t,z) = \frac{2}{r_0^2} \int_0^{r_0} rc(t,r,z)\mathrm{d}r.$$

$$\text{(II.6)}$$

The functions $u(r,z)$, $v(r,z)$, $c(t,r,z)$ in (II.5) can be presented with the help of the average functions (II.6):

$$u(r,z) = \bar{u}(z)\,\tilde{u}(r,z), \quad v(r,z) = \bar{v}(z)\,\tilde{v}(r,z),$$
$$c(t,r,z) = \bar{c}(t,z)\,\tilde{c}(t,r,z), \quad \text{(II.7)}$$

where $\tilde{u}(r,z)$, $\tilde{v}(r,z)$ and $\tilde{c}(t,r,z)$ present the radial non-uniformity of the velocity and concentration and satisfy the following conditions:

$$\frac{2}{r_0^2}\int_0^{r_0} r\tilde{u}(r,z)\mathrm{d}r = 1, \quad \frac{2}{r_0^2}\int_0^{r_0} r\tilde{v}(r,z)\mathrm{d}r = 1, \quad \frac{2}{r_0^2}\int_0^{r_0} r\tilde{c}(t,r,z)\mathrm{d}r = 1. \quad \text{(II.8)}$$

The average-concentration model may be obtained when putting (II.7) into (II.5), multiplying by r and integrating over r in the interval $[0, r_0]$. As a result, the following is obtained:

$$\frac{\partial\bar{c}}{\partial t} + \alpha(t,z)\bar{u}\frac{\partial\bar{c}}{\partial z} + \beta(t,z)\bar{u}\,\bar{c} + \gamma(t,z)\bar{v}\,\bar{c} = D\frac{\partial^2\bar{c}}{\partial z^2} - k\bar{c};$$
$$t = 0, \quad \bar{c}(0,z) \equiv c^0; \quad z = 0, \quad \bar{c}(t,0) \equiv \bar{c}(t,l), \quad \frac{\partial\bar{c}}{\partial z} \equiv 0, \quad \text{(II.9)}$$

where

$$\alpha(t,z) = \frac{2}{r_0^2}\int_0^{r_0} r\tilde{u}\tilde{c}\,\mathrm{d}r, \quad \beta(t,z) = \frac{2}{r_0^2}\int_0^{r_0} r\tilde{u}\frac{\partial\tilde{c}}{\partial z}\mathrm{d}r, \quad \gamma(t,z) = \frac{2}{r_0^2}\int_0^{r_0} r\tilde{v}\frac{\partial\tilde{c}}{\partial r}\mathrm{d}r. \quad \text{(II.10)}$$

The average radial velocity component $\bar{v}$ can be obtained from the continuity equation in (II.5) if it is multiplied by r^2 and then integrated with respect to r over the interval $[0, r_0]$:

$$\bar{v} = \delta\frac{\mathrm{d}\bar{u}}{\mathrm{d}z} + \frac{\mathrm{d}\delta}{\mathrm{d}z}\bar{u}, \quad \delta(z) = \frac{2}{r_0^2}\int_0^{r_0} r^2\tilde{u}\mathrm{d}r. \quad \text{(II.11)}$$

If (II.11) is put into (II.9), the average-concentration model assumes the form:

$$\frac{\partial\bar{c}}{\partial t} + \alpha\bar{u}\frac{\partial\bar{c}}{\partial z} + \left(\beta + \frac{\mathrm{d}\delta}{\mathrm{d}z}\right)\bar{u}\,\bar{c} + \gamma\delta\bar{c}\frac{\mathrm{d}\bar{u}}{\mathrm{d}z} = D\frac{\partial^2\bar{c}}{\partial z^2} - k\bar{c};$$
$$t = 0, \quad \bar{c}(0,z) \equiv c^0; \quad z = 0, \quad \bar{c}(t,0) \equiv \bar{c}(t,l), \quad \frac{\partial\bar{c}}{\partial z} \equiv 0. \quad \text{(II.12)}$$

Practically, the cross-sectional area surface in the columns is a constant $(r_0 = \text{const})$, i.e.,

$$\frac{\mathrm{d}\bar{u}}{\mathrm{d}z} = 0, \quad u = u(r). \quad \text{(II.13)}$$

In many practical cases, $\frac{\partial \tilde{u}}{\partial z} = 0$ and from (II.7), (II.10), and (II.13) follows:

$$\bar{v} = \tilde{v} = \gamma = \frac{d\delta}{dz} = 0, \quad \beta = \frac{\partial \alpha}{\partial z}. \tag{II.14}$$

As a result from (II.12) is obtained:

$$\frac{\partial \bar{c}}{\partial t} + \alpha \bar{u} \frac{\partial \bar{c}}{\partial z} + \frac{\partial \alpha}{\partial z} \bar{u} \bar{c} = D \frac{\partial^2 \bar{c}}{\partial z^2} - k\bar{c};$$

$$t = 0, \quad \bar{c}(0, z) \equiv c^0; \quad z = 0, \quad \bar{c}(t, 0) \equiv \bar{c}(t, l), \quad \frac{\partial \bar{c}}{\partial z} \equiv 0. \tag{II.15}$$

In the model (II.15), $\bar{u}$ is the average velocity of the laminar or turbulent flow in the column, and D is the diffusivity or the turbulent diffusivity (as a result of the small scale pulsations). The model parameter α is related to the radial non-uniformity of the velocity distribution and shows the influence of the column radius on the mass transfer kinetics. The parameter k may be obtained beforehand as a result of the chemical kinetics modeling.

The parameters in the model (II.15) show the influence of the scale-up (column radius increase) on the mass transfer kinetics if there exists a radial non-uniformity of the velocity distribution.

The presented theoretical analysis shows that in the convection–diffusion and average-concentration models, the velocity components and average velocity are:

$$u = u(r), \quad v = 0, \quad \bar{u} = \text{const.} \tag{II.16}$$

References

1. Boyadjiev CB (2006) Diffusion models and scale-up. Int J Heat Mass Transfer 49:796–799
2. Boyadjiev CB (2010) Theoretical chemical engineering. Modeling and simulation. Springer, Berlin, Heidelberg
3. Doichinova CB, Boyadjiev Chr (2012) On the column apparatuses modeling. Int J Heat Mass Transfer 55:6705–6715
4. Boyadjiev CB (2013) A new approach for the column apparatuses modeling in chemical engineering. J Pure Appl Math: Adv Appl 10(2):131–150

Chapter 5
Column Reactors Modeling

The theoretical procedure (II.5–II.15) presented in the Part II will be used for creation of average-concentration models of simple and complex chemical processes in one-phase column apparatuses. On this basis, the effect of the velocity radial non-uniformity will be analyzed and methods for model parameter identification [1–3] proposed.

The convection–diffusion model of the one-phase systems has the form (2.1.11):

$$u\frac{\partial c_i}{\partial z} = D_i\left(\frac{\partial^2 c_i}{\partial z^2} + \frac{1}{r}\frac{\partial c_i}{\partial r} + \frac{\partial^2 c_i}{\partial r^2}\right) + Q_i(c_1, c_2);$$

$$r = 0, \quad \frac{\partial c_i}{\partial r} \equiv 0; \quad r = r_0, \quad \frac{\partial c_i}{\partial r} \equiv 0; \tag{5.0.1}$$

$$z = 0, \quad c_i \equiv c_i^0, \quad u^0 c_i^0 \equiv u c_i^0 - D_i\frac{\partial c_i}{\partial z}; \quad i = 1, 2.$$

The average values of the velocity and concentration at the column cross-sectional area in one-phase systems follow from (II.3):

$$\bar{u} = \frac{2}{r_0^2}\int_0^{r_0} r u(r)\mathrm{d}r, \quad \bar{c}_i(z) = \frac{2}{r_0^2}\int_0^{r_0} r c_i(r, z)\mathrm{d}r \quad i = 1, 2. \tag{5.0.2}$$

The functions $u(r), c_i(r, z)$ can be presented with the help of the average functions (5.0.2):

$$u(r) = \bar{u}\tilde{u}(r), \quad c_i(r, z) = \bar{c}_i(z)\tilde{c}_i(r, z), \quad i = 1, 2, \tag{5.0.3}$$

where $\tilde{u}(r)$ and $\tilde{c}_i(r, z)$ represent the radial non-uniformity of the velocity and concentration and satisfy the following conditions:

© Springer International Publishing AG, part of Springer Nature 2018
C. Boyadjiev et al., *Modeling of Column Apparatus Processes,*
Heat and Mass Transfer, https://doi.org/10.1007/978-3-319-89966-4_5

$$\frac{2}{r_0^2} \int_0^{r_0} r\tilde{u}(r)\mathrm{d}r = 1, \quad \frac{2}{r_0^2} \int_0^{r_0} r\tilde{c}_i(r,z)\mathrm{d}r = 1, \quad i = 1,2. \qquad (5.0.4)$$

The average-concentration model may be obtained if (5.0.3) is put into (5.0.1), multiplied by r and integrated over r in the interval $[0, r_0]$. As a result, the average-concentration model has the form:

$$\alpha_i \bar{u} \frac{\mathrm{d}\bar{c}_i}{\mathrm{d}z} + \frac{\mathrm{d}\alpha_i}{\mathrm{d}z} \bar{u}\, \bar{c}_i = D_i \frac{\mathrm{d}^2 \bar{c}_i}{\mathrm{d}z^2} + \frac{2}{r_0^2} \int_0^{r_0} rQ_i(c_1,c_2)\mathrm{d}r;$$

$$\qquad (5.0.5)$$

$$z = 0, \quad \bar{c}_i = c_i^0, \quad \left(\frac{\mathrm{d}c_i}{\mathrm{d}z}\right)_{z=0} = 0; \quad i = 1,2.$$

where

$$\alpha_i(z) = \frac{2}{r_0^2} \int_0^{r_0} r\tilde{u}(r)\tilde{c}_i(r,z)\mathrm{d}r, \quad i = 1,2. \qquad (5.0.6)$$

5.1   Simple Chemical Reactions

Let us consider the stationary simple chemical reaction case

$$u\frac{\partial c}{\partial z} = D\left(\frac{\partial^2 c}{\partial z^2} + \frac{1}{r}\frac{\partial c}{\partial r} + \frac{\partial^2 c}{\partial r^2}\right) - kc;$$

$$r = 0, \quad \frac{\partial c}{\partial r} \equiv 0; \quad r = r_0, \quad \frac{\partial c}{\partial r} \equiv 0; \qquad (5.1.1)$$

$$z = 0, \quad c \equiv c^0, \quad u^0 c^0 \equiv uc^0 - D\frac{\partial c}{\partial z}.$$

5.1.1   Average-Concentration Model

From (II.3) follow the average values of the velocity and concentration at the column cross-sectional area:

$$\bar{u} = \frac{2}{r_0^2} \int_0^{r_0} ru(r)\mathrm{d}r, \quad \bar{c}(z) = \frac{2}{r_0^2} \int_0^{r_0} rc(r,z)\mathrm{d}r. \qquad (5.1.2)$$

The functions $u(r), c(r, z)$ in (5.1.1) can be presented with the average functions (5.1.2):

$$u(r) = \bar{u}\tilde{u}(r), \quad c(r, z) = \bar{c}(z)\tilde{c}(r, z), \tag{5.1.3}$$

where $\tilde{u}(r)$ and $\tilde{c}(r, z)$ represent the radial non-uniformity of the velocity and concentration and satisfy the following conditions:

$$\frac{2}{r_0^2} \int_0^{r_0} r\tilde{u}(r)\mathrm{d}r = 1, \quad \frac{2}{r_0^2} \int_0^{r_0} r\tilde{c}(r, z)\mathrm{d}r = 1. \tag{5.1.4}$$

The average-concentration model may be obtained if (5.1.3) is put into (5.1.1), multiplied by r and integrated over r in the interval $[0, r_0]$. As a result, the average-concentration model has the form:

$$\alpha\bar{u}\frac{\mathrm{d}\bar{c}}{\mathrm{d}z} + \frac{\mathrm{d}\alpha}{\mathrm{d}z}\bar{u}\,\bar{c} = D\frac{\mathrm{d}^2\bar{c}}{\mathrm{d}z^2} - k\bar{c};$$

$$z = 0, \quad \bar{c}(0) = c^0, \quad \frac{\mathrm{d}\bar{c}}{\mathrm{d}z} = 0, \tag{5.1.5}$$

where

$$\alpha(z) = \frac{2}{r_0^2} \int_0^{r_0} r\tilde{u}(r)\tilde{c}(r, z)\mathrm{d}r \tag{5.1.6}$$

represents effect of the radial non-uniformity of the velocity.

The use of the generalized variables

$$r = r_0 R, \quad z = lZ, \quad u(r) = \bar{u}U(R), \quad \tilde{u}(r) = \frac{u(r)}{\bar{u}} = U(R),$$

$$c(r, z) = c^0 C(R, Z), \quad \bar{c}(z) = c^0\overline{C}(Z), \quad \tilde{c}(r, z) = \frac{c(r, z)}{\bar{c}(z)} = \frac{C(R, Z)}{\overline{C}(Z)},$$

$$\overline{C}(Z) = 2\int_0^1 RC(R, Z)\mathrm{d}R, \quad \alpha(z) = \alpha(lZ) = A(Z) = 2\int_0^1 RU(R)\frac{C(R, Z)}{\overline{C}(Z)}\mathrm{d}R,$$

$$\tag{5.1.7}$$

leads to:

$$A(Z)\frac{\mathrm{d}\overline{C}}{\mathrm{d}Z} + \frac{\mathrm{d}A}{\mathrm{d}Z}\overline{C} = Pe^{-1}\frac{\mathrm{d}^2\overline{C}}{\mathrm{d}Z^2} - Da\,\overline{C}$$

$$Z = 0, \quad \overline{C} = 1, \quad \frac{\mathrm{d}\overline{C}}{\mathrm{d}Z} = 0, \tag{5.1.8}$$

where Pe and Da are the Peclet and Damkohler numbers, respectively:

$$Pe = \frac{\bar{u}l}{D}, \quad Da = \frac{kl}{\bar{u}}. \tag{5.1.9}$$

The case of parabolic velocity distribution (Poiseuille flow) will be presented as an example:

$$u = \bar{u}\left(2 - 2\frac{r^2}{r_0^2}\right), \quad \bar{u} = u^0, \quad U(R) = 2 - 2R^2. \tag{5.1.10}$$

The use of the velocity distribution (5.1.10) permits to obtain the function $A(Z)$ in (5.1.7), where $C(R, Z)$ is the solution of the model (1.1.15) for short ($\varepsilon = 10^{-1}$) columns [4]. Figure 5.1 displays the function $A(Z)$ for $Fo = 0.1$, $Da = 1$ showing that the function can be presented [3, 4] as linear approximation $A = a_0 + a_1 Z$ ($a_0 = 1, a_1 = 0.254$). As a result, the model (5.1.8) assumes the form:

$$(a_0 + a_1 Z)\frac{\mathrm{d}\overline{C}}{\mathrm{d}Z} + a_1\overline{C} = Pe^{-1}\frac{\mathrm{d}^2\overline{C}}{\mathrm{d}Z^2} - Da\,\overline{C};$$

$$Z = 0, \quad \overline{C} = 1, \quad \frac{\mathrm{d}\overline{C}}{\mathrm{d}Z} = 0. \tag{5.1.11}$$

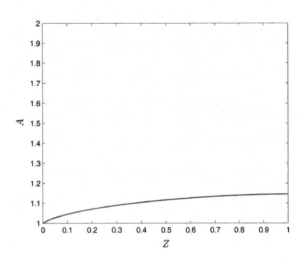

Fig. 5.1 Function $A(Z)$ for $Da = 1$, $Fo = 0.1$, $\varepsilon = 0.1$

5.1.2 Effect of the Velocity Radial Non-uniformity

In the cases of absence of radial non-uniformity of the velocity distribution at the column cross-sectional area (plug flow cases) $u = \bar{u}, U(R) \equiv 1$ and from (5.1.7) follows that $A(Z) \equiv 1$, i.e., the radial non-uniformity of the velocity distribution leads to $A(Z) > 1$.

The equation in (5.1.8) can be modified as

$$\frac{d\overline{C}}{dZ} = [A(Z)]^{-1}\left[Pe^{-1}\frac{d^2\overline{C}}{dZ^2} - \left(Da + \frac{dA}{dZ}\right)\overline{C}\right], \qquad (5.1.12)$$

i.e., the radial non-uniformity of the velocity distribution leads $(A(Z) > 1)$ to a decrease of the axial gradient of the average concentration $(d\overline{C}/dZ)$ and the conversion degree, because the conversion degree is possible to be presented as $G = \overline{C}(0) - \overline{C}(1)$.

5.1.3 Model Parameter Identification

Here (until the end), methods for the model parameter identification will use "artificial experimental data."

The solution of the model (1.1.27) for short $(\varepsilon = 10^{-1})$ columns [5], in the case $Fo = 0.1, Da = 1, Pe^{-1} = \varepsilon Fo = 0.05$, permits to $C(Z_n, R)$ be obtained for different $Z_n = 0.1n, n = 1, 2, \ldots, 10$ and average concentrations:

$$\overline{C}(Z_n) = 2\int_0^1 RC(Z_n, R)dR, \quad n = 1, \ldots, 10. \qquad (5.1.13)$$

As a result, it is possible to obtain "artificial experimental data" for different values of Z:

$$\overline{C}_{\text{exp}}^m(Z_n) = (0.95 + 0.1B_m)\overline{C}(Z_n), \quad m = 1, \ldots 10, \quad Z_n = 0.1n, \quad n = 1, 2, \ldots, 10, \qquad (5.1.14)$$

where $0 \leq B_m \leq 1, m = 1, \ldots, 10$ are obtained with a generator of random numbers. The obtained artificial experimental data (5.1.14) are used for illustration of the parameters' (a_0, a_1) identification in the average-concentration models (5.1.11) by minimization of the least-squares functions for different values of Z:

$$Q_n(a_{0n}, a_{1n}) = \sum_{m=1}^{10} \left[\overline{C}(Z_n, a_{0n}, a_{1n}) - \overline{C}_{\exp}^m(Z_n) \right]^2, \quad Z_n = 0.1n, \quad n = 1, 3, 5,$$

$$(5.1.15)$$

where the values of $\overline{C}(Z_n, a_{0n}, a_{1n})$ are obtained as solutions of (5.1.11) for different $Z_n = 0.1n, n = 1, 3, 5$. For the solution of (5.1.11) in the cases of short columns ($Fo = 0.1, Da = 1, \varepsilon = 10^{-1}, Pe^{-1} = \varepsilon Fo = 0.01$), the perturbation method is to be used (see Chap. 7 and [5]).

The solutions (a_{0n}, a_{1n}), $n = 1, 3, 5$, of the inverse problem for the parameter identification in the two-parameter average-concentrations model (5.1.11) for different values of Z_n, $n = 1, 3, 5$, after the minimization of (5.1.15), are obtained in [4]. These parameter values are used for the calculations of the average concentration in the model (5.1.11). The obtained values $\overline{C}(Z_n, a_{0n}, a_{1n})$, $Z_n = 0.1n, n = 1, 3, 5$ (the points) are compared (see Fig. 5.2) with the "exact" function (5.1.7) of the average concentration $\overline{C}(Z)$ (the line) obtained after solution of the model equation (2.1.27).

From Fig. 5.2, it is evident that the experimental data, obtained in a short column ($Z = 0.1$) with real diameter, are useful for the model parameter identification.

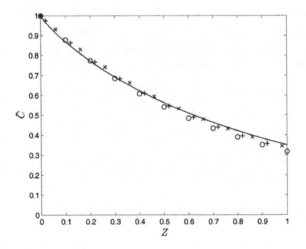

Fig. 5.2 Average concentration $\overline{C}(Z)$ for $Fo = 0.1, Da = 1, \varepsilon = 10^{-1}, Pe^{-1} = 0.01$:
line—$\overline{C}(Z)$—(1.1.27), (5.1.7)
O—$Z = 0.1, a_{01} = 1, a_{11} = 0.3519, \overline{C}(Z, a_{01}, a_{11})$—(5.1.11)
+—$Z = 0.3, a_{03} = 1, a_{13} = 0.2707, \overline{C}(Z, a_{03}, a_{13})$—(5.1.11)
×—$Z = 0.5, a_{05} = 1, a_{15} = 0.2162, \overline{C}(Z, a_{05}, a_{15})$—(5.1.11)

5.2 Complex Chemical Reaction

The theoretical procedure (II.5–II.15) is possible to be used for the creation of an average-concentration model of the complex chemical processes in one-phase column apparatuses. The base is the convection–diffusion model:

$$u\frac{\partial c_i}{\partial z} = D_i\left(\frac{\partial^2 c_i}{\partial z^2} + \frac{1}{r}\frac{\partial c_i}{\partial r} + \frac{\partial^2 c_i}{\partial r^2}\right) - kc_1^m c_2^n;$$

$$r = 0, \quad \frac{\partial c_i}{\partial r} \equiv 0; \quad r = r_0, \quad \frac{\partial c_i}{\partial r} \equiv 0; \qquad (5.2.1)$$

$$z = 0, \quad c = c_i^0, \quad u^0 c_i^0 \equiv uc_i^0 - D_i\frac{\partial c_i}{\partial z}, \quad i = 1, 2.$$

From (II.3) follow the average values of the velocity and concentration functions in (5.2.1) at the column cross-sectional area:

$$\bar{u} = \frac{2}{r_0^2}\int_0^{r_0} ru(r)dr, \quad \bar{c}_1(z) = \frac{2}{r_0^2}\int_0^{r_0} rc_1(r,z)dr, \quad \bar{c}_2(z) = \frac{2}{r_0^2}\int_0^{r_0} rc_2(r,z)dr.$$

$$(5.2.2)$$

The functions $u(r), c_1(r,z), c_2(r,z)$ in (4.1.2) can be presented with the help of the average functions (5.2.2):

$$u(r) = \bar{u}\tilde{u}(r), \quad c_1(r,z) = \bar{c}_1(z)\tilde{c}_1(r,z), \quad c_2(r,z) = \bar{c}_2(z)\tilde{c}_2(r,z) \qquad (5.2.3)$$

where

$$\frac{2}{r_0^2}\int_0^{r_0} r\tilde{u}(r)dr = 1, \quad \frac{2}{r_0^2}\int_0^{r_0} r\tilde{c}_1(r,z)dr = 1, \quad \frac{2}{r_0^2}\int_0^{r_0} r\tilde{c}_2(r,z)dr = 1. \qquad (5.2.4)$$

The average-concentration model may be obtained when (5.2.3) is put into (5.2.1), multiplied by r and integrated over r in the interval $[0, r_0]$. As a result, the average-concentration model has the form:

$$\alpha_i\bar{u}\frac{d\bar{c}_i}{dz} + \frac{d\alpha_i}{dz}\bar{u}\,\bar{c}_i = D_i\frac{d^2\bar{c}_i}{dz^2} - \delta k\bar{c}_1^m\bar{c}_2^n;$$

$$z = 0, \quad \bar{c}_i(0) = c_i^0, \quad \frac{d\bar{c}_i}{dz} = 0, \quad i = 1, 2, \qquad (5.2.5)$$

where

$$
\alpha_i(z) = \frac{2}{r_0^2} \int_0^{r_0} r \bar{u}(r) \tilde{c}_i(r,z) dr, \quad i = 1, 2; \quad \delta(z) = \frac{2}{r_0^2} \int_0^{r_0} r \tilde{c}_1^m(r,z) \tilde{c}_2^n(r,z) dr.
$$

$$(5.2.6)$$

The using of the generalized variables

$$
r = r_0 R, \quad z = lZ, \quad u(r) = \bar{u}U(R), \quad \tilde{u}(r) = \frac{u(r)}{\bar{u}} = U(R),
$$

$$
c_i(r,z) = c_i^0 C_i(R,Z), \quad \bar{c}_i(z) = c_i^0 \overline{C}_i(Z), \quad \overline{C}_i(Z) = 2 \int_0^1 R C_i(R,Z) dR,
$$

$$
\tilde{c}_i(r,z) = \frac{c_i(r,z)}{\bar{c}_i(z)} = \frac{C_i(R,Z)}{\overline{C}_i(Z)},
$$

$$(5.2.7)$$

$$
\alpha_i(z) = \alpha_i(lZ) = A_i(Z) = 2 \int_0^1 R U(R) \frac{C_i(R,Z)}{\overline{C}_i(Z)} dR, \quad i = 1, 2,
$$

$$
\delta(z) = \delta(lZ) = \Delta = 2 \int_0^1 R \left[\frac{C_1(R,Z)}{\overline{C}_1(Z)} \right]^m \left[\frac{C_2(R,Z)}{\overline{C}_2(Z)} \right]^n dR,
$$

leads to:

$$
A_i(Z) \frac{d\overline{C}_i}{dZ} + \frac{dA_i}{dZ} \overline{C}_i = Pe_i^{-1} \frac{d^2\overline{C}_i}{dZ^2} - \Delta(Z) Da_i \, \overline{C}_1^m \overline{C}_2^n;
$$

$$
Z = 0, \quad \overline{C}_i = 1, \quad \frac{d\overline{C}_i}{dZ} = 0; \quad i = 1, 2,
$$

$$(5.2.8)$$

where Pe and Da are the Peclet and Damkohler numbers, respectively:

$$
Pe_i = \frac{\bar{u}l}{D_i}, \quad Da_i = \theta^{i-1} Da, \quad Da = \frac{kl}{\bar{u}} \left(c_1^0 \right)^{m-1} \left(c_2^0 \right)^n, \quad \theta = \frac{c_1^0}{c_2^0}; \quad i = 1, 2.
$$

$$(5.2.9)$$

The model (2.1.26) for the high column ($\varepsilon = 0$) has the form:

$$
U \frac{\partial C_i}{\partial Z} = Fo_i \left(\frac{1}{R} \frac{\partial C_i}{\partial R} + \frac{\partial^2 C_i}{\partial R^2} \right) - Da_i \, C_1^m C_2^n;
$$

$$
R = 0, \quad \frac{\partial C_i}{\partial R} \equiv 0; \quad R = 1, \quad \frac{\partial C_i}{\partial R} \equiv 0; \quad Z = 0, \quad C_i \equiv 1.
$$

$$(5.2.10)$$

The solution of (5.2.10) for $m = n = 1, Fo_i = 0.1, Da_i = 1, i = 1, 2$, permits to calculate the functions $\overline{C}_i(Z), A_i(Z), i = 1, 2, \Delta(Z)$ in (5.2.7). The functions $\overline{C}_i(Z), i = 1, 2$, are presented on the Figs. 5.3 and 5.4. The functions $A_i(Z), i = 1, 2, \Delta(Z)$ are presented on Figs. 5.5, 5.6, and 5.7, where it is seen that linear approximations are possible to be used:

$$A_i = a_{0i} + a_{1i}Z, \quad i = 1, 2, \quad \Delta = \Delta_0 + \Delta_1 Z \qquad (5.2.11)$$

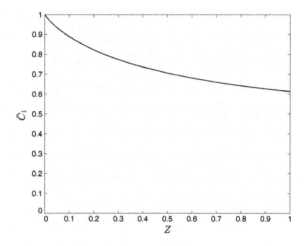

Fig. 5.3 Average concentration $\overline{C}_1(Z)$

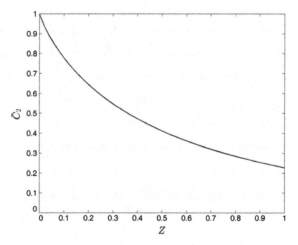

Fig. 5.4 Average concentration $\overline{C}_2(Z)$

Fig. 5.5 Function $A_1(Z)$

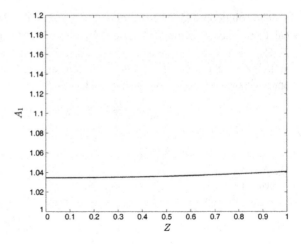

Fig. 5.6 Function $A_2(Z)$

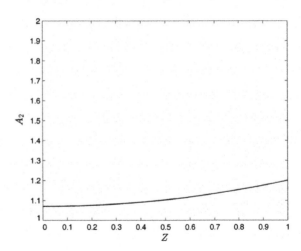

Fig. 5.7 Function $\Delta(Z)$

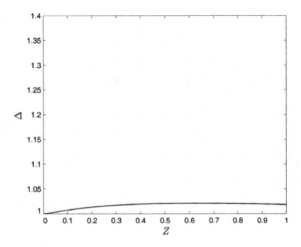

and the values of the parameters are:

$$a_{01} = 1.0346, \quad a_{11} = 0.0063, \quad a_{02} = 1.0708,$$
$$a_{12} = 0.1297, \quad \Delta_0 = 1.0095, \quad \Delta_1 = 0.0148. \tag{5.2.12}$$

5.3 Examples

5.3.1 Non-isothermal Chemical Reactors

Let us consider a non-isothermal chemical reactor, where simultaneous mass and heat transfer processes (2.4.9), (2.4.10) take place:

$$u \frac{\partial c}{\partial z} = D \left(\frac{\partial^2 c}{\partial z^2} + \frac{1}{r} \frac{\partial c}{\partial r} + \frac{\partial^2 c}{\partial r^2} \right) - kc;$$

$$r = 0, \quad \frac{\partial c}{\partial r} \equiv 0; \quad r = r_0, \quad \frac{\partial c}{\partial r} \equiv 0; \tag{5.3.1}$$

$$z = 0, \quad c \equiv c^0, \quad \bar{u}c^0 \equiv uc^0 - D \frac{\partial c}{\partial z}.$$

$$u \frac{\partial \theta}{\partial z} = \frac{\lambda}{\rho c_p} \left(\frac{\partial^2 \theta}{\partial z^2} + \frac{1}{r} \frac{\partial \theta}{\partial r} + \frac{\partial^2 \theta}{\partial r^2} \right) + \frac{q}{\rho c_p} kc;$$

$$r = 0, \quad \frac{\partial \theta}{\partial r} \equiv 0; \quad r = r_0, \quad -\lambda \frac{\partial \theta}{\partial r} \equiv k_0(\theta - \theta^*); \tag{5.3.2}$$

$$z = 0, \quad \theta \equiv \theta_0, \quad \bar{u}\theta_0 \equiv u\theta_0 - \frac{\lambda}{\rho c_p} \frac{\partial \theta}{\partial z},$$

where θ^* is the column wall temperature.

The average values of the velocity, concentration, and temperature at the column cross-sectional area are:

$$\bar{u} = \frac{2}{r_0^2} \int_0^{r_0} ru(r)dr, \quad \bar{c}(z) = \frac{2}{r_0^2} \int_0^{r_0} rc(r,z)dr, \quad \bar{\theta}(z) = \frac{2}{r_0^2} \int_0^{r_0} r\theta(r,z)dr. \tag{5.3.3}$$

The functions $u(r), c(r,z), \theta(r,z)$ in (5.3.1) and (5.3.2) can be presented with the help of the average functions (5.3.3):

$$u(r) = \bar{u}\tilde{u}(r), \quad c(r,z) = \bar{c}(z)\tilde{c}(r,z), \quad \theta(r,z) = \bar{\theta}(z)\tilde{\theta}(r,z), \tag{5.3.4}$$

where $\tilde{u}(r), \tilde{c}(r,z)$ and $\tilde{\theta}(r,z)$ present the radial non-uniformity of the velocity, concentration, and temperature distributions and satisfy the next conditions:

$$\frac{2}{r_0^2} \int_0^{r_0} r\tilde{u}(r)\mathrm{d}r = 1, \quad \frac{2}{r_0^2} \int_0^{r_0} r\tilde{c}(r,z)\mathrm{d}r = 1, \quad \frac{2}{r_0^2} \int_0^{r_0} r\tilde{\theta}(r,z)\mathrm{d}r = 1. \quad (5.3.5)$$

The average-concentration model may be obtained when putting (5.3.4) into (5.3.1) and (5.3.2), multiplying by r and integrating over r in the interval $[0, r_0]$. As a result, the average-concentration and temperature models have the forms:

$$\alpha\bar{u}\frac{\mathrm{d}\bar{c}}{\mathrm{d}z} + \frac{\mathrm{d}\alpha}{\mathrm{d}z}\bar{u}\,\bar{c} = D\frac{\mathrm{d}^2\bar{c}}{\mathrm{d}z^2} - k\bar{c};$$

$$z = 0, \quad \bar{c}(0) \equiv c^0, \quad \frac{\mathrm{d}\bar{c}}{\mathrm{d}z} \equiv 0; \qquad (5.3.6)$$

$$\alpha_\theta\bar{u}\frac{\mathrm{d}\bar{\theta}}{\mathrm{d}z} + \frac{\mathrm{d}\alpha_\theta}{\mathrm{d}z}\bar{u}\bar{\theta} = \frac{\lambda}{\rho c_p}\frac{\mathrm{d}^2\bar{\theta}}{\mathrm{d}z^2} - \frac{2k_0}{\rho c_p r_0}\left[\bar{\theta}\tilde{\theta}(r_0, z) - \theta^*\right] + \frac{qk}{\rho c_p}\bar{c};$$

$$z = 0, \quad \bar{\theta}(0) \equiv \theta_0, \quad \frac{\mathrm{d}\bar{\theta}}{\mathrm{d}z} \equiv 0; \qquad (5.3.7)$$

where

$$\alpha(z) = \frac{2}{r_0^2} \int_0^{r_0} r\tilde{u}(r)\tilde{c}(r,z)\mathrm{d}r, \quad \alpha_\theta(z) = \frac{2}{r_0^2} \int_0^{r_0} r\tilde{u}(r)\tilde{\theta}(r,z)\mathrm{d}r. \qquad (5.3.8)$$

The use of generalized variables

$$z = lZ, \quad \bar{c}(z) = \bar{c}(lZ) = c^0\overline{C}(Z), \quad \bar{\theta}(z) = \bar{\theta}(lZ) = \theta_0\overline{\Theta}(Z),$$

$$\overline{C}(Z) = 2\int_0^1 RC(R,Z)\mathrm{d}R, \quad \tilde{c}(r,z) = \frac{c(r,z)}{\bar{c}(z)} = \frac{C(R,Z)}{\overline{C}(Z)},$$

$$\alpha(z) = \alpha(lZ) = A(Z) = 2\int_0^1 RU(R)\frac{C(R,Z)}{\overline{C}(Z)}\mathrm{d}R,$$

$$\overline{\Theta}(Z) = 2\int_0^1 R\Theta(R,Z)\mathrm{d}R, \quad \tilde{\theta}(r,z) = \frac{\theta(r,z)}{\bar{\theta}(z)} = \frac{\Theta(R,Z)}{\overline{\Theta}(Z)}, \quad \theta^* = \theta_0\Theta^*,$$

$$\alpha_\theta(z) = \alpha_\theta(lZ) = A_\theta(Z) = 2\int_0^1 RU(R)\frac{\Theta(R,Z)}{\overline{\Theta}(Z)}\mathrm{d}R$$

$$(5.3.9)$$

leads (see (5.1.7), (5.1.8)) to

$$A(Z)\frac{\mathrm{d}\overline{C}}{\mathrm{d}Z} + \frac{\mathrm{d}A}{\mathrm{d}Z}\overline{C} = Pe^{-1}\frac{\mathrm{d}^2\overline{C}}{\mathrm{d}Z^2} - Da\,\overline{C};$$

$$Z = 0, \quad \overline{C} = 1, \quad \frac{\mathrm{d}\overline{C}}{\mathrm{d}Z} = 0. \tag{5.3.10}$$

$$A_\theta\frac{\mathrm{d}\overline{\Theta}}{\mathrm{d}Z} + \frac{\mathrm{d}A_\theta}{\mathrm{d}Z}\overline{\Theta} = \frac{\lambda}{\rho c_p \bar{u} l}\frac{\mathrm{d}^2\overline{\Theta}}{\mathrm{d}Z^2} - \frac{2k_0 l}{\rho c_p \bar{u} r_0}\left[\overline{\Theta}\tilde{\theta}(r_0, lZ) - \Theta^*\right] + \frac{qklc^0}{\rho c_p \bar{u}\theta_0}\overline{C};$$

$$Z = 0, \quad \overline{\Theta}(0) \equiv 1, \quad \frac{\mathrm{d}\overline{\Theta}}{\mathrm{d}Z} \equiv 0.$$

$$\tag{5.3.11}$$

The parameter identification of the models (5.3.10), (5.3.11) is possible to be made in a way similar to (5.1.8).

References

1. Boyadjiev C (2006) Diffusion models and scale-up. Int J Heat Mass Transf 49:796–799
2. Boyadjiev C (2009) Modeling of column apparatuses. Trans Acad 3:7–22
3. Boyadjiev C (2010) Theoretical chemical engineering. Modeling and simulation. Springer, Berlin
4. Doichinova M, Chr Boyadjiev (2012) On the column apparatuses modeling. Int J Heat Mass Transf 55:6705–6715
5. Boyadjiev B, Doichinova M, Boyadjiev C (2015) Computer modeling of column apparatuses. 3. Perturbation method approach. J Eng Thermophys 24(4):371–380

Chapter 6
Interphase Mass Transfer Process Modeling

The theoretical procedure (II.5–II.15) presented in Part II will be used for the creation of average-concentration models of absorption, adsorption, and catalytic processes in two-phase systems.

The convection–diffusion model of the two-phase systems [1–3] has the form (3.0.1):

$$u_j \frac{\partial c_{ij}}{\partial z_j} = D_{ij} \left(\frac{\partial^2 c_{ij}}{\partial z_j^2} + \frac{1}{r} \frac{\partial c_{ij}}{\partial r} + \frac{\partial^2 c_{ij}}{\partial r^2} \right) + Q_{ij}(c_{ij});$$

$$r = 0, \quad \frac{\partial c_{ij}}{\partial r} \equiv 0; \quad r = r_0, \quad \frac{\partial c_{ij}}{\partial r} \equiv 0;$$

$$z_j = 0, \quad c_{ij} \equiv c_{ij}^0, \quad u_j^0 c_{ij}^0 \equiv u_j c_{ij}^0 - D_{ij} \left(\frac{\partial c_{ij}}{\partial z_j} \right)_{z_j = 0}; \quad (6.0.1)$$

$$i = 1, 2, \ldots, i_0; \quad j = 1, 2 = 1, 3 = 2, 3.$$

The average values of the velocities and concentrations at the column cross-sectional area in two-phase systems follow from (II.3):

$$\bar{u}_j = \frac{2}{r_0^2} \int_0^{r_0} r u_j(r) \mathrm{d}r, \quad \bar{c}_{ij}(z) = \frac{2}{r_0^2} \int_0^{r_0} r c_{ij}(r, z) \mathrm{d}r, \quad (6.0.2)$$

$$i = 1, 2, \ldots, i_0, \quad j = 1, 2 = 1, 3 = 2, 3.$$

The functions $u_j(r), c_{ij}(r, z)$ in (6.0.1) can be presented by the average functions (6.0.2):

$$u_j(r) = \bar{u}_j \tilde{u}_j(r), \quad c_{ij}(r, z) = \bar{c}_{ij}(z) \tilde{c}_{ij}(r, z),$$
$$i = 1, 2, \ldots, i_0, \quad j = 1, 2 = 1, 3 = 2, 3, \quad (6.0.3)$$

© Springer International Publishing AG, part of Springer Nature 2018
C. Boyadjiev et al., *Modeling of Column Apparatus Processes*,
Heat and Mass Transfer, https://doi.org/10.1007/978-3-319-89966-4_6

where $\tilde{u}_j(r)$ and $\tilde{c}_{ij}(r,z)$ present the radial non-uniformity of the velocity and concentration and satisfy the conditions

$$\frac{2}{r_0^2}\int_0^{r_0} r\tilde{u}_j(r)\mathrm{d}r = 1, \quad \frac{2}{r_0^2}\int_0^{r_0} r\tilde{c}_{ij}(r,z)\mathrm{d}r = 1,$$

$$i = 1,2,\ldots,i_0, \quad j = 1,2 = 1,3 = 2,3. \tag{6.0.4}$$

The average-concentration model may be obtained when putting (6.0.3) into (6.0.1), multiplying by r, and integrating over r in the interval $[0, r_0]$. As a result, the average-concentration model has the form:

$$\alpha_{ij}\bar{u}_j \frac{\mathrm{d}\bar{c}_{ij}}{\mathrm{d}z_j} + \frac{\mathrm{d}\alpha_{ij}}{\mathrm{d}z_j}\bar{u}_j\bar{c}_{ij} = D_{ij}\frac{\mathrm{d}^2\bar{c}_{ij}}{\mathrm{d}z_j^2} + \frac{2}{r_0^2}\int_0^{r_0} rQ_{ij}\mathrm{d}r;$$

$$z_j = 0, \quad \bar{c}_{ij} = c_{ij}^0, \quad \left(\frac{\mathrm{d}c_{ij}}{\mathrm{d}z_j}\right)_{z_j=0} = 0; \tag{6.0.5}$$

$$i = 1,2,\ldots,i_0, \quad j = 1,2 = 1,3 = 2,3,$$

where

$$\alpha_{ij}(z) = \frac{2}{r_0^2}\int_0^{r_0} r\tilde{u}_j(r)\tilde{c}_{ij}(r,z)\mathrm{d}r, \quad i = 1,2,\ldots,i_0, \quad j = 1,2 = 1,3 = 2,3. \tag{6.0.6}$$

6.1 Absorption Process Modeling

6.1.1 Physical Absorption

The convection–diffusion model of the physical absorption ($i_0 = 1$ and the substance index i is possible to be ignored, $j = 1,2$) in a countercurrent column [4, 5] has the form (3.1.2), (3.1.3):

$$u_j\frac{\partial c_j}{\partial z_j} = D_j\left(\frac{\partial^2 c_j}{\partial z_j^2} + \frac{1}{r}\frac{\partial c_j}{\partial r} + \frac{\partial^2 c_j}{\partial r^2}\right) + (-1)^j k_0(c_1 - \chi c_2); \quad j = 1,2;$$

$$r = 0, \quad \frac{\partial c_1}{\partial r} = \frac{\partial c_2}{\partial r} \equiv 0; \quad r = r_0, \quad \frac{\partial c_1}{\partial r} = \frac{\partial c_2}{\partial r} \equiv 0;$$

$$z_1 = 0, \quad c_1(r,0) \equiv c_1^0, \quad u_1^0 c_1^0 \equiv u_1(r)c_1^0 - D_1\left(\frac{\partial c_1}{\partial z_1}\right)_{z_1=0};$$

$$z_2 = 0, \ c_2(r,0) \equiv c_2^0, \quad u_2^0 c_2^0 \equiv u_2(r)c_2^0 - D_2\left(\frac{\partial c_2}{\partial z_2}\right)_{z_2=0}, \qquad (6.1.1)$$

where $z_1 + z_2 = l$ (l is the column active zone height in the co-current column $z_1 = z_2 = z$) and $c_2^0 = 0$, practically.

The use of the averaging procedure (6.0.1)–(6.0.5) leads to the average-concentration model of the physical absorption:

$$\alpha_j \bar{u}_j \frac{d\bar{c}_j}{dz_j} + \frac{d\alpha_j}{dz_j}\bar{u}_j\bar{c}_j = D_j \frac{d^2\bar{c}_j}{dz_j^2} + (-1)^j k_0(\bar{c}_1 - \chi\bar{c}_2); \quad j = 1,2;$$

$$z_1 = 0, \quad \bar{c}_1(r,0) = c_1^0, \quad \left(\frac{dc_1}{dz_1}\right)_{z_1=0} = 0; \qquad (6.1.2)$$

$$z_2 = 0, \quad \bar{c}_2(r,0) = 0, \quad \left(\frac{dc_2}{dz_2}\right)_{z_2=0} = 0,$$

where

$$\alpha_j(z_j) = \frac{2}{r_0^2}\int_0^{r_0} r\tilde{u}_j(r)\tilde{c}_j(r,z_j)dr, \quad j = 1,2. \qquad (6.1.3)$$

For a theoretical analysis of the physical absorption, the following dimensionless (generalized) variables have to be used:

$$Z_1 = \frac{z_1}{l}, \quad Z_2 = \frac{z_2}{l}, \quad \overline{C}_1 = \frac{\bar{c}_1}{c_1^0}, \quad \overline{C}_2 = \frac{\bar{c}_2\chi}{c_1^0}. \qquad (6.1.4)$$

If (6.1.4) is put into (6.1.2), the model in generalized variables takes the form:

$$A_j \frac{d\overline{C}_j}{dZ_j} + \frac{dA_j}{dZ_j}\overline{C}_j = Pe_j^{-1}\frac{d^2\overline{C}_j}{dZ_j^2} - (-1)^{j-1}K_{0j}(\overline{C}_1 - \overline{C}_2);$$

$$Z_1 = 0, \quad \overline{C}_1 = 1, \quad \left(\frac{d\overline{C}_1}{dZ_1}\right)_{Z_1=0} = 0; \quad Z_2 = 0, \quad \overline{C}_2 = 0, \quad \left(\frac{d\overline{C}_2}{dZ_2}\right)_{Z_2=0} = 0,$$

$$\qquad (6.1.5)$$

where

$$Pe_j = \frac{\bar{u}_j l}{D_j}, \quad K_{0j} = \frac{k_0 l}{\bar{u}_j}\chi^{j-1}, \quad j = 1,2. \qquad (6.1.6)$$

From (6.0.2), (6.0.3), (6.0.6), (6.1.3), and (6.1.4) follow the expressions:

$$\tilde{u}_j(r) = \frac{u_j(r_0 R)}{\bar{u}_j} = U_j(R), \quad \tilde{c}_j(r, z_j) = \frac{c_j(r_0 R, lZ_j)}{\bar{c}_j(lZ_j)} = \frac{C_j(R, Z_j)}{\bar{C}_j(Z_j)},$$

$$\bar{C}_j(Z_j) = 2 \int_0^1 R C_j(R, Z_j) dR,$$

$$\alpha_j(z_j) = \alpha_j(lZ_j) = A_j(Z_j) = 2 \int_0^1 R U_j(R) \frac{C_j(R, Z_j)}{\bar{C}_j(Z_j)} dR, \quad j = 1, 2.$$

(6.1.7)

The case of parabolic velocity distribution (Poiseuille flow, where the difference between the phase velocities is in the average velocities only) will be considered as an example:

$$U_1 = U_2 = 2 - 2R^2.$$

(6.1.8)

The solution of the model equations (2.1.8) for a high column ($0 = \varepsilon \leq 10^{-2}$, $0 = Pe_j^{-1} = \varepsilon Fo_j \leq 10^{-2}$ for $Fo_j \leq 1, j = 1, 2, Fo_1 = 0.1, Fo_2 = 0.01, K_{01} = 1$, $K_{02} = 0.1$) using the iterative algorithm [5] in Chap. 8 and (6.1.7) permits to obtain the average concentrations $\bar{C}_j(Z_j)$ (Figs. 6.1 and 6.2) and the functions $A_j(Z_j)$ $j = 1, 2$ (Figs. 6.3 and 6.4).

The functions $A_j = (Z_j), j = 1, 2$ presented in Figs. 6.3 and 6.4 show that linear approximations are possible to be used:

$$A_j = a_{0j} + a_{1j} Z_j, \quad j = 1, 2.$$

(6.1.9)

The obtained ("theoretical") parameter values are presented in Table 6.1.

In the case of high columns ($Pe_j^{-1} = \varepsilon = 0$), the average-concentration model of a countercurrent physical absorption process has the form:

$$(a_{01} + a_{11} Z_1) \frac{d\bar{C}_1}{dZ_1} + a_{11} \bar{C}_1 = -K_{01} (\bar{C}_1 - \bar{C}_2); \quad Z_1 = 0, \quad \bar{C}_1(0) = 1.$$

$$(a_{02} + a_{12} Z_2) \frac{d\bar{C}_2}{dZ_2} + a_{12} \bar{C}_2 = K_{02} (\bar{C}_1 - \bar{C}_2); \quad Z_2 = 0, \quad \bar{C}_2(0) = 0.$$

(6.1.10)

The obtained average concentrations $\bar{C}_j(Z_j), j = 1, 2$ (Figs. 6.1 and 6.2) permit to obtain "artificial experimental data" for different values of $Z_j, j = 1, 2$:

$$\bar{C}_{j\,exp}^m(Z_{j1n}) = (0.95 + 0.1 S_m) \bar{C}_j(Z_{jn}), \quad m = 1, \dots, 10,$$

$$Z_{jn} = 0.1 n, \quad n = 1, 2, \dots, 10, \quad j = 1, 2,$$

(6.1.11)

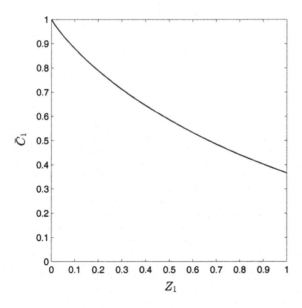

Fig. 6.1 Average concentration $\overline{C}_1(Z_1)$ for $Fo_1 = 0.1, Fo_2 = 0.01, K_{01} = 1, K_{02} = 0.1$

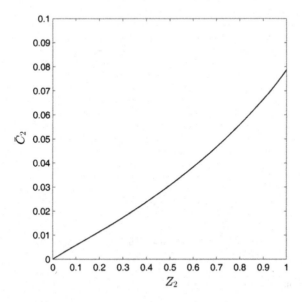

Fig. 6.2 Average concentration $\overline{C}_2(Z_2)$ for $Fo_1 = 0.1, Fo_2 = 0.01, K_{01} = 1, K_{02} = 0.1$

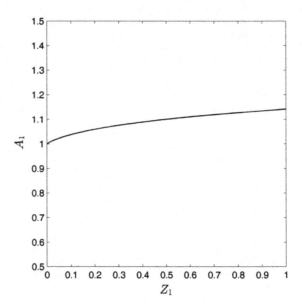

Fig. 6.3 Function $A_1(Z_1)$ for $Fo_1 = 0.1, Fo_2 = 0.01, K_{01} = 1, K_{02} = 0.1$

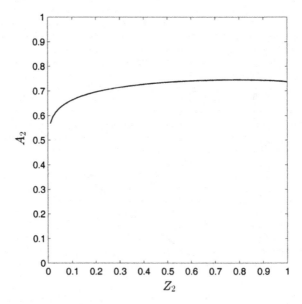

Fig. 6.4 Function $A_2(Z_2)$ for $Fo_1 = 0.1, Fo_2 = 0.01, K_{01} = 1, K_{02} = 0.1$

Table 6.1 Parameter values of $A_j(Z), j = 1, 2$ for $Fo_1 = 0.1, Fo_2 = 0.01, K_{01} = 1, K_{02} = 0.1$

"Theoretical" values	"Experimental" values		
$a_{01} = 1.0316$	$a_{01}^0 = 1.2242$	$a_{01}^1 = 0.9264$	$a_{01}^2 = 0.8888$
$a_{11} = 0.1225$	$a_{11}^0 = 0.4759$	$a_{11}^1 = 0.1564$	$a_{11}^2 = 0.0798$
$a_{02} = 0.6664$	$a_{02}^0 = 0.7191$	$a_{02}^1 = 0.5863$	$a_{02}^2 = 0.6021$
$a_{12} = 0.1036$	$a_{12}^0 = 0.0223$	$a_{12}^1 = 0.1096$	$a_{12}^2 = 0.1289$

where $0 \leq S_m \leq 1, m = 1, \ldots, 10$ are obtained by means of a generator of random numbers. The obtained "artificial experimental data" (6.1.11) are used as illustration of the parameter identification in the average-concentration models (6.1.10) by minimization of the least-squares functions $Q_n, n = 1, 2, \ldots, 10$ and Q:

$$Q_n\left(Z_n, a_{01}^n, a_{11}^n, a_{02}^n, a_{12}^n\right) = \sum_{m=1}^{10} \left[\overline{C}_1\left(Z_{1n}, a_{01}^n, a_{11}^n\right) - \overline{C}_{1\,\exp}^m(Z_{1n})\right]^2$$

$$+ \sum_{m=1}^{10} \left[\overline{C}_2\left(Z_{2n}, a_{02}^n, a_{12}^n\right) - \overline{C}_{2\,\exp}^m(Z_{2n})\right]^2, \qquad (6.1.12)$$

$$Z_n = Z_{1n} = Z_{2n} = 0.1n, \quad n = 1, 2;$$

$$Q\left(a_{01}^0, a_{11}^0, a_{02}^0, a_{12}^0\right) = \sum_{n=1}^{10} Q_n\left(Z_n, a_{01}^0, a_{11}^0, a_{02}^0, a_{12}^0\right),$$

where the values of $\overline{C}_j\left(Z_{jn}, a_{01}^n, a_{11}^n, a_{02}^n, a_{12}^n\right)$ are obtained as solutions of (6.1.10) for different $Z_{jn} = 0.1n, n = 1, 2, \ldots, 10, j = 1, 2$. The obtained ("experimental") values $\left(a_{01}^0, a_{11}^0, a_{02}^0, a_{12}^0\right), \left(a_{01}^1, a_{11}^1, a_{02}^1, a_{12}^1\right)$ and $\left(a_{01}^2, a_{11}^2, a_{02}^2, a_{12}^2\right)$ are presented in Table 6.1. They are used for the calculation of the functions $\overline{C}_1^0\left(Z_1, a_{01}^0, a_{11}^0\right)$, $\overline{C}_1^1\left(Z_1, a_{01}^1, a_{11}^1\right), \overline{C}_1^2\left(Z_1, a_{01}^2, a_{11}^2\right)$, (the lines in Fig. 6.5), and $\overline{C}_2^0\left(Z_2, a_{02}^0, a_{12}^0\right)$, $\overline{C}_2^1\left(Z_2, a_{02}^1, a_{12}^1\right), \overline{C}_2^2\left(Z_2, a_{02}^2, a_{12}^2\right)$, (the lines in Fig. 6.6). The points in Figs. 6.5 and 6.6 are the "artificial experimental data" (6.1.11).

The comparison of the functions (lines) and "artificial experimental data" (points) in Figs. 6.5 and 6.6 shows that the experimental data obtained from the column with real radius and small height ($Z_j = 0.1, j = 1, 2$) are useful for the parameter identification.

Fig. 6.5 Comparison of
concentration distributions
(6.1.10): (1)
$-\overline{C}_1^1(Z_1, a_{01}^1, a_{11}^1)$; (2)
$-\overline{C}_1^2(Z_1, a_{01}^2, a_{11}^2)$; (3)
$-\overline{C}_1^0(Z_1, a_{01}^0, a_{11}^0)$; $\circ$
—"artificial experimental
data" (6.1.11)

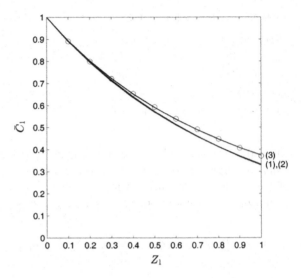

Fig. 6.6 Comparison of
concentration distributions
(6.1.10):
(1)—$\overline{C}_2^1(Z_2, a_{02}^1, a_{12}^1)$;
(2)—$\overline{C}_2^2(Z_2, a_{02}^2, a_{12}^2)$;
(3)—$\overline{C}_2^0(Z_2, a_{02}^0, a_{12}^0)$;
$\circ$—"artificial experimental
data" (6.1.11)

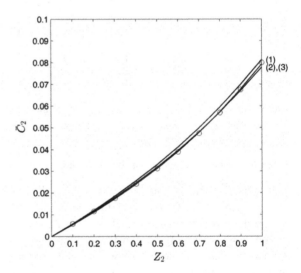

6.1.2   Chemical Absorption

The chemical absorption $(i_0 = 2)$ is a result of the chemical reaction between the
absorbed substance in the liquid phase $(i = 1, j = 2)$ and a reagent in the liquid
phase $(i = j = 2)$. If the kinetic model of the chemical reaction is $kc_{12}c_{22}$, the
convection–diffusion model of the chemical absorption in a co-current column has
the form (3.1.11), (3.1.12):

$$u_1 \frac{\partial c_{11}}{\partial z} = D_{11} \left(\frac{\partial^2 c_{11}}{\partial z^2} + \frac{1}{r} \frac{\partial c_{11}}{\partial r} + \frac{\partial^2 c_{11}}{\partial r^2} \right) - k_0(c_{11} - \chi c_{12});$$

$$u_2 \frac{\partial c_{12}}{\partial z} = D_{12} \left(\frac{\partial^2 c_{12}}{\partial z^2} + \frac{1}{r} \frac{\partial c_{12}}{\partial r} + \frac{\partial^2 c_{12}}{\partial r^2} \right) + k_0(c_{11} - \chi c_{12}) - k c_{12} c_{22};$$

$$u_2 \frac{\partial c_{22}}{\partial z} = D_{22} \left(\frac{\partial^2 c_{22}}{\partial z^2} + \frac{1}{r} \frac{\partial c_{22}}{\partial r} + \frac{\partial^2 c_{22}}{\partial r^2} \right) - k c_{12} c_{22};$$

$$r = 0, \quad \frac{\partial c_{11}}{\partial r} = \frac{\partial c_{12}}{\partial r} = \frac{\partial c_{22}}{\partial r} \equiv 0; \quad r = r_0, \quad \frac{\partial c_{11}}{\partial r} = \frac{\partial c_{12}}{\partial r} = \frac{\partial c_{22}}{\partial r} = 0;$$

$$z = 0, \quad c_{11} \equiv c_{11}^0, \quad c_{12} \equiv c_{12}^0, \quad c_{22} \equiv c_{22}^0, \quad u_1^0 c_{11}^0 \equiv u_1(r)c_{11}^0 - D_{11} \left(\frac{\partial c_{11}}{\partial z} \right)_{z=0},$$

$$u_2^0 c_{12}^0 \equiv u_2(r)c_{12}^0 - D_{12} \left(\frac{\partial c_{12}}{\partial z} \right)_{z=0}, \quad u_2^0 c_{22}^0 \equiv u_2(r)c_{22}^0 - D_{22} \left(\frac{\partial c_{22}}{\partial z} \right)_{z=0},$$

$$(6.1.13)$$

where $u_j^0, c_{ij}^0, \quad i = 1,2, j = 1,2$, are the inlet velocities and concentrations in the gas and liquid phases ($c_{12}^0 = 0$, practically).

From (II.3) follow the average values of the velocity and concentration in (6.1.13) at the column cross-sectional area:

$$\bar{u}_1 = \frac{2}{r_0^2} \int_0^{r_0} r u_1(r) dr, \quad \bar{u}_2 = \frac{2}{r_0^2} \int_0^{r_0} r u_2(r) dr, \quad \bar{c}_{11}(z) = \frac{2}{r_0^2} \int_0^{r_0} r c_{11}(r,z) dr,$$

$$(6.1.14)$$

$$\bar{c}_{12}(z) = \frac{2}{r_0^2} \int_0^{r_0} r c_{12}(r,z) dr, \quad \bar{c}_{22}(z) = \frac{2}{r_0^2} \int_0^{r_0} r c_{22}(r,z) dr.$$

The functions in (6.1.13) can be presented by the average functions (6.1.14):

$$u_1(r) = \bar{u}_1 \tilde{u}_1(r), \quad u_2(r) = \bar{u}_2 \tilde{u}_2(r), \quad c_{11}(r,z) = \bar{c}_{11}(z)\tilde{c}_{11}(r,z),$$
$$c_{12}(r,z) = \bar{c}_{12}(z)\tilde{c}_{12}(r,z), \quad c_{22}(r,z) = \bar{c}_{22}(z)\tilde{c}_{22}(r,z).$$

$$(6.1.15)$$

where

$$\frac{2}{r_0^2} \int_0^{r_0} r \tilde{u}_1(r) dr = 1, \quad \frac{2}{r_0^2} \int_0^{r_0} r \tilde{u}_2(r) dr = 1,$$

$$\frac{2}{r_0^2} \int_0^{r_0} r \tilde{c}_{11}(r,z) dr = 1, \quad \frac{2}{r_0^2} \int_0^{r_0} r \tilde{c}_{12}(r,z) dr = 1, \quad \frac{2}{r_0^2} \int_0^{r_0} r \tilde{c}_{22}(r,z) dr = 1.$$

$$(6.1.16)$$

The use of the averaging procedure (6.0.1)–(6.0.5) leads to the average-concentration model of the chemical absorption:

$$\alpha_{11}\bar{u}_1\frac{d\bar{c}_{11}}{dz} + \frac{d\alpha_{11}}{dz}\bar{u}_1\bar{c}_{11} = D_{11}\frac{d^2\bar{c}_{11}}{dz^2} - k_0(\bar{c}_{11} - \chi\bar{c}_{12});$$

$$\alpha_{12}\bar{u}_2\frac{d\bar{c}_{12}}{dz} + \frac{d\alpha_{12}}{dz}\bar{u}_2\bar{c}_{12} = D_{12}\frac{d^2\bar{c}_{12}}{dz^2} + k_0(\bar{c}_{11} - \chi\bar{c}_{12}) - \delta k\bar{c}_{12}\bar{c}_{22};$$

$$\alpha_{22}\bar{u}_2\frac{d\bar{c}_{22}}{dz} + \frac{d\alpha_{22}}{dz}\bar{u}_2\bar{c}_{22} = D_{22}\frac{d^2\bar{c}_{22}}{dz^2} - \delta k\bar{c}_{12}\bar{c}_{22}; \qquad (6.1.17)$$

$$z = 0, \quad \bar{c}_{11}(0) = c_{11}^0, \quad \bar{c}_{12}(0) = 0, \quad \bar{c}_{22}(0) = c_{22}^0,$$

$$\left(\frac{d\bar{c}_{11}}{dz}\right)_{z=0} = 0, \quad \left(\frac{d\bar{c}_{12}}{dz}\right)_{z=0} = 0, \quad \left(\frac{d\bar{c}_{22}}{dz}\right)_{z=0} = 0,$$

where

$$\alpha_{11}(z) = \frac{2}{r_0^2}\int_0^{r_0} r\tilde{u}_1(r)\tilde{c}_{11}(r,z)dr,$$

$$\alpha_{12}(z) = \frac{2}{r_0^2}\int_0^{r_0} r\tilde{u}_2(r)\tilde{c}_{12}(r,z)dr,$$

$$\alpha_{22}(z) = \frac{2}{r_0^2}\int_0^{r_0} r\tilde{u}_2(r)\tilde{c}_{22}(r,z)dr, \qquad (6.1.18)$$

$$\delta(z) = \frac{2}{r_0^2}\int_0^{r_0} r\tilde{c}_{12}(r,z)\tilde{c}_{22}(r,z)dr.$$

The use of dimensionless (generalized) variables

$$Z = \frac{z}{l}, \quad \overline{C}_{11} = \frac{\bar{c}_{11}}{c_{11}^0}, \quad \overline{C}_{12} = \frac{\bar{c}_{12}\chi}{c_{11}^0}, \quad \overline{C}_{22} = \frac{\bar{c}_{22}}{c_{22}^0}. \qquad (6.1.19)$$

leads to

$$A_{11}\frac{d\overline{C}_{11}}{dZ} + \frac{dA_{11}}{dZ}\overline{C}_{11} = Pe_{11}^{-1}\frac{d^2\overline{C}_{11}}{dZ^2} - K_{01}(\overline{C}_{11} - \overline{C}_{12});$$

$$A_{12}\frac{d\overline{C}_{12}}{dZ} + \frac{dA_{12}}{dZ}\overline{C}_{12} = Pe_{12}^{-1}\frac{d^2\overline{C}_{12}}{dZ^2} + K_{02}(\overline{C}_{11} - \overline{C}_{12}) - \Delta Kc_{22}^0\overline{C}_{12}\overline{C}_{22};$$

$$A_{22}\frac{d\overline{C}_{22}}{dZ} + \frac{dA_{22}}{dZ}\overline{C}_{22} + = Pe_{22}^{-1}\frac{d^2\overline{C}_{22}}{dZ^2} - \Delta K\frac{c_{11}^0}{\chi}\overline{C}_{12}\overline{C}_{22};$$

$$Z = 0, \quad \overline{C}_{11} = 1, \quad \overline{C}_{12} = 0, \quad \overline{C}_{22} = 1, \tag{6.1.20}$$

$$\left(\frac{d\overline{C}_{11}}{dZ}\right)_{Z=0} = 0, \quad \left(\frac{d\overline{C}_{12}}{dZ}\right)_{Z=0} = 0, \quad \left(\frac{d\overline{C}_{22}}{dZ}\right)_{Z=0} = 0,$$

where

$$Pe_{11} = \frac{\bar{u}_1 l}{D_{11}}, \quad Pe_{12} = \frac{\bar{u}_2 l}{D_{12}}, \quad Pe_{22} = \frac{\bar{u}_2 l}{D_{22}}, \tag{6.1.21}$$

$$K_{0j} = \frac{k_0 l}{\bar{u}_j}\chi^{j-1}, \quad j = 1, 2, \quad K = \frac{kl}{\bar{u}_2}.$$

From (6.0.2), (6.0.3), (6.0.6), (6.1.3), and (6.1.4) for the co-current flows ($z_1 = z_2 = z$) follow the expressions:

$$\tilde{c}_{1j}(r, z) = \frac{c_{1j}(r_0 R, lZ)}{\bar{c}_{1j}(lZ)} = \frac{C_{1j}(R, Z)}{\overline{C}_{1j}(Z)}, \quad \overline{C}_{1j}(Z) = 2\int_0^1 RC_{1j}(R, Z)dR,$$

$$\tilde{c}_{22}(r, z) = \frac{c_{22}(r_0 R, lZ)}{\bar{c}_{22}(lZ)} = \frac{C_{22}(R, Z)}{\overline{C}_{22}(Z)}, \quad \overline{C}_{22}(Z) = 2\int_0^1 RC_{22}(R, Z)dR,$$

$$\tag{6.1.22}$$

$$\alpha_{1j}(z) = \alpha_{1j}(lZ) = A_{1j}(Z) = 2\int_0^1 RU_j(R)\frac{C_{1j}(R, Z)}{\overline{C}_{1j}(Z)}dR, \quad j = 1, 2,$$

$$\delta(z) = \delta(lZ) = \Delta(Z) = 2\int_0^1 R\frac{C_{12}(R, Z)}{\overline{C}_{12}(Z)}\frac{C_{22}(R, Z)}{\overline{C}_{22}(Z)}dR.$$

A practical case leads to have the following orders of magnitude of the model parameters in (4.1.14):

$$Fo_s = 0.1, \quad s = 11, 12, 22, \quad K_0 = K_{01} = 1, \quad Da = K\frac{c_{11}^0}{\chi} = 1,$$

$$\tag{6.1.23}$$

$$K_{02} = K_0\frac{u_1^0\chi}{u_2^0} = 1, \quad \frac{u_1^0\chi}{u_2^0} = 1, \quad \frac{c_{22}^0\chi}{c_{11}^0} = 2.$$

The use of the parameter values (6.1.23) for solution of the model equations (3.1.20) permits to obtain the functions $C_{11}(R, Z), C_{12}(R, Z), C_{22}(R, Z)$ and after that the functions $\overline{C}_{11}(Z), \overline{C}_{12}(Z), \overline{C}_{22}(Z), A_{11}(Z), A_{12}(Z), \Delta(Z)$ in (6.1.22) (Figs. 6.7, 6.8 and 6.9), where $U_j = 2 - 2R^2, j = 1, 2$. From Figs. 6.7, 6.8 and 6.9,

Fig. 6.7 Function $A_{11}(Z)$

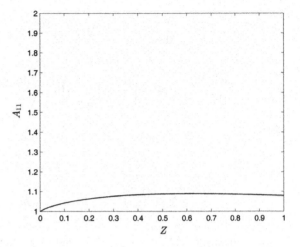

Fig. 6.8 Function $A_{12}(Z)$

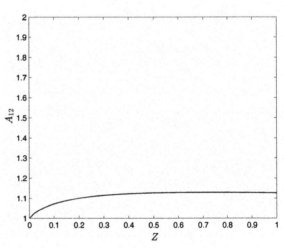

Fig. 6.9 Function $\Delta(Z)$

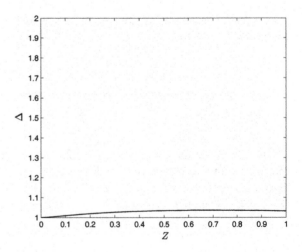

it is seen that the functions $A_{11}(Z), A_{12}(Z), \Delta(Z)$ can be presented as linear approximations

$$A_{11}(Z) = a_{110} + a_{111}Z, \quad A_{12}(Z) = a_{120} + a_{121}Z, \quad \Delta(Z) = \delta_0 + \delta_1 Z \quad (6.1.24)$$

with the approximation parameters being

$$a_{110} = 1.04, \quad a_{111} = 0.05, \quad a_{120} = 1.07, \quad a_{121} = 0.08, \quad \delta_0 = 1.01, \quad \delta_1 = 0.03. \quad (6.1.25)$$

From (6.1.25), it is seen that the maximal effect of the velocity non-uniformity on the function $\Delta(Z)$ is about 4% and cannot be registered experimentally, i.e., $\Delta(Z) \equiv 1$.

The parameter identification in (6.1.24) is possible to be realized similar to the case of the physical absorption.

6.2 Adsorption Process Modeling

6.2.1 Physical Adsorption

The convection–diffusion model of the non-stationary physical adsorption in the column apparatuses [6, 7] has the form (3.2.6), (3.2.7):

$$\frac{\partial c_{11}}{\partial t} + u_1 \frac{\partial c_{11}}{\partial z} = D_{11} \left(\frac{\partial^2 c_{11}}{\partial z^2} + \frac{1}{r} \frac{\partial c_{11}}{\partial r} + \frac{\partial^2 c_{11}}{\partial r^2} \right) - k_0(c_{11} - c_{13});$$

$$\frac{dc_{13}}{dt} = k_0(c_{11} - c_{13}) - b_0 k_1 c_{13} \frac{c_{23}}{c_{23}^0} + k_2 c_{23}^0 \left(1 - \frac{c_{23}}{c_{23}^0} \right);$$

$$\frac{dc_{23}}{dt} = -b_0 k_1 c_{13} \frac{c_{23}}{c_{23}^0} + k_2 c_{23}^0 \left(1 - \frac{c_{23}}{c_{23}^0} \right); \qquad (6.2.1)$$

$$t = 0, \quad c_{11} \equiv c_{11}^0, \quad c_{13} \equiv 0, \quad c_{23} \equiv c_{23}^0;$$

$$r = 0, \quad \frac{\partial c_{11}}{\partial r} \equiv 0; \quad r = r_0, \quad \frac{\partial c_{11}}{\partial r} \equiv 0;$$

$$z = 0, \quad c_{11} \equiv c_{11}^0, \quad u_1^0 c_{11}^0 \equiv u_1(r) c_{11}^0 - D_{11} \left(\frac{\partial c_{11}}{\partial z} \right)_{z=0}.$$

From (II.3) follow the average values of the velocity and concentration functions in (6.2.1) at the column cross-sectional area:

$$\bar{u}_1 = \frac{2}{r_0^2} \int_0^{r_0} r u_1(r) dr, \quad \bar{c}_{11}(t,z) = \frac{2}{r_0^2} \int_0^{r_0} r c_{11}(t,r,z) dr,$$

$$\bar{c}_{13}(t,z) = \frac{2}{r_0^2} \int_0^{r_0} r c_{13}(t,r,z) dr, \quad \bar{c}_{23}(t,z) = \frac{2}{r_0^2} \int_0^{r_0} r c_{23}(t,r,z) dr. \tag{6.2.2}$$

The functions in (6.2.1) can be presented by the average functions (6.2.2):

$$u_1(r) = \bar{u}_1 \tilde{u}_1(r), \quad c_{11}(t,r,z) = \bar{c}_{11}(t,z) \tilde{c}_{11}(t,r,z),$$
$$c_{13}(t,r,z) = \bar{c}_{13}(t,z) \tilde{c}_{13}(t,r,z), \quad c_{23}(t,r,z) = \bar{c}_{23}(t,z) \tilde{c}_{23}(t,r,z), \tag{6.2.3}$$

where

$$\frac{2}{r_0^2} \int_0^{r_0} r \tilde{u}_1(r) dr = 1, \quad \frac{2}{r_0^2} \int_0^{r_0} r \tilde{c}_{11}(t,r,z) dr = 1,$$

$$\frac{2}{r_0^2} \int_0^{r_0} r \tilde{c}_{13}(t,r,z) dr = 1, \quad \frac{2}{r_0^2} \int_0^{r_0} r \tilde{c}_{23}(t,r,z) dr = 1. \tag{6.2.4}$$

The use of the averaging procedure (6.0.1)–(6.0.5) leads to the average-concentration model of the physical adsorption:

$$\frac{\partial \bar{c}_{11}}{\partial t} + \alpha \bar{u}_1 \frac{\partial \bar{c}_{11}}{\partial z} + \frac{\partial \alpha}{\partial z} \bar{u}_1 \bar{c}_{11} = D_{11} \frac{\partial^2 \bar{c}_{11}}{\partial z^2} - k_0 (\bar{c}_{11} - \bar{c}_{13});$$

$$\frac{d \bar{c}_{13}}{dt} = k_0 (\bar{c}_{11} - \bar{c}_{13}) - b_0 k_1 \beta \bar{c}_{13} \frac{\bar{c}_{23}}{c_{23}^0} + k_2 c_{23}^0 \left(1 - \frac{\bar{c}_{23}}{c_{23}^0}\right);$$

$$\frac{d \bar{c}_{23}}{\partial t} = -b_0 k_1 \beta \bar{c}_{13} \frac{\bar{c}_{23}}{c_{23}^0} + k_2 c_{23}^0 \left(1 - \frac{\bar{c}_{23}}{c_{23}^0}\right); \tag{6.2.5}$$

$$t = 0, \quad \bar{c}_{11} \equiv c_{11}^0, \quad \bar{c}_{13} \equiv 0, \quad \bar{c}_{23} \equiv c_{23}^0;$$

$$z = 0, \quad \bar{c}_{11} \equiv c_{11}^0, \quad \left(\frac{\partial \bar{c}_{11}}{\partial z}\right)_{z=0} \equiv 0.$$

where

$$\alpha = \alpha(t,z) = \frac{2}{r_0^2} \int_0^{r_0} r \tilde{u}_1(r) \tilde{c}_{11}(t,r,z) dr,$$

$$\beta = \beta(t,z) = \frac{2}{r_0^2} \int_0^{r_0} r \tilde{c}_{13}(t,r,z) \tilde{c}_{23}(t,r,z) dr. \tag{6.2.6}$$

The use of the generalized variables

$$T = \frac{t}{t^0}, \quad Z = \frac{z}{l}, \quad \overline{C}_{11} = \frac{\overline{c}_{11}}{c_{11}^0}, \quad \overline{C}_{13} = \frac{\overline{c}_{13}}{c_{11}^0}, \quad \overline{C}_{23} = \frac{\overline{c}_{23}}{c_{23}^0}, \tag{6.2.7}$$

leads to:

$$\gamma \frac{\partial \overline{C}_{11}}{\partial T} + A \frac{\partial \overline{C}_{11}}{\partial Z} + \frac{\partial A}{\partial Z} \overline{C}_{11} = Pe^{-1} \frac{\partial^2 \overline{C}_{11}}{\partial Z^2} - K_0 (\overline{C}_{11} - \overline{C}_{13});$$

$$\frac{d\overline{C}_{13}}{dT} = K_3 (\overline{C}_{11} - \overline{C}_{13}) - BK_1 \overline{C}_{13} \overline{C}_{23} + K_2 \frac{c_{23}^0}{c_{11}^0} (1 - \overline{C}_{23});$$

$$\frac{d\overline{C}_{23}}{dT} = -BK_1 \frac{c_{11}^0}{c_{23}^0} \overline{C}_{13} \overline{C}_{23} + K_2 (1 - \overline{C}_{23}); \tag{6.2.8}$$

$$T = 0, \quad \overline{C}_{11} \equiv 1, \quad \overline{C}_{13} \equiv 0, \quad \overline{C}_{23} \equiv 1;$$

$$Z = 0, \quad \overline{C}_{11} \equiv 1, \quad \left(\frac{\partial \overline{C}_{11}}{\partial Z} \right)_{Z=0} \equiv 0,$$

where

$$K_0 = \frac{k_0 l}{u_1^0}, \quad K_1 = k_1 t^0 b_0, \quad K_2 = k_2 t^0 \quad K_3 = k_0 t^0,$$

$$A(T, Z) = \alpha(t_0 T, lZ) = \alpha(t, z) = 2 \int_0^1 RU(R) \frac{C_{11}(T, R, Z)}{\overline{C}_{11}(T, Z)} dR,$$

$$B(T, Z) = \beta(t_0 T, lZ) = \beta(t, z) = 2 \int_0^1 R \frac{C_{13}(T, R, Z)}{\overline{C}_{13}(T, Z)} \frac{C_{23}(T, R, Z)}{\overline{C}_{23}(T, Z)} dR,$$

$$\overline{C}_{11}(T, Z) = 2 \int_0^1 RC_{11}(T, R, Z) dR, \quad \overline{C}_{13}(T, Z) = 2 \int_0^1 RC_{13}(T, R, Z) dR,$$

$$\overline{C}_{23}(T, Z) = 2 \int_0^1 RC_{23}(T, R, Z) dR. \tag{6.2.9}$$

In (6.2.8), Z is a parameter in $\overline{C}_{13}(T, Z)$, $\overline{C}_{23}(T, Z)$ and T is a parameter in $\overline{C}_{11}(T, Z)$.

Practically, for lengthy (long-term) processes $0 \leq \gamma \leq 10^{-2}$ and high columns $(0 = \varepsilon \leq 10^{-2}, 0 = Pe^{-1} = \varepsilon Fo \leq 10^{-2}$ for $Fo \leq 1)$, the problem (6.2.8) has the form:

$$A\frac{d\overline{C}_{11}}{dZ} + \frac{dA}{dZ}\overline{C}_{11} = -K_0\left(\overline{C}_{11} - \overline{C}_{13}\right);$$

$$\frac{d\overline{C}_{13}}{dT} = K_3\left(\overline{C}_{11} - \overline{C}_{13}\right) - BK_1\overline{C}_{13}\overline{C}_{23} + K_2\frac{c_{23}^0}{c_{11}^0}\left(1 - \overline{C}_{23}\right);$$

$$\frac{d\overline{C}_{23}}{dT} = -BK_1\frac{c_{11}^0}{c_{23}^0}\overline{C}_{13}\overline{C}_{23} + K_2\left(1 - \overline{C}_{23}\right); \qquad (6.2.10)$$

$$T = 0, \quad \overline{C}_{11} \equiv 1, \quad \overline{C}_{13} \equiv 0, \quad \overline{C}_{23} \equiv 1; \quad Z = 0, \quad \overline{C}_{11} \equiv 1.$$

The solution of the model equations (3.2.11), using the multistep algorithm (see Chap. 9 and [7]), for the case $0 = \varepsilon \le 10^{-2}, 0 = Pe^{-1} = \varepsilon Fo \le 10^{-2}$, $Fo = 10^{-1}, K_0 = K_1 = K_3 = c_{23}^0 = 1, K_2 = 10^{-3}, c_{11}^0 = 10^{-2}$, permits to obtain the concentrations $C_{11}(T, R, Z), C_{13}(T, R, Z), C_{23}(T, R, Z)$ and the functions $\overline{C}_{11}(T, Z)$, $\overline{C}_{13}(T, Z), \overline{C}_{23}(T, Z), A(T, Z), B(T, Z)$ in (6.2.9). The results for $A(T, Z), B(T, Z)$ show that $B(T, Z) \equiv 1$ and $A(T, Z)$ are possible to be presented as a linear approximation:

$$A = a_0 + a_z Z + a_t T. \qquad (6.2.11)$$

The obtained ("theoretical") parameter values are $a_0 = 1.0471, a_z = 0.09025$, $a_t = -0.03770$ (see Table 6.2). The functions $\overline{C}_{11}(T, Z), A(T, Z)$ are presented in Figs. 6.10, 6.11, 6.12, and 6.13.

In Figs. 6.12 and 6.13 are compared the function $\overline{C}_{11}(T, Z)$ obtained in (6.2.9) (the lines) and the function $\overline{C}_{11}(T, Z)$ obtained as a solution of the (6.2.10) (the dotted lines), where $K_0 = K_1 = K_3 = c_{23}^0 = 1, K_2 = 10^{-3}, c_{11}^0 = 10^{-2}$ and $A = 1.047 + 0.0902Z - 0.0377T, B = 1$.

The concentration $C_{11}(0.6, R, Z)$ obtained as a solution of the problem (3.2.13)–(3.2.15) for the case $Fo = 10^{-1}, K_0 = K_1 = K_3 = c_{23}^0 = 1, K_2 = 10^{-3}, c_{11}^0 = 10^{-2}$ permits to obtain the average concentration $\overline{C}_{11}(0.6, Z)$ in (6.2.9) and "artificial experimental data" for different values of Z:

$$\overline{C}_{\exp}^m(Z_n) = (0.95 + 0.1S_m)\overline{C}_{11}(0.6, Z_n),$$

$$m = 1, \ldots 10, \quad Z_n = 0.1n, \quad n = 1, 2, \ldots, 10, \qquad (6.2.12)$$

Table 6.2 Parameter values

"Theoretical values"		"Experimental values"			
		Q	Q_1	Q_2	Q_3
a_0	1.0471	2.2291	0.7962	0.8721	0.9005
a_z	9.9246×10^{-2}	0.6849	7.3048×10^{-4}	4.4452×10^{-4}	3.1391×10^{-4}
a_T	-3.7701×10^{-2}	-0.7892	2.7259×10^{-4}	1.8971×10^{-4}	2.0352×10^{-4}

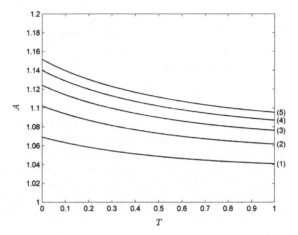

Fig. 6.10 Function $A(T, Z)$:
(1) $Z = 0.2$; (2) $Z = 0.4$;
(3) $Z = 0.6$; (4) $Z = 0.8$;
(5) $Z = 1.0$

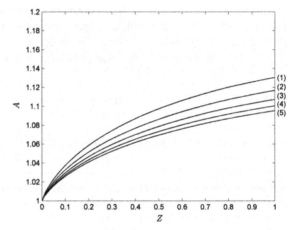

Fig. 6.11 Function $A(T, Z)$:
(1) $T = 0.2$; (2) $T = 0.4$;
(3) $T = 0.6$; (4) $T = 0.8$;
(5) $T = 1.0$

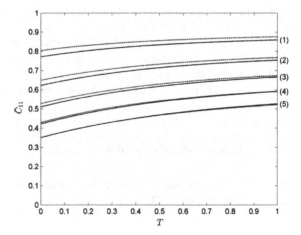

Fig. 6.12 Function
$\overline{C}_{11}(T, Z)$ in (6.2.9):
(1) $T = 0.2$; (2) $T = 0.4$;
(3) $T = 0.6$; (4) $T = 0.8$;
(5) $T = 1.0$; dotted lines are
solution of (6.2.10)

Fig. 6.13 Function
$\overline{C}_{11}(T,Z)$ in (6.2.9):
(1) $Z = 0.2$; (2) $Z = 0.4$;
(3) $Z = 0.6$; (4) $Z = 0.8$;
(5) $Z = 1.0$; dotted lines are
solution of (6.2.10)

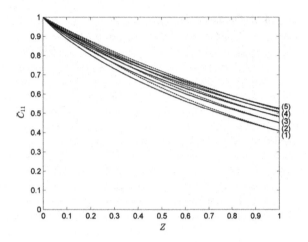

where $0 \leq S_m \leq 1, m = 1, \ldots, 10$ are obtained by means of a generator of random numbers. The obtained "artificial experimental" data (6.2.12) are used for the illustration of the parameter identification in the average-concentration models (6.2.10) by minimization of the least-squares functions Q_n and Q:

$$Q_n(Z_n, a_0, a_z, a_t) = \sum_{m=1}^{10} \left[\overline{C}_{11}(0.6, Z_n, a_0, a_z, a_t) - \overline{C}_{\exp}^m(Z_n) \right]^2,$$

$$Z_n = 0.1n, \quad n = 1, 2, \ldots, 10; \quad Q(a_0, a_z, a_t) = \sum_{n=1}^{10} Q_n(Z_n, a_0, a_z, a_t),$$

(6.2.13)

where the values of $\overline{C}_{11}(0.6, Z_n, a_0, a_z, a_t)$ are obtained as solutions of (6.2.10) for different $Z_n = 0.1n$, $n = 1, 2, \ldots, 10$.

The obtained ("experimental") values of a_0, a_z, a_t by minimization of Q, Q_1, Q_2, Q_3 are presented in Table 6.2.

In Fig. 6.14 are compared the average concentration $\overline{C}_{11}(Z)$ (the lines) as a solution of (6.2.10) for the parameter values a_0, a_z, a_t obtained by the minimization of Q_1 and Q in (5.2.13) with the "artificial experimental data" (6.2.12) (the points). The result presented shows that the parameter identification problems of the average-concentration models is possible to be solved using experimental data obtained in a short column ($Z = 0.1$) with a real diameter.

6.2.2 Chemical Adsorption

The convection–diffusion model of the non-stationary chemical adsorption [6, 7] has the form (3.2.18), (3.2.19):

Fig. 6.14 Function $\overline{C}_{11}$: 1—minimization of Q_1; 2—minimization of Q; ∘ —"artificial experimental data" (6.2.12)

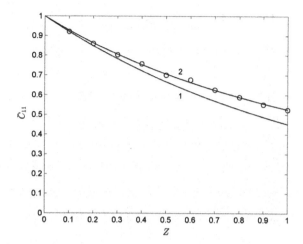

$$\frac{\partial c_{11}}{\partial t} + u_1 \frac{\partial c_{11}}{\partial z} = D_{11}\left(\frac{\partial^2 c_{11}}{\partial z^2} + \frac{1}{r}\frac{\partial c_{11}}{\partial r} + \frac{\partial^2 c_{11}}{\partial r^2}\right) - k_0(c_{11} - c_{13}),$$

$$\frac{dc_{13}}{dt} = k_0(c_{11} - c_{13}) - kc_{13}c_{23} = 0, \quad \frac{dc_{23}}{dt} = -kc_{13}c_{23};$$

$$t = 0, \quad c_{11} \equiv c_{11}^0, \quad c_{13} \equiv 0, \quad c_{23} \equiv c_{23}^0; \tag{6.2.14}$$

$$r = 0, \quad \frac{\partial c_{11}}{\partial r} \equiv 0; \quad r = r_0, \quad \frac{\partial c_{11}}{\partial r} \equiv 0;$$

$$z = 0, \quad c_{11} \equiv c_{11}^0, \quad u_1^0 c_{11}^0 \equiv u_1(r)c_{11}^0 - D_{11}\left(\frac{\partial c_{11}}{\partial z}\right)_{z=0}.$$

The use of the expressions (6.2.2)–(6.2.4) and averaging procedure (6.0.1)–(6.0.5) leads to the average-concentration model of the chemical absorption:

$$\frac{\partial \bar{c}_{11}}{\partial t} + \alpha \bar{u}_1 \frac{\partial \bar{c}_{11}}{\partial z} + \frac{\partial \alpha}{\partial z} \bar{u}_1 \bar{c}_{11} = \varepsilon_1 D_{11} \frac{\partial^2 \bar{c}_{11}}{\partial z^2} - k_0(\bar{c}_{11} - \bar{c}_{13});$$

$$\frac{d\bar{c}_{13}}{dt} = k_0(\bar{c}_{11} - \bar{c}_{13}) - \beta k \bar{c}_{13}\bar{c}_{23};$$

$$\frac{\partial \bar{c}_{23}}{\partial t} = -\beta k \bar{c}_{13}\bar{c}_{23}; \tag{6.2.15}$$

$$t = 0, \quad \bar{c}_{11} \equiv c_{11}^0, \quad \bar{c}_{13} \equiv 0, \quad \bar{c}_{23} \equiv c_{23}^0;$$

$$z = 0, \quad \bar{c}_{11} \equiv c_{11}^0, \quad \left(\frac{\partial \bar{c}_{11}}{\partial z}\right)_{z=0} \equiv 0.$$

where $\alpha = \alpha(t, z)$ and $\beta = \beta(t, z)$ are presented in (6.2.6).

The use of the generalized variables (6.2.7) leads to:

$$\gamma \frac{\partial \overline{C}_{11}}{\partial T} + A \frac{\partial \overline{C}_{11}}{\partial Z} + \frac{\partial A}{\partial Z} \overline{C}_{11} = Pe^{-1} \frac{\partial^2 \overline{C}_{11}}{\partial Z^2} - K_0 (\overline{C}_{11} - \overline{C}_{13});$$

$$\frac{d\overline{C}_{13}}{dT} = K_3 (\overline{C}_{11} - \overline{C}_{13}) - BK c_{23}^0 \overline{C}_{13} \overline{C}_{23};$$

$$\frac{d\overline{C}_{23}}{dT} = -BK c_{11}^0 \overline{C}_{13} \overline{C}_{23};$$

$$T = 0, \quad \overline{C}_{11} \equiv 1, \quad \overline{C}_{13} \equiv 0, \quad \overline{C}_{23} \equiv 1; \quad Z = 0, \quad \overline{C}_{11} \equiv 1, \quad \left(\frac{\partial \overline{C}_{11}}{\partial Z} \right)_{Z=0} \equiv 0,$$

$$\text{(6.2.16)}$$

where

$$K = kt^0, \quad K_0 = \frac{k_0 l}{u_1^0}, \quad K_3 = k_0 t^0,$$

$$A(T, Z) = \alpha(t_0 T, lZ) = \alpha(t, z) = 2 \int_0^1 RU(R) \frac{C_{11}(T, R, Z)}{\overline{C}_{11}(T, Z)} dR,$$

$$B(T, Z) = \beta(t_0 T, lZ) = \beta(t, z) = 2 \int_0^1 R \frac{C_{13}(T, R, Z)}{\overline{C}_{13}(T, Z)} \frac{C_{23}(T, R, Z)}{\overline{C}_{23}(T, Z)} dR, \qquad \text{(6.2.17)}$$

$$\overline{C}_{11}(T, Z) = 2 \int_0^1 RC_{11}(T, R, Z) dR, \quad \overline{C}_{13}(T, Z) = 2 \int_0^1 RC_{13}(T, R, Z) dR,$$

$$\overline{C}_{23}(T, Z) = 2 \int_0^1 RC_{23}(T, R, Z) dR.$$

In (6.2.8), Z is a parameter in $\overline{C}_{13}(T, Z)$, $\overline{C}_{23}(T, Z)$ and T is a parameter in $\overline{C}_{11}(T, Z)$.

Practically, for lengthy (long-term) processes $0 \leq \gamma \leq 10^{-2}$ and high columns $(0 = \varepsilon \leq 10^{-2}, 0 = Pe^{-1} = \varepsilon Fo \leq 10^{-2})$, the problem (6.2.8) has the form:

$$A \frac{d\overline{C}_{11}}{dZ} + \frac{dA}{dZ} \overline{C}_{11} = -K_0 (\overline{C}_{11} - \overline{C}_{13});$$

$$\frac{d\overline{C}_{13}}{dT} = K_3 (\overline{C}_{11} - \overline{C}_{13}) - BK c_{23}^0 \overline{C}_{13} \overline{C}_{23}; \qquad \text{(6.2.18)}$$

$$\frac{d\overline{C}_{23}}{dT} = -BK c_{11}^0 \overline{C}_{13} \overline{C}_{23};$$

$$T = 0, \quad \overline{C}_{13} = 0, \quad \overline{C}_{23} \equiv 1; \quad Z = 0, \quad \overline{C}_{11} \equiv 1.$$

The solution of the model equations (3.2.23) for the case $0 = \varepsilon \le 10^{-2}$, $0 = Pe^{-1} = \varepsilon Fo \le 10^{-2}, Fo = 10^{-1}, K = K_0 = K_3 = c_{23}^0 = 1; c_{11}^0 = 10^{-2}$, permits to obtain the average concentrations $\overline{C}_{11}(T,Z), \overline{C}_{13}(T,Z), \overline{C}_{23}(T,Z)$ and the functions $A(T,Z), B(T,Z)$ in (6.2.9). The results for $A(T,Z), B(T,Z)$ show that $B(T,Z) \equiv 1$ and $A(T,Z)$ are possible to be presented as a linear approximation:

$$A = a_0 + a_z Z + a_t T. \tag{6.2.19}$$

The obtained ("theoretical") parameter values are $a_0 = 1.0471, a_z = 9.9247 \times 10^{-2}$, $a_t = -3.7696 \times 10^{-2}$. The function $A(T,Z)$ is presented in Figs. 6.15 and 6.16.

The parameter identification of the chemical adsorption models is similar to the physical adsorption case.

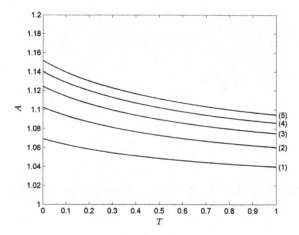

Fig. 6.15 Function $A(T, Z)$: (1) $Z = 0.2$; (2) $Z = 0.4$; (3) $Z = 0.6$; (4) $Z = 0.8$; (5) $Z = 1.0$

Fig. 6.16 Function $A(T, Z)$: (1) $T = 0.2$; (2) $T = 0.4$; (3) $T = 0.6$; (4) $T = 0.8$; (5) $T = 1.0$

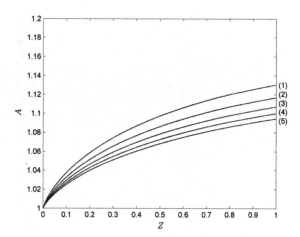

6.3   Catalytic Process Modeling

6.3.1   Physical Adsorption Mechanism

The convection–diffusion model of the catalytic processes in the column apparatuses [8] in the cases of physical adsorption mechanism has the form (3.3.3):

$$
\begin{aligned}
& u_1 \frac{\partial c_{11}}{\partial z} = D_{11}\left(\frac{\partial^2 c_{11}}{\partial z^2} + \frac{1}{r}\frac{\partial c_{11}}{\partial r} + \frac{\partial^2 c_{11}}{\partial r^2}\right) - k_{01}(c_{11} - c_{13}); \\[4pt]
& u_1 \frac{\partial c_{21}}{\partial z} = D_{21}\left(\frac{\partial^2 c_{21}}{\partial z^2} + \frac{1}{r}\frac{\partial c_{21}}{\partial r} + \frac{\partial^2 c_{21}}{\partial r^2}\right) - k_{02}(c_{21} - c_{23}); \\[4pt]
& k_{01}(c_{11} - c_{13}) - b_0 k_1 c_{13}\frac{c_{33}}{c_{33}^0} + k_2 c_{33}^0\left(1 - \frac{c_{33}}{c_{33}^0}\right) = 0; \\[4pt]
& k_{02}(c_{21} - c_{23}) - k c_{23}(c_{33}^0 - c_{33}) = 0; \\[4pt]
& - b_0 k_1 c_{13}\frac{c_{33}}{c_{33}^0} + k_2 c_{33}^0\left(1 - \frac{c_{33}}{c_{33}^0}\right) + k c_{23}(c_{33}^0 - c_{33}) = 0; \\[4pt]
& r = 0, \quad \frac{\partial c_{11}}{\partial r} = \frac{\partial c_{21}}{\partial r} \equiv 0; \quad r = r_0, \quad \frac{\partial c_{11}}{\partial r} = \frac{\partial c_{21}}{\partial r} \equiv 0; \\[4pt]
& z = 0, \quad c_{11} \equiv c_{11}^0, \quad u_1^0 c_{11}^0 \equiv u_1(r)c_{11}^0 - D_{11}\left(\frac{\partial c_{11}}{\partial z}\right)_{z=0}, \\[4pt]
& c_{21} \equiv c_{21}^0, \quad u_1^0 c_{21}^0 \equiv u_1(r)c_{21}^0 - D_{21}\left(\frac{\partial c_{21}}{\partial z}\right)_{z=0}.
\end{aligned}
\tag{6.3.1}
$$

From (II.3) follow the average values of the velocity and the concentration
functions in (6.3.1) at the column cross-sectional area:

$$
\bar{u}_1 = \frac{2}{r_0^2}\int_0^{r_0} r u_1(r)\,dr, \quad \bar{c}_{11}(z) = \frac{2}{r_0^2}\int_0^{r_0} r c_{11}(r,z)\,dr,
$$

$$
\bar{c}_{21}(z) = \frac{2}{r_0^2}\int_0^{r_0} r c_{21}(r,z)\,dr, \quad \bar{c}_{13}(z) = \frac{2}{r_0^2}\int_0^{r_0} r c_{13}(r,z)\,dr,
\tag{6.3.2}
$$

$$
\bar{c}_{23}(z) = \frac{2}{r_0^2}\int_0^{r_0} r c_{23}(r,z)\,dr, \quad \bar{c}_{33}(z) = \frac{2}{r_0^2}\int_0^{r_0} r c_{33}(r,z)\,dr.
$$

The functions in (6.3.1) can be presented by the average functions (6.3.2):

$$
\begin{aligned}
u_1(r) &= \bar{u}_1 \tilde{u}_1(r), \quad c_{11}(r,z) = \bar{c}_{11}(z)\tilde{c}_{11}(r,z), \\
c_{21}(r,z) &= \bar{c}_{21}(z)\tilde{c}_{21}(r,z), \quad c_{13}(r,z) = \bar{c}_{13}(z)\tilde{c}_{13}(r,z), \\
c_{23}(r,z) &= \bar{c}_{23}(z)\tilde{c}_{23}(r,z), \quad c_{33}(r,z) = \bar{c}_{33}(z)\tilde{c}_{33}(r,z).
\end{aligned}
\tag{6.3.3}
$$

where

$$
\frac{2}{r_0^2}\int_0^{r_0} r\tilde{u}_1(r)\,\mathrm{d}r = 1, \quad \frac{2}{r_0^2}\int_0^{r_0} r\tilde{c}_{11}(r,z)\,\mathrm{d}r = 1, \quad \frac{2}{r_0^2}\int_0^{r_0} r\tilde{c}_{21}(r,z)\,\mathrm{d}r = 1,
$$

$$
\frac{2}{r_0^2}\int_0^{r_0} r\tilde{c}_{13}(r,z)\,\mathrm{d}r = 1, \quad \frac{2}{r_0^2}\int_0^{r_0} r\tilde{c}_{23}(r,z)\,\mathrm{d}r = 1, \quad \frac{2}{r_0^2}\int_0^{r_0} r\tilde{c}_{33}(r,z)\,\mathrm{d}r = 1.
\tag{6.3.4}
$$

The use of (6.3.2), (6.3.3), (6.3.4) and the averaging procedure (6.0.1)–(6.0.5) leads to the average-concentration model of the catalytic processes in the column apparatuses in the cases of physical adsorption mechanism:

$$
\begin{aligned}
&\alpha_1 \bar{u}_1 \frac{\mathrm{d}\bar{c}_{11}}{\mathrm{d}z} + \frac{\mathrm{d}\alpha_1}{\mathrm{d}z}\bar{u}_1\bar{c}_{11} = D_{11}\frac{\mathrm{d}^2\bar{c}_{11}}{\mathrm{d}z^2} - k_{01}(\bar{c}_{11} - \bar{c}_{13}); \\
&\alpha_2 \bar{u}_1 \frac{\mathrm{d}\bar{c}_{21}}{\mathrm{d}z} + \frac{\mathrm{d}\alpha_2}{\mathrm{d}z}\bar{u}_1\bar{c}_{21} = D_{21}\frac{\mathrm{d}^2\bar{c}_{21}}{\mathrm{d}z^2} - k_{02}(\bar{c}_{21} - \bar{c}_{23}); \\
&k_{01}(\bar{c}_{11} - \bar{c}_{13}) - \beta b_0 k_1 \bar{c}_{13}\frac{\bar{c}_{33}}{c_{33}^0} + k_2 c_{33}^0\left(1 - \frac{\bar{c}_{33}}{c_{33}^0}\right) = 0; \\
&k_{02}(\bar{c}_{21} - \bar{c}_{23}) - k\bar{c}_{23}c_{33}^0 + \gamma k\bar{c}_{23}\bar{c}_{33} = 0; \\
&-\beta b_0 k_1 \bar{c}_{13}\frac{\bar{c}_{33}}{c_{33}^0} + k_2 c_{33}^0\left(1 - \frac{\bar{c}_{33}}{c_{33}^0}\right) + k\bar{c}_{23}c_{33}^0 - \gamma k\bar{c}_{23}\bar{c}_{33} = 0; \\
&z = 0, \quad \bar{c}_{11} = c_{11}^0, \quad \left(\frac{\mathrm{d}\bar{c}_{11}}{\mathrm{d}z}\right)_{z=0} = 0, \quad \bar{c}_{21} = c_{21}^0, \quad \left(\frac{\mathrm{d}\bar{c}_{21}}{\mathrm{d}z}\right)_{z=0} = 0.
\end{aligned}
\tag{6.3.5}
$$

where

$$\alpha_1 = \alpha_1(z) = \frac{2}{r_0^2} \int_0^{r_0} r \tilde{u}_1(r) \tilde{c}_{11}(r,z) \mathrm{d}r,$$

$$\alpha_2 = \alpha_2(z) = \frac{2}{r_0^2} \int_0^{r_0} r \tilde{u}_1(r) \tilde{c}_{21}(r,z) \mathrm{d}r,$$

$$\beta = \beta(z) = \frac{2}{r_0^2} \int_0^{r_0} r \tilde{c}_{13}(r,z) \tilde{c}_{33}(r,z) \mathrm{d}r, \tag{6.3.6}$$

$$\gamma = \gamma(z) = \frac{2}{r_0^2} \int_0^{r_0} r \tilde{c}_{23}(r,z) \tilde{c}_{33}(r,z) \mathrm{d}r.$$

The use of the generalized variables

$$Z = \frac{z}{l}, \quad \overline{C}_{11} = \frac{\overline{c}_{11}}{c_{11}^0}, \quad \overline{C}_{21} = \frac{\overline{c}_{21}}{c_{21}^0}, \quad \overline{C}_{13} = \frac{\overline{c}_{13}}{c_{11}^0}, \quad \overline{C}_{23} = \frac{\overline{c}_{23}}{c_{21}^0}, \quad \overline{C}_{33} = \frac{\overline{c}_{33}}{c_{33}^0},$$

$$\tilde{C}_{11} = \frac{\tilde{c}_{11}}{c_{11}^0}, \quad \tilde{C}_{21} = \frac{\tilde{c}_{21}}{c_{21}^0}, \quad \tilde{C}_{13} = \frac{\tilde{c}_{13}}{c_{11}^0}, \quad \tilde{C}_{23} = \frac{\tilde{c}_{23}}{c_{21}^0}, \quad \tilde{C}_{33} = \frac{\tilde{c}_{33}}{c_{33}^0},$$

$$\tag{6.3.7}$$

leads to:

$$A_1 \frac{\mathrm{d}\overline{C}_{11}}{\mathrm{d}Z} + \frac{\mathrm{d}A_1}{\mathrm{d}Z}\overline{C}_{11} = Pe_1^{-1} \frac{\mathrm{d}^2\overline{C}_{11}}{\mathrm{d}Z^2} - K_{01}\left(\overline{C}_{11} - \overline{C}_{13}\right);$$

$$A_2 \frac{\mathrm{d}\overline{C}_{21}}{\mathrm{d}Z} + \frac{\mathrm{d}A_2}{\mathrm{d}Z}\overline{C}_{21} = Pe_2^{-1} \frac{\mathrm{d}^2\overline{C}_{21}}{\mathrm{d}Z^2} - K_{02}\left(\overline{C}_{21} - \overline{C}_{23}\right); \tag{6.3.8}$$

$$Z = 0, \quad \overline{C}_{11} = 1, \quad \left(\frac{\mathrm{d}\overline{C}_{11}}{\mathrm{d}Z}\right)_{Z=0} = 0, \quad \overline{C}_{21} = 1, \quad \left(\frac{\mathrm{d}\overline{C}_{21}}{\mathrm{d}Z}\right)_{Z=0} = 0.$$

$$\overline{C}_{13} = \frac{\overline{C}_{11} + K_1\left(1 - \overline{C}_{33}\right)}{1 + BK_2\overline{C}_{33}}, \quad \overline{C}_{23} = \frac{\overline{C}_{21}}{1 + K_3\left(1 - G\overline{C}_{33}\right)},$$

$$\overline{C}_{33} = \frac{K_5 + \overline{C}_{23}}{BK_4\overline{C}_{13} + K_5 + G\overline{C}_{23}}. \tag{6.3.9}$$

The parameters in (6.3.8), (6.3.9) and the new functions have the forms:

$$K_{0i} = \frac{k_{0i}l}{u_1^0}, \quad Pe_{i1} = \frac{u_1^0 l}{D_{i0}}, \quad i = 1,2; \quad K_1 = \frac{k_2\,c_{33}^0}{k_{01}\,c_{11}^0},$$

$$K_2 = \frac{b_0 k_1}{k_{01}}, \quad K_3 = \frac{k_{23}c_{33}^0}{k_{02}}, \quad K_4 = \frac{b_0 k_1}{k_{23}c_{21}^0}\frac{c_{11}^0}{c_{33}^0}, \quad K_5 = \frac{k_2}{k_{23}c_{21}^0}. \tag{6.3.10}$$

$$A_i(Z) = \alpha_i(lZ) = \alpha_i(z) = 2 \int_0^1 RU(R) \frac{C_{i1}(R,Z)}{\overline{C}_{i1}(Z)} dR, \quad i = 1, 2,$$

$$B(Z) = \beta(lZ) = \beta(z) = 2 \int_0^1 R \frac{C_{13}(R,Z)}{\overline{C}_{13}(Z)} \frac{C_{33}(R,Z)}{\overline{C}_{33}(Z)} dR,$$

$$G(Z) = \gamma(lZ) = \gamma(z) = 2 \int_0^1 R \frac{C_{23}(R,Z)}{\overline{C}_{23}(Z)} \frac{C_{33}(R,Z)}{\overline{C}_{33}(Z)} dR,$$

$$\overline{C}_{11}(Z) = 2 \int_0^1 RC_{11}(R,Z) dR, \quad \overline{C}_{21}(Z) = 2 \int_0^1 RC_{21}(R,Z) dR,$$

$$\overline{C}_{13}(Z) = 2 \int_0^1 RC_{13}(R,Z) dR, \quad \overline{C}_{23}(Z) = 2 \int_0^1 RC_{23}(R,Z) dR,$$

$$\overline{C}_{33}(Z) = 2 \int_0^1 RC_{33}(R,Z) dR.$$

$$(6.3.11)$$

The use of (6.3.11) and $C_{11}(R,Z), C_{21}(R,Z), C_{13}(R,Z), C_{23}(R,Z), C_{33}(R,Z)$ as a solution of the problem (3.3.8), (3.3.12), (3.3.12) for the case (3.3.14) permits to obtain the average concentrations $\overline{C}_{11}(Z), \overline{C}_{21}(Z), \overline{C}_{13}(Z), \overline{C}_{23}(Z), \overline{C}_{33}(Z)$ and the functions $A_i(Z), i = 1, 2, B(Z), G(Z)$. They are presented in Figs. 6.17 and 6.18, where it is seen that the functions $A_i(Z), i = 1, 2, B(Z), G(Z)$ can be presented as linear approximations:

$$A_i(Z) = a_{0i} + a_{1i}Z, \quad i = 1, 2, \quad B(Z) = b_{0i} + b_{1i}Z, \quad G(Z) = g_{0i} + g_{1i}Z.$$

$$(6.3.12)$$

The approximation ("theoretical") parameter values are presented in Table 6.3, where it is seen that $B \equiv 1, G \equiv 1$, practically.

For high columns ($0 = \varepsilon \leq 10^{-2}, 0 = Pe_i^{-1} = \varepsilon.Fo_{i1} \leq 10^{-2}, Fo_{i1} \leq 1, i = 1, 2$), the problem (5.3.8) takes the form:

$$A_1 \frac{d\overline{C}_{11}}{dZ} + \frac{dA_1}{dZ} \overline{C}_{11} = -K_{01}(\overline{C}_{11} - \overline{C}_{13});$$

$$A_2 \frac{d\overline{C}_{21}}{dZ} + \frac{dA_2}{dZ} \overline{C}_{21} = -K_{02}(\overline{C}_{21} - \overline{C}_{23}); \quad Z = 0, \quad \overline{C}_{11} \equiv 1, \quad \overline{C}_{21} \equiv 1.$$

$$(6.3.13)$$

Fig. 6.17 Average functions
$\overline{C}(Z)$

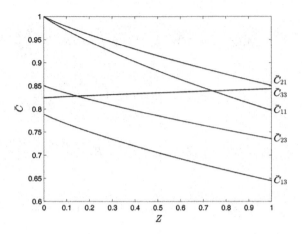

Fig. 6.18 Functions
$A_i(Z), i = 1, 2, B(Z), G(Z)$

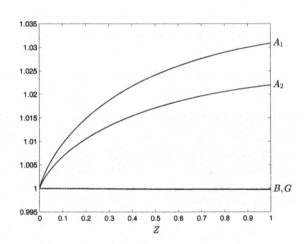

Table 6.3 Parameter values

A_1	A_2	B	G
$a_{01} = 1.0090$	$a_{02} = 1.0063$	$b_0 = 1.0000$	$g_0 = 1.0000$
$a_{11} = 0.0257$	$a_{12} = 0.0183$	$b_1 = -0.0003$	$g_1 = -0.0002$

The solution of (5.3.13) depends on the two functions:

$$\overline{C}_{13} = \frac{\overline{C}_{11} + K_1\left(1 - \overline{C}_{33}\right)}{1 + BK_2\overline{C}_{33}}, \quad \overline{C}_{23} = \frac{\overline{C}_{21}}{1 + K_3\left(1 - G\overline{C}_{33}\right)}, \tag{6.3.14}$$

where $\overline{C}_{33}$ is the solution of the cubic equation:

$$\bar{\omega}_3\left(\overline{C}_{33}\right)^3 + \bar{\omega}_2\left(\overline{C}_{33}\right)^2 + \bar{\omega}_1\overline{C}_{33} + \bar{\omega}_0 = 0,$$
$$\bar{\omega}_3 = BGK_3(K_1K_4 - K_2K_5),$$
$$\bar{\omega}_2 = K_5(BK_2 + 2BK_2K_3 - GK_3)$$
$$- K_4\left(BK_1 + BK_1K_3 + BGK_1K_3 + BGK_3\overline{C}_{11}\right) + BGK_2\overline{C}_{21}, \qquad (6.3.15)$$
$$\bar{\omega}_1 = BK_4\left(\overline{C}_{11} + K_1\right)(1 + K_3)$$
$$+ K_5(1 + K_3 + GK_3 - BK_2 - BK_2K_3) + (G - BK_2)\overline{C}_{21},$$
$$\bar{\omega}_0 = -\overline{C}_{21} - K_3K_5 - K_5$$

For solving (6.3.15), $0 \leq \overline{C}_{33} \leq 1$ has to be used.
The solution of (6.3.13)–(6.3.15) is obtained [8] as five-vector forms:

$$\overline{C}_{11}(Z) = \left|\overline{C}_{11(zeta)}\right|, \quad \overline{C}_{21}(Z) = \left|\overline{C}_{21(zeta)}\right|, \quad \overline{C}_{13}(Z) = \left|\overline{C}_{13(zeta)}\right|,$$
$$\overline{C}_{23}(Z) = \left|\overline{C}_{23(zeta)}\right|, \quad \overline{C}_{33}(Z) = \left|\overline{C}_{33(zeta)}\right|, \quad Z = \frac{\zeta - 1}{\zeta^0 - 1}, \quad \zeta = 1, 2, \ldots, \zeta^0.$$

$$(6.3.16)$$

For the case (3.3.14), Fig. 6.19 provides comparison of the functions $\overline{C}_{11}(Z), \overline{C}_{21}(Z)$ obtained as solutions of (6.3.13)–(6.3.15) using Table 6.3 (the dotted lines) with the solution of (3.3.8), (3.3.12), (3.3.13), using (6.3.11) (the lines).

Fig. 6.19 Functions $\overline{C}_{11}(Z), \overline{C}_{21}(Z)$: dotted lines —solution of (6.3.13)–(6.3.15) using Table 6.3; lines —solution of (3.3.8), (3.3.12), (3.3.13) using (6.3.11)

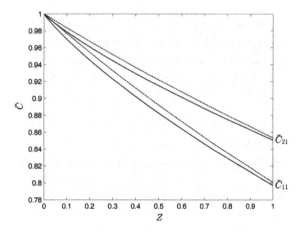

The obtained concentrations $\overline{C}_{11}(Z), \overline{C}_{21}(Z)$ for the case (3.3.14) after solution of (3.3.8), (3.3.12), (3.3.13) using (6.3.11) allow to obtain "artificial experimental data" for different values of Z:

$$\overline{C}_{11\,\exp}^{m}(Z_n) = (0.95 + 0.1 S_m)\overline{C}_{11}(Z_n),$$
$$\overline{C}_{21\,\exp}^{m}(Z_n) = (0.95 + 0.1 S_m)\overline{C}_{21}(Z_n), \tag{6.3.17}$$
$$m = 1,\ldots 10, \quad Z_n = 0.1 n, \quad n = 1, 2, \ldots, 10,$$

where $0 \leq S_m \leq 1, m = 1, \ldots, 10$ are obtained by means of a generator of random numbers. The obtained "artificial experimental data" (6.3.17) are used for illustration of the parameter identification in the average-concentration model (6.3.13)–(6.3.15) by minimization of the least-squares functions Q_n and Q:

$$Q_n\left(Z_n, a_{01}^n, a_{11}^n, a_{02}^n, a_{12}^n\right) = \sum_{m=1}^{10}\left[\overline{C}_{11}\left(Z_n, a_{01}^n, a_{11}^n, a_{02}^n, a_{12}^n\right) - \overline{C}_{11\,\exp}^{m}(Z_n)\right]^2$$

$$+ \sum_{m=1}^{10}\left[\overline{C}_{21}\left(Z_n, a_{01}^n, a_{11}^n, a_{02}^n, a_{12}^n\right) - \overline{C}_{21\,\exp}^{m}(Z_n)\right]^2,$$

$$Z_n = 0.1 n, \quad n = 1, 2, \ldots, 10;$$

$$Q\left(a_{01}^0, a_{11}^0, a_{02}^0, a_{12}^0\right) = \sum_{n=1}^{10} Q_n\left(Z_n, a_{01}^0, a_{11}^0, a_{02}^0, a_{12}^0\right),$$

$$\tag{6.3.18}$$

where the values of $\overline{C}_{11}\left(Z_n, a_{01}^n, a_{11}^n, a_{02}^n, a_{12}^n\right)$ and $\overline{C}_{21}\left(Z_n, a_{01}^n, a_{11}^n, a_{02}^n, a_{12}^n\right)$ are obtained as solutions of (5.3.13)–(5.3.15) for different values of Z: $Z_n = 0.1 n, \quad n = 1, 2, \ldots, 10$.

The obtained ("experimental") parameter values of $a_{01}^0, a_{11}^0, a_{02}^0, a_{12}^0, a_{01}^1, a_{11}^1, a_{02}^1, a_{12}^1, a_{01}^2, a_{11}^2, a_{02}^2, a_{12}^2$ are presented in Table 6.4. They are used for calculation of the functions $\overline{C}_{11}\left(Z, a_{01}^n, a_{11}^n, a_{02}^n, a_{12}^n\right), \overline{C}_{21}\left(Z, a_{01}^n, a_{11}^n, a_{02}^n, a_{12}^n\right), n = 0, 1, 2$ in the case (3.3.14) as solutions of (6.3.13)–(6.3.15) (the lines in Fig. 6.20), where the points are the "artificial experimental data" (6.3.17) (average values for every Z).

Table 6.4 Parameter values

"Theoretical values"	"Experimental" values —Q min	"Experimental" values —Q_1 min	"Experimental" values—Q_2 min
$a_{01} = 1.0090$	$a_{01}^0 = 1.0000$	$a_{01}^1 = 0.9984$	$a_{01}^2 = 0.9988$
$a_{11} = 0.0257$	$a_{11}^0 = 0.0397$	$a_{11}^1 = 0.1032$	$a_{11}^2 = 0.0779$
$a_{02} = 1.0063$	$a_{02}^0 = 1.0000$	$a_{02}^1 = 0.8865$	$a_{02}^2 = 0.9206$
$a_{12} = 0.0183$	$a_{12}^0 = 0.0316$	$a_{12}^1 = 0.0688$	$a_{12}^2 = 0.0499$

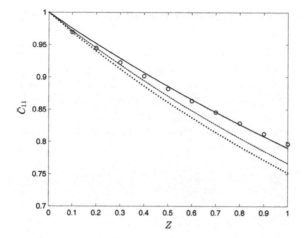

Fig. 6.20 Concentration distributions $\overline{C}_{11}\left(Z, a_{01}^n, a_{11}^n, a_{02}^n, a_{12}^n\right), n = 0, 1, 2$: lines—solutions of (6.3.13)–(6.3.15) in the case (3.3.14); points—the "artificial experimental data" (6.3.17) (average values for every Z)

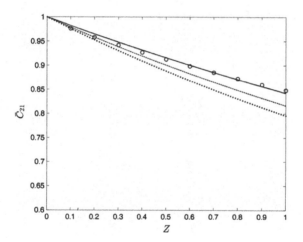

Fig. 6.21 Concentration distributions $\overline{C}_{21}\left(Z, a_{01}^n, a_{11}^n, a_{02}^n, a_{12}^n\right), n = 0, 1, 2$: lines—solutions of (6.3.13)–(6.3.15) in the case (3.3.14); points—the "artificial experimental data" (6.3.17) (average values for every Z)

The comparison of the functions (lines) with the "artificial experimental data" (points) in Figs. 6.20 and 6.21 shows that the experimental data obtained from a column with real radius and small height $(Z = 0.1)$ are useful for parameter identifications.

6.3.2   Chemical Adsorption Mechanism

The convection–diffusion model of the heterogeneous catalytic chemical reaction, in the case of chemical adsorption mechanism [8], has the form (3.3.11), (3.3.12), where the average values of the velocity and concentration functions at the column cross-sectional area have the forms (6.3.2)–(6.3.4). The use of (3.3.11), (3.3.12) and the averaging procedure (6.0.1)–(6.0.5) leads to the average-concentration model of the catalytic processes in the column apparatuses in the cases of chemical adsorption mechanism:

$$
\alpha_1 \bar{u}_1 \frac{d\bar{c}_{11}}{dz} + \frac{d\alpha_1}{dz} \bar{u}_1 \bar{c}_{11} = D_{11} \frac{d^2\bar{c}_{11}}{dz^2} - k_{01}(\bar{c}_{11} - \bar{c}_{13});
$$

$$
\alpha_2 \bar{u}_1 \frac{d\bar{c}_{21}}{dz} + \frac{d\alpha_2}{dz} \bar{u}_1 \bar{c}_{21} = D_{21} \frac{d^2\bar{c}_{21}}{dz^2} - k_{02}(\bar{c}_{21} - \bar{c}_{23}); \tag{6.3.19}
$$

$$
z = 0, \quad \bar{c}_{11} = c_{11}^0, \quad \left(\frac{d\bar{c}_{11}}{dz}\right)_{z=0} = 0, \quad \bar{c}_{21} = c_{21}^0, \quad \left(\frac{d\bar{c}_{21}}{dz}\right)_{z=0} = 0.
$$

$$
k_{01}(\bar{c}_{11} - \bar{c}_{13}) - \beta k_{13}\bar{c}_{13}\bar{c}_{33} = 0, \quad k_{02}(\bar{c}_{21} - \bar{c}_{23}) - k_{23}\bar{c}_{23}(c_{33}^0 - \gamma\bar{c}_{33}) = 0,
$$

$$
- \beta k_{13}\bar{c}_{13}\bar{c}_{33} + k_{23}\bar{c}_{23}(c_{33}^0 - \gamma\bar{c}_{33}) = 0. \tag{6.3.20}
$$

The new functions in (6.3.19), (6.3.20) are

$$
\alpha_i = \alpha_i(z) = \frac{2}{r_0^2} \int_0^{r_0} r\tilde{u}_1(r)\tilde{c}_{i1}(r,z)\,dr, \quad i = 1,2,
$$

$$
\beta = \beta(z) = \frac{2}{r_0^2} \int_0^{r_0} r\tilde{c}_{13}(r,z)\tilde{c}_{33}(r,z)\,dr, \tag{6.3.21}
$$

$$
\gamma = \gamma(z) = \frac{2}{r_0^2} \int_0^{r_0} r\tilde{c}_{23}(r,z)\tilde{c}_{33}(r,z)\,dr.
$$

The use of the generalized variables

$$
Z = \frac{z}{l}, \quad \overline{C}_{11} = \frac{\bar{c}_{11}}{c_{11}^0}, \quad \overline{C}_{21} = \frac{\bar{c}_{21}}{c_{21}^0}, \quad \overline{C}_{13} = \frac{\bar{c}_{13}}{c_{11}^0}, \quad \overline{C}_{23} = \frac{\bar{c}_{23}}{c_{21}^0}, \quad \overline{C}_{33} = \frac{\bar{c}_{33}}{c_{33}^0},
$$

$$
\tilde{C}_{11} = \frac{\tilde{c}_{11}}{c_{11}^0}, \quad \tilde{C}_{21} = \frac{\tilde{c}_{21}}{c_{21}^0}, \quad \tilde{C}_{13} = \frac{\tilde{c}_{13}}{c_{11}^0}, \quad \tilde{C}_{23} = \frac{\tilde{c}_{23}}{c_{21}^0}, \quad \tilde{C}_{33} = \frac{\tilde{c}_{33}}{c_{33}^0},
$$

$$
\tag{6.3.22}
$$

leads to:

$$A_1 \frac{d\overline{C}_{11}}{dZ} + \frac{dA_1}{dZ}\overline{C}_{11} = Pe_{11}^{-1}\frac{d^2\overline{C}_{11}}{dZ^2} - K_{01}(\overline{C}_{11} - \overline{C}_{13});$$

$$A_2 \frac{d\overline{C}_{21}}{dZ} + \frac{dA_2}{dZ}\overline{C}_{21} = Pe_{21}^{-1}\frac{d^2\overline{C}_{21}}{dZ^2} - K_{02}(\overline{C}_{21} - \overline{C}_{23}); \qquad (6.3.23)$$

$$Z = 0, \quad \overline{C}_{11} = 1, \quad \left(\frac{d\overline{C}_{11}}{dZ}\right)_{Z=0} = 0, \quad \overline{C}_{21} = 1, \quad \left(\frac{d\overline{C}_{21}}{dZ}\right)_{Z=0} = 0.$$

$$\overline{C}_{13} = \frac{\overline{C}_{11}}{1 + BK_1\overline{C}_{33}}, \quad \overline{C}_{23} = \frac{\overline{C}_{21}}{1 + K_2(1 - G\overline{C}_{33})}, \quad \overline{C}_{33} = \frac{\overline{C}_{23}}{G\overline{C}_{23} + BK_3\overline{C}_{13}},$$

$$(6.3.24)$$

where $K_1, K_2, K_3, A_i, \quad i = 1, 2, \quad B, G$ are presented in (3.3.20), (6.3.11).

The use of (6.3.11) and $C_{11}(R, Z), C_{21}(R, Z), C_{13}(R, Z), C_{23}(R, Z), C_{33}(R, Z)$, as a solution of (3.3.18), (3.3.19), and (3.3.21) for the case (3.3.22), permits to obtain the average concentrations $\overline{C}_{11}(Z), \overline{C}_{21}(Z), \overline{C}_{13}(Z), \overline{C}_{23}(Z), \overline{C}_{33}(Z)$ and the functions $A_i(Z), \quad i = 1, 2, \quad B(Z), \quad G(Z)$. They are presented in Figs. 6.22 and 6.23, where it is seen that the functions $A_i(Z), \quad i = 1, 2, \quad B(Z), \quad G(Z)$ are possible to be presented as linear approximations (6.3.12). The approximation ("theoretical") values of the parameters are presented in Table 6.5, where it is seen that $B \equiv 1, G \equiv 1$, practically.

For high columns $\left(0 = \varepsilon \leq 10^{-2}, \quad 0 = Pe_i^{-1} = \varepsilon Fo_{i1} \leq 10^{-2}, \quad Fo_{i1} \leq 1, \quad i = 1, 2\right)$, the problem (6.3.23) has the form (6.3.13).

Fig. 6.22 Average functions $\overline{C}(Z)$

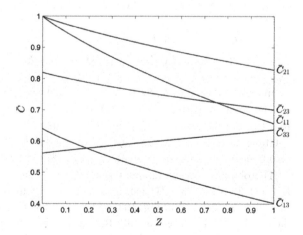

Fig. 6.23 Functions
$A_i(Z), i = 1, 2, B(Z), G(Z)$

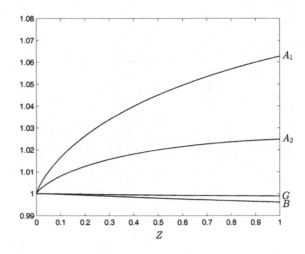

Table 6.5 Parameter's values

A_1	A_2	B	G
$a_{01} = 1.0143$	$a_{02} = 1.0078$	$b_0 = 1.0001$	$g_0 = 1.0000$
$a_{11} = 0.0544$	$a_{12} = 0.0204$	$b_1 = -0.0041$	$g_1 = -0.0012$

The solution of (6.3.13) depends on the two functions $(\overline{C}_{13}, \overline{C}_{23})$:

$$\overline{C}_{13} = \frac{\overline{C}_{11}}{1 + BK_1\overline{C}_{33}}, \quad \overline{C}_{23} = \frac{\overline{C}_{21}}{1 + K_2(1 - G\overline{C}_{33})}, \tag{6.3.25}$$

where $\overline{C}_{33}$ is the solution of the quadratic equation

$$\begin{aligned} &BG\big(\overline{C}_{21}K_1 - \overline{C}_{11}K_2K_3\big)\big(\overline{C}_{33}\big)^2 \\ &+ \big(G\overline{C}_{21} + B\overline{C}_{11}K_3 + B\overline{C}_{11}K_2K_3 - B\overline{C}_{21}K_1\big)\overline{C}_{33} - \overline{C}_{21} = 0. \end{aligned} \tag{6.3.26}$$

In order to solve (6.3.26), $0 \leq \overline{C}_{33} \leq 1$ has to be used.

The solution of (6.3.13), (6.3.25), (6.3.26) is obtained [8] as five-vector forms (6.3.16). For the case (3.3.22), Fig. 6.24 compares the functions $\overline{C}_{11}(Z), \overline{C}_{21}(Z)$ as solutions of (6.3.13), (6.3.25), (6.3.26) using Table 6.5 (the dotted lines) with the results of the solution of (3.3.18), (3.3.19), (3.3.21) using (6.3.11) (the lines).

The obtained concentrations $\overline{C}_{11}(Z), \overline{C}_{21}(Z)$ for the case (3.3.22) after solving (3.3.18), (3.3.19), (3.3.21) using (6.3.11) permit to obtain the "artificial experimental data" (6.3.17) for different values of Z. The obtained "artificial experimental data" (6.3.17) are used as illustration of the parameter identification in the average-concentration model (6.3.13), (6.3.25), (6.3.26) by minimization of the least-squares functions (6.3.18). The values of $\overline{C}_{11}\big(Z_n, a_{01}^n, a_{11}^n, a_{02}^n, a_{12}^n\big)$ and $\overline{C}_{21}\big(Z_n, a_{01}^n, a_{11}^n, a_{02}^n, a_{12}^n\big)$ are obtained for the case (3.3.22) as solutions of (6.3.13),

Fig. 6.24 Functions
$\overline{C}_{11}(Z), \overline{C}_{21}(Z)$: dotted lines
—solution of (6.3.13),
(6.3.25), (6.3.26) using
Table 6.5; solid lines—
solution of (3.3.18), (3.3.19),
(3.3.21) using (6.3.11)

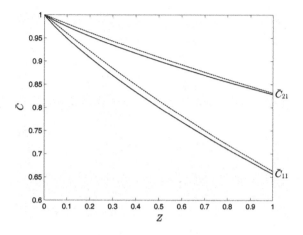

Table 6.6 Parameter's values

"Theoretical" values	"Experimental" values —Q min	"Experimental" values—Q_1 min	"Experimental" values—Q_2 min
$a_{01} = 1.0143$	$a_{01}^0 = 1.0000$	$a_{01}^1 = 0.9946$	$a_{01}^2 = 0.9978$
$a_{11} = 0.0544$	$a_{11}^0 = 0.0643$	$a_{11}^1 = 0.1007$	$a_{11}^2 = 0.0981$
$a_{02} = 1.0078$	$a_{02}^0 = 1.0000$	$a_{02}^1 = 0.9081$	$a_{02}^2 = 0.9159$
$a_{12} = 0.0204$	$a_{12}^0 = 0.0383$	$a_{12}^1 = 0.1024$	$a_{12}^2 = 0.0648$

(6.3.25), (6.3.26) for different $Z_n = 0.1n$, $n = 1, 2, \ldots, 10$. The obtained ("experimental") values of $a_{01}^0, a_{11}^0, a_{02}^0, a_{12}^0, a_{01}^1, a_{11}^1, a_{02}^1, a_{12}^1, a_{01}^2, a_{11}^2, a_{02}^2, a_{12}^2$ are presented in Table 6.6. They are used for the calculation of the functions $\overline{C}_{11}\left(Z, a_{01}^n, a_{11}^n, a_{02}^n, a_{12}^n\right)$, $\overline{C}_{21}\left(Z, a_{01}^n, a_{11}^n, a_{02}^n, a_{12}^n\right)$, $n = 0, 1, 2$ in the case (3.3.22) as solutions of (6.3.13), (6.3.25), (6.3.26) (the lines in Fig. 6.25), where the points are the "artificial experimental data" (6.3.17) (average values for every value of Z).

The comparison of the functions (lines) and experimental data (points) in Figs. 6.25 and 6.26 shows that the experimental data obtained from a column with a real radius and a small height $(Z = 0.1)$ are useful for the parameter's identifications.

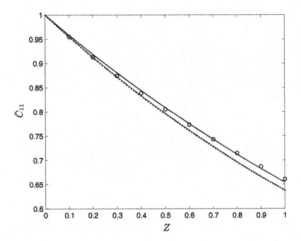

Fig. 6.25 Concentration distributions $\overline{C}_{11}\left(Z, a_{01}^n, a_{11}^n, a_{02}^n, a_{12}^n\right), n = 0, 1, 2$: lines—solutions of (6.3.13), (6.3.25), (6.3.26) in the case (3.3.22); points—the "artificial experimental data" (6.3.17) (average values for every Z)

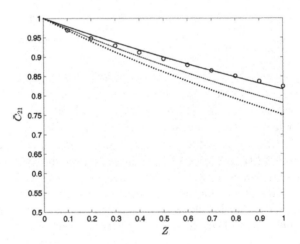

Fig. 6.26 Concentration distributions $\overline{C}_{21}\left(Z, a_{01}^n, a_{11}^n, a_{02}^n, a_{12}^n\right), n = 0, 1, 2$: lines—solutions of (6.3.13), (6.3.25), (6.3.26) in the case (3.3.22); points—the "artificial experimental data" (6.3.17) (average values for every Z)

6.4 Examples

6.4.1 Airlift Reactor Modeling

The airlift reactor modeling is possible to be made on the base of the average-concentration-type model [9], if the average velocities and concentrations are introduced in the model Eqs. (3.2.4)–(3.4.8):

$$
\bar{u}_1 = \frac{2}{r_0^2} \int_0^{r_0} r u_1(r) dr, \quad \bar{u}_2 = \frac{2}{r_0^2} \int_0^{r_0} r u_2(r) dr, \quad \bar{\hat{u}}_1 = \frac{2}{r_0^2} \int_0^{r_0} r \hat{u}_1(r) dr,
$$

$$
\bar{\hat{u}}_2 = \frac{2}{r_0^2} \int_0^{r_0} r \hat{u}_2(r) dr, \quad \bar{c}_{11}(t, z_1) = \frac{2}{r_0^2} \int_0^{r_0} r c_{11}(t, r, z_1) dr,
$$

$$
\bar{c}_{12}(t, z_1) = \frac{2}{r_0^2} \int_0^{r_0} r c_{12}(t, r, z_1) dr, \quad \bar{c}_{22}(t, z_1) = \frac{2}{r_0^2} \int_0^{r_0} r c_{22}(t, r, z_1) dr, \tag{6.4.1}
$$

$$
\bar{\hat{c}}_{12}(t, z_2) = \frac{2}{r_0^2} \int_0^{r_0} r \hat{c}_{12}(t, r, z_2) dr, \quad \bar{\hat{c}}_{22}(t, z_2) = \frac{2}{r_0^2} \int_0^{r_0} r \hat{c}_{22}(t, r, z_2) dr,
$$

using the expressions:

$$
\begin{aligned}
&u_1(r) = \bar{u}_1 \tilde{u}_1(r), \quad u_2(r) = \bar{u}_2 \tilde{u}_2(r), \quad \hat{u}_1(r) = \bar{\hat{u}}_1 \tilde{\hat{u}}_1(r), \\
&\hat{u}_2(r) = \bar{\hat{u}}_2 \tilde{\hat{u}}_2(r), \quad c_{11}(t, r, z) = \bar{c}_{11}(t, z) \tilde{c}_{11}(t, r, z), \\
&c_{12}(t, r, z) = \bar{c}_{12}(t, z) \tilde{c}_{12}(t, r, z), \quad c_{22}(t, r, z) = \bar{c}_{22}(t, z) \tilde{c}_{22}(t, r, z), \\
&\hat{c}_{12}(t, r, z) = \bar{\hat{c}}_{12}(t, z) \tilde{\hat{c}}_{12}(t, r, z), \quad \hat{c}_{22}(t, r, z) = \bar{\hat{c}}_{22}(t, z) \tilde{\hat{c}}_{22}(t, r, z).
\end{aligned} \tag{6.4.2}
$$

Applying procedure (6.0.1)–(6.0.6) for averaging Eqs. (3.2.4)–(3.4.8) in the cross section of the column leads to:

$$
\frac{\partial \bar{c}_{11}}{\partial t} + \alpha_{11} \bar{u}_1 \frac{\partial \bar{c}_{11}}{\partial z_1} + \frac{\partial \alpha_{11}}{\partial z_1} \bar{u}_1 \bar{c}_{11} = D_{11} \frac{\partial^2 \bar{c}_{11}}{\partial z_1^2} - k_0 (\bar{c}_{11} - \chi \bar{c}_{12}),
$$

$$
\frac{\partial \bar{c}_{12}}{\partial t} + \alpha_{12} \bar{u}_2 \frac{\partial \bar{c}_{12}}{\partial z_1} + \frac{\partial \alpha_{12}}{\partial z_1} \bar{u}_2 \bar{c}_{12} = D_{12} \frac{\partial^2 \bar{c}_{12}}{\partial z_1^2} + k_0 (\bar{c}_{11} - \chi \bar{c}_{12})
$$

$$
- k\beta \bar{c}_{12}^{\alpha_1} \bar{c}_{22}^{\alpha_2}, \tag{6.4.3}
$$

$$
\frac{\partial \bar{c}_{22}}{\partial t} + \alpha_{22} \bar{u}_2 \frac{\partial \bar{c}_{22}}{\partial z_1} + \frac{\partial \alpha_{22}}{\partial z_1} \bar{u}_2 \bar{c}_{22} = D_{22} \frac{\partial^2 \bar{c}_{22}}{\partial z_1^2} - k\beta \bar{c}_{12}^{\alpha_1} \bar{c}_{22}^{\alpha_2}.
$$

$$\frac{\partial \bar{c}_{12}}{\partial t} + \hat{\alpha}_{12}\bar{u}_2 \frac{\partial \bar{c}_{12}}{\partial z_2} + \frac{\partial \hat{\alpha}_{12}}{\partial z_2}\bar{u}_2\bar{c}_{12} = D_{12}\frac{\partial^2 \bar{c}_{12}}{\partial z_2^2} - k\beta\bar{c}_{12}^{\bar{\alpha}_1}\bar{c}_{22}^{\bar{\alpha}_2},$$

$$\frac{\partial \bar{c}_{22}}{\partial t} + \hat{\alpha}_{22}\bar{u}_2 \frac{\partial \bar{c}_{22}}{\partial z_2} + \frac{\partial \hat{\alpha}_{22}}{\partial z_2}\bar{u}_2\bar{c}_{22} = D_{22}\frac{\partial^2 \bar{c}_{22}}{\partial z_2^2} - k\hat{\beta}\bar{c}_{12}^{\bar{\alpha}_1}\bar{c}_{22}^{\bar{\alpha}_2}.$$

$$(6.4.4)$$

The initial conditions of (6.4.3), (6.4.4) will be formulated for the case, when at $t = 0$ the process starts with the beginning of the gas motion:

$$t = 0, \quad \bar{c}_{11} \equiv c_{11}^0, \quad \bar{c}_{12} \equiv 0, \quad \bar{c}_{22} \equiv c_{22}^0, \quad \bar{\hat{c}}_{12} \equiv 0, \quad \bar{\hat{c}}_{22} \equiv c_{22}^0, \quad (6.4.5)$$

where c_{11}^0 and c_{22}^0 are the initial concentrations of the reagents in the two phases.

The boundary conditions are equalities of the concentrations and mass fluxes at the two ends of the working zones—$z_1 = 0(z_2 = l)$ and $z_1 = l(z_2 = 0)$. The boundary conditions for $\bar{c}_{11}, \bar{c}_{22}, \bar{c}_{12}$ in the riser are:

$$z_1 = 0, \quad \bar{c}_{11} \equiv c_{11}^0, \quad \left(\frac{\partial \bar{c}_{11}}{\partial z_1}\right)_{z_1=0} \equiv 0,$$

$$\bar{c}_{22} \equiv c_{22}^0, \quad \left(\frac{\partial \bar{c}_{22}}{\partial z_1}\right)_{z_1=0} \equiv 0, \quad \bar{c}_{12} \equiv c_{12}^0, \quad \left(\frac{\partial \bar{c}_{12}}{\partial z_1}\right)_{z_1=0} \equiv 0.$$

$$(6.4.6)$$

The boundary conditions for $\bar{\hat{c}}_{12}, \bar{\hat{c}}_{22}$ in the downcomer are:

$$z_2 = 0, \quad \bar{\hat{c}}_{12} \equiv \hat{c}_{12}^0, \quad \left(\frac{\partial \bar{\hat{c}}_{12}}{\partial z_2}\right)_{z_2=0} \equiv 0, \quad \bar{\hat{c}}_{22} \equiv \hat{c}_{22}^0, \quad \left(\frac{\partial \bar{\hat{c}}_{22}}{\partial z_2}\right)_{z_2=0} \equiv 0. \quad (6.4.7)$$

In (6.4.3), (6.4.4), the functions

$$\alpha_{11}(t, z_1) = \frac{2}{r_0^2} \int_0^{r_0} r\tilde{u}_1(r)\tilde{c}_{11}(t, r, z_1)dr,$$

$$\alpha_{12}(t, z_1) = \frac{2}{r_0^2} \int_0^{r_0} r\tilde{u}_1(r)\tilde{c}_{12}(t, r, z_1)dr,$$

$$\alpha_{22}(t, z_1) = \frac{2}{r_0^2} \int_0^{r_0} r\tilde{u}_1(r)\tilde{c}_{22}(t, r, z_1)dr,$$

$$\hat{\alpha}_{12}(t, z_2) = \frac{2}{r_0^2} \int_0^{r_0} r\tilde{u}_2(r)\hat{\tilde{c}}_{12}(t, r, z_2)dr,$$

$$\hat{\alpha}_{22}(t, z_2) = \frac{2}{r_0^2} \int_0^{r_0} r\tilde{u}(r)\tilde{\hat{c}}_{22}(t, r, z_2)dr,$$

$$\beta(t, z_1) = \frac{2}{r_0^2} \int_0^{r_0} r\tilde{c}_{12}(t, r, z_1)\tilde{c}_{22}(t, r, z_1)dr, \qquad (6.4.8)$$

$$\hat{\beta}(t, z_2) = \frac{2}{r_0^2} \int_0^{r_0} r\tilde{c}_{12}(t, r, z_2)\tilde{\hat{c}}_{22}(t, r, z_2)dr,$$

have to be obtained using experimental data.

The holdup coefficient of the liquid phase ε_2 is possible to be obtained using

$$\varepsilon_2 = \frac{F_2 l_0}{l(F_1 + F_2)}, \qquad (6.4.9)$$

where l and l_0 are the liquid levels in the riser with and without gas motion, and F_1 and F_2 are the gas and liquid flow rates.

6.4.2 Moisture Adsorption Modeling

The presented theoretical analysis (3.4.32)–(3.4.36) shows that in the practical cases the model of a non-stationary physical adsorption process (for moisture absorption in new composite sorbent bed column [10, 11]) has a convective form:

$$u\frac{dc_{11}}{dz} = -k_0(c_{11} - c_{13}); \quad z = 0, \quad c_{11} \equiv c_{11}^0. \qquad (6.4.10)$$

$$\frac{dc_{13}}{dt} = k_0(c_{11} - c_{13}) - b_0 k_1 c_{13} \frac{c_{23}}{c_{23}^0} + k_2 c_{23}^0 \left(1 - \frac{c_{23}}{c_{23}^0}\right); \quad t = 0, \quad c_{13} \equiv 0.$$

$$(6.4.11)$$

$$\frac{dc_{23}}{dt} = -b_0 k_1 c \frac{c_{23}}{c_{23}^0} + k_2 c_{23}^0 \left(1 - \frac{c_{23}}{c_{23}^0}\right); \quad t = 0, \quad c_{23} \equiv c_{23}^0. \qquad (6.4.12)$$

The average values of the velocity and concentration in the column's cross-sectional area are possible to be obtained [1–3] using the expressions:

$$\bar{u}_1 = \frac{2}{r_0^2} \int_0^{r_0} r u_1(r) dr = u_1^0, \quad \bar{c}_{11}(t, z) = \frac{2}{r_0^2} \int_0^{r_0} r c_{11}(t, z, r) dr,$$

$$\bar{c}_{13}(t, z) = \frac{2}{r_0^2} \int_0^{r_0} r c_{13}(t, z, r) dr, \quad \bar{c}_{23}(t, z) = \frac{2}{r_0^2} \int_0^{r_0} r c_{23}(t, z, r) dr. \tag{6.4.13}$$

The velocity and concentration distributions in (6.4.10), (6.4.11), (6.4.12) can be represented by the average functions (6.4.13):

$$u_1(r) = \bar{u}_1 \tilde{u}_1(r), \quad c_{11}(t, z, r) = \bar{c}_{11}(t, z) \tilde{c}_{11}(t, r, z),$$

$$c_{13}(t, z, r) = \bar{c}_{13}(t, z) \tilde{c}_{13}(t, r, z), \quad c_{23}(t, z, r) = \bar{c}_{23}(t, z) \tilde{c}_{23}(t, r, z). \tag{6.4.14}$$

Here, $\tilde{u}(r)$, $\tilde{c}_1(t, r, z)$, $\tilde{c}(t, r, z)$, and $\tilde{c}_0(t, r, z)$ represent the radial non-uniformity of the velocity and the concentration distributions satisfying the conditions:

$$\frac{2}{r_0^2} \int_0^{r_0} r \tilde{u}_1(r) dr = 1, \quad \frac{2}{r_0^2} \int_0^{r_0} r \tilde{c}_{11}(t, r, z) dr = 1,$$

$$\frac{2}{r_0^2} \int_0^{r_0} r \tilde{c}_{13}(t, r, z) dr = 1, \quad \frac{2}{r_0^2} \int_0^{r_0} r \tilde{c}_{23}(t, r, z) dr = 1. \tag{6.4.15}$$

The use of the averaging procedure (6.0.1)–(6.0.5) leads to the average-concentration model of the moisture absorption:

$$\alpha \bar{u}_1 \frac{d\bar{c}_{11}}{dz} + \frac{d\alpha}{dz} \bar{u}_1 \bar{c}_{11} = -k_0(\bar{c}_{11} - \bar{c}_{13}); \quad z = 0, \quad \bar{c}_{11} \equiv c_{11}^0. \tag{6.4.16}$$

$$\frac{d\bar{c}_{13}}{dt} = k_0(\bar{c}_{11} - \bar{c}_{13}) - b_0 k_1 \beta \bar{c}_{13} \frac{\bar{c}_{23}}{c_{23}^0} + k_2 c_{23}^0 \left(1 - \frac{\bar{c}_{23}}{c_{23}^0}\right); \quad t = 0, \quad \bar{c}_{13} \equiv 0. \tag{6.4.17}$$

$$\frac{d\bar{c}_{23}}{dt} = -b_0 k_1 \beta \bar{c}_{13} \frac{\bar{c}_{23}}{c_{23}^0} + k_2 c_{23}^0 \left(1 - \frac{\bar{c}_{23}}{c_{23}^0}\right); \quad t = 0, \quad \bar{c}_{23} \equiv c_{23}^0, \tag{6.4.18}$$

where

$$\alpha = \alpha(t,z) = \frac{2}{r_0^2} \int_0^{r_o} r\tilde{u}_1(r)\tilde{c}_{11}(t,r,z)dr,$$

$$\beta = \beta(t,z) = \frac{2}{r_0^2} \int_0^{r_o} r\tilde{c}_{13}(t,r,z)\tilde{c}_{23}(t,r,z)dr. \qquad (6.4.19)$$

The use of the generalized variables

$$T = \frac{t}{t_0}, \quad R = \frac{r}{r_0}, \quad Z = \frac{z}{h}, \quad \overline{C}_1 = \frac{\tilde{c}_{11}}{c_{11}^0}, \quad \overline{C} = \frac{\tilde{c}_{13}}{c_{11}^0}, \quad \overline{C}_0 = \frac{\tilde{c}_{23}}{c_{23}^0},$$

$$\alpha(t,z) = \alpha(t_0 T, hZ) = A(T,Z), \quad \beta(t,z) = \beta(t_0 T, hZ) = B(T,Z), \qquad (6.4.20)$$

leads to

$$A\frac{d\overline{C}_1}{dZ} + \frac{dA}{dZ}\overline{C}_1 = -K_0(\overline{C}_1 - \overline{C}); \quad Z = 0, \quad \overline{C}_1 \equiv 1. \qquad (6.4.21)$$

$$\frac{d\overline{C}}{dT} = K_1(\overline{C}_1 - \overline{C}) - K_2 B\overline{C}\,\overline{C}_0 + K_3(1 - \overline{C}_0); \quad T = 0, \quad \overline{C} \equiv 0. \qquad (6.4.22)$$

$$\frac{d\overline{C}_0}{dT} = -K_4 B\overline{C}\,\overline{C}_0 + K_5(1 - \overline{C}_0); \quad T = 0, \quad \overline{C}_0 \equiv 1, \qquad (6.4.23)$$

where

$$K_0 = \frac{k_0 l}{\bar{u}_1}, \quad K_1 = k_0 t^0, \quad K_2 = k_1 t^0 b_0,$$

$$K_3 = k_2 t^0 \frac{c_{23}^0}{c_{11}^0}, \quad K_4 = k_1 t^0 b_0 \frac{c_{11}^0}{c_{23}^0}, \quad K_5 = k_2 t^0. \qquad (6.4.24)$$

For the functions $A(Z)$ and $B(Z)$, it is possible to use [4] the approximations

$$A(T,Z) = 1 + aZ + a_t T, \quad B(T,Z) = 1 + bZ + b_t T \qquad (6.4.25)$$

where the average concentration $\overline{C}(Z)$ obtained from experimental data has to be used for the determination of the parameters a, b.

At the end, the model equations (6.4.21), (6.4.22), (6.4.23) assume the form:

$$(1 + aZ + a_t T)\frac{d\overline{C}_1}{dZ} + a\overline{C}_1 = -K_0(\overline{C}_1 - \overline{C}); \quad Z = 0, \quad \overline{C}_1 \equiv 1. \qquad (6.4.26)$$

$$\frac{d\overline{C}}{dT} = K_1(\overline{C}_1 - \overline{C}) - K_2(1 + bZ + b_t T)\overline{C}\,\overline{C}_0 + K_3(1 - \overline{C}_0);$$
$$T = 0, \quad \overline{C} \equiv 0. \tag{6.4.27}$$

$$\frac{d\overline{C}_0}{dT} = -K_4(1 + bZ + b_t T)\overline{C}\,\overline{C}_0 + K_5(1 - \overline{C}_0); \quad T = 0, \quad \overline{C}_0 \equiv 1, \tag{6.4.28}$$

where T is a parameter in (6.4.26), while Z is a parameter in (6.4.27), (6.4.28).

The solution of the model equations (6.4.26), (6.4.27), (6.4.28) is possible to be obtained as three matrix forms

$$\overline{C}_1(T, Z) = \left\| C_{(1)\tau\zeta} \right\|, \quad \overline{C}(T, Z) = \left\| \overline{C}_{\tau\zeta} \right\|, \quad \overline{C}_0(T, Z) = \left\| \overline{C}_{(0)\tau\zeta} \right\|;$$
$$T = 0.01\tau, \quad \tau = 1, 2, \ldots, 100; \quad Z = 0.01\zeta, \quad \zeta = 1, 2, \ldots, 100; \tag{6.4.29}$$

using a multistep algorithm (see Chap. 9 and [7]).

The parameters $a, b, K_0, K_1, K_2, K_3, K_4, K_5$ in the model equations (6.4.26), (6.4.27), (6.4.28), where

$$K_1 = \frac{u_1^0}{l} K_0 = 1.565 K_0, \quad K_4 = \frac{c_{11}^0}{c_{23}^0} K_2 = 10^{-2} K_2,$$
$$K_5 = \frac{c_{11}^0}{c_{23}^0} K_3 = 10^{-2} K_3, \tag{6.4.30}$$

can be obtained using the following algorithm:

1. Minimization of the function:

$$F_1(K_0, K_2, K_3) = \left(\overline{C}_{(1)100,100} - 1\right)^2 + \left(\overline{C}_{(0)100,100}\right)^2 \tag{6.4.31}$$

   after solving (6.4.27), (6.4.28), (6.4.29) for $a = b = 0$. The obtained values K_0, K_2, K_3 have to be used for minimization of the function F_2 (6.4.32).

2. Minimization of the function:

$$F_2(a, b) = \sum_{\tau_0=1}^{10} \left(\overline{C}_{(1)\tau_0 100} - C_{(1)\tau_0 100}^{exp}\right)^2, \quad \tau_0 = 0.1\tau, \tag{6.4.32}$$

   where $C_{(1)\tau_0 100}^{exp}$ are experimental data.

3. The obtained parameter values a, b have to be used for minimization of the function $F_1(K_0, K_2, K_3)$ in (6.4.31), etc.

6.4.3 Three-Phase Process Modeling

The theoretical procedure (II.5–II.15) presented in Part II is possible to be used for the creation of three-phase average-concentration models in the cases of two-phase absorbent processes and absorption–adsorption processes.

The theoretical analysis of the three-phase processes provided in Chap. 4 shows that in the cases of $CaCO_3/H_2O$ absorbent and $Ca(OH)_2/H_2O$ absorbent the processes are physical (4.1.4) and chemical (4.1.7) absorption and the average-concentration models are (6.1.2) and (6.1.17), respectively.

In the case of an absorption–adsorption process, an intensive liquid-adsorbent particle-gas bubble flow takes place and an ideal mixing regime is created in the liquid–solid phase in the column, i.e., all concentrations are averaged over the cross-sectional area of the column. As a result, the models (4.2.3)–(4.2.6) and (4.2.18)–(4.2.21) are average-concentration models.

References

1. Boyadjiev C (2010) Theoretical chemical engineering. Modeling and simulation. Springer, Berlin
2. Boyadjiev C (2006) Diffusion models and scale-up. Int J Heat Mass Transf 49:796–799
3. Boyadjiev C (2013) A new approach for the column apparatuses modeling in chemical engineering. J Pure Appl Math: Adv Appl 10(2):131–150
4. Doichinova M, Boyadjiev C (2012) On the column apparatuses modeling. Int J Heat Mass Transf 55:6705–6715
5. Boyadjiev B, Doichinova M, Boyadjiev C (2015) Computer modeling of column apparatuses. 1. Two coordinate systems approach. J Eng Thermophys 24(3):247–258
6. Boyadjiev C, Boyadjiev B, Popova-Krumova P, Doichinova M (2015) An innovative approach for adsorption column modeling. Chem Eng Technol 38(4):675–682
7. Boyadjiev B, Doichinova M, Boyadjiev C (2015) Computer modeling of column apparatuses. 2. Multi-steps modeling approach. J Eng Thermophys 24(4):362–370
8. Boyadjiev B, Boyadjiev C (2015) A new approach for the catalytic processes modeling in columns apparatuses. Int J Mod Trends Eng Res 2(8):152–167
9. Boyadjiev C (2006) On the modeling of an airlift reactor. Int J Heat Mass Transf 49:2053–2057
10. Aristov Y, Mezentsev I, Mukhin V (2006) New approach to regenerate heat and moisture in a ventilation system: 2. Prototype of real unit. J Eng Thermophys 79:151–157
11. Aristov Y, Mezentsev I, Mukhin V, Boyadjiev C, Doichinova M, Popova P (2006) New approach to regenerate heat and moisture in a ventilation system: experiment. In: Proceedings of 11th workshop on "transport phenomena in two-phase flow", Bulgaria, pp 77–85

Part III
Computer Calculation Problems

Calculation Algorithms

In many cases, the computer modeling of the processes in column apparatuses, made on the base of the new approach using the convection–diffusion type model and average-concentration type model, does not allow a direct use of the MATLAB program. In these cases, it is necessary to create combinations of appropriate algorithms.

Practically, the new type models are characterized by the presence of small parameters at the highest derivates. As a result, the use of the conventional software for solving the model differential equations is difficult. This difficulty may be eliminated by an appropriate combination of MATLAB and perturbation method [1].

In the cases of countercurrent gas–liquid or liquid–liquid processes, the mass transfer process models are presented in two-coordinate systems, because in a one-coordinate system one of the equations has no solution by reason of the negative value in the equation Laplacian. Thus, a combination of an iterative algorithm and MATLAB has to be used for solving the equations set in different coordinate systems [2].

In the practical cases of non-stationary adsorption in gas–solid systems, the presence of mobile (gas) and immobile (solid) phases in the conditions of lengthy (long-term) processes leads to a non-stationary process in the immobile phase and a stationary process in the mobile phase. As a result, different coordinate systems must be used for the gas- and solid-phase models. A combination of a multi-step algorithm and MATLAB has to be used for the solutions of the equations set in different coordinate systems [3].

References

1. Boyadjiev B, Doichinova M, Boyadjiev Chr (2015) Computer modeling of column apparatuses. 3. Perturbations method approach. J Eng Thermophys 24(4):371–380
2. Boyadjiev B, Doichinova M, Boyadjiev Chr (2015) Computer modeling of column apparatuses. 1. Two coordinates systems approach. J Eng Thermophys 24(3):247–258
3. Boyadjiev B, Doichinova M, Boyadjiev Chr (2015) Computer modeling of column apparatuses. 2. Multi-step modeling approach. J Eng Thermophys 24(4):362–370

Chapter 7
Perturbation Method Approach

A new approach for the column apparatuses modeling uses convection–diffusion-type models and average-concentration models. All these new types of models [1–3] are characterized by the presence of small parameters at the highest derivatives. As a result, the model equations have no exact solutions and approximate (asymptotic) solutions have to be obtained [4–6]. In these cases, the use of the conventional software (MATLAB) for solving the model differential equations is difficult and this difficulty may be eliminated by an appropriate combination with the perturbation method.

7.1 Perturbation Method

Let ε is a small parameter and $y = \varphi(t, \varepsilon)$ is the solution of the ordinary differential equation [4, 5]

$$y' = F(y, \varepsilon) \tag{7.1.1}$$

in the finite interval

$$t_0 \leq t \leq T, \quad 0 \leq \varepsilon \leq \varepsilon_0, \tag{7.1.2}$$

where ε_0 is a small numeral. The exact solution of (7.1.1) is possible to be presented (like Taylor series expansion) as a power series expansion with respect to the small parameter ε:

$$\varphi(t, \varepsilon) = \sum_{s=0}^{\infty} \varepsilon^s \varphi_s(t), \tag{7.1.3}$$

© Springer International Publishing AG, part of Springer Nature 2018

C. Boyadjiev et al., *Modeling of Column Apparatus Processes*,

Heat and Mass Transfer, https://doi.org/10.1007/978-3-319-89966-4_7

where $\varphi_0(t)$ is the solution of the ordinary differential equation

$$y' = F(y, 0). \qquad (7.1.4)$$

The exact solution (7.1.3) is valid [3, 4] in the finite interval (7.1.2), only.
In the case of existence of small parameters at the highest derivate

$$\varepsilon y' = f(y, z), \quad z' = g(y, z), \qquad (7.1.5)$$

a new variable $\theta = t/\varepsilon$ has to be used:

$$y' = \frac{dy}{d\theta}\frac{1}{\varepsilon}, \quad z' = \frac{dz}{d\theta}\frac{1}{\varepsilon}, \quad \frac{dy}{d\theta} = f(y, z, \varepsilon), \quad \frac{dz}{d\theta} = \varepsilon g(y, z, \varepsilon), \qquad (7.1.6)$$

but these equations set has no exact solution

$$\varphi(\theta, \varepsilon) = \sum_{s=0}^{\infty} \varepsilon^s \varphi_s(\theta), \quad \gamma(\theta, \varepsilon) = \sum_{s=0}^{\infty} \varepsilon^s \gamma_s(\theta), \qquad (7.1.7)$$

because

$$\frac{t_0}{\varepsilon} \le \theta \le \frac{T}{\varepsilon} \to \infty, \quad \varepsilon \to 0, \qquad (7.1.8)$$

i.e., the interval (7.1.8) is not finite [3, 4].

In the case of (7.1.5), an approximate solution $y = \bar{\varphi}(t, \varepsilon)$ has to be sought if

$$|\varphi(t, \varepsilon) - \bar{\varphi}(t, \varepsilon)| \le \delta, \qquad (7.1.9)$$

where practically $\delta \sim 10^{-2}$ because the relative error in the experimental measurements are typically more than 1% (all mathematical operators which represent very small $\le 10^{-2}$ physical effects must be neglected, because they are not possible to be measured experimentally). This asymptotic solution is possible to be presented (like Taylor series expansion) as a power series expansion with respect to the small parameter ε:

$$\bar{\varphi}(t, \varepsilon) = \sum_{s=0}^{s_0} \varepsilon^s \varphi_s(t), \quad s_0 = s_0(\varepsilon, \delta). \qquad (7.1.10)$$

Let us consider the function $y = \varphi(t, \varepsilon)$ in the interval $0 \le t \le 1$ as a solution of the differential equation

$$\varepsilon y'' = y' + y, \quad y(0) = 1, \quad y'(0) = 0. \qquad (7.1.11)$$

An approximate (asymptotic) solution $y = \bar{\varphi}(t, \varepsilon)$ of (7.1.11) is possible to be presented as

$$\bar{\varphi}(t, \varepsilon) = \bar{\varphi}_0(t) + \varepsilon \bar{\varphi}_1(t) + \varepsilon^2 \bar{\varphi}_2(t). \tag{7.1.12}$$

The introduction of (7.1.12) in (7.1.11) and grouping of members with the same power of ε and their equalization to zero leads to individual differential equations for the functions in (7.1.12):

$$\begin{aligned}
\bar{\varphi}_0' + \bar{\varphi}_0 &= 0, & \bar{\varphi}_0 &= 1; \\
\bar{\varphi}_1'' + \bar{\varphi}_1 &= \bar{\varphi}_0'', & \bar{\varphi}_1 &= 0; \\
\bar{\varphi}_2'' + \bar{\varphi}_2 &= \bar{\varphi}_1'', & \bar{\varphi}_2 &= 0.
\end{aligned} \tag{7.1.13}$$

7.2   Convection–Diffusion-Type Models

Let us consider a model of the column apparatuses with pseudo-first-order chemical reaction (2.1.27), where the fluid flow is of Poiseuille type:

$$\begin{aligned}
(2 - 2R^2)\frac{\partial C}{\partial Z} &= Fo\left(\varepsilon\frac{\partial^2 C}{\partial Z^2} + \frac{1}{R}\frac{\partial C}{\partial R} + \frac{\partial^2 C}{\partial R^2}\right) - Da\,C; \\
R = 0, \quad \frac{\partial C}{\partial R} &= 0; \quad R = 1, \quad \frac{\partial C}{\partial R} = 0; \\
Z = 0, \quad C &= 1, \quad 1 = U - Pe^{-1}\frac{\partial C}{\partial Z}.
\end{aligned} \tag{7.2.1}$$

The convection–diffusion-type model (7.2.1) is of elliptical type. In the case of a short column, ε is a small parameter and the perturbation method [4–6] can be used, i.e., the substitution of an elliptical equation by a set of parabolic equations. A computer realization of this method will be presented as an example of the chemical reactor column modeling [2, 7].

7.2.1   Short Columns Model

For short columns, ε is a small parameter and if $\varepsilon < 0.3$ the problem (7.2.1) is possible to be solved using the following approximation of the perturbation method [6]

$$C(R, Z) = C^{(0)}(R, Z) + \varepsilon C^{(1)}(R, Z) + \varepsilon^2 C^{(2)}(R, Z) + \varepsilon^3 C^{(3)}(R, Z) \tag{7.2.2}$$

where $C^{(0)}, C^{(1)}$ and $C^{(2)}$ are solutions of the next problems:

$$\left(2 - 2R^2\right)\frac{\partial C^{(0)}}{\partial Z} = Fo\left(\frac{1}{R}\frac{\partial C^{(0)}}{\partial R} + \frac{\partial^2 C^{(0)}}{\partial R^2}\right) - Da\, C^{(0)};$$

$$R = 0, \quad \frac{\partial C^{(0)}}{\partial R} = 0; \quad R = 1, \quad \frac{\partial C^{(0)}}{\partial R} = 0; \quad Z = 0, \quad C^{(0)} = 1.$$

(7.2.3)

$$\left(2 - 2R^2\right)\frac{\partial C^{(s)}}{\partial Z} = Fo\left(\frac{1}{R}\frac{\partial C^{(s)}}{\partial R} + \frac{\partial^2 C^{(s)}}{\partial R^2}\right) - Da\, C^{(s)} + Fo\frac{\partial^2 C^{(s-1)}}{\partial Z^2};$$

$$R = 0, \quad \frac{\partial C^{(s)}}{\partial R} = 0; \quad R = 1, \quad \frac{\partial C^{(s)}}{\partial R} = 0; \quad Z = 0, \quad C^{(s)} = 0, \quad s = 1, \dots, 3.$$

(7.2.4)

In (7.2.2), the individual effects (mathematical operators) and their relative role (influence) in the overall process (model) must be greater than 10^{-2} ($\varepsilon^4 < 10^{-2}$), because the accuracy of the experimental measurements is greater than 1%.

7.2.2 Calculation Problem

The numerical solution of the equations set (7.2.3), (7.2.4) is possible if MATLAB and a four-step procedure are used, the functions $C^{(s)}(R, Z)$, $s = 0, 1, 2, 3$ being obtained in four matrix forms:

$$C^{(s)}(R, Z) = \left\|a_{\rho\zeta}^s\right\|, \quad s = 0, 1, 2, 3, \quad \rho = 1, 2, \dots, \rho^0, \quad \zeta = 1, 2, \dots, \zeta^0,$$

$$0 \le R \le 1, \quad 0 \le Z \le 1, \quad R = \frac{\rho - 1}{\rho^0 - 1}, \quad Z = \frac{\zeta - 1}{\zeta^0 - 1}, \quad \rho^0 = \zeta^0.$$

(7.2.5)

The first step is the solution of (7.2.3), i.e., elements calculation of the matrix:

$$C^{(0)}(R, Z) = \left\|a_{\rho\zeta}^0\right\|, \quad \rho = 1, 2, \dots, \rho^0, \quad \zeta = 1, 2, \dots, \zeta^0,$$

$$0 \le R \le 1, \quad 0 \le Z \le 1, \quad R = \frac{\rho - 1}{\rho^0 - 1}, \quad Z = \frac{\zeta - 1}{\zeta^0 - 1}, \quad \rho^0 = \zeta^0.$$

(7.2.6)

The next step is a polynomial approximation of the function $C^{(0)}(R, Z)$:

$$C^{(0)}(R, Z) = \left\|a_{\rho\zeta}^0\right\| = \left|\alpha_{0\rho}^0\right| + \left|\alpha_{1\rho}^0\right|Z + \left|\alpha_{2\rho}^0\right|Z^2 + \left|\alpha_{3\rho}^0\right|Z^3 + \left|\alpha_{4\rho}^0\right|Z^4,$$

$$\rho = 1, 2, \dots, \rho^0, \quad 0 \le R \le 1, \quad R = \frac{\rho - 1}{\rho^0 - 1}$$

(7.2.7)

and the determination of the second derivative

$$\frac{\partial^2 C^{(0)}}{\partial Z^2} = \left\| g^0_{\rho\zeta} \right\| = 2\left|\alpha^0_{2\rho}\right| + 6\left|\alpha^0_{3\rho}\right|Z + 12\left|\alpha^0_{4\rho}\right|Z^2,$$

$$\rho = 1, 2, \ldots, \rho^0, \quad 0 \leq R \leq 1, \quad R = \frac{\rho - 1}{\rho^0 - 1}. \tag{7.2.8}$$

The next step is the solution of (7.2.4) for $s = 1$ using (7.2.8), i.e., elements calculation of the matrix:

$$C^{(1)}(R, Z) = \left\| a^1_{\rho\zeta} \right\|, \quad \rho = 1, 2, \ldots, \rho^0, \quad \zeta = 1, 2, \ldots, \zeta^0,$$

$$0 \leq R \leq 1, \quad 0 \leq Z \leq 1, \quad R = \frac{\rho - 1}{\rho^0 - 1}, \quad Z = \frac{\zeta - 1}{\zeta^0 - 1}, \quad \rho^0 = \zeta^0. \tag{7.2.9}$$

Then follows the polynomial approximation of the function $C^{(1)}(R, Z)$

$$C^{(1)}(R, Z) = \left\| a^1_{\rho\zeta} \right\| = \left|\alpha^1_{0\rho}\right| + \left|\alpha^1_{1\rho}\right|Z + \left|\alpha^1_{2\rho}\right|Z^2 + \left|\alpha^1_{3\rho}\right|Z^3 + \left|\alpha^1_{4\rho}\right|Z^4,$$

$$\rho = 1, 2, \ldots, \rho^0, \quad 0 \leq R \leq 1, \quad R = \frac{\rho - 1}{\rho^0 - 1} \tag{7.2.10}$$

and the determination of the second derivative

$$\frac{\partial^2 C^{(1)}}{\partial Z^2} = \left\| g^1_{\rho\zeta} \right\| = 2\left|\alpha^1_{2\rho}\right| + 6\left|\alpha^1_{3\rho}\right|Z + 12\left|\alpha^1_{4\rho}\right|Z^2,$$

$$\rho = 1, 2, \ldots, \rho^0, \quad 0 \leq R \leq 1, \quad R = \frac{\rho - 1}{\rho^0 - 1}. \tag{7.2.11}$$

The next step is the solution of (7.2.4) for $s = 2$ using (7.2.8), i.e., elements calculation of the matrix:

$$C^{(2)}(R, Z) = \left\| a^2_{\rho\zeta} \right\|, \quad \rho = 1, 2, \ldots, \rho^0, \quad \zeta = 1, 2, \ldots, \zeta^0,$$

$$0 \leq R \leq 1, \quad 0 \leq Z \leq 1, \quad R = \frac{\rho - 1}{\rho^0 - 1}, \quad Z = \frac{\zeta - 1}{\zeta^0 - 1}, \quad \rho^0 = \zeta^0. \tag{7.2.12}$$

The next step is the polynomial approximation of the function $C^{(2)}(R, Z)$:

$$C^{(2)}(R, Z) = \left\| a^2_{\rho\zeta} \right\| = \left|\alpha^2_{0\rho}\right| + \left|\alpha^2_{1\rho}\right|Z + \left|\alpha^2_{2\rho}\right|Z^2 + \left|\alpha^2_{3\rho}\right|Z^3 + \left|\alpha^2_{4\rho}\right|Z^4,$$

$$\rho = 1, 2, \ldots, \rho^0, \quad 0 \leq R \leq 1, \quad R = \frac{\rho - 1}{\rho^0 - 1} \tag{7.2.13}$$

and the determination of the second derivative

$$\frac{\partial^2 C^{(2)}}{\partial Z^2} = \left\| g_{\rho\zeta}^2 \right\| = 2\left| \alpha_{2\rho}^2 \right| + 6\left| \alpha_{3\rho}^2 \right| Z + 12\left| \alpha_{4\rho}^2 \right| Z^2,$$

$$\rho = 1, 2, \ldots, \rho^0, \quad 0 \le R \le 1, \quad R = \frac{\rho - 1}{\rho^0 - 1}. \tag{7.2.14}$$

The last step is the solution of (7.2.4) for $s = 3$ using (6.2.14).

The solution of the problem (7.2.3), (7.2.4) was obtained using MATLAB program. It solves Eqs. (7.2.3) and (7.2.4) using the built-in MATLAB function *pdepe*, which solves initial–boundary value problems for parabolic partial differential equations. The second derivatives $\frac{\partial^2 C^{(s)}}{\partial Z^2}, s = 0, 1, 2$ are obtained with a polynomial approximation using the functions *polyfit* and *polyder* of MATLAB and then are introduced in the partial differential equations (7.2.3) and (7.2.4) using the built-in MATLAB function *interp2*.

7.2.3   Concentration Distributions

The solutions of the problem (7.2.1) obtained for the cases $Fo = 0.5, Da = 1, \varepsilon = 0.1, 0.3$ and concentration distributions $C(R, Z)$ in (7.2.2) for $Pe^{-1} = \varepsilon Fo = 0.05, 0.15$ and $Z = 0.2, 0.5, 0.8, 1.0$ are presented in Figs. 7.1 and 7.2.

Fig. 7.1 Solution of (7.2.1) for $\varepsilon = 0.1$

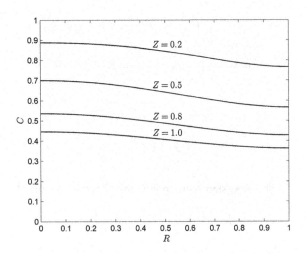

Fig. 7.2 Solution of (7.2.1) for $\varepsilon = 0.3$

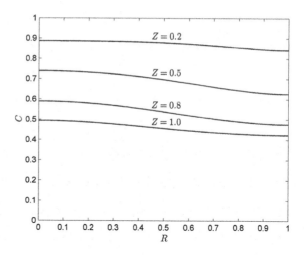

7.3   Average-Concentration Models

In the cases where the velocity distribution in the column is unknown, an average-concentration model (5.1.7) is possible to be used for the chemical reaction modeling:

$$A(Z)\frac{d\overline{C}}{dZ} + \frac{dA}{dZ}\overline{C} = Pe^{-1}\frac{d^2\overline{C}}{dZ^2} - Da\,\overline{C}; \quad Z = 0, \quad \overline{C} = 1, \quad \frac{d\overline{C}}{dZ} = 0; \quad (7.3.1)$$

where

$$A(Z) = 2\int_0^1 RU(R)\frac{C(R,Z)}{\overline{C}(Z)}\,dR, \quad U(R) = 2 - 2R^2, \quad \overline{C}(Z) = 2\int_0^1 RC(R,Z)\,dR.$$

$$(7.3.2)$$

The solution of (2.1.27) and (7.3.2) in the case $Fo = 0.5, Da = 1, Pe^{-1} = \varepsilon Fo$, $\varepsilon = 0.1, 0.3$ permits to obtain the functions $\overline{C}(Z), A(Z)$: They are presented in Figs. 7.3 and 7.4.

It is seen from Fig. 7.4 that the function $A(Z)$ can be presented [2] as a linear approximation $A = a_0 + a_1 Z$ and the (theoretical) values of the parameters a_0, a_1 are presented in Table 7.1. As a result, the model (7.3.1) has the form:

$$(a_0 + a_1 Z)\frac{d\overline{C}}{dZ} + a_1\overline{C} = Pe^{-1}\frac{d^2\overline{C}}{dZ^2} - Da\,\overline{C};$$

$$Z = 0, \quad \overline{C}(0) \equiv 1, \quad \left(\frac{d\overline{C}}{dZ}\right)_{Z=0} \equiv 0. \quad (7.3.3)$$

Fig. 7.3 Average
concentration $\overline{C}(Z)$ (7.3.2):
(*1*) $\varepsilon = 0.1$, (*2*) $\varepsilon = 0.3$

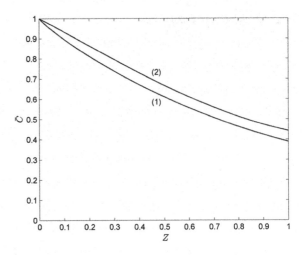

Fig. 7.4 Function $A(Z)$
(7.3.2): (*1*) $\varepsilon = 0.1$, (*2*)
$\varepsilon = 0.3$

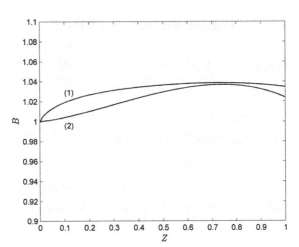

Table 7.1 Model parameter values

ε	a_0	a_1	a_0^0	a_1^0	a_0^1	a_1^1	a_0^2	a_1^2
0.1	1.0126	0.0075	1.0589	0.0863	0.9579	0.0003	0.9330	0.0005
0.3	0.9937	0.0288	1.0743	0.1663	1.2018	−0.0010	0.9299	0.0005

In (7.3.3) for $Fo = 0.5, Da = 1, \varepsilon = 0.3$, the small parameter is $Pe^{-1} = \varepsilon Fo = \theta = 0.15$, $\theta^3 < 10^{-2}$; i.e., the perturbation method is possible to be used:

$$\overline{C}(Z) = \overline{C}^{(0)}(Z) + \theta\overline{C}^{(1)}(Z) + \theta^2\overline{C}^{(2)}(Z) \tag{7.3.4}$$

and from (7.3.3) and (7.3.4) follows:

$$(a_0 + a_1 Z) \frac{d\overline{C}^{(0)}}{dZ} + a_1 \overline{C}^{(0)} = -Da\,\overline{C}^{(0)}; \quad Z = 0, \quad \overline{C}^{(0)} = 1. \tag{7.3.5}$$

$$(a_0 + a_1 Z) \frac{d\overline{C}^{(s)}}{dZ} + a_1 \overline{C}^{(s)} = \frac{d^2 \overline{C}^{(s-1)}}{dZ^2} - Da\,\overline{C}^{(s)}; \quad Z = 0, \quad \overline{C}^{(s)}(0) = 0, \quad s = 1, 2. \tag{7.3.6}$$

7.3.1 Calculation Problem

The numerical solution of the equations set (7.3.5), (7.3.6) is possible if MATLAB and a three-step procedure are used, where the functions $\overline{C}^{(s)}(Z), s = 0, 1, 2$ will be obtained in four vectors forms:

$$\overline{C}^{(s)}(Z) = |\bar{a}_\zeta^s|, \quad s = 0, 1, 2,$$
$$0 \le Z \le 1, \quad Z = \frac{\zeta - 1}{\zeta^0 - 1}, \quad \zeta = 1, 2, \dots, \zeta^0. \tag{7.3.7}$$

The main problem in solving the equations set (7.3.5), (7.3.6) is the calculation of the second derivatives $\frac{d^2 \overline{C}^{(s)}}{dZ^2}(Z), s = 0, 1, 2$. A circumvention of this problem may be the application of one of the following two algorithms.

Algorithm 1

The equations set (7.3.5), (7.3.6) permits to obtain the expression for the derivatives $(k = 1, \dots, (4 - s))$ of the functions $\overline{C}^{(s)}(Z) \ (s = 0, 1, \dots, 4)$:

$$\frac{d^k \overline{C}^{(s)}}{dZ^k} = \frac{\frac{d^{(k+1)} \overline{C}^{(s-1)}}{dZ^{(k+1)}} - (Da + kb_1) \frac{d^{(k-1)} \overline{C}^{(s)}}{dZ^{(k-1)}}}{(b_0 + b_1 Z)}, \quad s = 0, 1, \dots, 4,$$

$$k = 1, \dots, (4 - s), \quad \overline{C}^{(-1)} = 0, \quad \frac{d^0 \overline{C}^{(s)}}{dZ^0} = \overline{C}^{(s)}. \tag{7.3.8}$$

The first step is the solution of (7.3.5)

$$\overline{C}^{(0)}(Z) = |\bar{a}_\zeta^0|, \quad 0 \le Z \le 1, \quad Z = \frac{\zeta - 1}{\zeta^0 - 1}, \quad \zeta = 1, 2, \dots, \zeta^0 \tag{7.3.9}$$

applying (7.3.8) $(s = 0, k = 1, 2)$ for calculating the elements of the vectors

$$\frac{d\overline{C}^{(0)}}{dZ} = |\bar{a}_\zeta'^0|, \quad \frac{d^2\overline{C}^{(0)}}{dZ^2} = |\bar{a}_\zeta''^0|,$$

$$0 \leq Z \leq 1, \quad Z = \frac{\zeta - 1}{\zeta^0 - 1}, \quad \zeta = 1, 2, \ldots, \zeta^0. \tag{7.3.10}$$

The next step is the solution of (7.3.6) using (7.3.10) for $s = 1$

$$\overline{C}^{(1)}(Z) = |\bar{a}_\zeta^1|, \quad 0 \leq Z \leq 1, \quad Z = \frac{\zeta - 1}{\zeta^0 - 1}, \quad \zeta = 1, 2, \ldots, \zeta^0 \tag{7.3.11}$$

and (7.3.8) ($s = 1$, $k = 1, 2$) for calculating the elements of the vectors:

$$\frac{d\overline{C}^{(1)}}{dZ} = |\bar{a}_\zeta'^1|, \quad \frac{d^2\overline{C}^{(1)}}{dZ^2} = |\bar{a}_\zeta''^1|,$$

$$0 \leq Z \leq 1, \quad Z = \frac{\zeta - 1}{\zeta^0 - 1}, \quad \zeta = 1, 2, \ldots, \zeta^0. \tag{7.3.12}$$

The last step is the solving of (7.3.6) using (7.3.12) for $s = 2$:

$$\overline{C}^{(2)}(Z) = |\bar{a}_\zeta^2|, \quad 0 \leq Z \leq 1, \quad Z = \frac{\zeta - 1}{\zeta^0 - 1}, \quad \zeta = 1, 2, \ldots, \zeta^0. \tag{7.3.13}$$

Algorithm 2

The first step is the solution of (7.3.5), i.e., elements calculation of the vector:

$$\overline{C}^{(0)}(Z) = |\bar{a}_\zeta^0|, \quad 0 \leq Z \leq 1, \quad Z = \frac{\zeta - 1}{\zeta^0 - 1}, \quad \zeta = 1, 2, \ldots, \zeta^0. \tag{7.3.14}$$

The next step is a polynomial approximation of the function $\overline{C}^{(0)}(Z)$

$$\overline{C}^{(0)}(Z) = |\bar{a}_\zeta^0| = \bar{\alpha}_0^0 + \bar{\alpha}_1^0 Z + \bar{\alpha}_2^0 Z^2 + \bar{\alpha}_3^0 Z^3 + \bar{\alpha}_4^0 Z^4 \tag{7.3.15}$$

and the determination of the second derivative

$$\frac{d^2\overline{C}^{(0)}}{dZ^2}(Z) = |\bar{g}_\zeta^0| = 2\bar{\alpha}_2^0 + 6\bar{\alpha}_3^0 Z + 12\bar{\alpha}_4^0 Z^2. \tag{7.3.16}$$

The next step is the solution of (7.3.6) using (7.3.16) for $s = 1$, i.e., elements calculation of the vector:

$$\overline{C}^{(1)}(Z) = |\bar{a}_\zeta^1|, \quad 0 \leq Z \leq 1, \quad Z = \frac{\zeta - 1}{\zeta^0 - 1}, \quad \zeta = 1, 2, \ldots, \zeta^0. \tag{7.3.17}$$

The next step is the polynomial approximation of the function $\overline{C}^{(1)}(Z)$:

$$\overline{C}^{(1)}(Z) = \left|\bar{a}_\zeta^1\right| = \bar{\alpha}_0^1 + \bar{\alpha}_1^1 Z + \bar{\alpha}_2^1 Z^2 + \bar{\alpha}_3^1 Z^3 + \bar{\alpha}_4^1 Z^4 \qquad (7.3.18)$$

and the determination of the second derivative

$$\frac{d^2\overline{C}^{(1)}}{dZ^2}(Z) = \left|\bar{g}_\zeta^1\right| = 2\bar{\alpha}_2^1 + 6\bar{\alpha}_3^1 Z + 12\bar{\alpha}_4^1 Z^2. \qquad (7.3.19)$$

The last step is the solution of (7.3.6) using (7.3.19) for $s = 2$.

The solution of the problem (7.3.3) was obtained using MATLAB program. It solves Eqs. (7.3.5) and (7.3.6) using the built-in MATLAB function *ode45*, which solves non-stiff differential equations. Two different algorithms are used to obtain the second derivatives, which are introduced in the differential equations (7.3.5) and (7.3.6), by using the built-in MATLAB function *interp1*.

7.3.2 Average-Concentration Distributions

The solutions of (7.3.3), for theoretical values of a_0, a_1 (see Table 7.1) and $\theta = 0.05, 0.15, Da = 1$, are obtained by applying the Algorithms 1 and 2. They are presented (dotted lines) in Figs. 7.5 and 7.6, where they are juxtaposed with the calculated average concentrations (7.3.2) (lines).

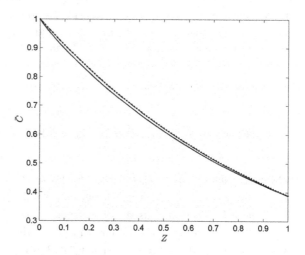

Fig. 7.5 Average concentrations for $\theta = 0.05$: solid line—calculated by (7.3.2), dotted line—solution of (7.3.3) (Algorithm 1), dashed line—solution of (7.3.3) (Algorithm 2)

Fig. 7.6 Average
concentrations for $\theta = 0.15$:
solid line—calculated by
(7.3.2), dotted line—solution
of (7.3.3) (Algorithm 1),
dashed line—solution of
(7.3.3) (Algorithm 2)

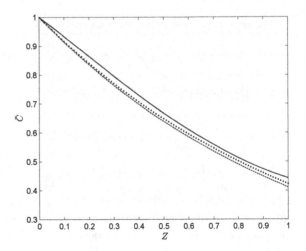

Fig. 7.6 Average concentrations for $\theta = 0.15$: solid line—calculated by (7.3.2), dotted line—solution of (7.3.3) (Algorithm 1), dashed line—solution of (7.3.3) (Algorithm 2)

7.3.3   Parameter Identification

The concentration $C(R, Z)$ in (7.2.1) obtained for the cases $Fo = 0.5, Da = 1$, $\varepsilon = 0.1, 0.3, Pe^{-1} = \varepsilon Fo = 0.05, 0.15$ allows to obtain the average concentrations $\overline{C}(Z)$ in (7.3.2) and "artificial experimental data" for different values of Z:

$$\overline{C}_{exp}^m(Z_n) = (0.95 + 0.1 S_m)\overline{C}(Z_n), \quad m = 1, \ldots 10,$$
$$Z_n = 0.1n, \quad n = 1, 2, \ldots, 10, \tag{7.3.20}$$

where $0 \leq S_m \leq 1, m = 1, \ldots, 10$ are obtained by a generator of random numbers. The obtained "artificial experimental data" (7.3.20) are used for illustration of the parameter identification in the average-concentration model (7.3.3) by minimization of the least-squares functions $Q_n, n = 1, 2$ and Q:

$$Q_n(Z_n, b_0^n, b_1^n) = \sum_{m=1}^{10} \left[\overline{C}(Z_n, b_0^n, b_1^n) - \overline{C}_{exp}^m(Z_n)\right]^2,$$
$$Q(b_0^0, b_1^0) = \sum_{n=1}^{10} Q_n(Z_n, b_0^0, b_1^0), \quad Z_n = 0.1n, \quad n = 1, 2, \ldots, 10, \tag{7.3.21}$$

where the values of $\overline{C}(Z_n, b_0^n, b_1^n)$ are obtained as solutions of (7.3.3) for different $Z_n = 0.1n, n = 1, 2, \ldots, 10$. The obtained values $\left(a_0^0, a_1^0; a_0^1, a_1^1; a_0^2, a_1^2\right)$ are presented in Table 7.1. They are used for calculation of the functions $\overline{C}(Z, a_0^0, a_1^0)$, $\overline{C}(Z, a_0^1, a_1^1)$, $\overline{C}(Z, a_0^2, a_1^2)$ as solutions of (7.3.3) (the lines in Fig. 7.7), where the points are the "artificial experimental data" (7.3.20).

Fig. 7.7 Comparison of the concentration distributions (7.3.3) and "artificial experimental data" (7.3.20) for $\theta = 0.05$: dashed line —$\overline{C}\left(Z, a_0^1, a_1^1\right)$; dotted line —$\overline{C}\left(Z, b_0^2, b_1^2\right)$; solid line —$\overline{C}\left(Z, a_0^0, a_1^0\right)$; circles —"artificial experimental data" (7.3.20)

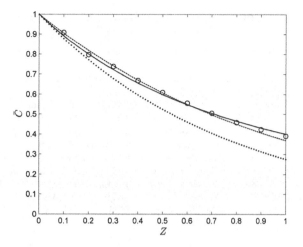

Fig. 7.8 Comparison of the concentration distributions (7.3.3) and "artificial experimental data" (7.3.20) for $\theta = 0.15$: dashed line —$\overline{C}\left(Z, a_0^1, a_1^1\right)$; dotted line —$\overline{C}\left(Z, a_0^2, a_1^2\right)$; solid line —$\overline{C}\left(Z, a_0^0, a_1^0\right)$; circles —"artificial experimental data" (7.3.20)

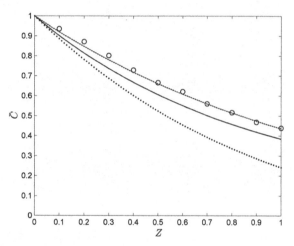

The comparison of the functions (lines) and experimental data (points) in Figs. 7.7 and 7.8 shows that the experimental data obtained from a column with real radius and small height ($Z = 0.1$) are useful for parameter identifications.

The computer modeling of the mass transfer processes in column apparatuses on the base of a new approach using a convection–diffusion-type model and an average-concentration-type model leads to calculation problems in the cases of presence of small parameters in the highest derivatives. This problem is solved by means of MATLAB and three algorithms, applying the perturbation method.

References

1. Boyadjiev C (2010) Theoretical chemical engineering. Modeling and simulation. Springer, Berlin
2. Doichinova M, Boyadjiev C (2012) On the column apparatuses modeling. Int J Heat Mass Transf 55:6705–6715
3. Boyadjiev C (2013) A new approach for the column apparatuses modeling in chemical engineering. J Pure Appl Math: Adv Appl 10(2):131–150
4. Мищенко ЕФ, Розов НХ (1975) Дифференциальные уравнения с малым параметром и релаксационные колебания. Изд. "Наука", Москва
5. O'Malley RE (1974) Introduction to singular perturbations. Academic Press, New York
6. Boyadjiev B, Doichinova M, Boyadjiev C (2015) Computer modeling of column apparatuses. 3. Perturbations method approach. J Eng Thermophys 24(4):371–380
7. Doichinova M, Boyadjiev C, Boyadjiev B (2015) Some problems in the column apparatuses modeling. Bulg Chem Commun 47(3):755–765

Chapter 8
Two-Coordinate Systems Problem

In the cases of physical absorption [1–4] in a high countercurrent gas–liquid column, the mass transfer process model has to be presented in two-coordinate systems (see 3.1.8):

$$U_j(R)\frac{\partial C_j}{\partial Z_j} = Fo_j\left(\frac{1}{R}\frac{\partial C_j}{\partial R} + \frac{\partial^2 C_j}{\partial R^2}\right) + (-1)^j K_j(C_1 - C_2);$$

$$R = 0, \quad \frac{\partial C_j}{\partial R} \equiv 0; \quad R = 1, \quad \frac{\partial C_j}{\partial R} \equiv 0; \quad j = 1, 2; \tag{8.0.1}$$

$$Z_1 = 0, \quad C_1 \equiv 1; \quad Z_2 = 0, \quad C_2 \equiv 0.$$

8.1 Convection–Diffusion-Type Model

Let us consider the convection–diffusion-type model (8.0.1), where the velocity distributions in the phases are of Poiseuille type [5], and the difference between the phase velocities is in the average velocities, only:

$$U_1 = U_2 = 2 - 2R^2. \tag{8.1.1}$$

From (8.0.1) and (8.1.1), it is possible to obtain the next form of the problem for computer modeling of the absorption processes in countercurrent column apparatuses:

$$(2 - 2R^2)\frac{\partial C_1}{\partial Z_1} = Fo_1\left(\frac{1}{R}\frac{\partial C_1}{\partial R} + \frac{\partial^2 C_1}{\partial R^2}\right) - K_1(C_1 - C_2);$$

$$R = 0, \quad \frac{\partial C_1}{\partial R} \equiv 0; \quad R = 1, \quad \frac{\partial C_1}{\partial R} \equiv 0; \quad Z_1 = 0, \quad C_1 \equiv 1. \tag{8.1.2}$$

$$(2 - 2R^2)\frac{\partial C_2}{\partial Z_2} = Fo_2\left(\frac{1}{R}\frac{\partial C_2}{\partial R} + \frac{\partial^2 C_2}{\partial R^2}\right) + K_2(C_1 - C_2);$$

$$R = 0, \quad \frac{\partial C_2}{\partial R} \equiv 0; \quad R = 1, \quad \frac{\partial C_2}{\partial R} \equiv 0; \quad Z_2 = 0, \quad C_2 \equiv 0. \qquad (8.1.3)$$

8.1.1 Calculation Problem

The numerical solution of the equations set (8.1.2), (8.1.3) is possible if an iterative procedure is used [6], where the concentration distributions in the column will be obtained in two matrix forms on every iteration step s:

$$C_1^s(R, Z_1) = \left\|a_{\rho\zeta_1}^s\right\|, \quad \rho = 1, 2, \ldots, \rho^0, \quad \zeta_1 = 1, 2, \ldots, \zeta^0,$$

$$0 \le R \le 1, \quad 0 \le Z_1 \le 1, \quad R = \frac{\rho - 1}{\rho^0 - 1}, \quad Z_1 = \frac{\zeta_1 - 1}{\zeta^0 - 1}, \quad \rho^0 = \zeta^0. \qquad (8.1.4)$$

$$C_2^s(R, Z_2) = \left\|b_{\rho\zeta_2}^s\right\|, \quad \rho = 1, 2, \ldots, \rho^0, \quad \zeta_2 = 1, 2, \ldots, \zeta^0,$$

$$0 \le R \le 1, \quad 0 \le Z_2 \le 1, \quad R = \frac{\rho - 1}{\rho^0 - 1}, \quad Z_2 = \frac{\zeta_2 - 1}{\zeta^0 - 1}. \qquad (8.1.5)$$

The iterative procedure starts with the zero step $s = 0$:

$$C_2^0(R, Z_2) = \left\|b_{\rho\zeta_2}^0\right\| \equiv 0, \quad \rho = 1, 2, \ldots, \rho^0, \quad \zeta_2 = 1, 2, \ldots, \zeta^0;$$

$$C_1^0(R, Z_1) = \left\|a_{\rho\zeta_1}^0\right\|, \quad \rho = 1, 2, \ldots, \rho^0, \quad \zeta_1 = 1, 2, \ldots, \zeta^0, \qquad (8.1.6)$$

where $C_1^0(R, Z_1)$ is a solution of the problem:

$$(2 - 2R^2)\frac{\partial C_1^0}{\partial Z_1} = Fo_1\left(\frac{1}{R}\frac{\partial C_1^0}{\partial R} + \frac{\partial^2 C_1^0}{\partial R^2}\right) - K_1 C_1^0;$$

$$R = 0, \quad \frac{\partial C_1^0}{\partial R} \equiv 0; \quad R = 1, \quad \frac{\partial C_1^0}{\partial R} \equiv 0; \quad Z_1 = 0, \quad C_1^0 \equiv 1. \qquad (8.1.7)$$

The solution of (8.1.7) permits to obtain a new function:

$$C_1^0(R, Z_1) = C_1^0(R, 1 - Z_2) = \widehat{C}_1^0(R, Z_2) = \left\|\hat{a}_{\rho\zeta_2}^0\right\|,$$

$$\rho = 1, 2, \ldots, \rho^0, \quad \zeta_2 = 1, 2, \ldots, \zeta^0. \qquad (8.1.8)$$

The iterative step s is the solution of the problem:

$$(2 - 2R^2)\frac{\partial C_2^s}{\partial Z_2} = Fo_2\left(\frac{1}{R}\frac{\partial C_2^s}{\partial R} + \frac{\partial^2 C_2^s}{\partial R^2}\right) + K_2\left(\widehat{C}_1^{(s-1)} - C_2^s\right);$$

$$R = 0, \quad \frac{\partial C_2^s}{\partial R} \equiv 0; \quad R = 1, \quad \frac{\partial C_2^s}{\partial R} \equiv 0; \quad Z_2 = 0, \quad C_2^s \equiv 0,$$

(8.1.9)

where

$$\widehat{C}_1^{(s-1)}(R, Z_2) = \left\|\widehat{a}_{\rho\zeta_2}^{(s-1)}\right\|, \quad \rho = 1, 2, \ldots, \rho^0, \quad \zeta_2 = 1, 2, \ldots, \zeta^0.$$

The solution of (8.1.9) permits to obtain a new function:

$$C_2^s(R, Z_2) = C_2^s(R, 1 - Z_1) = \widehat{C}_2^s(R, Z_1) = \left\|\widehat{b}_{\rho\zeta_1}^s\right\|,$$

$$\rho = 1, 2, \ldots, \rho^0, \quad \zeta_1 = 1, 2, \ldots, \zeta^0,$$

(8.1.10)

which will be used for solving (8.1.2) at the sth iterative step:

$$(2 - 2R^2)\frac{\partial C_1^s}{\partial Z_1} = Fo_1\left(\frac{1}{R}\frac{\partial C_1^s}{\partial R} + \frac{\partial^2 C_1^s}{\partial R^2}\right) - K_1\left(C_1^s - \widehat{C}_2^s\right);$$

$$R = 0, \quad \frac{\partial C_1^s}{\partial R} \equiv 0; \quad R = 1, \quad \frac{\partial C_1^s}{\partial R} \equiv 0; \quad Z_1 = 0, \quad C_1^s \equiv 1.$$

(8.1.11)

The solution of the problem (8.1.2), (8.1.3) is possible to be obtained using MATLAB program. It solves Eqs. (8.1.9), (8.1.11) through iterative procedure, using the built-in MATLAB function *pdepe*, which solves the initial–boundary value problems for parabolic partial differential equations. The obtained matrices $\widehat{C}_1^{s-1}$ (from 8.1.11) and $\widehat{C}_2^s$ (from 8.1.9) are introduced in (8.1.9) and (8.1.11), respectively, using the built-in MATLAB function *interp2*.

The stop criterion of the iterative procedure is the condition:

$$\left|\frac{a_{\rho\zeta_1}^s - a_{\rho\zeta_1}^{(s-1)}}{a_{\rho\zeta_1}^s}\right| \leq 10^{-3}, \quad \rho = 1, 2, \ldots, \rho^0, \quad \zeta_1 = 1, 2, \ldots, \zeta^0.$$

(8.1.12)

8.1.2 Concentration Distributions

A solution of the problem (8.1.2), (8.1.3) is obtained for the case $Fo_1 = 0.1$, $Fo_2 = 0.01$, $K_1 = 1$, $K_2 = 0.1$ and the concentration distributions $C_j(R, Z_j)$ for $Z_j = 0.2, 0.5, 0.8, 1.0, j = 1, 2$ are presented in Figs. 8.1 and 8.2. These results permit to obtain $\overline{C}_j(Z_j)$, $A_j(Z_j)$, $j = 1, 2$ in (6.1.7) (Figs. 6.1–6.4) and "theoretical" parameter values (6.1.9) presented in Table 6.1.

Fig. 8.1 Concentration distributions $C_1(R, Z_1)$ at $Fo_1 = 0.1$, $K_1 = 1$: (*1*) $C_1(R, 0.2)$; (*2*) $C_1(R, 0.5)$; (*3*) $C_1(R, 0.8)$; (*4*) $C_1(R, 1)$

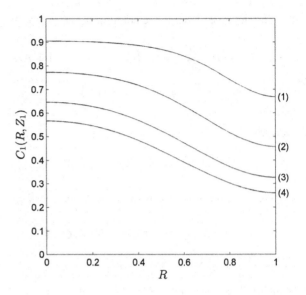

Fig. 8.2 Concentration distributions $C_2(R, Z_2)$ at $Fo_2 = 0.01$, $K_2 = 0.1$: (*1*) $C_2(R, 0.2)$; (*2*) $C_2(R, 0.5)$; (*3*) $C_2(R, 0.8)$; (*4*) $C_2(R, 1)$

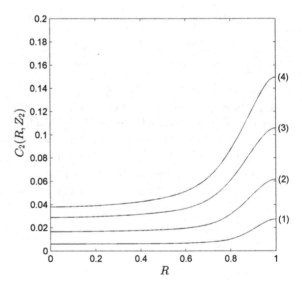

Table 8.1 Absorption degree

$Fo_1 = 0.1, K_1 = 1$ $Fo_2 = 0.01, K_2 = 0.1$	$U_1 = U_2 = 2 - 2R^2$	$U_1 = U_2 \equiv 1$
G	0.5814	0.6336

8.1.3 Absorption Process Efficiency

The solution of the problem (8.1.2), (8.1.3) permits to obtain the absorption efficiency g and the gas absorption degree G in the column using the inlet and outlet average convective mass flux at the cross-sectional area surface in the column:

$$g = u_1^0 c_1^0 - \frac{2}{r_0^2} \int_0^{r_0} r u_1(r) c_1(r, l) dr, \quad G = \frac{g}{u_1^0 c_1^0}. \tag{8.1.13}$$

The absorption degree in generalized variables (3.1.4) has the form:

$$G = 1 - 2 \int_0^1 R U_1(R) C_1(R, 1) dR. \tag{8.1.14}$$

In the cases of absence of the velocity radial non-uniformity ($U_1 = U_2 \equiv 1$), the absorption degree has the form:

$$G_0 = 1 - 2 \int_0^1 R C_1(R, 1) dR \tag{8.1.15}$$

and the reduction in the process efficiency due to the radial non-uniformity of the velocity is shown in Table 8.1.

8.2 Average-Concentration Model

In the cases of unknown velocity distribution in high countercurrent columns, the average-concentration model (6.1.10) is possible to be used for the physical absorption modeling:

$$
\begin{aligned}
(a_{01} + a_{11}Z_1)\frac{d\overline{C}_1}{dZ_1} + a_{11}\overline{C}_1 &= -K_{01}(\overline{C}_1 - \overline{C}_2); \quad Z_1 = 0, \quad \overline{C}_1(0) = 1. \\
(a_{02} + a_{12}Z_2)\frac{d\overline{C}_2}{dZ_2} + a_{12}\overline{C}_2 &= K_{02}(\overline{C}_1 - \overline{C}_2); \quad Z_2 = 0, \quad \overline{C}_2(0) = 0,
\end{aligned}
\tag{8.2.1}
$$

where a_{0j}, a_{1j}, $j = 1, 2$ are the "theoretical" parameter values presented in Table 6.1.

8.2.1 Calculation Problem

The numerical solution of the equation set (8.2.1) is possible if MATLAB, and an iterative procedure is used [6], where the average-concentration distributions in the column will be obtained in two vectors forms on every iteration step s:

$$\overline{C}_1^s(Z_1) = \left\| m_{\zeta_1}^s \right\|, \quad \zeta_1 = 1, 2, \ldots, \zeta^0, \quad 0 \leq Z_1 \leq 1, \quad Z_1 = \frac{\zeta_1 - 1}{\zeta^0 - 1}. \quad (8.2.2)$$

$$\overline{C}_2^s(Z_2) = \left\| n_{\zeta_2}^s \right\|, \quad \zeta_2 = 1, 2, \ldots, \zeta^0, \quad 0 \leq Z_2 \leq 1, \quad Z_2 = \frac{\zeta_2 - 1}{\zeta^0 - 1}. \quad (8.2.3)$$

The iterative procedure starts with the zero step $s = 0$:

$$\overline{C}_2^0(Z_2) = \left\| n_{\zeta_2}^0 \right\| \equiv 0, \quad \zeta_2 = 1, 2, \ldots, \zeta^0; \quad \overline{C}_1^0(Z_1) = \left\| m_{\zeta_1}^0 \right\|, \quad \zeta_1 = 1, 2, \ldots, \zeta^0,$$
$$(8.2.4)$$

where $\overline{C}_1^0(Z_1) = \left\| m_{\zeta_1}^0 \right\|$ is solution of the problem:

$$(a_{01} + a_{11}Z_1) \frac{d\overline{C}_1^0}{dZ_1} + a_{11}\overline{C}_1^0 = -K_1\overline{C}_1^0; \quad Z_1 = 0, \quad \overline{C}_1^0(0) \equiv 1. \quad (8.2.5)$$

As a result is possible to obtain

$$\hat{C}_1^0(Z_2) = \overline{C}_1^0(Z_1 = 1 - Z_2). \quad (8.2.6)$$

The iterative procedure s is the sequentially solving the equations:

$$(a_{02} + a_{12}Z_2) \frac{d\overline{C}_2^s}{dZ_2} + a_{12}\overline{C}_2^s = K_2\left(\hat{C}_1^{(s-1)} - \overline{C}_2^s \right); \quad Z_2 = 0, \quad \overline{C}_2^s(0) \equiv 0; \quad (8.2.7)$$

$$(a_{01} + a_{11}Z_1) \frac{d\overline{C}_1^s}{dZ_1} + a_{11}\overline{C}_1^s = -K_1\left(\overline{C}_1^s - \hat{C}_2^s \right); \quad Z_1 = 0, \quad \overline{C}_1^s(0) \equiv 1, \quad (8.2.8)$$

where

$$\widehat{C}_1^{(s-1)}(Z_2) = \overline{C}_1^{(s-1)}(Z_1 = 1 - Z_2), \quad \widehat{C}_2^s(Z_1) = \overline{C}_2^s(Z_2 = 1 - Z_1). \qquad (8.2.9)$$

The stop criterion of the iterative procedure is the condition:

$$\left| \frac{m_{\zeta_1}^s - m_{\zeta_1}^{(s-1)}}{m_{\zeta_1}^s} \right| \leq 10^{-3}, \quad \zeta_1 = 1, 2, \ldots, \zeta^0. \qquad (8.2.10)$$

The solving of the problem (8.2.7), (8.2.8) was obtained by MATLAB program, using iterative algorithm. First it solves Eq. (8.2.7) using $\widehat{C}_1^{(s-1)}(Z_2) = \overline{C}_1^{(s-1)}(Z_1 = 1 - Z_2)$ and the built-in MATLAB function $ode45$, which solves non-stiff differential equations by medium order method. The obtained matrix $\widehat{C}_2^s(Z_1) = \overline{C}_2^s(Z_2 = 1 - Z_1)$ is introduced in (8.2.8) using the built-in MATLAB interpolation function $interp1$.

The presented approach is used for the parameter identification in Chap. 6.

References

1. Chr Boyadjiev (2006) Diffusion models and scale-up. Int J Heat Mass Transfer 49:796–799
2. Boyadjiev C (2009) Modeling of column apparatuses. Trans Academenergo 3:7–22
3. Boyadjiev C (2010) Theoretical chemical engineering. Modeling and simulation. Springer, Berlin
4. Doichinova M, Boyadjiev C (2012) On the column apparatuses modeling. Int J Heat Mass Transfer 55:6705–6715
5. Chr Boyadjiev (2013) A new approach for the column apparatuses modeling in chemical engineering. J Pure Appl Math: Adv Appl 10(2):131–150
6. Boyadjiev B, Doichinova M, Boyadjiev C (2015) Computer modeling of column apparatuses. 1. Two coordinates systems approach. J Eng Thermophys 24(3):247–258

Chapter 9
Multi-step Modeling Algorithms

In the cases of a non-stationary chemical adsorption in gas–solid systems, the presence of mobile (gas) and immobile (solid) phases in lengthy processes leads to a non-stationary process in the immobile phase and stationary process in the mobile phase, practically. As a result, different coordinate systems have to be used in the gas and the solid phase model. A combination of multi-step algorithms and MATLAB has been used for solving the equations set in the different coordinate systems [1].

9.1 Convection–Diffusion-Type Model

Let us consider the convection–diffusion-type model of a non-stationary chemical adsorption in gas–solid systems [2–5]. In the case of a prolonged process in a high column from (3.2.23) follows $\gamma = \varepsilon = 0$; i.e., the model of chemical adsorption has the form:

$$U(R)\frac{\partial C_{11}}{\partial Z} = \text{Fo}\left(\frac{1}{R}\frac{\partial C_{11}}{\partial R} + \frac{\partial^2 C_{11}}{\partial R^2}\right) - K_0(C_{11} - C_{13});$$

$$\frac{dC_{13}}{dT} = K_3(C_{11} - C_{13}) - Kc_{23}^0 C_{13}C_{23};$$

$$\frac{dC_{23}}{dT} = -Kc_{11}^0 C_{13}C_{23}; T = 0, \quad C_{13} \equiv 0, \quad C_{23} \equiv 1; \tag{9.1.1}$$

$$R = 0, \quad \frac{\partial C_{11}}{\partial R} \equiv 0; \quad R = 1, \quad \frac{\partial C_{11}}{\partial R} \equiv 0; Z = 0, \quad C_{11} \equiv 1,$$

where T is a parameter in the function $C_{11}(T, R, Z)$, while R, Z are parameters in the functions $C_{13}(T, R, Z)$, $C_{23}(T, R, Z)$. As an example, the velocity distributions in the gas phase will be of Poiseuille type [2]:

$$U(R) = 2 - 2R^2. \tag{9.1.2}$$

For convenience, the functions of Eq. (9.1.1) can be denoted as:

$$C_{11}(T, R, Z) = C(T, R, Z), \quad C_{13}(T, R, Z) = C_1(T, R, Z),$$
$$C_{23}(T, R, Z) = C_0(T, R, Z) \tag{9.1.3}$$

and from (9.1.1) and (9.1.2) one obtains:

$$(2 - 2R^2)\frac{\partial C}{\partial Z} = \text{Fo}\left(\frac{1}{R}\frac{\partial C}{\partial R} + \frac{\partial^2 C}{\partial R^2}\right) - K_0(C - C_1);$$

$$R = 0, \quad \frac{\partial C}{\partial R} \equiv 0; \quad R = 1, \quad \frac{\partial C}{\partial R} \equiv 0; \quad Z = 0, \quad C \equiv 1. \tag{9.1.4}$$

$$\frac{dC_1}{dT} = K_3(C - C_1) - Kc_{23}^0 C_1 C_0; \quad T = 0, \quad C_1 \equiv 0. \tag{9.1.5}$$

$$\frac{dC_0}{dT} = -Kc_{11}^0 C_1 C_0; \quad T = 0, \quad C_0 \equiv 1. \tag{9.1.6}$$

9.1.1 Calculation Problem

For solving (9.1.4), (9.1.5), and (9.1.6), a multi-step approach [1] for different values of $T = \frac{\tau - 1}{\tau^0 - 1}$, $(\tau = 1, 2, \dots, \tau^0)$, $R = \frac{\rho - 1}{\rho^0 - 1}$, $(\rho = 1, 2, \dots, \rho^0)$, $Z = \frac{\zeta - 1}{\zeta^0 - 1}$, $(\zeta = 1, 2, \dots, \zeta^0)$, $\rho^0 = \zeta^0$ will be used and an upper index s $(s = 1, 2, \dots, \tau^0)$ will be the step number. At each step $s = 1, 2, \dots, \tau^0$, the solutions of (9.1.4), (9.1.5), (9.1.6) as three matrix forms will be obtained:

$$C^s(T, R, Z) = \left\|C_{\tau\rho\zeta}^s\right\|, \quad C_0^s(T, R, Z) = \left\|C_{(0)\tau\rho\zeta}^s\right\|, \quad C_1^s(T, R, Z) = \left\|C_{(1)\tau\rho\zeta}^s\right\|;$$

$$T = \frac{\tau - 1}{\tau^0 - 1}, \quad \tau = 1, 2, \dots, \tau^0; \quad R = \frac{\rho - 1}{\rho^0 - 1} \quad \rho = 1, 2, \dots, \rho^0;$$

$$Z = \frac{\zeta - 1}{\zeta^0 - 1}, \quad \zeta = 1, 2, \dots, \zeta^0; \quad 0 \leq T \leq 1, \quad 0 \leq R \leq 1, \quad 0 \leq Z \leq 1.$$

$$\tag{9.1.7}$$

As a zero step $(s = 0)$ will be used

$$C^0(T,R,Z) = \left\|C^0_{\tau\rho\zeta}\right\| \equiv C^0(R,Z), \quad C^0_0(T,R,Z) = \hat{C}^0_0 = \left\|C^0_{(0)\tau\rho\zeta}\right\| \equiv 1,$$

$$C^0_1(T,R,Z) = \hat{C}^0_1 = \left\|C^0_{(1)\tau\rho\zeta}\right\| \equiv 0,$$

$$(9.1.8)$$

where $C^0(R,Z)$ is a solution of (9.1.4) for $C_1 \equiv 0$, i.e.

$$(2 - 2R^2)\frac{\partial C^0}{\partial Z} = \mathrm{Fo}\left(\frac{1}{R}\frac{\partial C^0}{\partial R} + \frac{\partial^2 C^0}{\partial R^2}\right) - KC^0;$$

$$R = 0, \quad \frac{\partial C^0}{\partial R} \equiv 0; \quad R = 1, \quad \frac{\partial C^0}{\partial R} \equiv 0; \quad Z = 0, \quad C^0 \equiv 1.$$

$$(9.1.9)$$

The step s is the solution of the equations set:

$$\frac{dC^s_1}{dT} = K_0\left(C^{(s-1)} - C^s_1\right) - K_3 c^0_{23} C^s_0 C^s_1; \quad T = 0, \quad C^s_1 \equiv \hat{C}^{(s-1)}_1(R,Z). \quad (9.1.10)$$

$$\frac{dC^s_0}{dT} = -K_3 c^0_{11} C^s_0 C^s_1; \quad T = 0, \quad C^s_0 \equiv \hat{C}^{(s-1)}_0(R,Z). \quad (9.1.11)$$

The solutions of (9.1.10), (9.1.11) permit to obtain:

$$\hat{C}^s_0(R,Z) = C^s_0\left(\frac{1}{\tau^0}, R, Z\right) = \left\|C^s_{(0)1\rho\zeta}\right\|;$$

$$\hat{C}^s_1(R,Z) = C^s_1\left(\frac{1}{\tau^0}, R, Z\right) = \left\|C^s_{(1)1\rho\zeta}\right\|.$$

$$(9.1.12)$$

The obtained function $\hat{C}^s_1(R,Z)$ allows to obtain $C^s(R,Z) = \left\|C^s_{1\rho\zeta}\right\|$ a solution of (9.1.4) at the s step:

$$(2 - 2R^2)\frac{\partial C^s}{\partial Z} = \mathrm{Fo}\left(\frac{1}{R}\frac{\partial C^s}{\partial R} + \frac{\partial^2 C^s}{\partial R^2}\right) - K\left(C^s - \hat{C}^s_1\right);$$

$$R = 0, \quad \frac{\partial C^s}{\partial R} \equiv 0; \quad R = 1, \quad \frac{\partial C^s}{\partial R} \equiv 0; \quad Z = 0, \quad C^s \equiv 1.$$

$$(9.1.13)$$

The multi-step computational procedure ends at $s = \tau^0$ and the solution of (9.1.4), (9.1.5), (9.1.6) is:

$$C(T,R,Z) = \left\| C_{\tau\rho\zeta}^{\tau^0} \right\|, \quad C_0(T,R,Z) = \left\| C_{(0)\tau\rho\zeta}^{\tau^0} \right\|, \quad C_1(T,R,Z) = \left\| C_{(1)\tau\rho\zeta}^{\tau^0} \right\|;$$

$$T = \frac{\tau - 1}{\tau^0 - 1}, \quad \tau = 1, 2, \ldots, \tau^0; \quad R = \frac{\rho - 1}{\rho^0 - 1}, \quad \rho = 1, 2, \ldots, \rho^0;$$

$$Z = \frac{\zeta - 1}{\zeta^0 - 1}, \quad \zeta = 1, 2, \ldots, \zeta^0.$$

$$\tag{9.1.14}$$

The solution of the problem (9.1.4), (9.1.5), and (9.1.6) was obtained using MATLAB program and applying a multi-step algorithm. First, it solves Eqs. (9.1.10), (9.1.11) using the built-in MATLAB function *ode45*, which solves non-stiff differential equations by medium-order method. The obtained matrix $\hat{C}_1$ is introduced in (9.1.13), and the built-in MATLAB interpolation function *interp2* is used. Then Eq. (9.1.13) is solved by means of the built-in MATLAB function *pdepe*, which solves initial-boundary value problems for parabolic partial differential equations.

9.1.2 Concentration Distributions

A solution of the problem (9.1.4), (9.1.5), and (9.1.6) is obtained for the case $Fo = 0.1$, $K = K_0 = K_3 = 1$, $c_{11}^0 = c_{23}^0 = 1$. The concentration distributions $C(0.6, R, Z)$ for $Z = 0.2$, 0.4, 0.6, 0.8, 1.0, and $C(T, 0.2, Z)$ for $T = 0.2$, 0.4, 0.6, 0.8, 1.0 are presented on Figs. 9.1 and 9.2. The concentration distributions $C_0(T, 0.5, Z)$ for $Z = 0.2$, 0.4, 0.6, 0.8, 1.0 and $C_0(T, 0.5, Z)$ for $T = 0.2$, 0.4, 0.6, 0.8, 1.0 are presented on Figs. 9.3 and 9.4.

Fig. 9.1 Concentration distributions $C(0.6, R, Z)$: (1) $Z = 0.2$; (2) $Z = 0.4$; (3) $Z = 0.6$; (4) $Z = 0.8$; (5) $Z = 1.0$

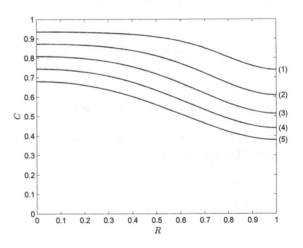

Fig. 9.2 Concentration
distributions $C(T, 0.2, Z)$:
(1) $T = 0.2$; (2) $T = 0.4$;
(3) $T = 0.6$; (4) $T = 0.8$;
(5) $T = 1.0$

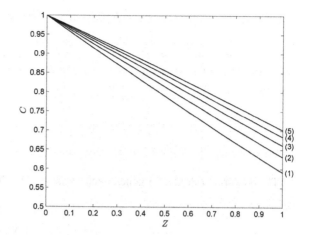

Fig. 9.3 Concentration
distributions $C_0(T, 0.5, Z)$:
(1) $Z = 0.2$; (2) $Z = 0.4$;
(3) $Z = 0.6$; (4) $Z = 0.8$;
(5) $Z = 1.0$

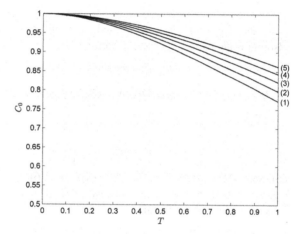

Fig. 9.4 Concentration
distributions $C_0(T, 0.5, Z)$:
(1) $T = 0.2$; (2) $T = 0.4$;
(3) $T = 0.6$; (4) $T = 0.8$;
(5) $T = 1.0$

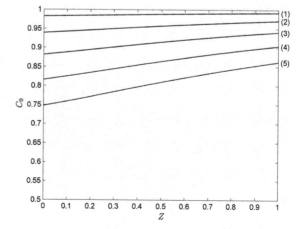

9.1.3 Adsorption Process Efficiency

The adsorption efficiency (g) and the adsorption degree (G) in the column are possible to be obtained using the inlet and outlet average convective mass flux at the cross-sectional area surface in the column:

$$g(t) = u_1^0 c_{11}^0 - \frac{2}{r_0^2} \int_0^{r_0} r u_1(r) c_{11}(t,r,l) dr, \quad G(t) = \frac{g}{u_1^0 c_{11}^0}. \qquad (9.1.15)$$

The adsorption degree in generalized variables (2.2.20) has the form:

$$\bar{G}(T) = 1 - 2 \int_0^1 R U(R) C(T,R,1) dR \qquad (9.1.16)$$

In the cases of velocity radial non-uniformity absence ($U(R) \equiv 1$), the adsorption degree $\bar{G}_0(T)$ has the form:

$$U = 1, \quad \bar{G}(T) = \bar{G}_0(T) = 1 - \bar{C}(T,1), \quad \bar{C}(T,Z) = 2 \int_0^1 R C(T,R,Z) dR. \qquad (9.1.17)$$

The adsorption degree (9.1.16), (9.1.17) is possible to be obtained using the solutions of the model Eqs. (9.1.4), (9.1.5), (9.1.6), and the results are presented on Fig. 9.5, where is seen that the adsorption degree decreases as a result of the velocity radial non-uniformity.

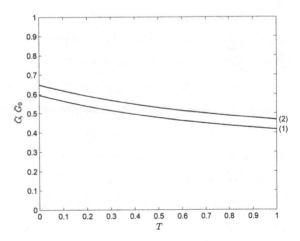

Fig. 9.5 Adsorption degree:
(1) $\bar{G}(T)$; (2) $\bar{G}_0(T)$.

9.2 Average-Concentration Model

The use of the convection–diffusion-type models for quantitative description of the chemical adsorption in column apparatuses is not possible because the velocity function in the convection–diffusion equation is unknown and an average-concentration model was obtained in Chap. 5. For lengthy processes and high columns, the problem (6.2.18) has the form (Figs. 9.6, 9.7).

Fig. 9.6 Function $\bar{C}(T, Z)$: (1) $Z = 0.2$; (2) $Z = 0.4$; (3) $Z = 0.6$; (4) $Z = 0.8$; (5) $Z = 1.0$; the dotted lines are solution of (9.2.4), (9.2.5), (9.2.6) with (theoretical) parameter values a_0, a_t, a_z.

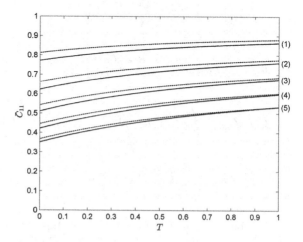

Fig. 9.7 Function $\bar{C}(T, Z)$: (1) $T = 0.2$; (2) $T = 0.4$; (3) $T = 0.6$; (4) $T = 0.8$; (5) $T = 1.0$; the dotted lines are solution of (9.2.4), (9.2.5), (9.2.6) with (theoretical) parameter values a_0, a_t, a_z.

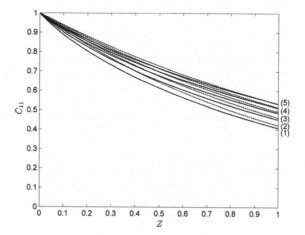

$$A\frac{d\bar{C}}{dZ} + \frac{dA}{dZ}\bar{C} = -K_0(\bar{C} - \bar{C}_1);$$

$$\frac{d\bar{C}_1}{dT} = K_3(\bar{C} - \bar{C}_1) - BKc_{23}^0\bar{C}_1\bar{C}_0; \quad \frac{d\bar{C}_0}{dT} = -BKc_{11}^0\bar{C}_1\bar{C}_0; \tag{9.2.1}$$

$$T = 0, \quad \bar{C}_1 \equiv 0, \quad \bar{C}_0 \equiv 1; \quad Z = 0, \quad \bar{C} \equiv 1.$$

In (9.2.1), Z is parameter in $\bar{C}_1(T,Z)$ and $\bar{C}_0(T,Z)$, while T is parameter in $\bar{C}(T,Z)$. The functions (6.2.17) are used in (9.2.1):

$$A(T,Z) = 2\int_0^1 RU(R)\frac{C(T,R,Z)}{\bar{C}(T,Z)}dR,$$

$$B(T,Z) = 2\int_0^1 R\frac{C_1(T,R,Z)}{\bar{C}_1(T,Z)}\frac{C_0(T,R,Z)}{\bar{C}_0(T,Z)}dR,$$

$$\bar{C}(T,Z) = 2\int_0^1 RC(T,R,Z)dR, \quad \bar{C}_0(T,Z) = 2\int_0^1 RC_0(T,R,Z)dR, \tag{9.2.2}$$

$$\bar{C}_1(T,Z) = 2\int_0^1 RC_1(T,R,Z)dR.$$

The obtained solutions from the model Eqs. (9.1.4), (9.1.5), (9.1.6) permit to find the functions $\bar{C}(T,Z)$, $\bar{C}_0(T,Z)$, $\bar{C}_1(T,Z)$, $A(T,Z)$, $B(T,Z)$, where $B(T,Z) \equiv 1$. The results are presented in Figs. 9.8 and 9.9, where is seen that the function $A(T,Z)$ is possible to be presented [3, 4] as linear approximation:

$$A = a_0 + a_t T + a_z Z. \tag{9.2.3}$$

The obtained ("theoretical") parameter values a_0, a_t, a_z, are presented in Table 9.1. As a result, from (9.2.1) is possible to be obtained:

$$(a_0 + a_t T + a_z Z)\frac{d\bar{C}}{dZ} + a_z\bar{C} = -K(\bar{C} - \bar{C}_1); \quad Z = 0, \quad \bar{C} \equiv 1. \tag{9.2.4}$$

$$\frac{d\bar{C}_1}{dT} = K_0(\bar{C} - \bar{C}_1) - K_3c_{23}^0\bar{C}_1\bar{C}_0; \quad T = 0, \quad \bar{C}_1 \equiv 0. \tag{9.2.5}$$

$$\frac{d\bar{C}_0}{dT} = -K_3c_{11}^0\bar{C}_1\bar{C}_0; \quad T = 0, \quad \bar{C}_0 \equiv 1. \tag{9.2.6}$$

Fig. 9.8 Function $A(T, Z)$:
(1) $Z = 0.2$; (2) $Z = 0.4$;
(3) $Z = 0.6$; (4) $Z = 0.8$;
(5) $Z = 1.0$

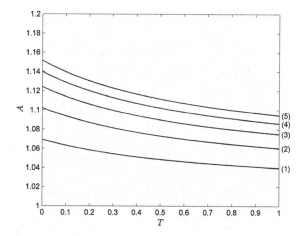

Fig. 9.9 Function $A(T, Z)$:
(1) $T = 0.2$; (2) $T = 0.4$;
(3) $T = 0.6$; (4) $T = 0.8$;
(5) $T = 1.0$

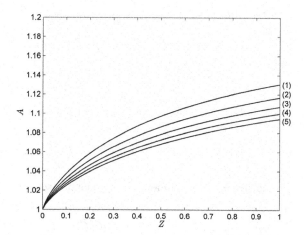

Table 9.1 Parameter values

	"Theoretical" values	"Experimental" values		
		Q	Q_1	Q_2
a_0	1.0473	3.507×10^{11}	0.8142	0.8991
a_z	0.0992	2.0671×10^{11}	7.0764×10^{-4}	3.2456×10^{-4}
a_t	-0.0387	-8.7123×10^{10}	2.4844×10^{-4}	1.8527×10^{-4}

9.2.1   Model Equations Solution

For solving (9.2.4), (9.2.5), (9.2.6), a multi-step approach for different values of $T = \frac{\tau - 1}{\tau^0 - 1}$, $(\tau = 1, 2, \ldots, \tau^0)$ and $Z = \frac{\zeta - 1}{\zeta^0 - 1}$, $(\zeta = 1, 2, \ldots, \zeta^0)$ will be used, where the upper index s $(s = 1, 2, \ldots, \tau^0)$ will be the step number. At each step, $s = 1, 2, \ldots, \tau^0$ the solution of (9.2.4), (9.2.5), (9.2.6) will be obtained as three matrix forms:

$$\bar{C}^s(T, Z) = \left\| \bar{C}^s_{\tau\zeta} \right\|, \quad \bar{C}^s_0(T, Z) = \left\| \bar{C}^s_{(0)\tau\zeta} \right\|, \quad \bar{C}^s_1(T, Z) = \left\| \bar{C}^s_{(1)\tau\zeta} \right\|;$$

$$T = \frac{\tau - 1}{\tau^0 - 1}, \quad \tau = 1, 2, \ldots, \tau^0; \quad Z = \frac{\zeta - 1}{\zeta^0 - 1}, \quad \zeta = 1, 2, \ldots, \zeta^0; \qquad (9.2.7)$$

$$0 \leq T \leq 1, \quad 0 \leq Z \leq 1.$$

As a zero step $(s = 0)$ will be used

$$\bar{C}^0(T, Z) = \left\| \bar{C}^0_{\tau\zeta} \right\| \equiv \bar{C}^0(Z), \quad \bar{C}^0_0(T, Z) = \left\| \bar{C}^0_{(0)\tau\zeta} \right\| \equiv 1,$$

$$\bar{C}^0_1(T, Z) = \left\| \bar{C}^0_{(1)\tau\zeta} \right\| \equiv 0, \qquad (9.2.8)$$

where $\bar{C}^0(Z)$ is solution of (9.2.4) for $\bar{C}_1 \equiv 0$, i.e.

$$(a_0 + a_t T + a_z Z)\frac{d\bar{C}^0}{dZ} + a_z \bar{C}^0 = -K\bar{C}^0, \quad Z = 0, \quad \bar{C}^0 \equiv 1. \qquad (9.2.9)$$

The step s is the solution of the equations set:

$$\frac{d\bar{C}^s_1}{dT} = K_0\left(\bar{C}^{(s-1)} - \bar{C}^s_1\right) - K_3 c^0_{23}\bar{C}^s_0\bar{C}^s_1; \quad T = 0, \quad \bar{C}^s_1 \equiv \hat{C}^{(s-1)}_1(Z). \qquad (9.2.10)$$

$$\frac{d\bar{C}^s_0}{dT} = -K_3 c^0_{11}\bar{C}^s_0\bar{C}^s_1; \quad T = 0, \quad \bar{C}^s_0 \equiv \hat{C}^{(s-1)}_0(Z). \qquad (9.2.11)$$

The solving of (9.2.10) and (9.2.11) leads to:

$$\hat{C}^s_0(Z) = \bar{C}^s_0\left(\frac{1}{\tau^0}, Z\right) = \left\| \bar{C}^s_{(0)1\zeta} \right\|; \quad \hat{C}^s_1(Z) = \bar{C}^s_1\left(\frac{1}{\tau^0}, Z\right) = \left\| \bar{C}^s_{(1)1\zeta} \right\|. \qquad (9.2.12)$$

The obtained function $\hat{C}^s_1(Z)$ permits to obtain $\bar{C}^s(Z) = \left\| \bar{C}^s_{1\zeta} \right\|$ as a solution of (9.2.4) at the s step:

$$(a_0 + a_t T + a_z Z) \frac{d\bar{C}^s}{dZ} + a_z \bar{C}^s = -K(\bar{C}^s - \hat{C}_1^s); \quad Z = 0, \quad \bar{C}^s \equiv 1. \quad (9.2.13)$$

The end of the multi-step computational procedure is $s = \tau^0$ and the solution of (9.2.4), (9.2.5), (9.2.6) is:

$$\bar{C}(T,Z) = \left\| \bar{C}_{\tau\zeta}^{\tau^0} \right\|, \quad \bar{C}_0(T,Z) = \left\| \bar{C}_{(0)\tau\zeta}^{\tau^0} \right\|, \quad \bar{C}_1(T,Z) = \left\| \bar{C}_{(1)\tau\zeta}^{\tau^0} \right\|;$$

$$T = \frac{\tau - 1}{\tau^0 - 1}, \quad \tau = 1, 2, \ldots, \tau^0; \quad Z = \frac{\zeta - 1}{\zeta^0 - 1}, \quad \zeta = 1, 2, \ldots, \zeta^0. \quad (9.2.14)$$

The solution of the problem (9.2.4), (9.2.5), (9.2.6) is possible to be obtained using the MATLAB program and applying a multi-step algorithm. First Eqs. (9.2.10) and (9.2.11) are solved using the built-in MATLAB function *ode45*, which solves non-stiff differential equations by medium-order method. The obtained matrix $\hat{C}_1$ is introduced in (9.2.13), and the built-in MATLAB interpolation function *interp1* is used. Then Eq. (9.2.13) is solved using again the *ode45* solver.

9.2.2   Parameter Identification

The obtained concentration $C(0.6, R, Z)$ as a solution of the problem (9.1.4), (9.1.5), and (9.1.6) for the case Fo $= 0.1$, $K = K_0 = K_3 = 1$, $c_{11}^0 = c_{23}^0 = 1$ permits to obtain the average concentration $\bar{C}(0.6, Z) = 2 \int\limits_0^1 RC(0.6, R, Z) dR$ and artificial experimental data for different values of Z:

$$\bar{C}_{exp}^m(Z_n) = (0.95 + 0.1 S_m)\bar{C}(0.6, Z_n),$$
$$m = 1, \ldots 10, \quad Z_n = 0.1n, \quad n = 1, 2, \ldots, 10, \quad (9.2.15)$$

where $0 \leq S_m \leq 1$, $m = 1, \ldots, 10$ are obtained by means of a random numbers generator. The obtained artificial experimental data (9.2.15) are used for illustration of the parameter identification in the average-concentrations model of the chemical adsorption (9.2.4), (9.2.5), (9.2.6) by minimization of the least-squares functions Q_n and Q:

$$Q_n(Z_n, a_0, a_z, a_t) = \sum_{m=1}^{10} \left[\bar{C}(0.6, Z_n, a_0, a_z, a_t) - \bar{C}_{exp}^m(Z_n) \right]^2,$$

$$Q(a_0, a_z, a_t) = \sum_{n=1}^{10} Q_n(Z_n, a_0, a_z, a_t), \quad Z_n = 0.1n, \quad n = 1, 2, \ldots, 10; \quad (9.2.16)$$

Fig. 9.10 Function $\bar{C}$:
(dotted line)—minimization
of Q_1; (dashed line)—
minimization of Q_2; (solid
line)—minimization of Q;
(circles)—experimental data
(8.2.15)

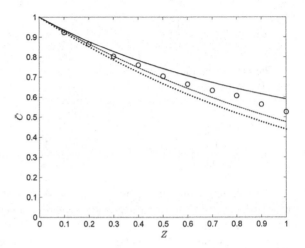

where the values of $\bar{C}(0.6, Z_n, a_0, a_z, a_t)$ are obtained as solutions of (9.2.4), (9.2.5), (9.2.6) for different $Z_n = 0.1n, \quad n = 1, 2, \ldots, 10$.

The obtained ("experimental") values a_0, a_z, a_t are presented in Table 9.1.

In Fig. 9.10 is compared the average concentration $\bar{C}(Z)$ (the lines), as a solution of (9.2.4), (9.2.5), (9.2.6) for the ("experimental") parameter values a_0, a_z, a_t obtained by the minimization of Q_1, Q_2, and Q in (9.2.16), with the artificial experimental data (9.2.15) (the points). The result displayed shows that problems of the parameter identification in the average-concentration model of the chemical adsorption are possible to be solved using experimental data obtained in a short column $(Z = 0.1)$ with a real diameter.

References

1. Boyadjiev B, Doichinova M, Boyadjiev C (2015) Computer modeling of column apparatuses. 2. Multi-step modeling approach. J Eng Thermophys 24(4):362–370
2. Boyadjiev C (2010) Theoretical chemical engineering. Modeling and simulation. Springer, Berlin
3. Doichinova M, Boyadjiev C (2012) On the column apparatuses modeling. Int J Heat Mass Transfer 55:6705–6715
4. Boyadjiev C (2013) A new approach for the column apparatuses modeling in chemical engineering. J Pure Appl Math: Adv Appl 10(2):131–150
5. Doichinova M, Boyadjiev C (2015) A new approach for the column apparatuses modeling in chemical and power engineering. Therm Sci 19(5):1747–1759

Part IV
Modeling of Processes in Industrial Column Apparatuses

Industrial Types of Convection–Diffusion and Average-Concentration Models

In Parts I–III were presented convection–diffusion and average-concentration models, where the radial velocity component is equal to zero in the cases of a constant axial velocity radial non-uniformity along the column height. There are many cases of industrial columns, where these conditions are not satisfied. In this Part, will be presented the models of chemical, absorption, adsorption, and catalytic processes in the cases of an axial modification of the radial non-uniformity of the axial velocity component and the radial velocity component is not equal to zero.

Chapter 10
Industrial Column Chemical Reactors

The new approach for the modeling of the processes in column apparatuses [1–3] presents the convection–diffusion and average-concentration models of the column chemical reactors (in Chaps. 2 and 5), where the radial velocity component is equal to zero in the cases of a constant axial velocity radial non-uniformity along the column height:

$$u = u(r), \quad v \equiv 0. \tag{10.0.1}$$

In the pseudo-first-order reactions case, these models (2.1.24, 2.1.27, 5.1.5) have the forms:

$$
\begin{aligned}
&u\frac{\partial c}{\partial z} = D\left(\frac{\partial^2 c}{\partial z^2} + \frac{1}{r}\frac{\partial c}{\partial r} + \frac{\partial^2 c}{\partial r^2}\right) - kc; \\
&r = 0, \quad \frac{\partial c}{\partial r} \equiv 0; \quad r = r_0, \quad \frac{\partial c}{\partial r} \equiv 0; \\
&z = 0, \quad c \equiv c^0, \quad u^0 c^0 \equiv uc^0 - D\left(\frac{\partial c}{\partial z}\right)_{z=0}.
\end{aligned}
\tag{10.0.2}
$$

$$
\begin{aligned}
&\alpha \bar{u}\frac{d\bar{c}}{dz} + \frac{d\alpha}{dz}\bar{u}\bar{c} = D\frac{d^2\bar{c}}{dz^2} - k\bar{c}; \\
&z = 0, \quad \bar{c}(0) = c^0, \quad \frac{d\bar{c}}{dz} = 0.
\end{aligned}
\tag{10.0.3}
$$

In (10.0.2), (10.0.3) is possible to be introduced the generalized variables (2.1.25, 5.1.7):

$$
\begin{aligned}
&r = r_0 R, \quad z = lZ, \quad u(r) = \bar{u}U(R), \quad \tilde{u}(r) = \frac{u(r)}{\bar{u}} = U(R), \\
&c(r,z) = c^0 C(R,Z), \quad \bar{c}(z) = c^0\overline{C}(Z), \\
&\overline{C}(Z) = 2\int_0^1 RC(R,Z)dR, \quad \tilde{c}(r,z) = \frac{c(r,z)}{\bar{c}(z)} = \frac{C(R,Z)}{\overline{C}(Z)}, \\
&\alpha(z) = \alpha(lZ) = A(Z) = 2\int_0^1 RU(R)\frac{C(R,Z)}{\overline{C}(Z)}dR
\end{aligned}
\tag{10.0.4}
$$

© Springer International Publishing AG, part of Springer Nature 2018
C. Boyadjiev et al., *Modeling of Column Apparatus Processes*,
Heat and Mass Transfer, https://doi.org/10.1007/978-3-319-89966-4_10

and as a result is obtained:

$$U\frac{\partial C}{\partial Z} = Fo\left(\varepsilon^2\frac{\partial^2 C}{\partial Z^2} + \frac{1}{R}\frac{\partial C}{\partial R} + \frac{\partial^2 C}{\partial R^2}\right) - Da\,C; \quad \varepsilon^2 = Fo^{-1}Pe^{-1};$$
$$R = 0, \quad \frac{\partial C}{\partial R} \equiv 0; \quad R = 1, \quad \frac{\partial C}{\partial R} \equiv 0; \quad (10.0.5)$$
$$Z = 0, \quad C \equiv 1, \quad 1 \equiv U - Pe^{-1}\frac{\partial C}{\partial Z}.$$

$$A(Z)\frac{d\overline{C}}{dz} + \frac{dA}{dz}\overline{C} = Pe^{-1}\frac{d^2\overline{C}}{dz^2} - Da\,\overline{C};$$
$$Z = 0, \quad \overline{C} = 1, \quad \frac{d\overline{C}}{dz} = 0. \quad (10.0.6)$$

In an industrial column $(l > 1\,(m))$, the average velocity is $\bar{u} > 1\,(m\ s^{-1})$ and the diffusivity is $(D < 10^{-4}\,m^2\ s^{-1})$. These conditions are possible to be obtained the order of magnitude of the parameter values:

$$Pe^{-1} < 10^{-4}, \quad \varepsilon = \frac{h_0}{l} > 10^{-1}, \quad Fo < 10^{-2} \quad (10.0.7)$$

and the models (10.0.5), (10.0.6) have convective forms:

$$U\frac{\partial C}{\partial Z} = -Da\,C; \quad Z = 0, \quad C \equiv 1. \quad (10.0.8)$$

$$A(Z)\frac{d\overline{C}}{dZ} + \frac{dA}{dZ}\overline{C} = -Da\,\overline{C}; \quad Z = 0, \quad \overline{C} = 1. \quad (10.0.9)$$

There are many cases of industrial columns, where the conditions (10.0.1) are not satisfied. In this chapter will be presented models of chemical processes in the cases of an axial modification of the radial non-uniformity of the axial velocity component [4], and the radial velocity component is not equal to zero [5].

10.1 Effect of the Axial Modification of the Radial Non-uniformity of the Axial Velocity Component

Very often in industrial conditions, an axial modification of the radial non-uniformity of the velocity is realized. The radial non-uniformity of the axial velocity component in the column apparatuses is caused by the fluid hydrodynamics at the column inlet, where it has a maximum and decreases along the column height as a result of the fluid viscosity. The theoretical determination of the change in the radial non-uniformity of the axial velocity component in a column is difficult in one-phase processes and practically impossible in two-phase and three-phase processes. The difficulty of the theoretical analysis of the effect of the axial modification of the radial non-uniformity of the velocity, can be circumvented by appropriate hydrodynamic model, where the average velocity at the cross section of the column is a constant, while the maximal velocity (and as a result the radial non-uniformity of the axial velocity component too) decreases along the column height.

Let us consider [4] the velocity distribution

$$u_n(r, z_n) = u^0 U_n(R, Z_n), \quad n = 0, 1, \ldots, 9, \tag{10.1.1}$$

where $u^0 = \text{const}$ is the inlet velocity and an axial step change of the radial non-uniformity of the axial velocity component in a column (Fig. 10.1):

$$
\begin{aligned}
&U_n(R, Z_n) = a_n - b_n R^2, \\
&a_n = 2 - 0.1n, \quad b_n = 2(1 - 0.1n), \\
&0.1n \leq Z_n \leq 0.1(n+1) \quad n = 0, 1, \ldots, 9, \quad 0 \leq R \leq 1,
\end{aligned}
\tag{10.1.2}
$$

where $U_n(R, Z_n)$ satisfy the equation:

$$2 \int_0^1 R U_n(R, Z_n) dR = 1 \tag{10.1.3}$$

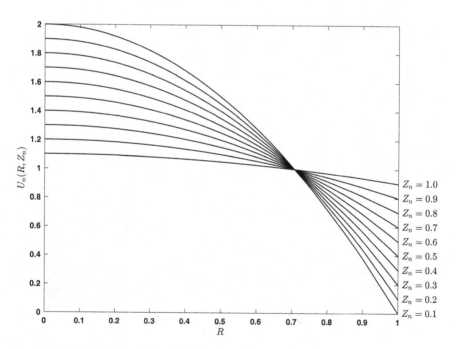

Fig. 10.1 Velocity distributions $U_n(R, Z_n)$, $Z_n = 0.1(n+1)$, $n = 0, 1, \ldots, 9$

i.e., $\bar{u} = u^0 = $ const.

If put (10.1.1), (10.1.2) in (10.0.8), the model has the form:

$$U_n \frac{\partial C_n}{\partial Z_n} = -Da\, C_n; \quad 0.1n \leq Z_n \leq 0.1(n+1);$$
$$Z_n = 0.1n, \quad C_n(R, Z_n) = C_{n-1}(R, Z_n); \quad n = 0, 1, \ldots, 9; \tag{10.1.4}$$
$$Z_0 = 0, \quad C_0(R, Z_0) \equiv 1.$$

10.1.1   Model Equation Solution

The solution of (10.1.4) $C(R, Z) = C_n(R, Z_n), Z_n = 0.1(n+1), n = 0, 1, \ldots, 9$ in $A(Z)$ the case $Da = 1$ is presented in Fig. 10.2. This solution $C(R, Z)$ permits to be obtained in (10.0.4) the average ("theoretical") concentration distribution $\overline{C}(Z) = \overline{C}_n(Z_n), Z_n = 0.1(n+1), n = 0, 1, \ldots, 9$ in the column (the points in Fig. 10.3) and function (the points in Fig. 10.4) on every step:

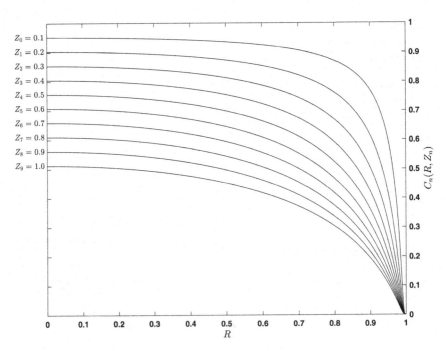

Fig. 10.2 Concentration distributions $C(Z) = C_n(Z_n), Z_n = 0.1(n+1), n = 0, 1, \ldots, 9$

$$\overline{C}(Z) = \overline{C}_n(Z_n) = 2\int_0^1 R C_n(R, Z_n)\,dR,$$

$$A(Z) = A_n(Z_n) = 2\int_0^1 R U_n(R) \frac{C_n(R, Z_n)}{\overline{C}_n(Z_n)}\,dR, \qquad (10.1.5)$$

$$Z_n = 0.1(n+1), n = 0, 1, \ldots, 9,$$

which are presented in Figs. 10.3 and 10.4. From Fig. 10.4 is seen that the function $A(Z)$ is possible to be presented as a quadratic approximation:

$$A(Z) = a_0 + a_1 Z + a_2 Z^2, \qquad (10.1.6)$$

where the ("theoretical") values of are presented in Table 10.1. As a result, in the case of axial modification of the radial non-uniformity of the velocity, the model (10.0.9) has the form:

$$\left(a_0 + a_1 Z + a_2 Z^2\right)\frac{d\overline{C}}{dZ} + (a_1 + 2a_2 Z)\overline{C} = -Da\overline{C}; \quad Z = 0, \quad \overline{C} = 1, \quad (10.1.7)$$

where the parameters a_0, a_1, a_2 must be obtained, using experimental data.

10.1.2 Parameter Identification

The obtained value of the function $\overline{C}(1)$ (Fig. 10.3) permits to be obtained the artificial experimental data $\overline{C}_{exp}^m(1)$ for the column end $(Z = 1)$:

$$\overline{C}_{exp}^m(1) = (0.95 + 0.1B_m)\overline{C}(1), \quad m = 1, \ldots, 10, \qquad (10.1.8)$$

where $0 \le B_m \le 1, m = 0, 1, \ldots, 10$ are obtained by a generator of random numbers. The obtained artificial experimental data (10.1.8) are used for the illustration of the parameters (a_0, a_1, a_2) identification in the average-concentration model (10.1.7) by the minimization of the least-squares function:

$$Q(a_0, a_1, a_2) = \sum_{m=1}^{10} \left[\overline{C}(1, a_0, a_1, a_2) - \overline{C}_{exp}^m(1)\right]^2, \qquad (10.1.9)$$

where the value of $\overline{C}(1, a_0, a_1, a_2)$ is obtained after the solution of (10.1.7) for $Z = 1$. The obtained "experimental" parameter values are presented in Table 10.1.

The obtained ("experimental") parameter values are used for the solution of (10.1.7), and the result (the line) is compared with the average ("theoretical") concentration values $\overline{C}(Z) = \overline{C}_n(Z_n), Z_n = 0.1(n+1), n = 0, 1, \ldots, 9$ (points) [as solution of (10.1.4) and (10.1.5)] in Fig. 10.3.

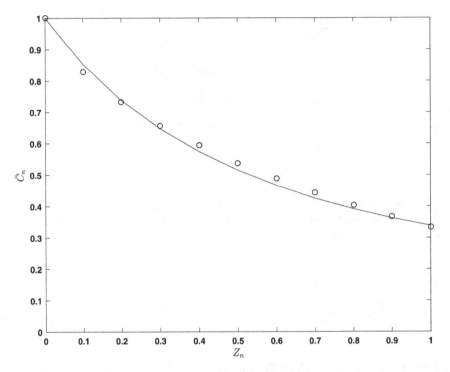

Fig. 10.3 Average-concentration distribution: "theoretical" values [as solution of (10.1.4) and (10.1.5)] $\overline{C}(Z) = \overline{C}_n(Z_n), Z_n = 0.1(n+1), n = 0, 1, \ldots, 9$ (points); $\overline{C}(Z)$ as a solution of (10.1.7) for "experimental" values of a_0, a_1, a_2 (line)

10.1.3 Influence of the Model Parameter

The model (10.1.7), with "experimental" parameter values of a_0, a_1, a_2 in Table 10.1, is used for the calculation the average concentrations in the case $Da = 2$ and the result (line) is compared (Fig. 10.5) with the average ("theoretical") concentration values $\overline{C}(Z) = \overline{C}_n(Z_n), Z_n = 0.1(n+1), n = 0, 1, \ldots, 9$ [as solutions of (10.1.4) and (10.1.5)] (points) for this case.

The presented numerical analysis of the industrial column chemical reactors shows [4] that average-concentration model, where the radial velocity component is equal to zero (in the cases of a constant velocity radial non-uniformity along the column height), is possible to be used in the cases of an axial modification of the radial non-uniformity of the axial velocity component. The use of experimental data, for the average concentration at the column end, for a concrete process, permits to be obtained the model parameters (a_0, a_1, a_2), related with the radial non-uniformity of the velocity. These parameter values permit to be used the average-concentration model for modeling of different processes (different values of

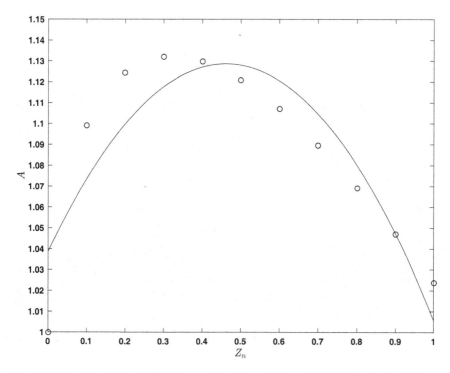

Fig. 10.4 Function $A(Z) = A_n(Z_n), Z_n = 0.1(n+1), n = 0, 1, \ldots, 9$ (10.1.5) (points); $A(Z)$ as a quadratic approximation (10.1.6) (line)

Table 10.1 Parameters a_0, a_1, a_2

Parameters	"Theoretical" values	"Experimental" values
a_0	1.0387	0.8582
a_1	0.3901	0.4505
a_2	−0.4230	−0.4343

the parameter Da, i.e., different values of the column height, average velocity, reagent diffusivity, and chemical reaction rate constant).

10.2 Effect of the Radial Velocity Component

The radial non-uniformity of the axial velocity component in a column apparatus is a result of the fluid hydrodynamics at the column inlet, where it is a maximum and decreases along the column height and as a result, a radial velocity component is initiated. The theoretical analysis of the change in the radial non-uniformity of the axial velocity component and the effect of the radial velocity component in a

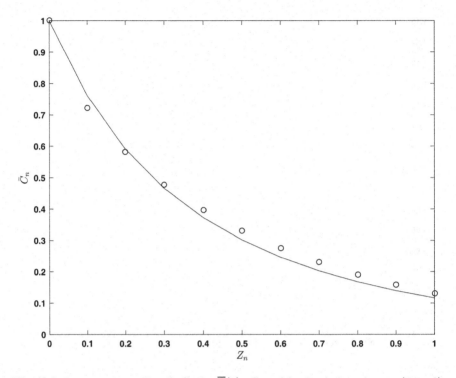

Fig. 10.5 Average-concentration distribution $\overline{C}(Z)$: effect of the chemical reaction rate $(Da = 2)$

column can be made by an appropriate hydrodynamic model, where the average velocity at the cross section of the column is a constant, while the maximal velocity (and as a result the radial non-uniformity of the axial velocity component, too) decreases along the column height.

A theoretical analysis of the effect of the radial velocity components in the industrial column chemical reactors will be presented in the case, when the radial velocity component is not equal to zero for pseudo-first-order chemical reactions. In the stationary case, the convection–diffusion model [1–3] has the form:

$$
\begin{aligned}
&u\frac{\partial c}{\partial z} + v\frac{\partial c}{\partial r} = D\left(\frac{\partial^2 c}{\partial z^2} + \frac{1}{r}\frac{\partial c}{\partial r} + \frac{\partial^2 c}{\partial r^2}\right) - kc;\\
&r = 0, \quad \frac{\partial c}{\partial r} \equiv 0; \quad r = r_0, \quad \frac{\partial c}{\partial r} \equiv 0; \\
&z = 0, \quad c \equiv c^0, \quad u^0 c^0 \equiv uc^0 - D\frac{\partial c}{\partial z}.
\end{aligned}
\tag{10.2.1}
$$

$$
\frac{\partial u}{\partial z} + \frac{\partial v}{\partial r} + \frac{v}{r} = 0; \quad r = r_0, \quad v(r_0, z) \equiv 0, \quad z = 0, \quad u = u(r, 0). \tag{10.2.2}
$$

In (10.2.1, 10.2.2) $c(r, z), D$, are the concentrations (kg mol m^{-3}) and the diffusivities (m^2 s^{-1}) of the reagents in the fluid, $u(r, z)$ and $v(r, z)$—the axial and radial velocity components (m s^{-1}), (r, z)—the radial and axial coordinates (m),

k—chemical reaction rate constant, u^0, c^0—input $(z = 0)$ velocity and concentrations.

The theoretical analysis of the model (10.2.1), (10.2.2) will be made, using generalized variables [1]:

$$r = r_0 R, \quad z = lZ, \quad \varepsilon = \frac{r_0}{l},$$
$$u(r, z) = u(r_0 R, lZ) = u^0 U(R, Z),$$
$$v(r, z) = v(r_0 R, lZ) = u^0 \varepsilon V(R, Z), \tag{10.2.3}$$
$$c(r, z) = c(r_0 R, lZ) = c^0 C(R, Z).$$

As a result from (10.2.1)–(10.2.3) is possible to be obtained:

$$U \frac{\partial C}{\partial Z} + V \frac{\partial C}{\partial R} = Fo \left(\varepsilon^2 \frac{\partial^2 C}{\partial Z^2} + \frac{1}{R} \frac{\partial C}{\partial R} + \frac{\partial^2 C}{\partial R^2} \right) - Da\, C;$$
$$R = 0, \quad \frac{\partial C}{\partial R} \equiv 0; \quad R = 1, \quad \frac{\partial C}{\partial R} \equiv 0; \tag{10.2.4}$$
$$Z = 0, \quad C \equiv 1, \quad 1 \equiv U - Pe^{-1} \frac{\partial C}{\partial Z}.$$

$$\frac{\partial U}{\partial Z} + \frac{\partial V}{\partial R} + \frac{V}{R} = 0; \quad R = 1, \quad V(1, Z) \equiv 0; \quad Z = 0, \quad U = U(R, 0). \tag{10.2.5}$$

In (10.2.4) are used the parameters:

$$Fo = \frac{D_i l}{u^0 r_0^2}, \quad Da = \frac{kl}{u^0}, \quad Pe = \frac{u^0 l}{D}, \tag{10.2.6}$$

where Fo, Da, and Pe are the Fourier, Damkohler, and Peclet numbers, respectively.

10.2.1 Axial and Radial Velocities

The theoretical analysis of the change in the radial non-uniformity of the axial velocity component (effect of the radial velocity component) in a column can be made by an appropriate hydrodynamic model, where the average velocity at the cross section of the column is a constant (inlet average axial velocity component), while the radial non-uniformity of the axial velocity component decreases along the column height, and as a result, a radial velocity component is initiated. In generalized variables, (10.2.3) is possible to be used the model:

$$U = (2 - 0.4Z) - 2(1 - 0.4Z)R^2, \quad V = 0.2(R - R^3), \tag{10.2.7}$$

where the velocity components satisfy Eq. (10.2.5), and the average velocity is a constant, and the inlet velocity distribution is the Poiseuille type:

$$\bar{u} = \frac{2}{r_0^2} \int_0^{r_0} ru(r,z)\mathrm{d}r = u^0, \quad \overline{U} = 2 \int_0^1 RU\mathrm{d}R = 1. \qquad (10.2.8)$$

The velocity components (10.2.7) are presented in Figs. 10.6 and 10.7.

10.2.2   Concentration Distributions

In industrial conditions $Fo < 10^{-2}, Pe > 10^2$, and the model (10.2.4) has the convective [2] form:

$$U\frac{\partial C}{\partial Z} + V\frac{\partial C}{\partial R} = -Da\,C; \quad R = 1, \quad \frac{\partial C}{\partial R} \equiv 0; \quad Z = 0, \quad C \equiv 1. \qquad (10.2.9)$$

From Fig. 10.7 is seen that $V < 0.1$ and must be presented with the help of a small parameter $0.1 = \alpha \ll 1$, i.e.,

$$V = \alpha V_0, \quad V_0 = 0.2(R - R^3)\alpha^{-1} \qquad (10.2.10)$$

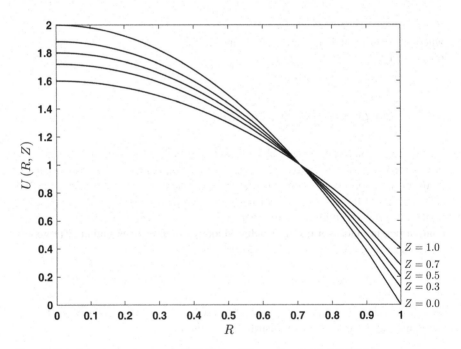

Fig. 10.6 Axial velocity component $U(R,Z)$ for different $Z = 0, 0.3, 0.5, 0.7, 1.0$

Fig. 10.7 Radial velocity component $V(R)$

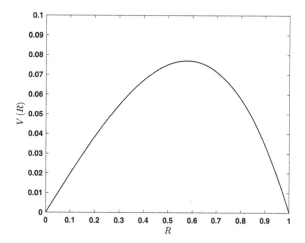

As a result, the problem (10.2.9) has the form:

$$U\frac{\partial C}{\partial Z} + \alpha V_0 \frac{\partial C}{\partial R} = -Da\,C; \quad R = 1, \quad \frac{\partial C}{\partial R} \equiv 0; \quad Z = 0, \quad C \equiv 1, \quad (10.2.11)$$

where $0.1 = \alpha \ll 1$ is a small parameter and (10.2.11) must be solved by the perturbation method [5], i.e., the concentration must be presented as

$$C(R,Z) = C_0(R,Z) + \alpha C_1(R,Z) + \alpha^2 C_2(R,Z) \qquad (10.2.12)$$

and the problem (10.2.11) must be replaced by a set of equations:

$$U\frac{\partial C_0}{\partial Z} = -Da\,C_0; \quad Z = 0, \quad C_0 \equiv 1. \qquad (10.2.13)$$

$$U\frac{\partial C_1}{\partial Z} = -Da\,C_1 - V_0\frac{\partial C_0}{\partial R}; \quad Z = 0, \quad C_1 \equiv 0. \qquad (10.2.14)$$

$$U\frac{\partial C_2}{\partial Z} = -Da\,C_2 - V_0\frac{\partial C_1}{\partial R}; \quad Z = 0, \quad C_2 \equiv 0. \qquad (10.2.15)$$

The solution of (10.2.13)–(10.2.15) is possible to be obtained, using cubic spline interpolations for $\frac{\partial C_0}{\partial R}$ and $\frac{\partial C_1}{\partial R}$.

The solutions of (10.2.13)–(10.2.15) in the case $Da = 1$, using (10.2.10) and $0.1 = \alpha \ll 1$, lead to $.C(R,Z)$ in (10.2.12), which is presented in Fig. 10.8.

Fig. 10.8 Concentration
distributions $C(R, Z)$ for
different
$Z = 0, 0.3, 0.5, 0.7, 1.0$

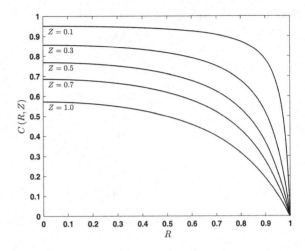

10.2.3   Average-Concentration Model

In [5] is possible to see that the average values of the velocity and concentrations at
the column cross-sectional area are:

$$\bar{u}(z) = \frac{2}{r_0^2} \int_0^{r_0} r u(r, z) \mathrm{d}r, \quad \bar{c}(z) = \frac{2}{r_0^2} \int_0^{r_0} r c(r, z) \mathrm{d}r, \tag{10.2.16}$$

where in a cylindrical columns $r_0 = \text{const}, \bar{u}(z) \equiv u^0 = \text{const}$.

The functions $u(r, z), v(r, z), c(r, z)$ in (10.2.1) can be presented with the help of
the average functions (10.2.16):

$$u(r, z) = \bar{u}U(R, Z), \quad v(r, z) = \varepsilon\bar{u}V(R), \quad c(r, z) = \bar{c}(z)\tilde{c}(r, z), \tag{10.2.17}$$

where $\tilde{c}(r, z)$ present the radial non-uniformity of the concentrations and satisfy the
following conditions:

$$\frac{2}{r_0^2} \int_0^{r_0} r\tilde{c}(r, z) \mathrm{d}r = 1. \tag{10.2.18}$$

The average-concentration model may be obtained when putting (10.2.17) into
(10.2.1), multiplying by r and integrating over r in the interval $[0, r_0]$. As a result,
the following is obtained:

$$\alpha(z)\bar{u}\frac{d\bar{c}}{dz} + [\beta(z) + \gamma(z)]\bar{u}\bar{c}_i = D\frac{d^2\bar{c}}{dz^2} - k\bar{c};$$
$$z = 0, \quad \bar{c} \equiv c^0, \quad \frac{d\bar{c}}{dz} \equiv 0, \tag{10.2.19}$$

where

$$\alpha(z) = \frac{2}{r_0^2}\int_0^{r_0} rU\tilde{c}\,dr, \quad \beta(z) = \frac{2}{r_0^2}\int_0^{r_0} rU\frac{\partial\tilde{c}}{\partial z}dr, \quad \gamma(z) = \frac{2}{r_0^2}\int_0^{r_0} rV\frac{\partial\tilde{c}}{\partial r}dr,$$
$$U = U(R, Z), \quad V = V(R), \quad \tilde{c}(r, z) = \tilde{C}(R, Z). \tag{10.2.20}$$

The theoretical analysis of the model (10.2.19) will be made, using the next generalized variables and functions:

$$z = lZ, \quad r = r_0 R, \quad \bar{c}(z) = c^0\overline{C}(Z),$$
$$\overline{C}(Z) = 2\int_0^1 RC(R, Z)dR, \quad \tilde{c}(r, z) = \frac{c(r,z)}{\bar{c}(z)} = \frac{C(R,Z)}{\overline{C}(Z)} = \tilde{C}(R, Z),$$
$$\alpha(z) = \alpha(lZ) = A(Z) = 2\int_0^1 RU(R, Z)\tilde{C}(R, Z)dR, \tag{10.2.21}$$
$$\beta(z) = \beta(lZ) = B(Z) = 2\int_0^1 RU(R, Z)\frac{\partial\tilde{C}}{\partial Z}dR,$$
$$\gamma(z) = \gamma(lZ) = G(Z) = 2\int_0^1 RV(R)\frac{\partial\tilde{C}}{\partial R}dR$$

and as a result

$$A(Z)\frac{d\overline{C}}{dz} + [B(Z) + G(Z)]\overline{C} = Pe^{-1}\frac{d^2\overline{C}}{dz^2} - Da\,\overline{C};$$
$$Z = 0, \quad \overline{C} = 1, \quad \frac{d\overline{C}}{dz} = 0. \tag{10.2.22}$$

In industrial conditions $Pe > 10^2$, and the model (10.2.22) has the convective [2] form:

$$A(Z)\frac{d\overline{C}}{dz} + [B(Z) + G(Z)]\overline{C} = -Da\,\overline{C};$$
$$Z = 0, \quad \overline{C} = 1. \tag{10.2.23}$$

The solution of (10.2.7), (10.2.9), and (10.2.21) permits to be obtained the average concentrations $\overline{C}(Z_n), Z_n = 0.1(n+1), n = 0, 1, \ldots, 9$ ("theoretical" values) and functions $A(Z_n), B(Z_n), G(Z_n), Z_n = 0.1(n+1), n = 0, 1, \ldots, 9$ in (10.2.21), which are presented (points) in Figs. 10.9, 10.10, and 10.11.

From Figs. 10.10 and 10.11 is seen that the functions $A(Z), B(Z), G(Z)$ are possible to be presented as the next approximations:

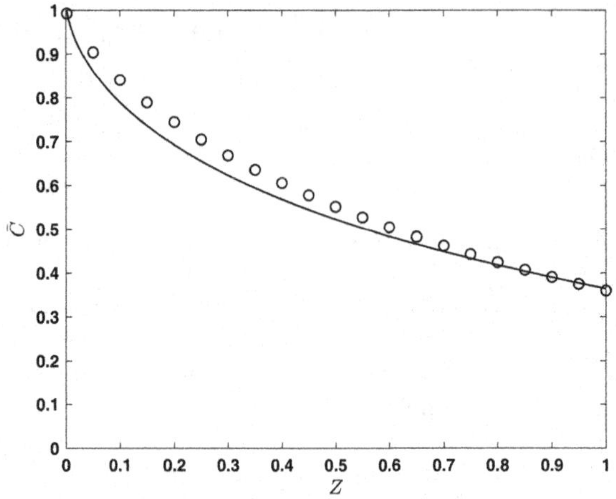

Fig. 10.9 Average concentrations $\overline{C}(Z)$: "theoretical" values $\overline{C}(Z_n), Z_n = 0.1(n+1), n = 0, 1, \ldots, 9$ (points); solution of (10.2.25) (lines)

Fig. 10.10 Functions $A(Z_n), G(Z_n), Z_n = 0.1(n+1), n = 0, 1, \ldots, 9$ (points) and theirs quadratic and linear approximations (10.2.24) (lines)

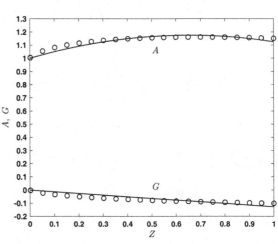

Fig. 10.11 Functions $B(Z_n), Z_n = 0.1(n+1), n = 0, 1, \ldots, 9$ (points) and its parabolic approximation (10.2.24) (line)

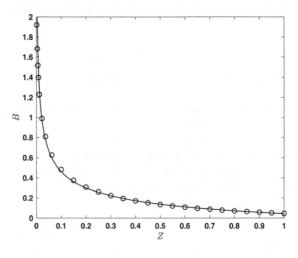

$$A(Z) = 1 + a_1 Z + a_2 Z^2, \quad B(Z) = b_0 + b_1 Z^{b_2}, \quad G(Z) = gZ \qquad (10.2.24)$$

and the "theoretical" parameter values $(a_1, a_2, b_0, b_1, b_2, g)$ are presented in Table 10.2. As a result, the model (10.2.23) has the form:

$$(1 + a_1 Z + a_2 Z^2) \frac{d\overline{C}}{dz} + (b_0 + b_1 Z^{b_2} + gZ)\overline{C} = -Da\,\overline{C}; \\ Z = 0, \quad \overline{C} = 1, \qquad (10.2.25)$$

where the parameters $P(a_1, a_2, b_0, b_1, b_2, g)$ must be obtained, using experimental data.

The parameters b_0, b_1, b_2 in the parabolic approximation of the function $B(Z)$ are obtained, using the least-squares method [1, 5].

10.2.4 Parameter Identification

The obtained from (10.2.9) and (10.2.21) value of the function $\overline{C}(1)$ (Fig. 10.9) permits to be obtained the artificial experimental data $\overline{C}_{exp}^{m}(1)$ for the column end $(Z = 1)$:

$$\overline{C}_{exp}^{m}(1) = (0.95 + 0.1 B_m)\overline{C}(1), \quad m = 1, \ldots, 10, \qquad (10.2.26)$$

where $0 \leq B_m \leq 1, m = 0, 1, \ldots, 10$ are obtained by a generator of random numbers.

The obtained artificial experimental data (10.2.26) are possible to be used for the illustration of the parameters P identification in the average-concentration model (10.2.27) by the minimization of the least-squares function:

$$Q(P) = \sum_{m=1}^{10} \left[\overline{C}(1, P) - \overline{C}_{exp}^{m}(1) \right]^2, \qquad (10.2.27)$$

where the values of $\overline{C}(1, P)$ are obtained after the solution of (10.2.25) for $Z = 1$.

From the "theoretical" values of the parameters in Table 10.2, it is seen that as starting parameter values in the minimization procedure of the parametric

Table 10.2 Parameters $P(a_1, a_2, b_0, b_1, b_2, g)$

Parameters	"Theoretical" values	"Experimental" values
a_1	0.537	0.532
a_2	−0.412	−0.516
b_0	−0.311	−0.571
b_1	0.354	0.385
b_2	−0.339	−0.487
g	−0.127	−0.109

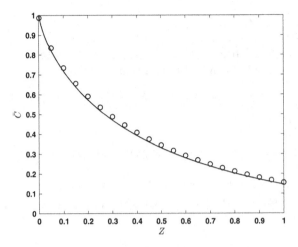

Fig. 10.12 Average-concentration distribution $\overline{C}(Z)$: effect of the chemical reaction rate ($Da = 2$)

identification, it is convenient to use $a_1 = 0.5, a_2 = -0.5, b_0 = -0.5, b_1 = 0.5, b_2 = -0.5, g = -0.1$.

The obtained "experimental" parameter values after the minimization procedure (10.2.27) are presented in Table 10.2.

The obtained ("experimental") parameter values are used for the solution of (10.2.25), and the results (the lines) are compared with the average ("theoretical") concentration values $\overline{C}(Z_n), Z_n = 0.1(n+1), n = 0, 1, \ldots, 9$ (points) [as solution of (10.2.7), (10.2.9) and (10.2.21)] in Fig. 10.9.

10.2.5   Influence of the Model Parameters

The model (10.2.25), with "experimental" parameter values of P in Table 10.2, is used for the calculation the average concentrations in the cases $Da = 2$, and the results (lines) are compared (Fig. 10.12) with the average ("theoretical") concentration values $\overline{C}(Z) = \overline{C}_n(Z_n), Z_n = 0.1(n+1), n = 0, 1, \ldots, 9$ [as solution of (10.2.7), (10.2.9), and (10.2.21)] (points) for this case.

References

1. Boyadjiev C (2010) Theoretical chemical engineering. Modeling and simulation. Springer, Berlin, Heidelberg
2. Doichinova M, Boyadjiev C (2012) On the column apparatuses modeling. Int J Heat Mass Transf 55:6705–6715

3. Boyadjiev C (2013) A new approach for the column apparatuses modeling in chemical engineering. J Pure Appl Math Adv Appl 10(2):131–150
4. Boyadjiev B, Boyadjiev C (2017) New models of industrial column chemical reactors. Bul Chem Commun 49(3):706–710
5. Boyadjiev B, Boyadjiev C (2018) A new approach to modeling of chemical processes in industrial column apparatuses. Chem Eng Tech (in press)

Chapter 11
Industrial Co-current Column Absorber

The new approach for the modeling of the processes in column apparatuses [1–3] presents the convection–diffusion and average-concentration models of the column chemical reactors [4], in the cases of an axial modification of the axial velocity radial non-uniformity along the column height (see Chap. 10). This problem will be solved in the cases of the absorption processes in a co-current column [5].

In the cases of a constant radial non-uniformity of the velocities along the column height $\left(u_j = u_j(r), j = 1, 2\right)$, the convection–diffusion and average-concentration models of a chemical absorption (with a pseudo-first-order chemical reaction in the liquid phase) in a co-current column (see Chaps. 3 and 6) have the forms:

$$u_1 \frac{\partial c_1}{\partial z} = D_1 \left(\frac{\partial^2 c_1}{\partial z^2} + \frac{1}{r} \frac{\partial c_1}{\partial r} + \frac{\partial^2 c_1}{\partial r^2} \right) - k(c_1 - \chi c_2);$$

$$u_2 \frac{\partial c_2}{\partial z} = D_2 \left(\frac{\partial^2 c_2}{\partial z^2} + \frac{1}{r} \frac{\partial c_2}{\partial r} + \frac{\partial^2 c_2}{\partial r^2} \right) + k(c_1 - \chi c_2) - k_0 c_2;$$

$$r = 0, \quad \frac{\partial c_j}{\partial r} \equiv 0; \quad r = r_0, \quad \frac{\partial c_j}{\partial r} \equiv 0; \quad j = 1, 2;$$

$$z = 0, \quad c_1 \equiv c_1^0, \quad c_2 \equiv 0, \quad u_1^0 c_1^0 \equiv u_1 c_1^0 - D_1 \left(\frac{\partial c_1}{\partial z} \right)_{z=0}, \quad \left(\frac{\partial c_2}{\partial z} \right)_{z=0} = 0.$$

$$\tag{11.0.1}$$

$$\alpha_1(z) \bar{u}_1 \frac{d\bar{c}_1}{dz} + \frac{d\alpha_1}{dz} \bar{u}_1 \bar{c}_1 = D_1 \frac{d^2 \bar{c}_1}{dz^2} - k(\bar{c}_1 - \chi \bar{c}_2);$$

$$\alpha_2(z) \bar{u}_2 \frac{d\bar{c}_2}{dz} + \frac{d\alpha_2}{dz} \bar{u}_2 \bar{c}_2 = D_2 \frac{d^2 \bar{c}_2}{dz^2} + k(\bar{c}_1 - \chi \bar{c}_2) - k_0 \bar{c}_2;$$

$$z = 0, \quad \bar{c}_1(0) \equiv c_1^0, \quad \bar{c}_2(0) \equiv 0, \quad \frac{d\bar{c}_1}{dz} \equiv 0, \quad \frac{d\bar{c}_2}{dz} \equiv 0; \tag{11.0.2}$$

$$\alpha_j(z) = \frac{2}{r_0^2} \int_0^{r_0} r \tilde{u}_j \tilde{c}_j dr, \quad \tilde{u}_j(r) = \frac{u_j(r)}{\bar{u}_j}, \quad \tilde{c}_j(r, z) = \frac{c_j(r, z)}{\bar{c}_j(z)}, \quad j = 1, 2.$$

In (11.0.1) and (11.0.2), $u_j(r), c_j(r, z), D_j, j = 1, 2$ are the velocities, concentrations, and diffusivities in the gas $(j = 1)$ and liquid $(j = 2)$ phases, (r, z)—the radial and axial coordinates (m), χ—Henry's number, k—volume interphase mass transfer

© Springer International Publishing AG, part of Springer Nature 2018
C. Boyadjiev et al., *Modeling of Column Apparatus Processes,*
Heat and Mass Transfer, https://doi.org/10.1007/978-3-319-89966-4_11

coefficient (s^{-1}), k_0—chemical reaction rate constant (s^{-1}), r_0, l—column radius and height (m), $u_j^0, c_j^0, j = 1, 2$—input $(z = 0)$ velocities and concentrations $\bar{u}_j, \bar{c}_j(z), j = 1, 2$—the average velocities and concentrations at the column cross-sectional area, $\tilde{u}_j(r), \tilde{c}_j(r, z), j = 1, 2$—the radial non-uniformities of the velocities and concentrations.

In (11.0.1) and (11.0.2), it is possible to introduce the generalized variables:

$$r = r_0 R, \quad z = lZ, \quad u_j(r) = \bar{u}_j U_j(R), \quad j = 1, 2,$$

$$c_1(r, z) = c_1^0 C_1(R, Z), \quad c_2(r, z) = \frac{c_1^0}{\chi} C_2(R, Z), \tag{11.0.3}$$

$$\bar{c}_1(z) = c_1^0 \overline{C}_1(Z), \quad \bar{c}_2(z) = \frac{c_1^0}{\chi} \overline{C}_2(Z),$$

and result is obtained:

$$U_1 \frac{\partial C_1}{\partial Z} = Fo_1 \left(\varepsilon^2 \frac{\partial^2 C_1}{\partial Z^2} + \frac{1}{R} \frac{\partial C_1}{\partial R} + \frac{\partial^2 C_1}{\partial R^2} \right) - K_1(C_1 - C_2);$$

$$U_2 \frac{\partial C_2}{\partial Z} = Fo_2 \left(\varepsilon^2 \frac{\partial^2 C_2}{\partial Z^2} + \frac{1}{R} \frac{\partial C_2}{\partial R} + \frac{\partial^2 C_2}{\partial R^2} \right) + K_2(C_1 - C_2) - DaC_2;$$

$$R = 0, \quad \frac{\partial C_j}{\partial R} \equiv 0; \quad R = 1, \quad \frac{\partial C_j}{\partial R} \equiv 0; \quad j = 1, 2; \tag{11.0.4}$$

$$Z = 0, \quad C_1 \equiv 1, \quad C_2 = 0, \quad 1 \equiv U_1 - Pe_1^{-1} \frac{\partial C_1}{\partial Z}, \quad \frac{\partial C_2}{\partial Z} \equiv 0.$$

$$A_1(Z) \frac{d\overline{C}_1}{dZ} + \frac{dA_1}{dZ} \overline{C}_1 = Pe_1^{-1} \frac{d^2\overline{C}_1}{dZ^2} - K_1(\overline{C}_1 - \overline{C}_2);$$

$$A_2(Z) \frac{d\overline{C}_2}{dZ} + \frac{dA_2}{dZ} \overline{C}_2 = Pe_2^{-1} \frac{d^2\overline{C}_2}{dZ^2} + K_2(\overline{C}_1 - \overline{C}_2) - Da\overline{C}_2; \tag{11.0.5}$$

$$Z = 0, \quad \overline{C}_1 = 1, \quad \overline{C}_2 = 0, \quad \frac{d\overline{C}_1}{dZ} = 0, \quad \frac{d\overline{C}_2}{dZ} = 0,$$

where Fo, Da, and Pe are the Fourier, Damkohler, and Peclet numbers, respectively:

$$Fo_j = \frac{D_j l}{\bar{u}_j r_0^2}, \quad Pe_j = \frac{\bar{u}_j l}{D_j}, \quad Da = \frac{k_0 l}{\bar{u}_2}, \quad \varepsilon^2 = \frac{r_0^2}{l^2} = Fo_j^{-1} Pe_j^{-1},$$

$$K_1 = \frac{kl}{\bar{u}_1}, \quad K_2 = \omega K_1, \quad \omega = \frac{\bar{u}_1 \chi}{\bar{u}_2}, \quad j = 1, 2. \tag{11.0.6}$$

In the cases of a physical absorption $Da = 0$ and $\omega \to 0 (\omega \to \infty)$ for highly (lightly) soluble gases.

In (11.0.5) are used the expressions:

$$\overline{C}_j(Z) = 2\int_0^1 RC_j(R,Z)dR, \quad \tilde{c}_j(r,z) = \frac{c_j(r,z)}{\bar{c}_j(z)} = \frac{C_j(R,Z)}{\overline{C}_j(Z)},$$

$$\alpha_j(z) = \alpha_j(lZ) = A_j(Z) = 2\int_0^1 RU_j(R)\frac{C_j(R,Z)}{\overline{C}_j(Z)}dR, \quad j = 1,2. \tag{11.0.7}$$

Let's consider the physical absorption $(Da = 0)$ of an average soluble gas $(\omega \sim 1)$ in an industrial absorption column $(l > 1)$ (m), in the cases of down co-current gas–liquid drops flow. The average gas velocity and the average liquid drops velocity are $\bar{u}_1 \sim 1, \bar{u}_2 \sim 1$ (m s^{-1}); the diffusivities in the gas (air) and the liquid (water) are $D_1 \sim 10^{-4}, D_2 \sim 10^{-9}$ (m^2 s^{-1}). In these conditions, it is possible to obtain the next order of magnitude of the parameter values:

$$Pe_1^{-1} \sim 10^{-4}, \quad \frac{h_0}{l} \sim 10^{-1}, \quad Fo_1 \sim 10^{-2}, \quad Pe_2^{-1} \sim 10^{-9}, \quad Fo_2 \sim 10^{-7} \tag{11.0.8}$$

and the model (11.0.4) has a convective form:

$$U_1\frac{dc_1}{dz} = -K_1(C_1 - C_2); \quad U_2\frac{dc_2}{dz} = \omega K_1(C_1 - C_2);$$

$$Z = 0, \quad C_1 \equiv 1, \quad C_2 = 0. \tag{11.0.9}$$

In the conditions (11.0.8), the average-concentration model (11.0.5) has the form:

$$A_1(Z)\frac{d\overline{C}_1}{dZ} + \frac{dA_1}{dZ}\overline{C}_1 = -K_1(\overline{C}_1 - \overline{C}_2);$$

$$A_2(Z)\frac{d\overline{C}_2}{dZ} + \frac{dA_2}{dZ}\overline{C}_2 = \omega K_1(\overline{C}_1 - \overline{C}_2); \tag{11.0.10}$$

$$Z = 0, \quad \overline{C}_1 = 1, \quad \overline{C}_2 = 0.$$

11.1 Effect of the Axial Modification of the Radial Non-uniformity of the Axial Velocity Component

The radial non-uniformity of the axial velocity components of the gas and liquid phases in a co-current column apparatus is the result of the fluid hydrodynamics at the column inlet, where they are maximums and decrease along the column height as a result of the fluid viscosities. For a theoretical analysis of the effect of the axial modification of the radial non-uniformities of the velocities in a two-phase, co-current column will be used a hydrodynamic model, where the average velocities in the phases at the cross section of the column are constants, while the maximal velocities (and as a result the radial non-uniformity of the axial velocities components too) decrease along the column height [6].

Let's considers [6] the velocity distributions

$$u_{jn}(r, z_n) = u_j^0 U_{jn}(R, Z_n), \quad j = 1, 2, \quad n = 0, 1, .., 9, \tag{11.1.1}$$

where $u_j^0 = \text{const}, j = 1, 2$ are the inlet velocities and an axial step change of the radial non-uniformity of the axial velocity components in the column:

$$
\begin{aligned}
&U_{jn}(R, Z_n) = a_{jn} - b_{jn}R^2, \\
&a_{jn} = 2 - 0.1n, \quad b_{jn} = 2(1 - 0.1n), \\
&0.1n \le Z_n \le 0.1(n+1), \quad n = 0, 1, \ldots, 9, \quad j = 1, 2,
\end{aligned}
\tag{11.1.2}
$$

where $U_{jn}(R, Z_n), j = 1, 2 n = 0, 1, \ldots, 9$ satisfy the equations:

$$2 \int_0^1 R U_{jn}(R, Z_n) dR = 1, \quad j = 1, 2, \quad n = 0, 1, \ldots, 9, \tag{11.1.3}$$

i.e. $\bar{u}_j = u_j^0 = \text{const}, j = 1.2$.

If put (11.1.2) in (11.0.9), the convection–diffusion model has the form:

$$
\begin{aligned}
&U_{1n} \frac{dc_{1n}}{dz_n} = -K_1(C_{1n} - C_{2n}); \quad 0.1n \le Z_n \le 0.1(n+1); \\
&U_{2n} \frac{dc_{2n}}{dz_n} = \omega K_1(C_{1n} - C_{2n}); \\
&Z_n = 0.1n, \quad C_{jn}(R, Z_n) = C_{j(n-1)}(R, Z_n); \\
&n = 0, 1, \ldots, 9; \quad j = 1, 2; \\
&Z_0 = 0, \quad C_{10}(R, Z_0) \equiv 1, \quad C_{20}(R, Z_0) = 0.
\end{aligned}
\tag{11.1.4}
$$

11.1.1   Model Equations Solution

The parameter ω in (11.1.4) is known beforehand. The solution of (11.1.4), for a concrete absorption process ($\omega = 1$) of an average soluble gas and "theoretical" value of $K_1 = 1$, permits to be obtained the concentration distributions $C_{jn}(R, Z_n), j = 1, 2$ for different $Z_n = 0.1(n+1), n = 0, 1, \ldots, 9$ (Fig. 11.1).

The solution of (11.1.4) (Fig. 11.1) and (11.0.7) permits to be obtained the "theoretical" average-concentration distributions $\bar{C}_{jn}(Z_n), j = 1, 2$ (the points in Fig. 11.2) and the functions $A_{jn}(Z_n), j = 1, 2$ (the points in Fig. 11.3) for different $Z_n = 0.1(n+1), n = 0, 1, \ldots, 9$.

Figure. 11.3 shows that the functions $A_{jn}(Z_n), n = 0, 1, \ldots, 4, j = 1, 2$ are possible to be presented as quadratic approximations:

$$A_1(Z) = a_{10} + a_{11}Z + a_{12}Z^2, \quad A_2(Z) = a_{20} + a_{21}Z + a_{22}Z^2, \tag{11.1.5}$$

where the ("theoretical") values of $a_{j0}, a_{j1}, a_{j2}, j = 1, 2$ are presented in Table 11.1.

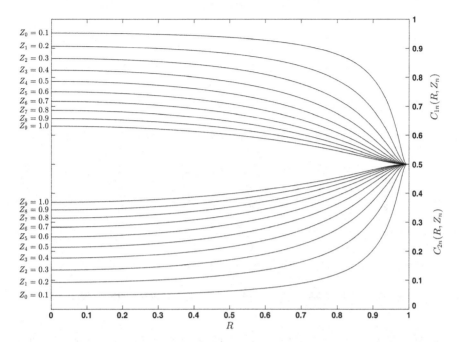

Fig. 11.1 Concentration distributions $C_{jn}(R, Z_n), j = 1, 2$ in the case $\omega = K_1 = 1$ for different $Z_n = 0.1(n+1), n = 0, 1, \ldots, 9$

As a result, in the case of axial modification of the radial non-uniformity of the velocity, the model (11.0.10) has the form:

$$
\begin{aligned}
(a_{10} + a_{11}Z + a_{12}Z^2)\frac{\mathrm{d}\overline{C}_1}{\mathrm{d}z} + (a_{11} + 2a_{12}Z)\overline{C}_1 &= -K_1(\overline{C}_1 - \overline{C}_2); \\
(a_{20} + a_{21}Z + a_{22}Z^2)\frac{\mathrm{d}\overline{C}_2}{\mathrm{d}z} + (a_{21} + 2a_{22}Z)\overline{C}_2 &= \omega K_1(\overline{C}_1 - \overline{C}_2); \\
Z = 0, \quad \overline{C}_1 = 1, \quad \overline{C}_2 = 0.
\end{aligned}
\tag{11.1.6}
$$

where, at an unknown velocity distribution in the two phases, ω is known beforehand for a concrete process, while the parameters $a_{j0}, a_{j1}, a_{j2}, j = 1, 2, K_1$ must be obtained, using experimental data.

11.1.2 Parameter Identification

The obtained values of the functions $\overline{C}_{jn}(Z_n), j = 1, 2$, for a concrete process ($\omega = 1$), "theoretical" value of $K_1 = 1$ and different $Z_n = 0.1(n+1), n = 0, 1, \ldots, 9$ (Fig. 11.2), permit to be obtained the values of $\overline{C}_j(1), j = 1, 2$ and the artificial experimental data:

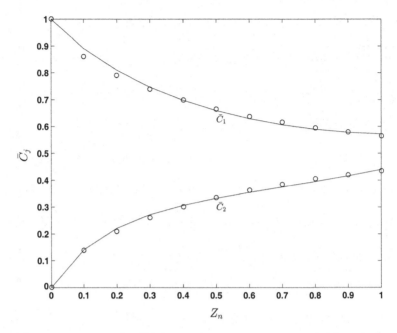

Fig. 11.2 Average concentration $\overline{C}_j(Z), j = 1, 2$ in the case $\omega = K_1 = 1$: "theoretical" values $\overline{C}_j(Z) = \overline{C}_{jn}(Z_n), j = 1, 2$ as solutions of (11.1.4) and (11.0.7) for different $Z_n = 0.1(n+1), n = 0, 1, \ldots, 9$ (points); $\overline{C}_j(Z), j = 1, 2$ as a solution of (11.1.6), using the "experimental" parameter values $a_{j0}, a_{j1}, a_{j2}, j = 1, 2, K_1$ (lines)

$$\overline{C}^m_{j\exp}(1) = (0.95 + 0.1B_m)\overline{C}_j(1), \quad j = 1, 2, \quad m = 1, \ldots, 10, \qquad (11.1.7)$$

where $0 \leq B_m \leq 1, m = 1, \ldots, 10$ are obtained by a generator of random numbers. The obtained artificial experimental data (11.1.7) are used for the illustration of the parameters $\left(a_{j0}, a_{j1}, a_{j2}, j = 1, 2, K_1\right)$ identification in the average-concentrations model (11.1.6) by the minimization of the least-squares function with respect to $a_{j0}, a_{j1}, a_{j2}, j = 1, 2, K_1$:

$$Q(a_{j0}, a_{j1}, a_{j2}, j = 1, 2, K_1) = \sum_{m=1}^{10} \left[\overline{C}_1(1, a_{j0}, a_{j1}, a_{j2}, j = 1, 2, K_1) - \overline{C}^m_{1\exp}(1)\right]^2$$

$$+ \sum_{m=1}^{10} \left[\overline{C}_2(1, a_{j0}, a_{j1}, a_{j2}, j = 1, 2, K_1) - \overline{C}^m_{2\exp}(1)\right]^2,$$

$$(11.1.8)$$

where the values of $\overline{C}_j(1, a_{j0}, a_{j1}, a_{j2}, j = 1, 2, K_1), j = 1, 2$ are obtained as solutions of (11.1.6). The obtained (after the minimization) "experimental" parameter values $a_{j0}, a_{j1}, a_{j2}, j = 1, 2, K_1$ are compared with the "theoretical" values in Table 11.1.

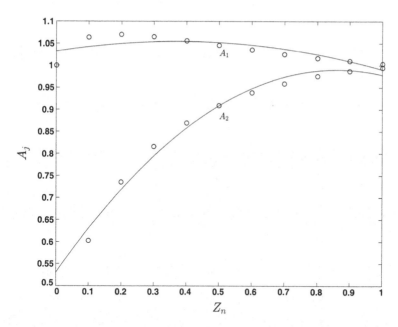

Fig. 11.3 Function $A_{jn}(Z_n), j = 1, 2$ in the case $\omega = K_1 = 1$: as a solution of (11.1.4) and (11.0.7) for different $Z_n = 0.1(n+1), n = 0, 1, \ldots, 9$ (points); $A_j(Z), j = 1, 2$ as a quadratic approximation (11.1.5) (line)

Table 11.1 Parameters $a_{j0}, a_{j1}, a_{j2}, K_j, j = 1, 2$ (physical absorption)

Parameters	"Theoretical" values	"Experimental" values
a_{10}	1.0318	0.9348
a_{11}	0.1226	0.1286
a_{12}	−0.1640	−0.1616
a_{20}	0.5301	0.5547
a_{21}	1.0671	1.0267
a_{22}	−0.6190	−0.6300
K_1	1	1.0637 (1.0720)
K_2	1	1.0637 (1.0681)

The obtained ("experimental") parameter values $a_{j0}, a_{j1}, a_{j2}, j = 1, 2, K_1$ are used for the solution of (11.1.6), and the results (the lines) are compared with the "theoretical" average-concentration values in Fig. 11.2.

The obtained "experimental" value of K_1 permits to be obtained the "experimental" value of the interphase mass transfer coefficient $k = K_1 \bar{u}_1 / l$.

In the same velocities distribution in the phases (the same "experimental" values of $a_{j0}, a_{j1}, a_{j2}, j = 1, 2, K_1$ in Table 11.1), for other concrete process ($\omega = 1.5$), the solution of (11.1.4) and (11.0.7) permits to be obtained the "theoretical" average concentrations $\overline{C}_j(Z), j = 1, 2$, which are compared (Fig. 11.4), with the solution of

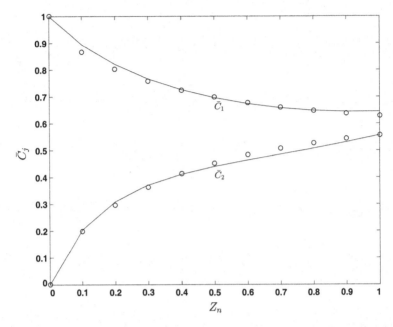

Fig. 11.4 Average concentration distribution $\overline{C}_j(Z), j = 1, 2$ in the case $\omega = 1.5$: "theoretical" values as solutions of (11.1.4) and (11.0.7) for different $Z_n = 0.1(n+1), n = 0, 1, \ldots, 9$ (points); as a solution of (11.1.6), using the "experimental" parameter values $a_{j0}, a_{j1}, a_{j2}, j = 1, 2, K_1$ (lines)

(11.1.6), using the "experimental" values of $a_{j0}, a_{j1}, a_{j2}, j = 1, 2, K_1$ in Table 11.1. From Fig. 11.4, it can be concluded that the values of the parameters $a_{j0}, a_{j1}, a_{j2}, j = 1, 2, K_1$ do not depend on ω (gas solubility).

11.1.3   Chemical Absorption

In the case of chemical absorption, from (11.0.4), (11.0.5), (11.0.8), (11.1.5) follow the models:

$$
\begin{aligned}
&U_{1n}\frac{dc_{1n}}{dz_n} = -K_1(C_{1n} - C_{2n}); \quad 0.1n \leq Z_n \leq 0.1(n+1); \\
&U_{2n}\frac{dc_{2n}}{dz_n} = \omega K_1(C_{1n} - C_{2n}) - DaC_{2n}; \\
&Z_n = 0.1n, \quad C_{jn}(R, Z_n) = C_{j(n-1)}(R, Z_n); \\
&n = 0, 1, \ldots, 9; \quad j = 1, 2; \\
&Z_0 = 0, \quad C_{10}(R, Z_0) \equiv 1, \quad C_{20}(R, Z_0) = 0.
\end{aligned}
\qquad (11.1.9)
$$

$$(a_{10} + a_{11}Z + a_{12}Z^2)\frac{d\overline{C}_1}{dz} + (a_{11} + 2a_{12}Z)\overline{C}_1 = -K_1(\overline{C}_1 - \overline{C}_2);$$

$$(a_{10} + a_{21}Z + a_{22}Z^2)\frac{d\overline{C}_2}{dz} + (a_{21} + 2a_{22}Z)\overline{C}_2 = \omega K_1(\overline{C}_1 - \overline{C}_2) - Da\overline{C}_2;$$

$$Z = 0, \quad \overline{C}_1 = 1, \quad \overline{C}_2 = 0.$$

$$(11.1.10)$$

The parameters ω, Da in (11.1.9) and (11.1.10) are known beforehand. The solution of (11.1.9) for a concrete process ($\omega = 1, Da = 1$) and a "theoretical" value of $K_1 = 1$ permits to be obtained the concentration distributions $C_{jn}(R, Z_n), j = 1, 2$ for different $Z_n = 0.1(n+1), n = 0, 1, \ldots, 9$ (Fig. 11.5).

The solution of (11.1.9) (Fig. 11.5) and (11.0.7) permits to be obtained the "theoretical" average concentration $\overline{C}_{jn}(Z_n), j = 1, 2$ (the points in Fig. 11.6) and the functions $A_{jn}(Z_n), j = 1, 2$ (the points in Fig. 11.7) for different $Z_n = 0.1(n+1), n = 0, 1, \ldots, 9$.

Figure. 11.7 shows that the functions $A_{jn}(Z_n), Z_n = 0.1(n+1), n = 0, 1, \ldots, 9$ are possible to be presented as quadratic approximations (11.1.5) and the "theoretical" values of $a_{j0}, a_{j1}, a_{j2}, j = 1, 2$, are presented in Table 11.2.

The "theoretical" average-concentration values $\overline{C}_j(1), j = 1, 2$ (Fig. 11.6) are used for to be obtained and the artificial experimental data (11.1.7). As a result,

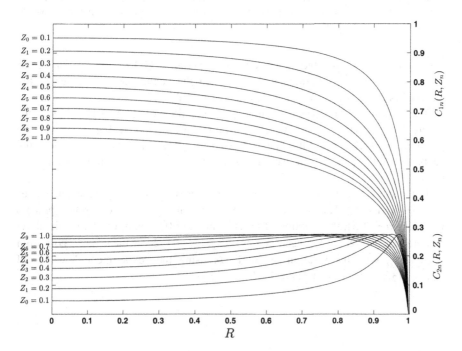

Fig. 11.5 Concentration distributions $C_{jn}(R, Z_n), j = 1, 2$ in the case $\omega = Da = K_1 = 1$ for different $Z_n = 0.1(n+1), n = 0, 1, \ldots, 9$ (chemical absorption)

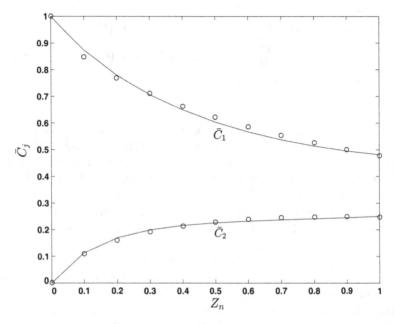

Fig. 11.6 Average concentration distributions $\overline{C}_{jn}(Z_n), j = 1, 2$ in the case $\omega = Da = K_1 = 1$: "theoretical" values as solutions of (11.1.9) and (11.0.7) for different $Z_n = 0.1(n+1), n = 0, 1, \ldots, 9$ (points); as a solution of (11.1 10), using the "experimental" parameter values $a_{j0}, a_{j1}, a_{j2}, j = 1, 2, K_1$ (lines)

the minimization of the least-squares function (11.1.8) with respect to $a_{j0}, a_{j1}, a_{j2}, j = 1, 2, K_1$ permits to be obtained the "experimental" values of the parameters $a_{j0}, a_{j1}, a_{j2}, j = 1, 2, K_1$, presented in Table 11.2.

The "theoretical" average concentration as solutions of (11.1.9) and (11.0.7) for different $Z_n = 0.1(n+1), n = 0, 1, \ldots, 9$ (points) are compared (Fig. 11.5) with the solution of (11.1.10) for the same case $(\omega = 1, Da = 1)$ (the line), where the "experimental" values of the parameters $a_{j0}, a_{j1}, a_{j2}, j = 1, 2, K_1$ in Table 11.2 are used.

In the same velocities distribution in the phases (the same "experimental" values of $a_{j0}, a_{j1}, a_{j2}, j = 1, 2, K_1$ in Table 11.2), for other concrete process $(\omega = 1, Da = 1.5)$, the solution of (11.1.9) and (11.0.7) permit to be obtained the "theoretical" average concentrations $\overline{C}_j(Z), j = 1, 2$, which are compared (Fig. 11.8), with the solution of (11.1.10), using the "experimental" values of $a_{j0}, a_{j1}, a_{j2}, j = 1, 2, K_1$ in Table 11.2.

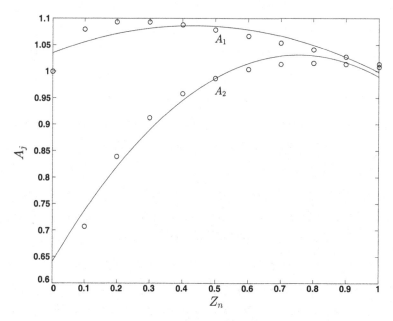

Fig. 11.7 Function $A_{jn}(Z_n), j = 1, 2$ in the case $\omega = Da = K_1 = 1$: as a solution of (11.1.9) and (11.0.7) for different $Z_n = 0.1(n+1), n = 0, 1, \ldots, 9$ (points);$A_j(Z), j = 1, 2$ as a quadratic approximation (11.1.5) (line)

Table 11.2 Parameters $a_{j0}, a_{j1}, a_{j2}, K_j, j = 1, 2$ (chemical absorption)

Parameters	"Theoretical" values	"Experimental" values
a_{10}	1.0346	0.8825
a_{11}	0.2378	0.2423
a_{12}	−0.2742	−0.2771
a_{20}	0.6405	0.6586
a_{21}	1.0359	1.1074
a_{22}	−0.6869	−0.6794
K_1	1	1.0684
K_2	1	1.0684

11.1.4 Physical Absorption of Highly Soluble Gas $(\omega = Da = 0)$

In the cases of physical absorption of highly soluble gas $(\omega = Da = 0)$, from (11.1.9), (11.1.10) follows $C_{2n} = \overline{C}_{2n} \equiv 0, n = 0, 1, \ldots, 9$, and as a result the models (11.1.9) and (11.1.10) have the forms:

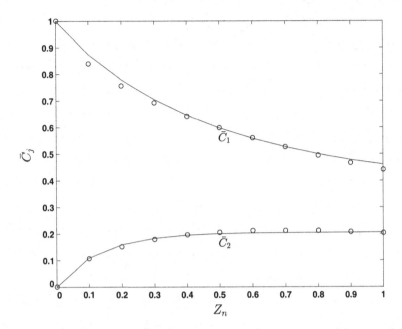

Fig. 11.8 Average concentration distribution $\overline{C}_j(Z), j = 1, 2$ in the case $\omega = 1, Da = 1.5$: "theoretical" values as solutions of (11.1.9) and (11.0.7) for different $Z_n = 0.1(n+1), n = 0, 1, \ldots, 9$ (points); as a solution of (11.1.10) (lines), using the "experimental" parameter values $a_{j0}, a_{j1}, a_{j2}, j = 1, 2, K_1$ in Table 11.2

$$U_{1n}\frac{dc_{1n}}{dz_n} = -K_1 C_{1n}; \quad 0.1n \leq Z_n \leq 0.1(n+1);$$
$$Z_n = 0.1n, \quad C_{1n}(R, Z_n) = C_{1(n-1)}(R, Z_n); \quad n = 0, 1, \ldots, 4; \qquad (11.1.11)$$
$$Z_0 = 0, \quad C_{10}(R, Z_0) \equiv 1.$$

$$(a_{10} + a_{11}Z + a_{12}Z^2)\frac{d\overline{c}_1}{dz} + (a_{11} + 2a_{12}Z)\overline{C}_1 = -K_1\overline{C}_1; \qquad (11.1.12)$$
$$Z = 0, \quad \overline{C}_1 = 1.$$

The solution of (11.1.11) for "theoretical" value of $K_1 = 1$ (dimensionless interphase mass transfer coefficient in the gas phase) permits to be obtained the "theoretical" concentration distributions $C_{1n}(R, Z_n)$ for different $Z_n = 0.1(n+1), n = 0, 1, \ldots, 9$. This solution of (11.1.11) and (11.0.7) permits to be obtained the "theoretical" average concentrations $\overline{C}_{1n}(Z_n)$ for different $Z_n = 0.1(n+1), n = 0, 1, \ldots, 9$ (the points in Fig. 11.9), to be obtained the value of $\overline{C}_1(1)$ and the artificial experimental data (11.1.7) for $j = 1$. They are used for the parameter (K_1) identification in the average-concentration model (11.1.12) by the minimization of the least-squares function with respect to K_1:

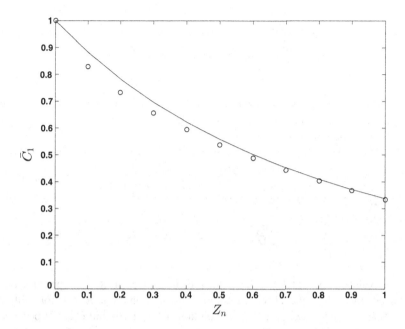

Fig. 11.9 Average concentration distribution $\overline{C}_1(Z)$ in the case $\omega = Da = 0$: "theoretical" values of $\overline{C}_{1n}(Z_n)$ for different $Z_n = 0.1(n+1), n = 0, 1, \ldots, 9$ (points); $\overline{C}_1(Z)$ as solutions of (11.1.12) for "experimental" values of a_{10}, a_{11}, a_{12}, and K_1 (in the brackets) in Table 11.1 (line)

$$Q_1(K_1) = \sum_{m=1}^{10} \left[\overline{C}_1(1, a_{10}, a_{11}, a_{12}, K_1) - \overline{C}_{1\,\text{exp}}^m(1) \right]^2, \qquad (11.1.13)$$

where the obtained "experimental" parameter value of K_1 (in the brackets) in Table 11.1 and the "experimental" parameter values a_{10}, a_{11}, a_{12}, in Table 11.1 are used for the solution of (11.1.12) (the line in Fig. 11.9).

11.1.5 Physical Absorption of Lightly Soluble Gas $\left(\omega^{-1} = 0 \right)$

The model (11.1.9) and (11.1.10) is possible to be presented as

$$
\begin{aligned}
&U_{1n} \frac{dc_{1n}}{dz_n} = -\omega^{-1} K_2 (C_{1n} - C_{2n}); \quad 0.1n \leq Z_n \leq 0.1(n+1); \\
&U_{2n} \frac{dc_{2n}}{dz_n} = K_2 (C_{1n} - C_{2n}); \\
&Z_n = 0.1n, \quad C_{jn}(R, Z_n) = C_{j(n-1)}(R, Z_n); \quad n = 0, 1, \ldots, 9; \quad j = 1, 2; \\
&Z_0 = 0, \quad C_{10}(R, Z_0) \equiv 1, \quad C_{20}(R, Z_0) = 0.
\end{aligned}
\qquad (11.1.14)
$$

$$(a_{10} + a_{11}Z + a_{12}Z^2)\frac{d\overline{C}_1}{dz} + (a_{11} + 2a_{12}Z)\overline{C}_1 = -\omega^{-1}K_2(\overline{C}_1 - \overline{C}_2);$$
$$(a_{20} + a_{21}Z + a_{22}Z^2)\frac{d\overline{C}_2}{dz} + (a_{21} + 2a_{22}Z)\overline{C}_2 = K_2(\overline{C}_1 - \overline{C}_2); \qquad (11.1.15)$$
$$Z = 0, \quad \overline{C}_1 = 1, \quad \overline{C}_2 = 0.$$

In the case of physical absorption of lightly soluble gas $(\omega^{-1} = 0)$, from (11.1.14), (11.1.15) follows $C_{1n} \equiv 1, n = 0, 1, \ldots, 9$ and as a result from (11.1.14), (11.1.15) is possible to be obtained:

$$U_{2n}\frac{dc_{2n}}{dz_n} = K_2(1 - C_{2n}); \quad 0.1n \leq Z_n \leq 0.1(n+1);$$
$$Z_n = 0.1n, \quad C_{2n}(R, Z_n) = C_{2(n-1)}(R, Z_n); \quad n = 0, 1, \ldots, 9; \qquad (11.1.16)$$
$$Z_0 = 0, \quad C_{20}(R, Z_0) = 0.$$

$$(a_{20} + a_{21}Z + a_{22}Z^2)\frac{d\overline{C}_2}{dz} + (a_{21} + 2a_{22}Z)\overline{C}_2 = K_2(1 - \overline{C}_2); \qquad (11.1.17)$$
$$Z = 0, \quad \overline{C}_2 = 0.$$

The solution of (11.1.16) for "theoretical" value of $K_2 = 1$ permits to be obtained the "theoretical" concentration distributions $C_{2n}(R, Z_n)$ for different $Z_n = 0.1(n+1), n = 0, 1, \ldots, 9$. This solution of (11.1.16) and (11.0.7) permits to be obtained the "theoretical" average-concentration values $\overline{C}_{2n}(Z_n)$ for different $Z_n = 0.1(n+1), n = 0, 1, \ldots, 9$ (the points in Fig. 11.10), to be obtained the value of $\overline{C}_2(1)$ and the artificial experimental data (11.1.7) for $j = 2$. They are used for the parameter (K_2) identification in the average-concentration model (11.1.17) by the minimization of the least-squares function with respect to K_2:

$$Q_2(K_2) = \sum_{m=1}^{10} \left[\overline{C}_2(1, a_{20}, a_{21}, a_{22}, K_2) - \overline{C}_{2\,\exp}^m(1)\right]^2, \qquad (11.1.18)$$

where the obtained "experimental" parameter value of K_2 (in the brackets) in Table 11.1 and the "experimental" parameter values a_{20}, a_{21}, a_{22}, in Table 11.1 are used for the solution of (11.1.17) (the line in Fig. 11.10).

The presented numerical analysis of a co-current absorption process in column apparatus shows that the average-concentration model, where the radial velocity components in the phases are equal to zero (in the cases of constant velocity radial non-uniformities along the column height), is possible to be used in the cases of an axial modification of the radial non-uniformities of the axial velocity components. The use of experimental data, for the average concentrations at column end, for a concrete process, permits to be obtained the model parameters $(a_{j0}, a_{j1}, a_{j2}, j = 1, 2)$, related with the radial non-uniformities of the velocities in the gas and liquid phases. These parameter values permit to be used the average-concentration model for modeling of physical and chemical absorption, absorption of highly and lightly soluble gases (different values of the parameters ω, Da, i.e., different values of the column height, average velocities, chemical reaction rate constant, and gas solubility).

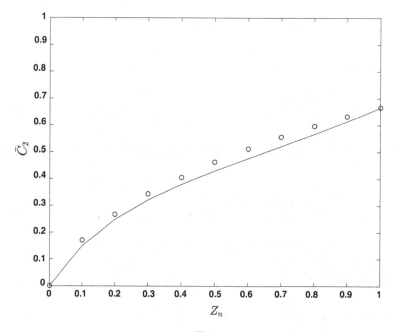

Fig. 11.10 Average concentration distribution $\overline{C}_2(Z)$ in the case $\omega^{-1} = 0$: "theoretical" values $\overline{C}_{2n}(Z_n)$ for different $Z_n = 0.1(n+1), n = 0, 1, \ldots, 9$ (points); $\overline{C}_2(Z)$ as solution of (11.1.17) (line) for "experimental" values of $a_{20}, a_{21}, a_{22}, K_2$ (K_2 in the brackets) in Table 11.1

11.2   Effect of the Radial Velocity Component

The numerical analysis of a co-current absorption process in column apparatus shows [5] that the average-concentration model where the radial velocity components in the phases are equal to zero is possible to be used in the cases of an axial modification of the radial non-uniformities of the axial velocity components. Here will be presented the case, when the radial velocity components in the phases are not equal to zero.

11.2.1   Convection–Diffusion Model

The new approach of the processes modeling in the column apparatuses [1–3] permits to be created the convection–diffusion model of the co-current chemical absorption process with a pseudo-first-order chemical reaction in the liquid phase:

$$u_j \frac{\partial c_j}{\partial z} + v_j \frac{\partial c_j}{\partial r} = D_j \left(\frac{\partial^2 c_j}{\partial z^2} + \frac{1}{r} \frac{\partial c_j}{\partial r} + \frac{\partial^2 c_j}{\partial r^2} \right) + (-1)^{(2-j)} k(c_1 - \chi c_2) - (j-1)k_0 c_2;$$

$$r = 0, \quad \frac{\partial c_j}{\partial r} \equiv 0; \quad r = r_0, \quad \frac{\partial c_j}{\partial r} \equiv 0; \quad j = 1, 2; \tag{11.2.1}$$

$$z = 0, \quad c_1 \equiv c_1^0, \quad c_2 \equiv 0, \quad u_1^0 c_1^0 \equiv u_1 c_1^0 - D_1 \left(\frac{\partial c_1}{\partial z} \right)_{z=0}, \quad \left(\frac{\partial c_2}{\partial z} \right)_{z=0} = 0.$$

$$\frac{\partial u_j}{\partial z} + \frac{\partial v_j}{\partial r} + \frac{v_j}{r} = 0;$$

$$r = r_0, \quad v_j(r_0, z) \equiv 0; \quad z = 0, \quad u_j = u_j(r, 0); \quad j = 1, 2. \tag{11.2.2}$$

In (11.2.1) and (11.2.2), it is possible to introduce the generalized variables:

$$r = r_0 R, \quad z = lZ, \quad u_j(r, z) = u_j(r_0 R, lZ) = u_j^0 U_j(R, Z),$$
$$v_j(r, z) = v_j(r_0 R, lZ) = u_j^0 \varepsilon V_j(R, Z), \quad j = 1, 2,$$
$$c_1(r, z) = c_1(r_0 R, lZ) = c_1^0 C_1(R, Z), \tag{11.2.3}$$
$$c_2(r, z) = c_2(r_0 R, lZ) = \frac{c_1^0}{\chi} C_2(R, Z)$$

and as a result is obtained:

$$U_j \frac{\partial C_j}{\partial Z} + V_j \frac{\partial C_j}{\partial R} = \mathrm{Fo}_j \left(\varepsilon^2 \frac{\partial^2 C_j}{\partial Z^2} + \frac{1}{R} \frac{\partial C_j}{\partial R} + \frac{\partial^2 C_j}{\partial R^2} \right)$$
$$+ (-1)^{(2-j)} K_j (C_1 - C_2) - (j-1)Da C_j;$$

$$R = 0, \quad \frac{\partial C_j}{\partial R} \equiv 0; \quad R = 1, \quad \frac{\partial C_j}{\partial R} \equiv 0; \quad j = 1, 2;$$

$$Z = 0, \quad C_1 \equiv 1, \quad C_2 = 0, \quad 1 \equiv U_1 - Pe_1^{-1} \frac{\partial C_1}{\partial Z}, \quad \frac{\partial C_2}{\partial Z} \equiv 0.$$

$$\frac{\partial U_j}{\partial Z} + \frac{\partial V_j}{\partial R} + \frac{V_j}{R} = 0; \quad R = 1, \quad V_j(1, Z) \equiv 0; \quad Z = 0, \quad U_j = U_j(R, 0).$$

$$\tag{11.2.4}$$

In (11.2.4), *Fo*, *Da*, and *Pe* are the Fourier, Damkohler, and Peclet numbers, respectively:

$$Fo_j = \frac{D_j l}{u_j^0 r_0^2}, \quad Da = \frac{k_0 l}{u_2^0}, \quad Pe_j = \frac{u_j^0 l}{D_j}, \quad \varepsilon^2 = \frac{r_0^2}{l^2} = Fo_j^{-1} Pe_j^{-1},$$

$$K_1 = \frac{kl}{u_1^0}, \quad K_2 = \omega K_1, \quad \omega = \frac{u_1^0 \chi}{u_2^0}, \quad j = 1, 2. \tag{11.2.5}$$

In the cases of a physical absorption $Da = 0$.

11.2.2 Axial and Radial Velocity Components

The radial non-uniformity of the gas and liquid axial velocity components in the absorption columns is a result of the fluid hydrodynamics at the column inlet and decreases along the column height as a result of the fluid viscosities. As a result, radial velocity components are initiated. The theoretical analysis of the change in the radial non-uniformity of the axial velocity components (effect of the radial velocity components) in a column can be made by an appropriate hydrodynamic model, where the average velocities at the cross section of the column are constants (inlet average axial velocity components). In generalized variables [1], as an example, is possible to be used the next velocity distributions, where the difference between the gas and liquid flows is in the average (inlet) velocities, only:

$$U_j = (2 - 0.4Z) - 2(1 - 0.4Z)R^2, \quad V_j = 0.2(R - R^3), \quad j = 1, 2, \quad (11.2.6)$$

where the velocity components satisfy Eq. (11.2.2) and the average axial velocity components are constants:

$$\bar{u}_j = \frac{2}{r_0^2} \int_0^{r_0} r u_j(r, z) \mathrm{d}r = u_j^0, \quad \overline{U}_j = 2 \int_0^1 R U_j \mathrm{d}R = 1, \quad j = 1, 2. \quad (11.2.7)$$

In Chap. 10, it is given that $V_j < 0.1, j = 1, 2$, and it must be presented with the help of a small parameter $0.1 = \alpha \ll 1$; i.e.,

$$V_j = \alpha V_{j0}, \quad V_{j0} = 0.2(R - R^3)\alpha^{-1}, \quad j = 1, 2. \quad (11.2.8)$$

11.2.3 Convective Type Model

In an industrial absorption column $[l > r_0 > 1 \text{ (m)}]$, in the cases of down co-current gas–liquid drops flow, the average gas velocity and the average liquid drops velocity are $\bar{u}_1 \sim 1, \bar{u}_2 \sim 1 \text{ (m s}^{-1})$, and the diffusivities in the gas (air) and the liquid (water) are $D_1 \sim 10^{-4}, D_2 \sim 10^{-9} \text{ (m}^2 \text{ s}^{-1})$. In these conditions, it is possible to obtain the next order of magnitude of the parameter values:

$$\varepsilon^2 \sim 10^{-2}, \quad Fo_1 \sim 10^{-2}, \quad Fo_2 \sim 10^{-7}, \quad Pe_1^{-1} \sim 10^{-4}, \quad Pe_2^{-1} \sim 10^{-9} \quad (11.2.9)$$

and the model (11.2.4) has the convective [1–3] form:

$$U_j \frac{\partial C_j}{\partial Z} + \alpha V_{j0} \frac{\partial C_j}{\partial R} = (-1)^{(2-j)} K_j (C_1 - C_2) - (j - 1) Da C_j;$$

$$R = 1, \quad \frac{\partial C_j}{\partial R} \equiv 0; \quad j = 1, 2; \quad (11.2.10)$$

$$Z = 0, \quad C_1 \equiv 1, \quad C_2 = 0.$$

11.2.4 Average-Concentration Model

In [3], it is possible to see that the average values of the velocity and concentrations at the column cross-sectional area are:

$$\bar{u}_j(z) = \frac{2}{r_0^2} \int_0^{r_0} r u_j(r, z) dr, \quad \bar{c}_j(z) = \frac{2}{r_0^2} \int_0^{r_0} r c_i(r, z) dr, \quad j = 1, 2. \qquad (11.2.11)$$

In the cylindrical columns,

$$r_0 = \text{const}, \quad \bar{u}_j(z) \equiv u_j^0 = \text{const}, \quad \frac{d\bar{u}_j}{dz} \equiv 0, \quad j = 1, 2. \qquad (11.2.12)$$

The functions $u_j(r, z), v_j(r, z), c_j(r, z), j = 1, 2$ in (11.2.1) can be presented with the help of the average functions (11.2.11):

$$\begin{aligned} u_j(r, z) &= \bar{u}_j U_j(R, Z), \quad v_j(r, z) = \varepsilon \bar{u}_j V_j(R, Z), \\ c_j(r, z) &= \bar{c}_j(z) \tilde{c}_j(r, z), \quad j = 1, 2, \end{aligned} \qquad (11.2.13)$$

where $\tilde{c}_j(r, z), j = 1, 2$ present the radial non-uniformity of the concentrations and satisfy the following conditions:

$$\frac{2}{r_0^2} \int_0^{r_0} r \tilde{c}_j(r, z) dr = 1, \quad j = 1, 2. \qquad (11.2.14)$$

The average-concentration model may be obtained when putting (11.2.13) into (11.2.1), multiplying by r and integrating over r in the interval $(0, r_0)$. As a result, the following is obtained:

$$\begin{aligned} \alpha_j(z) \bar{u}_j \frac{d\bar{c}_j}{dz} &+ [\beta_j(z) + \gamma_j(z)] \bar{u}_j \bar{c}_j \\ &= D_j \frac{d^2 \bar{c}_j}{dz^2} + (-1)^{(2-j)} k(\bar{c}_1 - \chi \bar{c}_2) - (j-1) k_0 \bar{c}_2; \qquad (11.2.15) \\ z = 0, \quad \bar{c}_j(0) &\equiv (2-j) c_j^0, \quad \frac{d\bar{c}_j}{dz} \equiv 0; \quad j = 1, 2. \end{aligned}$$

where

$$\alpha_j(z) = \frac{2}{r_0^2} \int_0^{r_0} r U_j \tilde{c}_j \mathrm{d}r, \quad \beta_j(z) = \frac{2}{r_0^2} \int_0^{r_0} r U_j \frac{\partial \tilde{c}_j}{\partial z} \mathrm{d}r,$$

$$\gamma_j(z) = \frac{2}{r_0^2} \int_0^{r_0} r V_j \frac{\partial \tilde{c}_j}{\partial r} \mathrm{d}r, \tag{11.2.16}$$

$$U_j = U_j(R, Z), \quad V_j = V_j(R), \quad \tilde{c}_j(r, z) = \tilde{C}_j(R, Z), \quad j = 1, 2.$$

The theoretical analysis of the model (11.2.15) will be made, using the next generalized variables and functions:

$$r = r_0 R, \quad z = lZ, \quad \bar{c}_j(z) = c_j^0 \overline{C}_j(Z), \quad c_2^0 = \frac{c_1^0}{\chi},$$

$$\overline{C}_j(Z) = 2 \int_0^1 R C_j(R, Z) \mathrm{d}R,$$

$$\tilde{c}_j(r, z) = \frac{c_j(r, z)}{\bar{c}_j(z)} = \frac{C_j(R, Z)}{\overline{C}_j(Z)} = \tilde{C}_j(R, Z),$$

$$\alpha_j(z) = \alpha_j(lZ) = A_j(Z) = 2 \int_0^1 R U_j(R, Z) \tilde{C}_j(R, Z) \mathrm{d}R, \tag{11.2.17}$$

$$\beta_j(z) = \beta_j(lZ) = B_j(Z) = 2 \int_0^1 R U_j(R, Z) \frac{\partial \tilde{C}_j}{\partial Z} \mathrm{d}R,$$

$$\gamma_j(z) = \gamma_j(lZ) = G_j(Z) = 2 \int_0^1 R V_j(R) \frac{\partial \tilde{C}_j}{\partial R} \mathrm{d}R, \quad j = 1, 2,$$

and as a result the model (11.2.15) has the form:

$$A_j(Z) \frac{\mathrm{d}\overline{C}_j}{\mathrm{d}z} + \left[B_j(Z) + G_j(Z) \right] \overline{C}_j$$
$$= Pe_j^{-1} \frac{\mathrm{d}^2 \overline{C}_j}{\mathrm{d}z^2} + (-1)^{(2-j)} K_j \left(\overline{C}_1 - \overline{C}_2 \right) - (j-1) Da \overline{C}_j; \tag{11.2.18}$$
$$Z = 0, \quad \overline{C}_1 \equiv 1, \quad \overline{C}_2 = 0, \quad \frac{\mathrm{d}\overline{C}_1}{\mathrm{d}z} = 0, \quad \frac{\mathrm{d}\overline{C}_2}{\mathrm{d}z} = 0; \quad j = 1, 2.$$

In industrial conditions (11.2.9), the model (11.2.18) has the convective [3] form:

$$A_j(Z) \frac{\mathrm{d}\overline{C}_j}{\mathrm{d}z} + \left[B_j(Z) + G_j(Z) \right] \overline{C}_j$$
$$= (-1)^{(2-j)} K_1 \omega^{(j-1)} \left(\overline{C}_1 - \overline{C}_2 \right) - (j-1) Da \overline{C}_j; \tag{11.2.19}$$
$$Z = 0, \quad \overline{C}_1 = 1, \quad \overline{C}_2 = 0; \quad j = 1, 2.$$

The presented models (11.2.10) and (11.2.19) permit to be analyzed the physical absorption $(0 = Da \leq 10^{-2})$ of highly soluble $(0 = \omega \leq 10^{-2})$, average soluble $(10^{-1} < \omega < 10)$, or lightly soluble $(0 = \omega^{-1} \leq 10^{-2})$ gases.

11.2.5 Physical Absorption of Highly Soluble Gas

In the cases of physical absorption of highly soluble gas ($\omega = K_2 = Da = 0$) and from (11.2.10), (11.2.19) follows $C_2 = \overline{C}_2 \equiv 0$, and as a result, the models (11.2.10) and (11.2.19) have the forms:

$$U_1 \frac{dC_1}{dZ} + \alpha V_{10} \frac{dC_1}{dR} = -K_1 C_1; \quad R = 1, \quad \frac{\partial C_1}{\partial R} \equiv 0; \quad Z = 0, \quad C_1 \equiv 1.$$

$$\tag{11.2.20}$$

$$A_1(Z) \frac{d\overline{C}_1}{dZ} + [B(Z) + G(Z)]\overline{C}_1 = -K_1 \overline{C}_1;$$
$$Z = 0, \quad \overline{C}_1 \equiv 1. \tag{11.2.21}$$

In (11.2.20), $0.1 = \alpha \ll 1$ is a small parameter and (11.2.20) must be solved by perturbation method (see Chap. 10); i.e., the concentration must be presented as

$$C_1(R, Z) = C_{10}(R, Z) + \alpha C_{11}(R, Z) + \alpha^2 C_{12}(R, Z), \tag{11.2.22}$$

where the functions $C_{10}(R, Z), C_{11}(R, Z), C_{12}(R, Z)$ are the solutions of the next set of equations:

$$U_1 \frac{\partial C_{10}}{\partial Z} = -K_1(C_{10}); \quad Z = 0, \quad C_{10} \equiv 1. \tag{11.2.23}$$

$$U_1 \frac{\partial C_{11}}{\partial Z} = -V_{10} \frac{\partial C_{10}}{\partial R} - K_1(C_{11}); \quad Z = 0, \quad C_{11} \equiv 0. \tag{11.2.24}$$

$$U_1 \frac{\partial C_{12}}{\partial Z} = -V_{10} \frac{\partial C_{11}}{\partial R} - K_1(C_{12}); \quad Z = 0, \quad C_{12} \equiv 0. \tag{11.2.25}$$

The solution of (11.2.23)–(11.2.25) is obtained, using cubic spline interpolations for $\frac{\partial C_{10}}{\partial R}$ and $\frac{\partial C_{11}}{\partial R}$.

The solution of (11.2.20) in the case $K_1 = 1$, using (11.2.6) and (11.2.8), is presented in Fig. 11.11.

The solution of (11.2.20) and (11.2.17) permits to be obtained the average concentrations $\overline{C}_1(Z_n), Z_n = 0.1(n+1), n = 0, 1, \ldots, 9$ ("theoretical" values, points in Fig. 11.12) and functions $A_1(Z_n), B_1(Z_n), G_2(Z_n), Z_n = 0.1(n+1), n = 0, 1, \ldots, 9$ in (11.2.17), which are presented (points) in Figs. 11.13 and 11.14.

The functions $A_1(Z), B_1(Z), G_1(Z)$ in Figs. 11.13 and 11.14 are possible to be presented as the next approximations:

$$A_1(Z) = 1 + a_{11}Z + a_{12}Z^2, \quad B_1(Z) = b_{10} + b_{11}Z^{b_{12}},$$
$$G_1(Z) = g_{11}Z, \tag{11.2.26}$$

where the ("theoretical") values of $a_{11}, a_{12}, b_{10}, b_{11}, b_{12}, g_{11}$ are presented in Table 11.3.

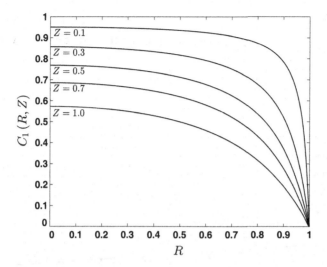

Fig. 11.11 Concentration distributions $C_1(R,Z)$ for different $Z = 0.1, 0.3, 0.5, 0.7, 1.0$

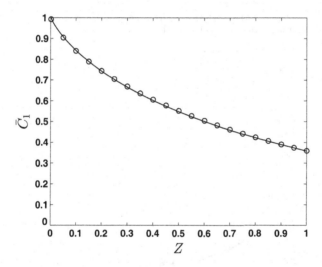

Fig. 11.12 Average concentrations $\overline{C}_1(Z)$: "theoretical" values $\overline{C}_1(Z_n), Z_n = 0.1(n+1), n = 0, 1, \ldots, 9$ (points); solution of (11.2.27), using the "experimental" parameter values in Table 11.3 (lines)

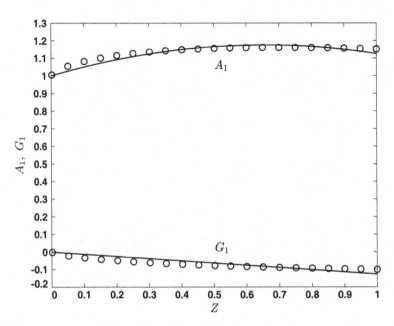

Fig. 11.13 Functions $A_1(Z), G_1(Z)$:$A_1(Z_n), G_1(Z_n), Z_n = 0.1(n+1), n = 0, 1, \ldots, 9$ (points); quadratic and linear approximations (11.2.26) (lines)

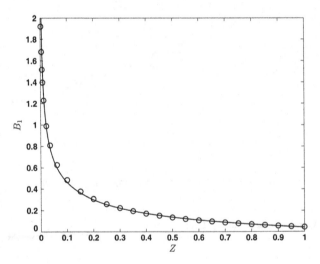

Fig. 11.14 Functions $B_1(Z)$: $B_1(Z_n), Z_n = 0.1(n+1), n = 0, 1, \ldots, 9$ (points); parabolic approximations (11.2.26) (lines)

Table 11.3 Parameters $a_{jo}, a_{j1}, a_{j2}, b_{jo}, b_{j1}, b_{j2}, g_{jo}, g_{j1}, K_j, j = 1, 2, K_2 = \omega K_1$ (physical absorption)

Parameters	"Theoretical" values	"Experimental" values	
a_{11}	0.537	0.552	(0.534)
a_{12}	−0.412	−0.519	(−0.497)
a_{20}	0.461	0.499	(0.523)
a_{21}	0.997	1.005	(1.121)
a_{22}	−0.574	−0.506	(−0.510)
b_{10}	−0.311	−0.542	(−0.551)
b_{11}	0.354	0.488	(0.435)
b_{12}	−0.339	−0.461	(−0.305)
b_{20}	−0.137	−0.506	(−0.549)
b_{21}	0.219	0.497	(0.518)
b_{22}	−0.733	−0.500	(−0.514)
g_{11}	−0.127	−0.104	(−0.095)
g_{20}	0.184	0.101	(0.099)
g_{21}	−0.149	−0.101	(−0.100)
K_1	1	0.821	(1.011)
K_2	1	1.010	–

As a result, the model (11.2.21) has the form:

$$(1 + a_{11}Z + a_{12}Z^2)\frac{d\overline{C}_1}{dz} + (b_{10} + b_{11}Z^{b_{12}} + g_{11}Z)\overline{C}_1 = -K_1\overline{C}_1;$$
$$Z = 0, \quad \overline{C}_1 \equiv 1, \tag{11.2.27}$$

where the parameters $P_1(a_{11}, a_{12}, b_{10}, b_{11}, b_{12}, g_{11}, K_1)$ must be obtained, using experimental data.

The values of the function $\overline{C}_1(Z_n)$, in the cases of physical absorption of highly soluble gas $(\omega = K_2 = Da = 0)$, for different $Z_n = 0.1(n + 1), n = 0, 1, \ldots, 9$, permit to be obtained the value of $\overline{C}_1(1)$ and the artificial experimental data:

$$\overline{C}^m_{j\,exp}(1) = (0.95 + 0.1B_m)\overline{C}_j(1), \quad j = 1, 2, \quad m = 1, \ldots, 10, \tag{11.2.28}$$

where $0 \le B_m \le 1, m = 1, \ldots, 10$ are obtained by a generator of random numbers.

The obtained artificial experimental data (11.2.28) are possible to be used for the parameters P_1 identification in the average-concentration model (11.2.27), by the minimization of the least-squares function with respect to P_1:

$$Q_1(P_1) = \sum_{m=1}^{10} \left[\overline{C}_1(1, P_1) - \overline{C}^m_{1\,exp}(1)\right]^2, \tag{11.2.29}$$

where the values of $\overline{C}_1(1, P_1)$ are obtained as solutions of (11.2.27).

From the "theoretical" values of the parameters in Table 11.3, it is seen that as starting parameter values in the minimization procedure (11.2.29) of the parametric identification, it is convenient to use $a_{11} = 0.5, a_{12} = -0.5, b_{10} = -0.5,$ $b_{11} = 0.5, .b_{12} = -0.5, g_{11} = -0.1, K_1 = 1.$

The obtained "experimental" parameter values after the minimization procedure (11.2.29) are presented in Table 11.3. They are used for the solution of (11.2.27), and the result (the line) is compared with the "theoretical" average-concentration values (points) in Fig. 11.12.

11.2.6 Physical Absorption of Lightly Soluble Gas

In the cases of physical absorption of lightly soluble gas ($\omega^{-1} = Da = 0$), from (11.2.9), (11.2.18) follows $C_1 = \overline{C}_1 \equiv 1$, and as a result, the models (11.2.9) and (11.2.18) have the forms:

$$U_2 \frac{dc_2}{dz} + V_2 \frac{dc_2}{dR} = -K_2(1 - C_2);$$
$$R = 1, \quad \frac{\partial C_2}{\partial R} \equiv 0; \quad Z = 0, \quad C_2 \equiv 0. \tag{11.2.30}$$

$$A_2(Z) \frac{d\overline{c}_2}{dz} + [B_2(Z) + G_2(Z)]\overline{C}_2 = K_2(1 - \overline{C}_2);$$
$$Z = 0, \quad \overline{C}_2 \equiv 0. \tag{11.2.31}$$

In (11.2.30) must be introduced (11.2.8):

$$U_2 \frac{dc_2}{dz} + \alpha V_{20} \frac{dc_2}{dR} = K_2(1 - C_2);$$
$$R = 1, \quad \frac{\partial C_2}{\partial R} \equiv 0; \quad Z = 0, \quad C_2 \equiv 0 \tag{11.2.32}$$

and to be used the perturbation method:

$$C_2(R, Z) = C_{20}(R, Z) + \alpha C_{21}(R, Z) + \alpha^2 C_{22}(R, Z) \tag{11.2.33}$$

$$U_2 \frac{\partial C_{20}}{\partial Z} = K_2(1 - C_{20}); \quad Z = 0, \quad C_{20} \equiv 0. \tag{11.2.34}$$

$$U_2 \frac{\partial C_{21}}{\partial Z} = -V_{20} \frac{\partial C_{20}}{\partial R} - K_2 C_{21}; \quad Z = 0, \quad C_{21} \equiv 0. \tag{11.2.35}$$

$$U_2 \frac{\partial C_{22}}{\partial Z} = -V_{20} \frac{\partial C_{21}}{\partial R} - K_2 C_{22}; \quad Z = 0, \quad C_{22} \equiv 0. \tag{11.2.36}$$

The solution of (11.2.35) and (11.2.36) is obtained, using cubic spline interpolations for $\frac{\partial C_{20}}{\partial R}$ and $\frac{\partial C_{21}}{\partial R}$.

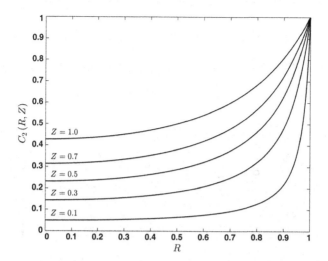

Fig. 11.15 Concentration distributions $C_2(R, Z)$ for different $Z = 0, 0.3, 0.5, 0.7, 1.0$

The solution of (11.2.32) in the case $K_2 = 1$, using (11.2.6) and (11.2.8), is presented in Fig. 11.15.

The solution of (11.2.32) permits to be obtained the average concentrations $\overline{C}_2(Z_n), Z_n = 0.1(n+1), n = 0, 1, \ldots, 9$ ("theoretical" values, points in Fig. 11.16) and functions $A_2(Z_n), B_2(Z_n), G_2(Z_n), Z_n = 0.1(n+1), n = 0, 1, \ldots, 9$ in (11.2.17), which are presented (points) in Figs. 11.17 and 11.18.

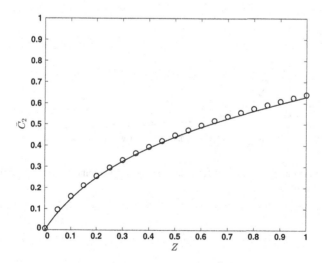

Fig. 11.16 Average concentrations $\overline{C}_2(Z)$: "theoretical" values $\overline{C}_2(Z_n), Z_n = 0.1(n+1)$, $n = 0, 1, \ldots, 9$ (points); solution of (11.2.37), using the "experimental" parameter values in Table 11.3 (lines)

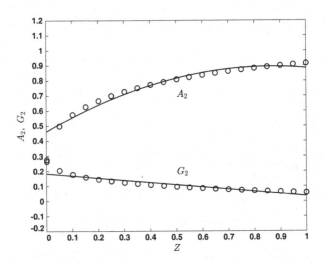

Fig. 11.17 Functions $A_2(Z_n), G_2(Z_n), Z_n = 0.1(n+1), n = 0, 1, \ldots, 9$ (points) and theirs quadratic and linear approximations (11.2.36) (lines)

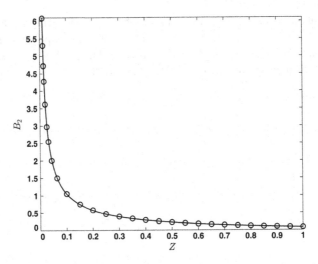

Fig. 11.18 Function $B_2(Z_n), Z_n = 0.1(n+1), n = 0, 1, \ldots, 9$ (points) and its parabolic approximation (11.2.36) (line)

The functions $A_2(Z), B_2(Z), G_2(Z)$ are possible to be presented as the next approximations:

$$A_2(Z) = a_{20} + a_{21}Z + a_{22}Z^2, \quad B_2(Z) = b_{20} + b_{21}Z^{b_{22}},$$
$$G_2(Z) = g_{20} + g_{21}Z, \tag{11.2.37}$$

As a result, the model (11.2.20) has the form:

$$(a_{20} + a_{21}Z + a_{22}Z^2)\frac{d\overline{C}_2}{dz}$$
$$+ \left[(b_{20} + b_{21}Z^{b_{22}}) + (g_{20} + g_{21}Z)\right]\overline{C}_2 = K_2(1 - \overline{C}_2); \tag{11.2.38}$$
$$Z = 0, \quad \overline{C}_2 \equiv 0.$$

The obtained "theoretical" average concentrations $\overline{C}_2(Z_n)$ for different $Z_n = 0.1(n+1), n = 0, 1, \ldots, 9$, permit to be obtained the value of $\overline{C}_2(1)$ and the artificial experimental data (11.2.28) for $j = 2$. They are used for the parameter $P_2(a_{20}, a_{21}, a_{22}, b_{20}, b_{21}, b_{22}, g_{20}, g_{21}, K_2)$ identification in the average-concentration model (11.2.38) by the minimization of the least-squares function with respect to P_2:

$$Q_2(P_2) = \sum_{m=1}^{10} \left[\overline{C}_2(1, P_2) - \overline{C}_{2\exp}^m(1)\right]^2, \tag{11.2.39}$$

where the values of $\overline{C}_2(1, P_2)$ are obtained as solutions of (11.2.38).

From the "theoretical" values of the parameters in Table 11.3, it is seen that as starting parameter values in the minimization procedure (11.2.39) of the parametric identification, it is convenient to use $a_{20} = 0.5, a_{21} = 1, a_{22} = -0.5, b_{20} = -0.5,$ $b_{21} = 0.5, b_{22} = -0.5, g_{20} = 0.1, g_{21} = -0.1, K_2 = 1.$

The obtained "experimental" parameter values after the minimization procedure (11.2.39) are presented in Table 11.3. They are used for the solution of (11.2.37), and the result (the line) is compared with the "theoretical" average-concentration values (points) in Fig. 11.16.

11.2.7 Physical Absorption of Average Soluble Gas

Let's consider the physical absorption $(Da = 0)$ of an average soluble gas $(\omega \sim 1)$ in an industrial absorption column (11.2.9). The convection–diffusion and average-concentration models (11.2.10) and (11.2.19) have the forms:

$$U_1 \frac{dc_1}{dz} + \alpha V_{10} \frac{dc_1}{dR} = -K_1(C_1 - C_2);$$
$$U_2 \frac{dc_2}{dz} + \alpha V_{20} \frac{dc_2}{dR} = \omega K_1(C_1 - C_2); \tag{11.2.40}$$
$$R = 1, \quad \frac{\partial C_j}{\partial R} \equiv 0, \quad j = 1, 2; \quad Z = 0, \quad C_1 \equiv 1, \quad C_2 = 0.$$

$$(1 + a_{11}Z + a_{12}Z^2)\frac{d\overline{C}_1}{dZ} + (b_{10} + b_{11}Z^{b_{12}} + g_1 Z)\overline{C}_1 = -K_1(\overline{C}_1 - \overline{C}_2);$$
$$Z = 0, \quad \overline{C}_1 \equiv 1.$$
$$(a_{20} + a_{21}Z + a_{22}Z^2)\frac{d\overline{C}_2}{dZ} + (b_{20} + b_{21}Z^{b_{22}} + g_{20} + g_{21}Z)\overline{C}_2$$
$$= \omega K_1(\overline{C}_1 - \overline{C}_2); \quad Z = 0, \quad \overline{C}_2 \equiv 0,$$
$$(11.2.41)$$

where the parameters $a_{j0}, a_{j1}, a_{j2}, b_{j0}, b_{j1}, b_{j2}, g_{j0}, g_{j1}, j = 1, 2, K_1$ must be obtained, using experimental data.

The solution of (11.2.40), in the case $K_1 = 1, \omega = 1$, is obtained with the help of (11.2.6), (11.2.8) and the perturbation method [2, 6]. For that purpose this solution is presented as (11.2.22), (11.2.33), where the functions are solutions of the sets of Eqs. (11.2.23)–(11.2.25), (11.2.34)–(11.2.36). The solution of (11.2.40) and (11.2.17) permits to be obtained the average concentrations in the phases $\overline{C}_j(Z_n), i = 1, 2, Z_n = 0.1(n+1), n = 0, 1, \ldots, 9$ ("theoretical" values, the points in Fig. 11.19).

The values of the functions $\overline{C}_j(Z_n), j = 1, 2$ for different $Z_n = 0.1(n+1), n = 0, 1, \ldots, 9$, permit to be obtained the values of $\overline{C}_j(1), j = 1, 2$ and the artificial experimental data (11.2.28).

The obtained artificial experimental data (11.2.28) are possible to be used for the parameters $P(a_{j0}, a_{j1}, a_{j2}, b_{j0}, b_{j1}, b_{j2}, g_{j0}, g_{j1}, j = 1, 2, K_1)$ identification in the average-concentration model (11.2.41) by the minimization of the least-squares function with respect to P:

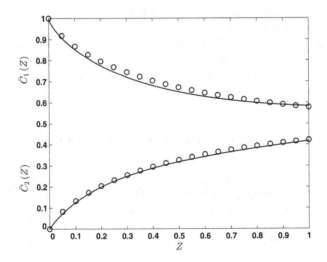

Fig. 11.19 Concentration distributions $C_j(R, Z), j = 1, 2$ for different $Z = 0, 0.3, 0.5, 0.7, 1.0$

$$Q(P) = \sum_{m=1}^{10} \left[\overline{C}_1(1,P) - \overline{C}_{1\,\mathrm{exp}}^m(1) \right]^2 + \sum_{m=1}^{10} \left[\overline{C}_2(1,P) - \overline{C}_{2\,\mathrm{exp}}^m(1) \right]^2, \quad (11.2.42)$$

where the values of $\overline{C}_j(1,P), j = 1,2$ are obtained as solutions of (11.2.41).

From the "theoretical" values of the parameters in Table 11.3, it is seen that as starting parameter values in the minimization procedure (11.2.42) of the parametric identification, it is convenient to use the parameter values in the cases of minimization of Q_1, Q_2 in (11.2.29) and (11.2.38). The obtained "experimental" parameter values after the minimization procedure (11.2.42) are presented in Table 11.3 (in brackets). They are used for the solution of (11.2.41), and the result (the line) is compared with the "theoretical" average-concentration values (points) in Fig. 11.19.

In the same velocities distribution in the phases (the same "experimental" values (in brackets) in Table 11.3), for other concrete process ($\omega = 2$), the solution of (11.2.40) and (11.2.17) permits to obtain the "theoretical" average concentrations $\overline{C}_j(Z), j = 1, 2$, which are compared (Fig. 11.20), with the solution of (11.2.41), using the "experimental" values (in brackets) in Table 11.3.

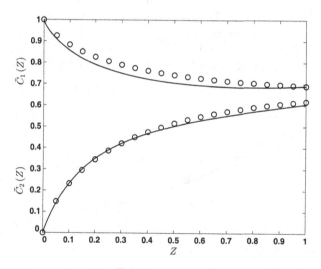

Fig. 11.20 Average concentrations $\overline{C}_j(Z), j = 1, 2, \omega = 2$: "theoretical" values $\overline{C}_j(Z_n), i = 1, 2, Z_n = 0.1(n+1), n = 0, 1, \ldots, 9$ (points); solution of (11.2.41), using the "experimental" parameter values $a_{j0}, a_{j1}, a_{j2}, b_{j0}, b_{j1}, b_{j2}, g_{j0}, g_{j1}, j = 1, 2, K_1$ (in brackets) in Table 11.3 (lines)

11.2.8   Chemical Absorption

In the case of chemical absorption, from (11.2.10), (11.2.19), (11.2.40), (11.2.41) follow the models:

$$U_1\frac{dC_1}{dZ} + \alpha V_{10}\frac{dC_1}{dR} = -K_1(C_1 - C_2);$$
$$U_2\frac{dC_2}{dZ} + \alpha V_{20}\frac{dC_2}{dR} = \omega K_1(C_1 - C_2) - DaC_2; \qquad (11.2.43)$$
$$R = 1, \quad \frac{\partial C_j}{\partial R} \equiv 0, \quad j = 1,2; \quad Z = 0, \quad C_1 \equiv 1, \quad C_2 = 0.$$

$$\left(1 + a_{11}Z + a_{12}Z^2\right)\frac{d\overline{C}_1}{dz} + \left(b_{10} + b_{11}Z^{b_{12}} + g_1Z\right)\overline{C}_1 = -K_1\left(\overline{C}_1 - \overline{C}_2\right);$$
$$Z = 0, \quad \overline{C}_1 \equiv 1.$$
$$\left(a_{20} + a_{21}Z + a_{22}Z^2\right)\frac{d\overline{C}_2}{dz} + \left(b_{20} + b_{21}Z^{b_{22}} + g_{20} + g_{21}Z\right)\overline{C}_2 = \qquad (11.2.44)$$
$$= \omega K_1\left(\overline{C}_1 - \overline{C}_2\right) - Da\overline{C}_2; \quad Z = 0, \quad \overline{C}_2 \equiv 0,$$

The parameters ω, Da in (11.2.43), (11.2.44) are known beforehand. The solution of (11.2.43) is possible to be made similar to the solution of (11.2.40). For a concrete process ($\omega = 1, Da = 1$) and a "theoretical" value of $K_1 = 1$ This solution permits to be obtained the concentration distributions $C_j(R, Z_n), j = 1, 2$ for different $Z_n = 0.1(n+1), n = 0, 1, \ldots, 9$. This solution of (11.2.17) permits to be obtained the "theoretical" average concentration $\overline{C}_{jn}(Z_n), j = 1, 2$ (the points in Fig. 11.21) for different $Z_n = 0.1(n+1), n = 0, 1, \ldots, 9$, which are compared with the solution of (11.2.44), using the "experimental" parameter values in Table 11.3 (in brackets).

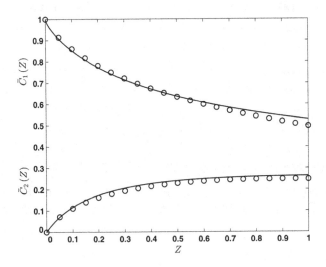

Fig. 11.21 Average concentration ("theoretical") $\overline{C}_j(R, Z_n), j = 1, 2$ in the case $\omega = Da = K_1 = 1$ as a solution of (11.2.43) and (11.2.17) for different $Z_n = 0.1(n+1), n = 0, 1, \ldots, 9$ (points); solution of (11.2.44), using the "experimental" parameter values (in brackets) in Table 11.3 (lines)

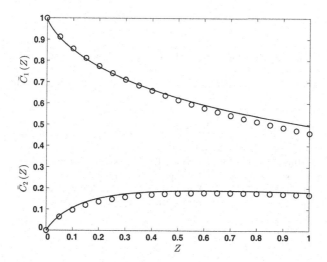

Fig. 11.22 Average concentration distribution $\overline{C}_j(Z), j = 1, 2$ in the case $\omega = 1, Da = 2$: "theoretical" values as solutions of (11.2.40) and (11.2.16) for different $Z_n = 0.1(n+1), n = 0, 1, \ldots, 9$ (points); as a solution of (11.2.41) (lines), using the "experimental" parameter values in Table 11.3

Using the same velocities distribution in the phases (the same "experimental" values (in brackets) in Table 11.3), for other concrete process ($\omega = 1, Da = 2$), the solution of (11.2.43) and (11.2.17) permits to be obtained the "theoretical" average concentrations $\overline{C}_j(Z), j = 1, 2$ (points), which are compared (Fig. 11.22), with the solution of (11.2.44) (lines), using the "experimental" values (in brackets) in Table 11.3.

The presented numerical analysis of a co-current absorption process in column apparatus shows that the average-concentration model, where the radial velocity components in the phases are not equal to zero, is possible to be used for modeling of the of co-current absorption process in column apparatus. The use of experimental data for the average concentrations at column end permits to be obtained the model parameters $a_{j0}, a_{j1}, a_{j2}, b_{j0}, b_{j1}, b_{j2}, g_{j0}, g_{j1}, g_{j2}, j = 1, 2, K_1$, related with the radial non-uniformities of the velocities in the gas and liquid phases. These parameter values permit to be used the average-concentration model for modeling of physical and chemical absorption, absorption of highly and lightly soluble gases (different values of the parameters ω, Da, i.e., different values of the column height, average velocities, chemical reaction rate constant, and gas solubility).

References

1. Boyadjiev C (2010) Theoretical chemical engineering. Modeling and simulation. Springer, Berlin, Heidelberg
2. Doichinova M, Boyadjiev C (2012) On the column apparatuses modeling. Int J Heat Mass Transf 55:6705–6715
3. Boyadjiev C (2013) A new approach for the column apparatuses modeling in chemical engineering. J Pure Appl Math: Adv Appl 10(2):131–150
4. Boyadjiev B, Boyadjiev C (2017) New models of industrial column chemical reactors. Bulgaria Chem Commun 49(3):706–710
5. Boyadjiev B, Boyadjiev C (2017) New models of industrial column absorbers. 1. Co-current absorption processes. Bulgaria Chem Commun 49(3):711–719
6. Boyadjiev B, Boyadjiev C (2017) A new approach to modeling of absorption in industrial column apparatuses. 1. Co-current process. Chem Eng Tech (in press)

Chapter 12
Industrial Counter-current Column Absorber

In Chap. 11 were presented the convection–diffusion and average-concentration models [1–3] of the gas absorption processes in the co-current columns, where the radial velocity component is not equal to zero, in the cases of an axial modification of the axial velocity radial non-uniformity along the column height [4]. This possibility will be used for modeling of the gas absorption processes in the counter-current columns, where the problem is complicated [5, 6], because the mass transfer process models have to be presented in two-coordinate systems (in a one-coordinate system, one of the equations has no solution due to the negative Laplacian value).

The modeling of the counter-current absorption processes uses two cylindrical coordinate systems (r, z_j), $j = 1, 2$ in the gas $(j = 1)$ and liquid $(j = 2)$ phases; i.e., the axial coordinates in the co-current models (11.0.1) and (11.0.2) must be replaced by $z = z_1, z_2; z_1 + z_2 = l$ (l is the column height) in the gas and liquid phases, respectively. As a result, the convection–diffusion and average-concentration models of the counter-current chemical absorption processes in the case, when the radial velocity component is equal to zero, have the forms:

$$u_1 \frac{\partial c_1}{\partial z_1} = D_1 \left(\frac{\partial^2 c_1}{\partial z_1^2} + \frac{1}{r} \frac{\partial c_1}{\partial r} + \frac{\partial^2 c_1}{\partial r^2} \right) - k(c_1 - \chi c_2);$$

$$u_2 \frac{\partial c_2}{\partial z_2} = D_2 \left(\frac{\partial^2 c_2}{\partial z_2^2} + \frac{1}{r} \frac{\partial c_2}{\partial r} + \frac{\partial^2 c_2}{\partial r^2} \right) + k(c_1 - \chi c_2) - k_0 c_2;$$

$$r = 0, \quad \frac{\partial c_j}{\partial r} \equiv 0; \quad r = r_0, \quad \frac{\partial c_j}{\partial r} \equiv 0; \quad j = 1, 2; \qquad (12.0.1)$$

$$z_1 = 0, \quad c_1 \equiv c_1^0, \quad u_1^0 c_1^0 \equiv u_1 c_1^0 - D_1 \left(\frac{\partial c_1}{\partial z} \right)_{z=0};$$

$$z_2 = 0, \quad c_2 \equiv 0, \quad \left(\frac{\partial c_2}{\partial z} \right)_{z=0} = 0.$$

$$\alpha_1(z)\bar{u}_1\frac{d\bar{c}_1}{dz_1} + \frac{d\alpha_1}{dz_1}\bar{u}_1\bar{c}_1 = D_1\frac{d^2\bar{c}_1}{dz_1^2} - k(\bar{c}_1 - \chi\bar{c}_2);$$

$$\alpha_2(z)\bar{u}_2\frac{d\bar{c}_2}{dz_2} + \frac{d\alpha_2}{dz_2}\bar{u}_2\bar{c}_2 = D_2\frac{d^2\bar{c}_2}{dz_2^2} + k(\bar{c}_1 - \chi\bar{c}_2) - k_0\bar{c}_2;$$

$$z_1 = 0, \quad \bar{c}_1(0) \equiv c_1^0, \quad \frac{d\bar{c}_1}{dz} \equiv 0; \quad z_2 = 0, \quad \bar{c}_2(0) \equiv 0, \quad \frac{d\bar{c}_2}{dz} \equiv 0; \quad (12.0.2)$$

$$\alpha_j(z_j) = \frac{2}{r_0^2}\int_0^{r_0} r\tilde{u}_j\tilde{c}_j dr, \quad \tilde{u}_j(r) = \frac{u_j(r)}{\bar{u}_j},$$

$$\tilde{c}_j(r,z_j) = \frac{c_j(r,z_j)}{\bar{c}_j(z)}, \quad j = 1,2.$$

In (12.0.1) and (12.0.2), it is possible to be introduced the generalized variables [1]:

$$r = r_0 R, \quad z_1 = lZ_1, \quad z_2 = lZ_2, \quad Z_1 + Z_2 = 1,$$
$$u_j(r) = \bar{u}_j U_j(R), \quad \tilde{u}_j(r) = \frac{u_j(r)}{\bar{u}_j} = U_j(R), \quad j = 1,2$$
$$c_1(r,z_1) = c_1^0 C_1(R,Z_1), \quad c_2(r,z_2) = \frac{c_1^0}{\chi}C_2(R,Z_2), \quad (12.0.3)$$
$$\bar{c}_1(z_1) = c_1^0\overline{C}_1(Z_1), \quad \bar{c}_2(z_2) = \frac{c_1^0}{\chi}\overline{C}_2(Z_2)$$

and as a result is obtained:

$$U_1\frac{\partial C_1}{\partial Z_1} = Fo_1\left(\varepsilon\frac{\partial^2 C_1}{\partial Z_1^2} + \frac{1}{R}\frac{\partial C_1}{\partial R} + \frac{\partial^2 C_1}{\partial R^2}\right) - K_1(C_1 - C_2);$$

$$U_2\frac{\partial C_2}{\partial Z_2} = Fo_2\left(\varepsilon\frac{\partial^2 C_2}{\partial Z_2^2} + \frac{1}{R}\frac{\partial C_2}{\partial R} + \frac{\partial^2 C_2}{\partial R^2}\right) + \omega K_1(C_1 - C_2) - DaC_2;$$

$$R = 0, \quad \frac{\partial C_j}{\partial R} \equiv 0; \quad R = 1, \quad \frac{\partial C_j}{\partial R} \equiv 0; \quad j = 1,2;$$

$$Z_1 = 0, \quad C_1 \equiv 1, \quad 1 \equiv U_1 - Pe_1^{-1}\frac{\partial C_1}{\partial Z_1}; \quad Z_2 = 0, \quad C_2 = 0, \quad \frac{\partial C_2}{\partial Z_2} \equiv 0.$$

$$(12.0.4)$$

$$A_1(Z_1)\frac{d\overline{C}_1}{dZ_1} + \frac{dA_1}{dZ_1}\overline{C}_1 = Pe_1^{-1}\frac{d^2\overline{C}_1}{dZ_1^2} - K_1(\overline{C}_1 - \overline{C}_2);$$

$$A_2(Z_2)\frac{d\overline{C}_2}{dZ_2} + \frac{dA_2}{dZ_2}\overline{C}_2 = Pe_2^{-1}\frac{d^2\overline{C}_2}{dZ_2^2} + \omega K_1(\overline{C}_1 - \overline{C}_2) - Da\overline{C}_2; \quad (12.0.5)$$

$$Z_1 = 0, \quad \overline{C}_1 = 1, \quad \frac{d\overline{C}_1}{dZ} = 0; \quad Z_2 = 0, \quad \overline{C}_2 = 0, \quad \frac{d\overline{C}_2}{dZ} = 0.$$

In (12.0.5), *Fo*, *Da*, and *Pe* are the Fourier, Damkohler, and Peclet numbers, respectively:

$$Fo_j = \frac{D_j l}{u_j^0 r_0^2}, \quad Pe_j = \frac{u_j^0 l}{D_j}, \quad \varepsilon^2 = Fo_j^{-1} Pe_j^{-1}, \quad j = 1, 2,$$

$$Da = \frac{k_0 l}{u_2^0}, \quad K_1 = \frac{kl}{u_1^0}, \quad \omega = \frac{u_1^0 \chi}{u_2^0}. \tag{12.0.6}$$

In the cases of a physical absorption, $Da = 0$.
In (12.0.5) are used the functions:

$$\overline{C}_j(Z_j) = 2 \int_0^1 RC_j(R, Z_j)\,dR, \tilde{c}_j(r, z_j) = \frac{c_j(r, z_j)}{\bar{c}_j(z_j)} = \frac{C_j(R, Z_j)}{\overline{C}_j(Z_j)},$$

$$\alpha_j(z_j) = A_j(Z_j) = 2 \int_0^1 RU_j(R) \frac{C_j(R, Z_j)}{\overline{C}_j(Z_j)}\,dR, \quad j = 1, 2. \tag{12.0.7}$$

The model parameters in (12.0.4) and (12.0.5) in the industrial absorption columns have very small values:

$$Fo_1 \sim 10^{-6}, \quad Fo_2 \sim 10^{-10}, \quad Pe_1^{-1} \sim 10^{-6}, \quad Pe_2^{-1} \sim 10^{-10} \tag{12.0.8}$$

and as a result, the models (12.0.4) and (12.0.5) have the convective forms:

$$U_1 \frac{dc_1}{dz_1} = -K_1(C_1 - C_2); \quad U_2 \frac{dc_2}{dz_2} = \omega K_1(C_1 - C_2) - DaC_2;$$

$$Z_1 = 0, \quad C_1 \equiv 1; \quad Z_2 = 0, \quad C_2 = 0. \tag{12.0.9}$$

$$A_1(Z_1) \frac{d\overline{C}_1}{dZ_1} + \frac{dA_1}{dZ_1} \overline{C}_1 = -K_1(\overline{C}_1 - \overline{C}_2);$$

$$A_2(Z_2) \frac{d\overline{C}_2}{dZ_2} + \frac{dA_2}{dZ_2} \overline{C}_2 = \omega K_1(\overline{C}_1 - \overline{C}_2) - Da\overline{C}_2; \tag{12.0.10}$$

$$Z_1 = 0, \quad \overline{C}_1 = 1; \quad Z_2 = 0, \quad \overline{C}_2 = 0.$$

12.1 Effect of the Axial Modification of the Radial Non-uniformity of the Velocity

Let us consider the velocity distributions [6]:

$$u_{jn}(r, z_{jn}) = \bar{u}_j \tilde{u}_{jn}(r, z_{jn}), \quad j = 1, 2 \tag{12.1.1}$$

and axial step changes of the radial non-uniformity of the axial velocity components in the column:

$$\tilde{u}_{jn}(r, z_{jn}) = \tilde{u}_{jn}(r_0 R, l Z_{jn}) = U_{jn}(R, Z_{jn}) = a_{jn} - b_{jn} R^2,$$

$$a_{jn} = 2 - 0.1n, \quad b_{jn} = 2(1 - 0.1n), \tag{12.1.2}$$

$$0.1n \le Z_{jn} \le 0.1(n+1), \quad n = 0, 1, \ldots, 9, \quad j = 1, 2,$$

where the average velocities at the cross section of the column are constants, while the maximal velocities (and as a result, the radial non-uniformity of the axial velocity components too) decrease along the column height.

12.1.1   Physical Absorption of the Average Soluble Gases

If put (12.1.2) in (12.0.8), the model, in the case of physical absorption ($Da = 0$), has the form:

$$U_{1n} \frac{dC_{1n}}{dz_{1n}} = -K_1(C_{1n} - C_{2n}), \quad 0.1n \le Z_{1n} \le 0.1(n+1);$$

$$U_{2n} \frac{dC_{2n}}{dz_{2n}} = \omega K_1(C_{1n} - C_{2n}), \quad 0.1n \le Z_{2n} \le 0.1(n+1);$$

$$Z_{jn} = 0.1n, \quad C_{jn}(R, Z_{jn}) = C_{j(n-1)}(R, Z_{jn}); \quad n = 0, 1, \ldots, 9; \quad j = 1, 2;$$

$$Z_{10} = 0, \quad C_{10}(R, Z_{10}) \equiv 1; \quad Z_{20} = 0, \quad C_{20}(R, Z_{20}) \equiv 0. \tag{12.1.3}$$

The solution of (12.1.3), using the method in [2, 6], for concrete process ($\omega = 1$) and "theoretical" value of $K_1 = 1$, permits to be obtained the concentration distributions $C_{jn}(R, Z_{jn}), j = 1, 2$ for different $Z_{jn} = 0.1(n+1), j = 1, 2$, $n = 0, 1, \ldots, 9$. These results and (12.0.6) permit to be obtained the "theoretical" average-concentration values $\overline{C}_{jn}(Z_{jn}), Z_{jn} = 0.1(n+1), j = 1, 2, n = 0, 1, \ldots, 9$ (the points in Fig. 12.1) and the function values $A_{jn}(Z_{jn}), j = 1, 2$ (the points in Fig. 12.2) for different $Z_{jn} = 0.1(n+1), j = 1, 2, n = 0, 1, \ldots, 9$.

From Fig. 12.2, it is seen that the functions $A_{jn}(Z_{jn})$, $n = 0, 1, \ldots, 9, j = 1, 2$ are possible to be presented as quadratic approximations:

$$A_j(Z_j) = a_{j0} + a_{j1} Z_j + a_{j2} Z_j^2, \quad j = 1, 2, \tag{12.1.4}$$

where the ("theoretical") values of $a_{j0}, a_{j1}, a_{j2}, j = 1, 2$ are presented in Table 12.1. As a result, the model (12.0.9) has the form:

$$\left(a_{10} + a_{11} Z_1 + a_{12} Z_1^2\right) \frac{d\overline{C}_1}{dZ_1} + (a_{11} + 2a_{12} Z_1)\overline{C}_1 = -K_1(\overline{C}_1 - \overline{C}_2);$$

$$\left(a_{20} + a_{21} Z_2 + a_{22} Z_2^2\right) \frac{d\overline{C}_2}{dZ_2} + (a_{21} + 2a_{22} Z_2)\overline{C}_2 = \omega K_1(\overline{C}_1 - \overline{C}_2); \tag{12.1.5}$$

$$Z_1 = 0, \quad \overline{C}_1 = 1; \quad Z_2 = 0, \quad \overline{C}_2 = 0,$$

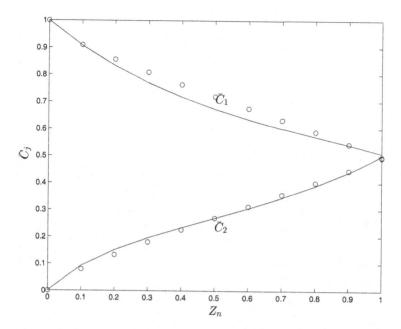

Fig. 12.1 Average concentration $\overline{C}_j(Z_j), j = 1,2$ in the case $\omega = K_1 = 1$: "theoretical" values $\overline{C}_j(Z_j) = \overline{C}_{jn}(Z_{jn}), j = 1,2, Z_{1n} = Z_n, Z_{2n} = 1 - Z_n$ as solutions of (12.1.3) and (12.0.6) for different $Z_{jn} = 0.1(n+1), j = 1,2, n = 0,1,\ldots,9$ (points); $\overline{C}_j(Z_j), j = 1,2, Z_1 = Z_n, Z_2 = 1 - Z_n$ as a solution of (12.1.5), using the "experimental" parameter values $a_{j0}, a_{j1}, a_{j2}, j = 1,2, K_1$ in Table 12.1 (lines)

where (at an unknown velocity distribution in two phases) $\omega = \frac{\overline{u}_1 \chi}{\overline{u}_2}$ is known beforehand for a concrete process, while the parameters $a_{j0}, a_{j1}, a_{j2}, j = 1,2, K_1$ must be obtained, using experimental data. For highly (lightly) soluble gases $\omega \to 0(\omega \to \infty)$.

12.1.2 Physical Absorption of Highly Soluble Gas $(\omega = Da = 0)$

In the cases of physical absorption of highly soluble gas ($\omega = Da = 0$), from (12.0.8), (12.0.9) follows $C_{2n} = \overline{C}_{2n} \equiv 0$, and as a result, the models (12.0.8) and (12.0.9) have the forms:

$$
\begin{aligned}
&U_{1n}\frac{dc_{1n}}{dz_{1n}} = -K_1 C_{1n}; \quad 0.1n \le Z_{1n} \le 0.1(n+1); \\
&Z_{1n} = 0.1n, \quad C_{1n}(R, Z_{1n}) = C_{1(n-1)}(R, Z_{1n}); \quad n = 0,1,\ldots,9; \quad (12.1.6) \\
&Z_{10} = 0, \quad C_{10}(R, Z_{10}) \equiv 1.
\end{aligned}
$$

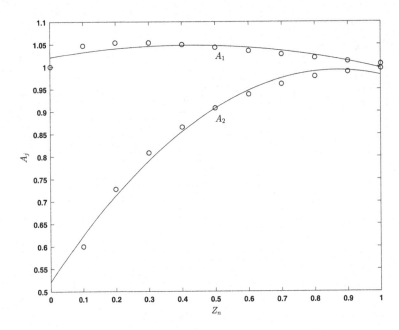

Fig. 12.2 Function $A_{jn}(Z_{jn}), Z_{1n} = Z_n, Z_{2n} = 1 - Z_n$, calculated from (12.0.6) for different $Z_{jn} = 0.1(n+1), j = 1, 2, n = 0, 1, \ldots, 9$(points); $A_j(Z_j), j = 1, 2, Z_1 = Z_n, Z_2 = 1 - Z_n$ as a quadratic approximation (12.1.4) (line)

Table 12.1 Parameters $a_{j0}, a_{j1}, a_{j2}, K_j, j = 1, 2$ (physical absorption)

Parameters	"Theoretical" values	"Experimental" values
a_{10}	0.919	0.567
a_{11}	0.420	0.443
a_{12}	−0.427	−0.420
a_{20}	0.433	0.414
a_{21}	1.105	0.910
a_{22}	−0.632	−0.766
K_1	1	1.077
$K_2 = \omega K_1$	1	1.029

$$\left(a_{10} + a_{11}Z_1 + a_{12}Z_1^2\right)\frac{d\overline{C}_1}{dZ_1} + (a_{11} + 2a_{12}Z_1)\overline{C}_1 = -K_1\overline{C}_1;$$

$$Z_1 = 0, \quad \overline{C}_1 = 1. \tag{12.1.7}$$

The solution of (12.1.6) for "theoretical" value of $K_1 = 1$ permits to be obtained the "theoretical" concentration distributions $C_{1n}(R, Z_{1n})$ for different $Z_{1n} = 0.1(n+1), n = 0, 1, \ldots, 9$. This solution of (12.1.6) and (12.0.6) permits to

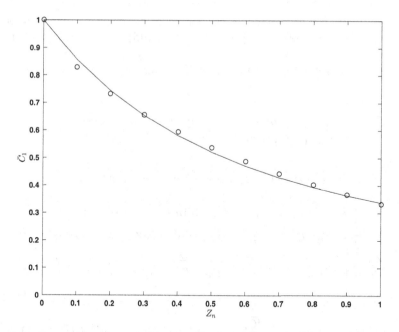

Fig. 12.3 Average-concentration distribution $\overline{C}_1(Z_1)$ in the case $\omega = Da = 0$: "theoretical" values of $\overline{C}_{1n}(Z_1 = Z_n)$ for different $Z_{1n} = 0.2(n+1)$, $n = 0, 1, \ldots, 9$ (points); $\overline{C}_1(Z_1 = Z_n)$ as solutions of (12.1.10) for "experimental" values of $a_{10}, a_{11}, a_{12}, K_1$ (line)

be obtained the "theoretical" average concentrations $\overline{C}_{1n}(Z_{1n})$ for different $Z_{1n} = 0.1(n+1)$, $n = 0, 1, \ldots, 9$ (the points in Fig. 12.3).

The obtained values of the function $\overline{C}_{1n}(Z_{1n})$, $Z_{1n} = 0.1(n+1)$, $n = 0, 1, \ldots, 9$ [using the solution of (12.1.6) and (12.0.6)], for the concrete process ($\omega = Da = 0$) and "theoretical" value of $K_1 = 1$, permit to be obtained the concentration values $\overline{C}_1(1)$ and the artificial experimental data:

$$\overline{C}_{1\,\text{exp}}^m(1) = (0.95 + 0.1B_m)\overline{C}_1(1), \quad m = 1, \ldots, 10, \qquad (12.1.8)$$

where $0 \leq B_m \leq 1, m = 1, \ldots, 10$ are obtained by a generator of random numbers.

The obtained artificial experimental data (12.1.8) are used for the parameter $(a_{10}, a_{11}, a_{12}, K_1)$ identification in the average-concentration model (12.1.7) by the minimization of the least-squares function Q_1 with respect to $a_{10}, a_{11}, a_{12}, K_1$:

$$Q_1(a_{10}, a_{11}, a_{12}, K_1) = \sum_{m=1}^{10}\left[\overline{C}_1(1, a_{10}, a_{11}, a_{12}, K_1) - \overline{C}_{1\,\text{exp}}^m(1)\right]^2, \qquad (12.1.9)$$

where the values of $\overline{C}_1(1, a_{10}, a_{11}, a_{12}, K_1)$ are obtained as solutions of (12.1.7), using the method in [5]. The obtained after the minimization "experimental" parameter values a_{10}, a_{11}, a_{12} are compared with the "theoretical" values on Table 12.1.

The obtained ("experimental") parameter values $a_{10}, a_{11}, a_{12}, K_1$ are used for the solution of (12.1.7) with the help of the method in [5], and the result (the line) is compared in Fig. 12.3 with the "theoretical" average concentrations $\overline{C}_{1n}(Z_{1n})$ for different $Z_{1n} = 0.1(n+1), n = 0, 1, \ldots, 9$ (the points).

12.1.3   Physical Absorption of Lightly Soluble Gas $(\omega^{-1} = 0, Da = 0)$

The models (12.1.3) and (12.1.5) are possible to be presented as

$$U_{1n}\frac{dc_{1n}}{dz_{1n}} = -\omega^{-1}K_2(C_{1n} - C_{2n}), \quad 0.1n \leq Z_{1n} \leq 0.1(n+1);$$
$$U_{2n}\frac{dc_{2n}}{dz_{2n}} = K_2(C_{1n} - C_{2n}) - DaC_{2n}, \quad 0.1n \leq Z_{2n} \leq 0.1(n+1);$$
$$Z_{jn} = 0.1n, \quad C_{jn}(R, Z_n) = C_{j(n-1)}(R, Z_{jn}); \quad n = 0, 1, \ldots, 9; \quad j = 1, 2;$$
$$Z_{10} = 0, \quad C_{10}(R, Z_{10}) \equiv 1; \quad Z_{20} = 0, \quad C_{20}(R, Z_0) = 0.$$

$$(12.1.10)$$

$$\left(a_{10} + a_{11}Z_1 + a_{12}Z_1^2\right)\frac{d\overline{C}_1}{dZ_1} + (a_{11} + 2a_{12}Z_1)\overline{C}_1 = -\omega^{-1}K_2(\overline{C}_1 - \overline{C}_2);$$
$$\left(a_{20} + a_{21}Z_2 + a_{22}Z_2^2\right)\frac{d\overline{C}_2}{dZ_2} + (a_{21} + 2a_{22}Z_2)\overline{C}_2 = K_2(\overline{C}_1 - \overline{C}_2);$$
$$Z_1 = 0, \quad \overline{C}_1 = 1; \quad Z_2 = 0, \quad \overline{C}_2 = 0.$$

$$(12.1.11)$$

In (12.1.10) and (12.1.11), $K_2 = \omega K_1$.

In the case of physical absorption of lightly soluble gas $(\omega^{-1} = 0, Da = 0)$ and from (12.1.10), (12.1.11) follows $C_{1n} \equiv 1, n = 0, 1, \ldots, 9$. As a result from (12.1.10), (12.1.11), it is possible to be obtained:

$$U_{2n}\frac{dc_{2n}}{dz_{2n}} = K_2(1 - C_{2n}); \quad 0.1n \leq Z_{2n} \leq 0.1(n+1);$$
$$Z_{2n} = 0.1n, \quad C_{2n}(R, Z_{2n}) = C_{2(n-1)}(R, Z_{2n}); \quad n = 0, 1, \ldots, 9; \quad (12.1.12)$$
$$Z_{20} = 0, \quad C_{20}(R, Z_{20}) = 0.$$

$$\left(a_{20} + a_{21}Z_2 + a_{22}Z_2^2\right)\frac{d\overline{C}_2}{dZ_2} + (a_{21} + 2a_{22}Z_2)\overline{C}_2 = K_2(1 - \overline{C}_2); \quad (12.1.13)$$
$$Z_2 = 0, \quad \overline{C}_2 = 0.$$

The solution of (12.1.12) for "theoretical" value of $K_2 = 1$ permits to be obtained the "theoretical" concentration distributions $C_{2n}(R, Z_{2n})$ for different $Z_{2n} = 0.1(n+1), n = 0, 1, \ldots, 9$. This solution of (12.1.12) and (12.0.6) permits to be obtained the "theoretical" average-concentration values $\overline{C}_{2n}(Z_{2n})$ for different $Z_{2n} = 0.1(n+1), n = 0, 1, \ldots, 9$ (the points in Fig. 12.4).

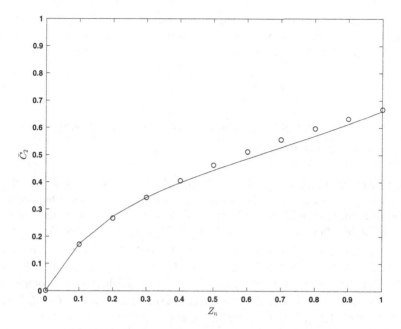

Fig. 12.4 Average-concentration distribution $\overline{C}_2(Z_2 = Z_n)$ in the case $\omega^{-1} = 0, Da = 0$:"theoretical" values of $\overline{C}_{2n}(Z_{2n})$ for different $Z_{2n} = 0.2(n+1)$, $n = 0, 1, \ldots, 9$ (points); $\overline{C}_2(Z_2)$ as solutions of (12.1.16) for "experimental" values of $a_{20}, a_{21}, a_{22}, K_2$ in Table 12.1 (line)

The obtained values of the function $\overline{C}_{2n}(Z_{2n}), Z_{2n} = 0.1(n+1), n = 0, 1, \ldots, 9$ (using the solution of (12.1.12) and (12.0.6)), for the concrete process ($\omega^{-1} = 0, Da = 0$) and "theoretical" value of $K_2 = 1$, permit to be obtained the concentration values $\overline{C}_2(1)$ and the artificial experimental data:

$$\overline{C}_{2\exp}^{m}(1) = (0.95 + 0.1B_m)\overline{C}_2(1), \quad m = 1, \ldots, 10, \qquad (12.1.14)$$

where $0 \leq B_m \leq 1, m = 1, \ldots, 10$ are obtained by a generator of random numbers.

The obtained artificial experimental data (12.1.14) are used for the parameter $(a_{20}, a_{21}, a_{22}, K_2)$ identification in the average-concentration model (12.1.13) by the minimization of the least-squares function Q_2 with respect to $a_{20}, a_{21}, a_{22}, K_2$:

$$Q_2(a_{20}, a_{21}, a_{22}, K_2) = \sum_{m=1}^{10} \left[\overline{C}_2(1, a_{20}, a_{21}, a_{22}, K_2) - \overline{C}_{2\exp}^{m}(1) \right]^2, \qquad (12.1.15)$$

where the values of $\overline{C}_2(1, a_{20}, a_{21}, a_{22}, K_2)$ are obtained as solutions of (12.1.13), using the method in [6]. The obtained after the minimization "experimental" parameter values a_{20}, a_{21}, a_{22} are compared with the "theoretical" values on Table 12.1.

The obtained ("experimental") parameter values $a_{20}, a_{21}, a_{22}, K_2$ are used for the solution of (12.1.13) with the help of the method in [6], and the result (the line) is compared in Fig. 12.4 with the "theoretical" average concentrations $\overline{C}_{2n}(Z_{2n})$ for different $Z_{2n} = 0.1(n+1), n = 0, 1, \ldots, 9$ (the points).

12.1.4 Parameter Identification in the Cases of Average Soluble Gases

The parameters $a_{j0}, a_{j1}, a_{j2}, j = 1, 2,$ in the model (12.1.5) are related to the velocity non-uniformity in the column, only. Their "experimental" values are obtained in the cases of absorption of highly and lightly gases (Table 12.1). The parameter K_1 in (18) must be obtained from the experimental values of K_1 and $K_2 = \omega K_1$ in Table 12.1 as $K_1 = \frac{K_1 + \omega^{-1} K_2}{2} = 1.05$. The obtained ("experimental") parameter values $a_{j0}, a_{j1}, a_{j2}, j = 1, 2,$ in Table 12.1 and $K_1 = 1.05$ are used for the solution of (12.1.5) with the help of the method [6], and the result (the lines) is compared in Fig. 12.1 with the "theoretical" average-concentration values $\overline{C}_{jn}(Z_{jn}), Z_{jn} = 0.1(n+1), j = 1, 2, n = 0, 1, \ldots, 9$ (the points).

The procedure for determining the concentrations in Fig. 12.1 is repeated for different values of Henry's number ($\omega = 0.5, 1.5$), and the results are presented in Figs. 12.5 and 12.6.

12.1.5 Chemical Absorption

In the case of chemical absorption, from (12.0.4), (12.0.5), (12.0.7), (12.1.4) follow the models:

$$
\begin{aligned}
&U_{1n}\frac{dc_{1n}}{dz_{1n}} = -K_1(C_{1n} - C_{2n}), \quad 0.1n \leq Z_{1n} \leq 0.1(n+1); \\
&U_{2n}\frac{dc_{2n}}{dz_{2n}} = \omega K_1(C_{1n} - C_{2n}) - DaC_{2n}, \quad 0.1n \leq Z_{2n} \leq 0.1(n+1); \\
&Z_{jn} = 0.1n, \quad C_{jn}(R, Z_n) = C_{j(n-1)}(R, Z_{jn}); \\
&n = 0, 1, \ldots, 9; \quad j = 1, 2; \\
&Z_{10} = 0, \quad C_{10}(R, Z_{10}) \equiv 1; \quad Z_{20} = 0, \quad C_{20}(R, Z_0) = 0.
\end{aligned}
\tag{12.1.18}
$$

$$
\begin{aligned}
&\left(a_{10} + a_{11}Z_1 + a_{12}Z_1^2\right)\frac{d\overline{c}_1}{dz_1} + (a_{11} + 2a_{12}Z_1)\overline{C}_1 = -K_1(\overline{C}_1 - \overline{C}_2); \\
&\left(a_{20} + a_{21}Z_2 + a_{22}Z_2^2\right)\frac{d\overline{c}_2}{dz_2} + (a_{21} + 2a_{22}Z_2)\overline{C}_2 \\
&\quad = \omega K_1(\overline{C}_1 - \overline{C}_2) - Da\overline{C}_2; \\
&Z_1 = 0, \quad \overline{C}_1 = 1; \quad Z_2 = 0, \quad \overline{C}_2 = 0.
\end{aligned}
\tag{12.1.19}
$$

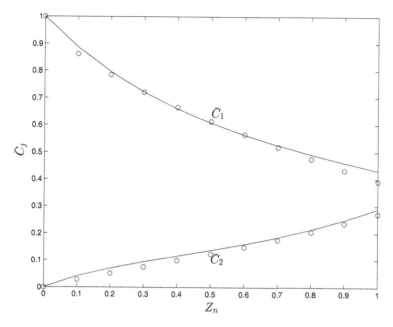

Fig. 12.5 Average concentration $\overline{C}_j(Z_j), j = 1, 2$ in the cases $\omega = 0.5, K_1 = 1$: "theoretical" values $\overline{C}_j(Z_j) = \overline{C}_{jn}(Z_{jn}), j = 1, 2, Z_{1n} = Z_n, Z_{2n} = 1 - Z_n$ as solutions of (12.1.3) and (12.0.6) for different $Z_{jn}, n = 0, 1, \ldots, 9$ (points); $\overline{C}_j(Z_j), j = 1, 2, Z_1 = Z_n, Z_2 = 1 - Z_n$ as solutions of (12.1.5) for "experimental" values of $a_{j0}, a_{j1}, a_{j2}, j = 1, 2, K_1$ in Table 12.1 (lines)

The solution of (12.1.18) for a concrete processes ($\omega = 1, Da = 1, 2$) and a "theoretical" value of $K_1 = 1$ permits to be obtained the concentration distributions $C_{jn}(R, Z_{jn}), j = 1, 2$ for different $Z_{jn} = 0.1(n + 1), j = 1, 2, n = 0, 1, \ldots, 9$. This solution of (12.1.18) and (12.0.6) permits to be obtained the "theoretical" average concentration $\overline{C}_{jn}(Z_{jn}), Z_{jn} = 0.1(n + 1), j = 1, 2, n = 0, 1, \ldots, 9$ (the points in Figs. 12.7, 12.8) and to be compared with $\overline{C}_j(Z_j), j = 1, 2$ (lines) as a solution of (12.1.19), using the "experimental" parameter values $a_{j0}, a_{j1}, a_{j2}, j = 1, 2$ in Table 12.1 and the "experimental" value of $K_1 = 1.05$.

12.2 Effect of the Radial Component of the Velocity

The new approach of the process modeling in the column apparatuses [1–3] permits to be created the convection–diffusion model of the counter-current chemical absorption process with a pseudo-first-order chemical reaction in the liquid phase in the cases of axial modification of the radial non-uniformities of the axial velocity components [6], when the radial velocity components in the phases are not equal to zero:

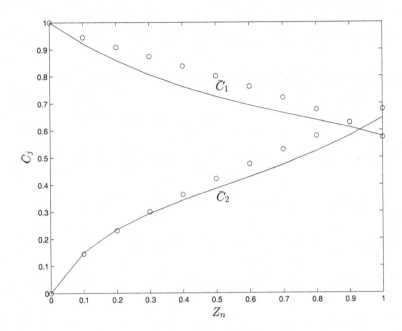

Fig. 12.6 Average concentration $\bar{C}_j(Z_j), j = 1, 2$ in the cases $\omega = 1.5, K_1 = 1$: "theoretical" values $\bar{C}_j(Z_j) = \bar{C}_{jn}(Z_{jn}), j = 1, 2, Z_{1n} = Z_n, Z_{2n} = 1 - Z_n$ as solutions of (12.1.3) and (12.0.6) for different $Z_{jn}, n = 0, 1, \ldots, 9$ (points); $\bar{C}_j(Z_j), j = 1, 2, Z_1 = Z_n, Z_2 = 1 - Z_n$ as solutions of (12.1.5) for "experimental" values of $a_{j0}, a_{j1}, a_{j2}, j = 1, 2, K_1$ in Table 12.1 (lines)

$$u_j \frac{\partial c_j}{\partial z_j} + v_j \frac{\partial c_j}{\partial r} = D_j \left(\frac{\partial^2 c_j}{\partial z_j^2} + \frac{1}{r} \frac{\partial c_j}{\partial r} + \frac{\partial^2 c_j}{\partial r^2} \right)$$
$$+ (-1)^{(2-j)} k(c_1 - \chi c_2) - (j - 1) k_0 c_2;$$
$$r = 0, \quad \frac{\partial c_j}{\partial r} \equiv 0; \quad r = r_0, \quad \frac{\partial c_j}{\partial r} \equiv 0; \quad j = 1, 2; \tag{12.2.1}$$
$$z_1 = 0, \quad c_1 \equiv c_1^0; \quad z_2 = 0, \quad c_2 \equiv 0;$$
$$u_1^0 c_1^0 \equiv u_1 c_1^0 - D_1 \left(\frac{\partial c_1}{\partial z_1} \right)_{z_1 = 0}, \quad \left(\frac{\partial c_2}{\partial z_2} \right)_{z_2 = 0} = 0.$$

$$\frac{\partial u_j}{\partial z_j} + \frac{\partial v_j}{\partial r} + \frac{v_j}{r} = 0; \tag{12.2.2}$$
$$r = r_0, \quad v_j(r_0, z_j) \equiv 0; \quad z_j = 0, \quad u_j = u_j(r, 0); \quad j = 1, 2.$$

In (12.2.1) and (12.2.2), $u_j(r, z_j), v_j(r, z_j), j = 1, 2$ are the axial and radial velocity components (m s^{-1}) in the phases.

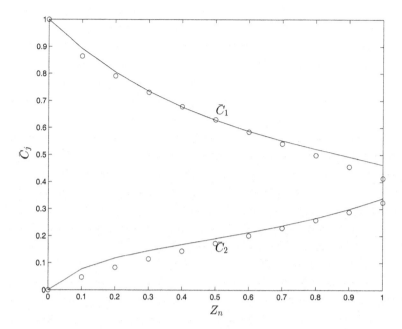

Fig. 12.7 Average concentration $\overline{C}_j(Z_j), j = 1, 2$ in the cases $\omega = 1, Da = 1$: "theoretical" values $\overline{C}_j(Z_j) = \overline{C}_{jn}(Z_{jn}), j = 1, 2, Z_{1n} = Z_n, Z_{2n} = 1 - Z_n$ as solutions of (12.1.18) and (12.0.6) for different $Z_{jn} = 0.1(n+1), j = 1, 2, n = 0, 1, \ldots, 9$ (points); $\overline{C}_j(Z_j), j = 1, 2, Z_1 = Z_n, Z_2 = 1 - Z_n$ as a solution of (12.1.19) (lines), using the "experimental" parameter values $a_{j0}, a_{j1}, a_{j2}, j = 1, 2$ in Table 12.1 and the "experimental" value of $K_1 = 1.05$

In (12.2.1) and (12.2.2), it is possible to be introduced the generalized variables:

$$
\begin{aligned}
&r = r_0 R, \quad z_j = lZ_j, \quad u_j(r, z_j) = u_j(r_0 R, lZ_j) = u_j^0 U_j(R, Z_j), \\
&v_j(r, z_j) = v_j(r_0 R, lZ_j) = u_j^0 \varepsilon V_j(R, Z_j), \\
&c_1(r, z_j) = c_1(r_0 R, lZ_j) = c_1^0 C_1(R, Z_j), \\
&c_2(r, z_j) = c_2(r_0 R, lZ_j) = \frac{c_1^0}{\chi} C_2(R, Z_j), \quad j = 1, 2
\end{aligned}
\tag{12.2.3}
$$

and as a result is obtained:

$$
\begin{aligned}
&U_j \frac{\partial C_j}{\partial Z} + V_j \frac{\partial C_j}{\partial R} = \mathrm{Fo}_j \left(\varepsilon^2 \frac{\partial^2 C_j}{\partial Z^2} + \frac{1}{R} \frac{\partial C_j}{\partial R} + \frac{\partial^2 C_j}{\partial R^2} \right) \\
&\quad + (-1)^{(2-j)} K_j (C_1 - C_2) - (j - 1) Da C_j; \\
&R = 0, \quad \frac{\partial C_j}{\partial R} \equiv 0; \quad R = 1, \quad \frac{\partial C_j}{\partial R} \equiv 0; \quad j = 1, 2; \\
&Z_1 = 0, \quad C_1 \equiv 1, \quad 1 \equiv U_1 - \mathrm{Pe}_1^{-1} \frac{\partial C_1}{\partial Z_1}; \\
&Z_2 = 0, \quad C_2 = 0, \quad \frac{\partial C_2}{\partial Z_2} \equiv 0.
\end{aligned}
\tag{12.2.4}
$$

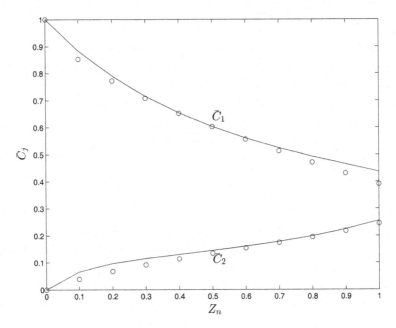

Fig. 12.8 Average concentration $\overline{C}_j(Z_j), j = 1, 2$ in the cases $\omega = 1, Da = 2$: "theoretical" values $\overline{C}_j(Z_j) = \overline{C}_{jn}(Z_{jn}), j = 1, 2, Z_{1n} = Z_n, Z_{2n} = 1 - Z_n$ as solutions of (12.1.18) and (12.0.6) for different $Z_{jn} = 0.1(n + 1), j = 1, 2, n = 0, 1, \ldots, 9$ (points); $\overline{C}_j(Z_j), j = 1, 2, Z_1 = Z_n, Z_2 = 1 - Z_n$ as a solution of (12.1.19) (lines), using the "experimental" parameter values $a_{j0}, a_{j1}, a_{j2}, j = 1, 2$ in Table 12.1 and the "experimental" value of $K_1 = 1.05$

$$\frac{\partial U_j}{\partial Z_j} + \frac{\partial V_j}{\partial R} + \frac{V_j}{R} = 0;$$

$$R = 1, \quad V_j(1, Z_j) \equiv 0; \quad Z = 0, \quad U_j = U_j(R, 0). \tag{12.2.5}$$

In (12.2.4), *Fo*, *Da*, and *Pe* are the Fourier, Damkohler, and Peclet numbers, respectively:

$$Fo_j = \frac{D_j l}{u_j^0 r_0^2}, \quad Pe_j = \frac{u_j^0 l}{D_j}, \quad \varepsilon^2 = Fo_j^{-1} Pe_j^{-1}, \quad j = 1, 2,$$

$$Da = \frac{k_0 l}{u_2^0}, \quad K_1 = \frac{kl}{u_1^0}, \quad K_2 = \omega K_1, \quad \omega = \frac{u_1^0 \chi}{u_2^0}. \tag{12.2.6}$$

In the cases of a physical absorption, $Da = 0$.

12.2.1 Axial and Radial Velocity Components

The radial non-uniformity of the axial velocity component in a column apparatus is a result of the fluid hydrodynamics at the column inlet and decreases along the column height as a result of the fluid viscosity. As a result, a radial velocity component is initiated. The theoretical analysis of the change in the radial non-uniformity of the axial velocity component (effect of the radial velocity component) in a column can be made by an appropriate hydrodynamic model, where the average velocity at the cross section of the column is a constant (inlet average axial velocity component). In generalized variables (12.2.3), as an example, it is possible to be used the next velocity distributions, where the difference between gas and liquid flows is in the average (inlet) velocities, only:

$$U_j = (2 - 0.4Z_j) - 2(1 - 0.4Z_j)R^2, \quad V_j = 0.2(R - R^3),$$

$$\overline{U}_j = 2 \int_0^1 RU_j dR = 1, \quad j = 1.2. \tag{12.2.7}$$

12.2.2 Convective-Type Model

The model parameters in (12.2.4) in the industrial absorption columns have very small values (12.0.8), and as a result, the model has a convective form:

$$U_j \frac{\partial C_j}{\partial Z_j} + V_j \frac{\partial C_j}{\partial R} = (-1)^{(2-j)} K_j (C_1 - C_2) - (j - 1)Da\, C_j;$$

$$R = 1, \quad \frac{\partial C_j}{\partial R} \equiv 0; \quad j = 1, 2; \tag{12.2.8}$$

$$Z_1 = 0, \quad C_1 \equiv 1; \quad Z_2 = 0, \quad C_2 = 0.$$

12.2.3 Average-Concentration Model

In Chap. 6, it is possible to see that the average values of the velocity and concentrations at the column cross-sectional area are:

$$\bar{u}_j(z_j) = \frac{2}{r_0^2} \int_0^{r_0} ru_j(r, z_j)dr, \quad \bar{c}_j(z_j) = \frac{2}{r_0^2} \int_0^{r_0} rc_i(r, z_j)dr, \quad j = 1, 2. \tag{12.2.9}$$

In the cylindrical columns, $r_0 = const, \bar{u}_j(z_j) \equiv u_j^0 = const, d\bar{u}_j/dz_j \equiv 0, j = 1, 2$, practically.

The functions $u_j(r, z_j), v_j(r, z_j), c_j(r, z_j), j = 1, 2$ in (12.2.1) can be presented with the help of the average functions (12.2.9):

$$u_j(r, z_j) = \bar{u}_j U_j(R, Z_j), \quad v_j(r, z_j) = \varepsilon \bar{u}_j V_j(R, Z_j),$$
$$c_j(r, z_j) = \bar{c}_j(z_j)\, \tilde{c}_j(r, z_j), \quad j = 1, 2, \tag{12.2.10}$$

where $\tilde{c}_j(r, z_j), j = 1, 2$ present the radial non-uniformity of the concentrations and satisfy the following conditions:

$$\frac{2}{r_0^2} \int_0^{r_0} r \tilde{c}_j(r, z_j)\, dr = 1, \quad j = 1, 2. \tag{12.2.11}$$

The average-concentration model may be obtained if put (12.2.10) into (12.2.1), multiplying by r and integrating over r in the interval $[0, r_0]$. As a result, the following is obtained:

$$\alpha_j(z_j)\bar{u}_j \frac{d\bar{c}_j}{dz_j} + [\beta_j(z_j) + \gamma_j(z_j)]\bar{u}_j\, \bar{c}_j$$
$$= D_j \frac{d^2\bar{c}_j}{dz_j^2} + (-1)^{(2-j)} k(\bar{c}_1 - \chi\bar{c}_2) - (j - 1)k_0\bar{c}_2; \tag{12.2.12}$$
$$z_j = 0, \quad \bar{c}_j(0) \equiv (2 - j)c_j^0, \quad \frac{d\bar{c}_j}{dz_j} \equiv 0; \quad j = 1, 2.$$

where

$$\alpha_j(z_j) = \frac{2}{r_0^2} \int_0^{r_0} r\, U_j \tilde{c}_j\, dr, \quad \beta_j(z_j) = \frac{2}{r_0^2} \int_0^{r_0} r U_j \frac{\partial \tilde{c}_j}{\partial z_j} dr,$$
$$\gamma_j(z_j) = \frac{2}{r_0^2} \int_0^{r_0} r V_j \frac{\partial \tilde{c}_j}{\partial r} dr, \quad U_j = U_j(R, Z_j), \quad V_j = V_j(R), \tag{12.2.13}$$
$$\tilde{c}_j(r, z_j) = \tilde{C}_j(R, Z_j), \quad j = 1, 2.$$

The theoretical analysis of the model (12.2.12) will be made, using the next generalized variables and functions:

$$r = r_0 R, \quad z_j = lZ_j, \quad u_j(r, z_j) = \bar{u}_j U_j(R, Z_j), \quad v_j(r) = \bar{u}_j \varepsilon V_j(R),$$
$$c_j(r, z_j) = c_j^0 C_j(R, Z_j), \quad \bar{c}_j(z_j) = c_j^0 \overline{C}_j(Z_j),$$
$$c_2^0 = \frac{c_1^0}{\chi}, \quad \overline{C}_j(Z_j) = 2 \int_0^1 RC_j(R, Z_j)dR, \tag{12.2.14}$$
$$\tilde{c}_j(r, z_j) = \frac{c_j(r, z_j)}{\bar{c}_j(z_j)} = \frac{C_j(R, Z_j)}{\overline{C}_j(Z_j)} = \tilde{C}_j(R, Z_j).$$

$$\alpha_j(z_j) = \alpha_j(lZ_j) = A_j(Z_j) = 2\int_0^1 RU_j(R,Z_j)\widetilde{C}_j(R,Z_j)\,dR,$$

$$\beta_j(z_j) = \beta_j(lZ_j) = B_j(Z_j) = 2\int_0^1 RU_j(R,Z_j)\frac{\partial\widetilde{C}_j}{\partial Z_j}\,dR, \qquad (12.2.15)$$

$$\gamma_j(z_j) = \gamma_j(lZ_j) = G_j(Z_j) = 2\int_0^1 RV_j(R)\frac{\partial\widetilde{C}_j}{\partial R}\,dR, \quad j = 1,2.$$

As a result, the model (12.2.12) has the form:

$$A_j(Z_j)\frac{d\overline{C}_j}{dZ_j} + [B_j(Z_j) + G_j(Z_j)]\overline{C}_j$$
$$= Pe_j^{-1}\frac{d^2\overline{C}_j}{dZ_j^2} + (-1)^{(2-j)}K_j(\overline{C}_1 - \overline{C}_2) - (j-1)Da\overline{C}_j; \qquad (12.2.16)$$
$$Z_j = 0, \quad \overline{C}_j = 2 - j, \quad \frac{d\overline{C}_j}{dZ_j} = 0; \quad j = 1,2.$$

In industrial conditions (12.0.7) the model (12.2.16) has the convective form:

$$A_j(Z_j)\frac{d\overline{C}_j}{dZ_j} + [B_j(Z_j) + G_j(Z_j)]\overline{C}_j =$$
$$= (-1)^{(2-j)}K_j(\overline{C}_1 - \overline{C}_2) - (j-1)Da\overline{C}_j; \qquad (12.2.17)$$
$$Z_j = 0, \quad \overline{C}_1 = 1, \quad \overline{C}_2 = 0; \quad j = 1,2.$$

The presented models (12.2.8) and (12.2.17) permit to be analyzed the physical absorption $(0 = Da \le 10^{-2})$ of highly soluble $(0 = \omega \le 10^{-2})$, average soluble $(10^{-1} < \omega < 10)$, or lightly soluble $(0 = \omega^{-1} \le 10^{-2})$ gases.

12.2.4 Physical Absorption of Highly Soluble Gas

In the cases of physical absorption of highly soluble gas $(\omega = K_2 = Da = 0)$ and from (12.2.8), (12.2.17) follows $C_2 = \overline{C}_2 \equiv 0$, and as a result, the models (12.2.8) and (12.2.17) have the forms:

$$U_1\frac{dC_1}{dZ_1} + V_1\frac{dC_1}{dR} = -K_1C_1; \quad R = 1, \quad \frac{\partial C_1}{\partial R} \equiv 0; \quad Z_1 = 0, \quad C_1 \equiv 1.$$
$$(12.2.18)$$

$$A_1(Z_1)\frac{d\overline{C}_1}{dZ} + [B(Z_1) + G(Z_1)]\overline{C}_1 = -K_1\overline{C}_1; \quad Z_1 = 0, \quad \overline{C}_1 \equiv 1. \qquad (12.2.19)$$

The solution of (12.2.7), (12.2.18) and (12.2.15) permits to be obtained the average concentrations $\overline{C}_1(Z_{1n}), Z_{1n} = 0.1(n+1), n = 0, 1, \ldots, 9$ and functions $A_1(Z_{1n}), B_1(Z_{1n}), G_2(Z_{1n}), Z_{1n} = 0.1(n+1), n = 0, 1, \ldots, 9$ in (12.2.15). The functions $A_1(Z_1), B_1(Z_1), G_1(Z_1)$ are possible to be presented as approximations that are similar to those used in Chap. 11.

12.2.5 Physical Absorption of Lightly Soluble Gas

In the cases of physical absorption of lightly soluble gas ($\omega^{-1} = Da = 0$), from (12.2.8) and (12.2.17) follows $C_1 = \overline{C}_1 \equiv 1$, and as a result, the models (12.2.8), (12.2.17) have the forms:

$$U_2 \frac{dC_2}{dZ_2} + V_2 \frac{dC_2}{dR} = -K_2(1 - C_2);$$

$$R = 1, \quad \frac{\partial C_2}{\partial R} \equiv 0; \quad Z_2 = 0, \quad C_2 \equiv 0.$$

(12.2.20)

$$A_2(Z_2) \frac{d\overline{C}_2}{dZ_2} + [B_2(Z_2) + G_2(Z_2)]\overline{C}_2 = K_2(1 - \overline{C}_2);$$

$$Z_2 = 0, \quad \overline{C}_2 \equiv 0.$$

(12.2.21)

The solution of (12.2.20), (12.2.7) and (12.2.15) permits to be obtained the average concentrations $\overline{C}_2(Z_{2n}), Z_{2n} = 0.1(n+1), n = 0, 1, \ldots, 9$ and functions $A_2(Z_{2n}), B_2(Z_{2n}), G_2(Z_{2n}), Z_{2n} = 0.1(n+1), n = 0, 1, \ldots, 9$ in (12.2.15). The functions $A_2(Z_2), B_2(Z_2), G_2(Z_2)$ are possible to be presented as approximations that are similar to those used in Chap. 11.

12.2.6 Physical Absorption of Average Soluble Gas

Let us consider the physical absorption ($Da = 0$) of an average soluble gas ($\omega \sim 1$) in an industrial absorption column (12.2.8). The convection–diffusion and average-concentration models (12.2.8) and (12.2.17) have the forms:

$$U_1 \frac{dC_1}{dZ} + V_1 \frac{dC_1}{dR} = -K_1(C_1 - C_2);$$

$$U_2 \frac{dC_2}{dZ} + V_2 \frac{dC_2}{dR} = \omega K_1(C_1 - C_2);$$

$$R = 1, \quad \frac{\partial C_j}{\partial R} \equiv 0; \quad j = 1, 2; \quad Z = 0, \quad C_1 \equiv 1, \quad C_2 = 0.$$

(12.2.22)

$$A_1(Z)\frac{d\overline{C}_1}{dZ} + [B_1(Z)+G_1(Z)]\overline{C}_1 = -K_1(\overline{C}_1 - \overline{C}_2);$$

$$A_2(Z)\frac{d\overline{C}_2}{dZ} + [B_2(Z)+G_2(Z)]\overline{C}_2 = \omega K_1(\overline{C}_1 - \overline{C}_2);$$

$$Z = 0, \quad \overline{C}_1 = 1, \quad \overline{C}_2 = 0,$$

(12.2.23)

where the functions $A_j(Z), B_j(Z), G_j(Z), j = 1, 2$ are presented as approximations that are similar to those used in Chap. 11.

12.2.7 Chemical Absorption

In the case of chemical absorption, from (12.2.8), (12.2.17) follow the models:

$$U_j\frac{\partial C_j}{\partial Z} + V_j\frac{\partial C_j}{\partial R} = (-1)^{(2-j)}K_1\omega^{(j-1)}(C_1 - C_2) - (j-1)Da\,C_j;$$

$$R = 1, \quad \frac{\partial C_j}{\partial R} \equiv 0; \quad j = 1, 2;$$

$$Z = 0, \quad C_1 \equiv 1, \quad C_2 \equiv 0.$$

(12.2.24)

$$A_j(Z_j)\frac{d\overline{C}_j}{dZ_j} + [B_j(Z_j)+G_j(Z_j)]\overline{C}_j$$

$$= (-1)^{(2-j)}K_j(\overline{C}_1 - \overline{C}_2) - (j-1)Da\,\overline{C}_j;$$

$$Z_j = 0, \quad \overline{C}_1 = 1, \quad \overline{C}_2 = 0; \quad j = 1, 2.$$

(11.2.25)

The theoretical analysis of the chemical absorption is similar to that in Chap. 11.

The presented numerical analysis of the counter-current absorption processes in the column apparatuses, in the cases of axial modification of the radial non-uniformities of the axial velocity components and radial modification of the radial velocity component, shows the possibility to be created convection–diffusion and average-concentration models. The hydrodynamic effects are presented in the average-concentration models as parameters. The use of experimental data, for the average concentrations at the column ends, for a concrete processes, permits to be obtained the model parameters $a_{j0}, a_{j1}, a_{j2}, b_{j0}, b_{j1}, b_{j2}, g_{j0}, g_{j1}, g_{j2}, j = 1, 2, K_1$, related to the radial non-uniformities of the velocities in the gas and liquid phases. These parameter values permit to be used the average-concentration model for modeling of physical and chemical absorption, absorption of highly, lightly, and average soluble gases (different values of the parameters ω, Da, i.e., different values of the column height, average velocities, chemical reaction rate constant, and gas solubility).

References

1. Boyadjiev C (2010) Theoretical Chemical Engineering. Modeling and simulation. Springer, Berlin
2. Doichinova M, Boyadjiev C (2012) On the column apparatuses modeling. Int J Heat Mass Transfer 55:6705–6715
3. Boyadjiev C (2013) A new approach for the column apparatuses modeling in chemical engineering. J Pure Appl Math: Adv Appl 10(2):131–150
4. Boyadjiev B, Boyadjiev C (2017) New models of industrial column absorbers. 1. Co-current absorption processes. Bulg Chem Commun 49(3):711–719
5. Boyadjiev B, Doichinova M, Boyadjiev C (2015) Computer modeling of column apparatuses. 1. Two coordinates systems approach. J Eng Thermophy 24(3):247–258
6. Boyadjiev B, Boyadjiev C (2017) New models of industrial column absorbers. 1. Counter-current absorption processes. Bulg Chem Commun 49(3):720–728

Chapter 13
Industrial Column Adsorber

In Chaps. 10–12 were presented convection–diffusion and average-concentration models of chemical [1], co-current absorption [2] and countercurrent absorption [3] processes in industrial column apparatuses, where an axial modification of the axial velocity radial non-uniformity along the column height exists and the radial velocity component is not equal to zero. This problem is solved in the cases of the physical and chemical adsorption processes in the industrial column apparatuses [4].

13.1 Effect of the Axial Modification of the Radial Non-uniformity of the Axial Velocity Component

13.1.1 Physical Adsorption Process

In Chaps. 3, 6 are presented convection-type (3.2.12) and average-concentration (6.2.10) models of the adsorption processes in the industrial column apparatuses:

$$
\begin{aligned}
&U(R)\frac{dC_{11}}{dZ} = -K(C_{11} - C_{13}); \\
&\frac{dC_{13}}{dT} = K\omega(C_{11} - C_{13}) - K_1 C_{13}C_{23} + K_2\theta^{-1}(1 - C_{23}); \\
&\frac{dC_{23}}{dT} = -K_1\theta C_{13}C_{23} + K_2(1 - C_{23}); \\
&T = 0, \quad C_{13} \equiv 0, \quad C_{23} \equiv 1; \quad Z = 0, \quad C \equiv 1.
\end{aligned}
\tag{13.1.1}
$$

© Springer International Publishing AG, part of Springer Nature 2018
C. Boyadjiev et al., *Modeling of Column Apparatus Processes,*
Heat and Mass Transfer, https://doi.org/10.1007/978-3-319-89966-4_13

$$A\frac{d\overline{C}_{11}}{dZ} + \frac{dA}{dZ}\overline{C}_{11} = -K(\overline{C}_{11} - \overline{C}_{13});$$

$$\frac{d\overline{C}_{13}}{dT} = K\omega(\overline{C}_{11} - \overline{C}_{13}) - BK_1\overline{C}_{13}\overline{C}_{23} + K_2\theta^{-1}(1 - \overline{C}_{23});$$

$$\frac{d\overline{C}_{23}}{dT} = -BK_1\theta\overline{C}_{13}\overline{C}_{23} + K_2(1 - \overline{C}_{23});$$ (13.1.2)

$$T = 0, \quad \overline{C}_{11} \equiv 1, \quad \overline{C}_{13} \equiv 0, \quad \overline{C}_{23} \equiv 1; \quad Z = 0, \quad \overline{C}_{11} \equiv 1,$$

where $\theta = \dfrac{c_{11}^0}{c_{23}^0}$, $\omega = \dfrac{u_1^0 t^0}{l}$. In (13.1.2) are used the functions:

$$A(T,Z) = 2\int_0^1 RU(R)\frac{C(T,R,Z)}{\overline{C}(T,Z)}dR,$$

$$B(T,Z) = 2\int_0^1 R\frac{C_1(T,R,Z)}{\overline{C}_1(T,Z)}\frac{C_0(T,R,Z)}{\overline{C}_0(T,Z)}dR,$$

$$\overline{C}(T,Z) = 2\int_0^1 RC(T,R,Z)dR,$$ (13.1.3)

$$\overline{C}_0(T,Z) = 2\int_0^1 RC_0(T,R,Z)dR,$$

$$\bar{C}_1(T,Z) = 2\int_0^1 RC_1(T,R,Z)dR,$$

where

$$C = C_{11}, \quad C_1 = C_{13}, \quad C_0 = C_{23}, \quad \overline{C} = \overline{C}_{11}, \quad \overline{C}_1 = \overline{C}_{13}, \quad \overline{C}_0 = \overline{C}c_{23}.$$ (13.1.4)

13.1.2 Modeling of the Industrial Column Adsorbers

Very often in industrial conditions, an axial modification of the radial
non-uniformity of the axial velocity component is realized. For a theoretical analysis of the effect of the axial modification of the radial non-uniformity of the
velocity in the industrial adsorbers will be used as an appropriate hydrodynamic
model (10.1.2), where the average velocity $\bar{u} = u_1^0$ at the cross section of the
column is a constant, while the maximal velocity (and as a result the radial

non-uniformity of the axial velocity component too) decreases (axial steps change) along the column height. As a result, the velocity and concentrations in (13.1.1)–(13.1.4) will be presented as:

$$
\begin{aligned}
&U = U_n(R, Z_n) = a_n - b_n R^2, \quad a_n = 2 - 0.1n, \quad b_n = 2(1 - 0.1n), \\
&C(T, R, Z) = C_n(T, R, Z_n), \quad C_1(T, R, Z) = C_{1n}(T, R, Z_n), \\
&C_0(T, R, Z) = C_{0n}(T, R, Z_n), \quad 0.1n \le Z_n \le 0.1(n+1) \quad n = 0, 1, \ldots, 9.
\end{aligned}
$$

$$(13.1.5)$$

If we put (13.1.5) and (13.1.4) in (13.1.1), the model has the form:

$$
\begin{aligned}
&U_n(R, Z_n)\frac{dC_n}{dZ_n} = -K(C_n - C_{1n}), \\
&0.1n \le Z_n \le 0.1(n+1), \quad n = 0, 1, \ldots, 9; \\
&Z_0 = 0, \quad C_0(R, Z_0) \equiv 1; \\
&Z_n = 0.1n, \quad C_n(R, Z_n) = C_{n-1}(R, Z_n), \quad n = 1, \ldots, 9.
\end{aligned}
$$

$$(13.1.6)$$

$$
\begin{aligned}
&\frac{dC_{1n}}{dT} = \omega K(C_n - C_{1n}) - K_1 C_{0n} C_{1n} + K_2 \theta^{-1}(1 - C_{0n}); \\
&\frac{dC_{0n}}{dT} = -K_1 \theta C_{0n} C_{1n} + K_2(1 - C_{0n}); \\
&T = 0, \quad C_{0n} \equiv 1, \quad C_{1n} \equiv 0.
\end{aligned}
$$

$$(13.1.7)$$

13.1.3 Model Equations Solution

For the solution of (13.1.6), (13.1.7), a multisteps approach [5–7] for different values of $T = \frac{\tau-1}{\tau^0-1}$, $(\tau = 1, 2, \ldots, \tau^0)$, $R = \frac{\rho-1}{\rho^0-1}$, $(\rho = 1, 2, \ldots, \rho^0)$, $Z = 0.01\zeta$, $\zeta = 1, 2, \ldots, 100$ will be used, where s, $(s = \tau = 1, 2, \ldots, \tau^0)$, will be the step number. At each step $s = \tau = 1, 2, \ldots, \tau^0$, the solutions of (13.1.6), (13.1.7) will be obtained as three matrix forms:

$$
\begin{aligned}
&C_n^s(T, R, Z_n) = \left\| C_{\tau\rho\zeta_n}^s \right\|, \quad C_{0n}^s(T, R, Z_n) = \left\| C_{(0)\tau\rho\zeta_n}^s \right\|, \\
&C_{1n}^s(T, R, Z_n) = \left\| C_{(1)\tau\rho\zeta_n}^s \right\|; \quad 0.1n \le Z_n \le 0.1(n+1); \\
&T = \frac{\tau-1}{\tau^0-1}, \quad s = \tau = 1, 2, \ldots, \tau^0; \quad R = \frac{\rho-1}{\rho^0-1} \quad \rho = 1, 2, \ldots, \rho^0; \\
&0 \le T \le 1, \quad 0 \le R \le 1; \\
&Z_n = 0.01\zeta_n, \quad \zeta_n = 10n + m, \quad n = 0, 1, \ldots, 9, \quad m = 1, 2, \ldots, 10.
\end{aligned}
$$

$$(13.1.8)$$

As a zero step ($s = \tau = 0$) will be used:

$$C_n^0(T, R, Z_n) = \left\| C_{0\rho\zeta_n}^0 \right\| \equiv C_n^0(R, Z_n),$$

$$C_{0n}^0(T, R, Z_n) = \widehat{C}_{0n}^0 = \left\| C_{(0)0\rho\zeta_n}^0 \right\| \equiv 1, \qquad (13.1.9)$$

$$C_{1n}^0(T, R, Z_n) = \widehat{C}_{1n}^0 = \left\| C_{(1)0\rho\zeta_n}^0 \right\| \equiv 0,$$

where $C_n^0(R, Z_n)$ is solution of (13.1.4) for $C_{1n} \equiv 0$, i.e.,

$$U_n(R, Z_n)\frac{dC_n^0}{dZ_n} = -KC_n^0, \quad 0.1n \leq Z_n \leq 0.1(n+1), \quad n = 0, 1, \ldots, 9;$$

$$Z_0 = 0, \quad C_0^0(R, Z_0) \equiv 1; \qquad (13.1.10)$$

$$Z_n = 0.1n, \quad C_n^0(R, Z_n) = C_{n-1}^0(R, Z_n), \quad n = 1, \ldots, 9.$$

The steps $s = \tau = 1, 2, \ldots, \tau^0$ are the solutions of the equations sets:

$$\frac{dC_{1n}^s}{dT} = \omega K\left(C_n^{(s-1)} - C_{1n}^s\right) - K_1 C_{0n}^s C_{1n}^s + K_2 \theta^{-1}\left(1 - C_{0n}^s\right);$$

$$\frac{dC_{0n}^s}{dT} = -K_1 \theta C_{0n}^s C_{1n}^s + K_2\left(1 - C_{0n}^s\right); \quad n = 1, \ldots, 9; \qquad (13.1.11)$$

$$T = \frac{\tau - 1}{\tau_0}, \quad C_{1n}^s \equiv \widehat{C}_{1n}^{(s-1)}(R, Z_n), \quad C_{0n}^s \equiv \widehat{C}_{0n}^{(s-1)}(R, Z_n).$$

The solutions of (13.1.11) permit to obtain:

$$\widehat{C}_{0n}^s(R, Z_n) = C_{0n}^s\left(\frac{\tau}{\tau_0}, R, Z_n\right) = \left\| C_{(0)n\tau\rho\zeta_n}^s \right\|,$$

$$\widehat{C}_{1n}^s(R, Z_n) = C_{1n}^s\left(\frac{\tau}{\tau_0}, R, Z_n\right) = \left\| C_{(1)n\tau\rho\zeta_n}^s \right\|, \quad n = 1, \ldots, 9. \qquad (13.11.12)$$

The obtained functions $\widehat{C}_{1n}^s(R, Z_n)$ permit to obtain $C_n^s(R, Z_n) = \left\| C_{n1\rho\zeta_n}^s \right\|$, as a solution of (13.1.6) at the s step:

$$U_n(R, Z_n)\frac{dC_n^s}{dZ_n} = -K\left(C_n^s - \widehat{C}_{1n}^s\right),$$

$$0.1n \leq Z_n \leq 0.1(n+1), \quad n = 0, 1, \ldots, 9; \qquad (13.1.13)$$

$$Z_0 = 0, \quad C_0^s(R, Z_0) \equiv 1;$$

$$Z_n = 0.1n, \quad C_n^s(R, Z_n) = C_{n-1}^s(R, Z_n), \quad n = 1, \ldots, 9.$$

The end of the multisteps computational procedure is $s = \tau = \tau^0$, and the solution of (13.1.6, 13.1.7) is:

$$C(T,R,Z) = \left\|C_{\tau\rho\zeta}\right\|, \quad C_0(T,R,Z) = \left\|C_{(0)\tau\rho\zeta}\right\|, \quad C_1(T,R,Z) = \left\|C_{(1)\tau\rho\zeta}\right\|;$$

$$T = \frac{\tau - 1}{\tau^0 - 1}, \quad \tau = 1,2,\ldots,\tau^0; \quad R = \frac{\rho - 1}{\rho^0 - 1}, \quad \rho = 1,2,\ldots,\rho^0;$$

$$Z = 0.01\zeta, \quad \zeta = 1,2,\ldots,100.$$

$$(13.1.14)$$

13.1.4 Concentration Distributions

The solution of the model Eqs. (13.1.6), (13.1.7), using the numerical multisteps algorithm [5–7] and calculation procedure (13.1.8)–(13.1.14), is obtained in the case:

$$K = K_1 = 1, \quad K_2 = 10, \quad \theta = 0.01, \quad \omega = 1 \qquad (13.1.15)$$

and the result is presented in Fig. 13.1.

The obtained solution of the model Eqs. (13.1.6), (13.1.7), and (13.1.3) permits to find the "theoretical" average concentrations $\overline{C}(T,Z)$, $\overline{C}_0(T,Z)$, $\overline{C}_1(T,Z)$ and the functions $A(T,Z)$, $B(T,Z)$ (the points in Figs. 13.2, 13.3).

The results for $A(T,Z)$, $B(T,Z)$ show that these functions are possible to be presented as the next approximations:

$$A = 1 + a_1 Z + a_2 Z^2 + a_t T, \quad B = 1 + b_1 Z + b_2 Z^2 + b_t T. \qquad (13.1.16)$$

The obtained "theoretical" parameter values a_1, a_2, a_t, b_1 b_2, b_t, $(K = 1)$ are presented in Table 13.1, where is seen that $B \equiv 1$. As a result, the average-concentration model (13.1.2) has the form:

$$\left(1 + a_1 Z + a_2 Z^2 + a_t T\right) \frac{d\overline{C}}{dZ} + (a_1 + 2a_2 Z)\overline{C} = -K\left(\overline{C} - \overline{C}_1\right);$$

$$Z = 0, \quad \overline{C} \equiv 1. \qquad (13.1.17)$$

$$\frac{d\overline{C}_1}{dT} = \omega K\left(\overline{C} - \overline{C}_1\right) - K_1\overline{C}_0\overline{C}_1 + K_2\theta^{-1}\left(1 - \overline{C}_0\right);$$

$$\frac{d\overline{C}_0}{dT} = -K_1\theta\overline{C}_0\overline{C}_1 + K_2\left(1 - \overline{C}_0\right);$$

$$T = 0, \quad \overline{C}_0 \equiv 1, \quad \overline{C}_1 \equiv 0. \qquad (13.1.18)$$

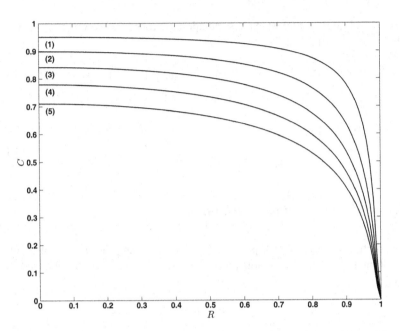

Fig. 13.1 Concentration $C(T, R, Z)$: $T = 0.8$; $Z = 0.2(1)$, $0.4(2)$, $0.6(3)$, $0.8(4)$, $1.0(5)$

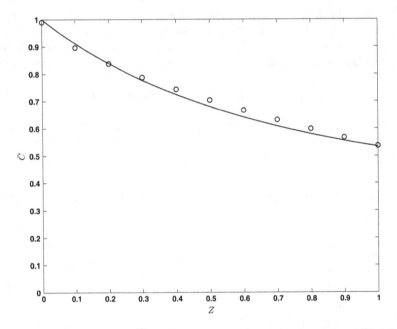

Fig. 13.2 Average concentrations $\overline{C}(T, Z)$: "theoretical" values, using the solution of (13.1.6) and (13.1.3) for $T = 0.8$, $Z = Z_n = 0.1n$, $n = 1, \ldots, 10$ (points); solution of (13.1.17), (13.1.18), using the "experimental" parameter values (Table 13.1) (line)

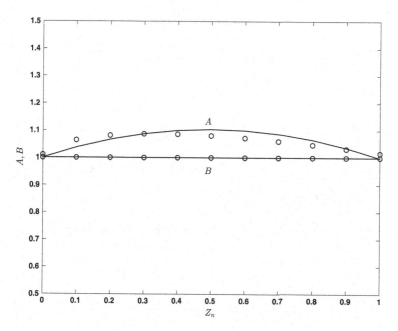

Fig. 13.3 Functions $A(T,Z)$, $B(T,Z)$ in (13.1.3) for $T = 0.8$, $Z = Z_n = 0.1n$, $n = 1, \ldots, 10$ (points); approximations (13.1.16) (lines)

Table 13.1 Model parameter values a_1, a_2, a_t, b_1, b_2, b_t, K (physical adsorption)		Theoretical values	Experimental values
	a_1	0.4160	0.4717
	a_2	−0.4171	−0.3213
	a_t	−0.0003	−0.0003
	b_1	-2.011×10^{-5}	–
	b_2	1.522×10^{-5}	–
	b_t	-2.311×10^{-5}	–
	K	1	1.1287

For the solution of (13.1.17), (13.1.18), a multisteps approach for different values of $T = \frac{\tau-1}{\tau^0-1}$, $(\tau = 1, 2, \ldots, \tau^0)$, $Z = 0.01\zeta$, $\zeta = 1, 2, \ldots, 100$ will be used, where s $(s = \tau = 1, 2, \ldots, \tau^0)$ will be the step number. At each step $s = \tau = 1, 2, \ldots, \tau^0$, the solutions of (13.1.17), (13.1.18) will be obtained as three matrix forms:

$$\overline{C}^s(T,Z) = \left\| \overline{C}^s_{\tau\zeta} \right\|, \quad \overline{C}^s_0(T,Z) = \left\| \overline{C}^s_{(0)\tau\zeta} \right\|, \quad C^s_1(T,Z) = \left\| \overline{C}^s_{(1)\tau\zeta} \right\|;$$

$$T = \frac{\tau-1}{\tau^0-1}, \quad \tau = 1, 2, \ldots, \tau^0; \quad Z = 0.01\zeta, \quad \zeta = 1, 2, \ldots, 100; \qquad (13.1.19)$$

$$0 \leq T \leq 1, \quad 0 \leq Z \leq 1.$$

As a zero step ($s = \tau = 0$) will be used:

$$\overline{C}^0(T,Z) = \left|\overline{C}^0_{0\zeta}\right| \equiv \overline{C}^0(Z), \quad \overline{C}^0_0(T,Z) = \hat{C}^0_0 = \left|\overline{C}^0_{(0)0\zeta}\right| \equiv 1,$$
$$\overline{C}^0_1(T,Z) = \hat{C}^0_1 = \left|\overline{C}^0_{(1)0\zeta}\right| \equiv 0,$$
(13.1.20)

where $\overline{C}^0(Z)$ is solution of (13.1.17) for $\overline{C}_1 \equiv 0$, i.e.,

$$\left(1 + a_1 Z + a_2 Z^2 + a_t T\right)\frac{d\overline{C}^0}{dZ} + (a_1 + 2a_2 Z)\overline{C}^0 = -K\overline{C}^0;$$
$$Z = 0, \quad \overline{C}^0 \equiv 1.$$
(13.1.21)

The steps $s = \tau = 1, 2, \ldots, \tau^0$ are the solutions of the equations sets:

$$\frac{d\overline{C}^s_1}{dT} = \omega K\left(\overline{C}^{(s-1)} - \overline{C}^s_1\right) - K_1\overline{C}^s_0\overline{C}^s_1 + K_2\theta^{-1}\left(1 - \overline{C}^s_0\right);$$
$$\frac{d\overline{C}^s_0}{dT} = -K_1\theta\overline{C}^s_0\overline{C}^s_1 + K_2\left(1 - \overline{C}^s_0\right);$$
$$T = \frac{\tau - 1}{\tau_0}, \quad \overline{C}^s_0 \equiv \overline{C}^{(s-1)}_0, \quad \overline{C}^s_1 \equiv \overline{C}^{(s-1)}_1.$$
(13.1.22)

The solutions of (13.1.22) permit to obtain:

$$\hat{C}^s_0(Z) = \overline{C}^s_0\left(\frac{\tau}{\tau^0}, Z\right) = \left\|\overline{C}^s_{(0)\tau\zeta}\right\|, \quad \hat{C}^s_1(Z) = \overline{C}^s_1\left(\frac{\tau}{\tau^0}, Z\right) = \left\|\overline{C}^s_{(1)\tau\zeta}\right\|,$$
$$s = \tau = 1, 2, \ldots, \tau^0$$
(13.1.23)

The obtained functions $\hat{C}^s_1(Z)$ permit to obtain $\overline{C}^s(Z) = \left\|\overline{C}^s_{\tau\zeta}\right\|$, as a solution of (13.1.17) at the s step:

$$\left(1 + a_1 Z + a_2 Z^2 + a_t\frac{\tau}{\tau^0}\right)\frac{d\overline{C}^s}{dZ} + (a_1 + 2a_2 Z)\overline{C}^s = -K\left(\overline{C}^s - \hat{C}^s_1\right);$$
$$Z = 0, \quad \overline{C}^s \equiv 1.$$
(13.1.24)

The end of the multisteps computational procedure is $s = \tau = \tau^0$, and the solution of (13.1.17, 13.1.18) is:

$$\overline{C}(T,Z) = \left\|\overline{C}_{\tau\zeta}\right\|, \quad \overline{C}_0(T,Z) = \left\|\overline{C}_{(0)\tau\zeta}\right\|, \quad \overline{C}_1(T,Z) = \left\|\overline{C}_{(1)\tau\zeta}\right\|;$$
$$T = \frac{\tau - 1}{\tau^0 - 1}, \quad \tau = 1, 2, \ldots, \tau^0; \quad Z = 0.01\zeta, \quad \zeta = 1, 2, \ldots, 100.$$
(13.1.25)

13.1.5 Parameter Identification

In Eqs. (13.1.17), (13.1.18), the values of the parameters K_1, K_2, θ, ω are known beforehand and only parameters a_1, a_2, a_t, K in (13.1.17) should be determined, using experimental data.

The solution of (13.1.6), (13.1.7) in the case (13.1.15) and Eq. (13.1.3) permits to obtain concentration $C(T, R, 1)$, the average concentrations $\overline{C}(T, 1)$, $T = 0.1m$, $m = 1, \ldots, 10$, and the artificial experimental data:

$$\overline{C}_{exp}^{m}(T, 1) = (0.95 + 0.1S_m)\overline{C}(T, 1), \quad T = 0.1m, \quad m = 1, \ldots, 10, \quad (13.1.26)$$

where $0 \le S_m \le 1$, $m = 1, \ldots, 10$ are obtained by a generator of random numbers. The obtained artificial experimental data (13.1.26) are used for the parameters a_1, a_2, a_t, K identification in the average-concentration model (13.1.17), (13.1.18) by the minimization of the least-squares function Q with respect to the parameters a_1, a_2, a_t, K:

$$Q(a_1, a_2, a_t, K) = \sum_{m=1}^{10} \left[\overline{C}(T_m, 1, a_1, a_2, a_t, K) - \overline{C}_{exp}^{m}(T_m, 1) \right]^2,$$
$$T_m = 0.1m, \quad m = 1, \ldots, 10, \tag{13.1.27}$$

where the values of $\overline{C}(T_m, 1, a_1, a_2, a_t, K)$ are obtained as solutions of (13.1.17), (13.1.18) for different $T_m = 0.1m$, $m = 1, 2, \ldots, 10$ and $Z = 1$.

The obtained (after the minimization) "experimental" values of a_1, a_2, a_t, K are presented in Table 13.1. They are used for the solution of (13.1.17), (13.1.18), and the result is presented in Fig. 13.2 (line).

13.1.6 Chemical Adsorption Process

The presence of chemical bonds between the active component (AC) molecules in the gas phase and active sites (AS) in the solid face leads to the expressions for the convection-type model (3.2.25) and average-concentration model (6.2.18) in the cases of the chemical adsorption.

If we put (13.1.5) in (3.2.25), the convection-type model has the form:

$$U_n(R, Z_n) \frac{dC_n}{dZ_n} = -K(C_n - C_{1n}),$$

$$0.1n \leq Z_n \leq 0.1(n+1), \quad n = 0, 1, \ldots, 9;$$

$$Z_0 = 0, \quad C_0(R, Z_0) \equiv 1;$$

$$Z_n = 0.1n, \quad C_n(R, Z_n) = C_{n-1}(R, Z_n); \quad n = 1, \ldots, 9. \tag{13.1.28}$$

$$\frac{dC_{1n}}{dT} = \omega K(C_n - C_{1n}) - K_3 C_{0n} C_{1n}; \quad \frac{dC_{0n}}{dT} = -K_3 \theta C_{0n} C_{1n};$$

$$T = 0, \quad C_{0n} \equiv 1, \quad C_{1n} \equiv 0,$$

where $K_3 = kt^0$.

The solution of the model Eq. (13.1.28), using the numerical multisteps algorithm [5–7] and calculation procedure (13.1.8)–(13.1.14), is obtained in the case:

$$K = K_3 = 1, \quad \theta = 0.01, \quad \omega = 1 \tag{13.1.29}$$

and is presented in Fig. 13.4.

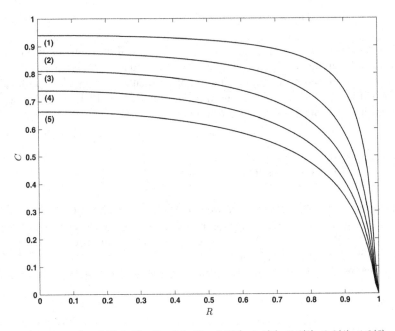

Fig. 13.4 Concentration $C(T, R, Z)$: $T = 0.8$; $Z = 0.2(1), 0.4(2), 0.6(3), 0.8(4), 1.0(5)$

The obtained solution of (13.1.28) and Eq. (13.1.3) permits to find the average concentrations $\overline{C}(T,Z)$, $\overline{C}_0(T,Z)$, $\overline{C}_1(T,Z)$ and the functions $A(T,Z)$, $B(T,Z)$ for different $Z = 0.1n$, $n = 1,\ldots,9$. They are presented in Figs. 13.5, 13.6 (points). The results for $A(T,Z)$, $B(T,Z)$ show that these functions are possible to be presented [5–7] as the approximations (13.1.16) and the obtained "theoretical" parameter values a_1, a_2, a_t, b_1, b_2, b_t are presented in Table 13.2, where is seen that $B \equiv 1$. As a result, the average-concentration model (6.2.18) has the form:

$$\left(1 + a_1 Z + a_2 Z^2 + a_t T\right)\frac{d\overline{C}}{dZ} + (a_1 + 2a_2 Z)\overline{C} = -K\left(\overline{C} - \overline{C}_1\right);$$

$$\frac{d\overline{C}_1}{dT} = \omega K\left(\overline{C} - \overline{C}_1\right) - K_3\overline{C}_0\overline{C}_1; \quad \frac{d\overline{C}_0}{dT} = -K_3\theta\overline{C}_0\overline{C}_1; \qquad (13.1.30)$$

$$T = 0, \quad \overline{C}_0 \equiv 1, \quad \overline{C}_1 \equiv 0; \quad Z = 0, \quad \overline{C} \equiv 1.$$

In Eq. (13.1.30), the values of the parameters K_3, θ, ω are known beforehand and only parameters a_1, a_2, a_t, K should be determined, using experimental data.

The obtained average concentrations $\overline{C}(T_m, 1)$, $T_m = 0.1m$, $m = 1,\ldots,10$, in the case (13.1.29), using the solution of (13.1.28) and Eq. (13.1.3), permit to obtain the artificial experimental data (13.1.26), which are used for the parameter identification in the average-concentration models (13.1.30) by the minimization of the

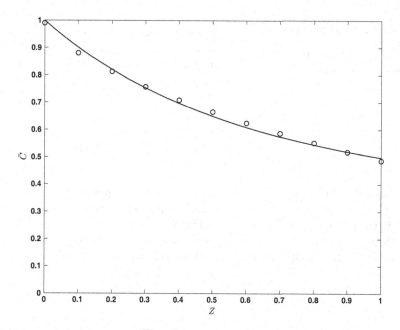

Fig. 13.5 Average concentrations $\overline{C}(T,Z)$: "theoretical" values, using the solution of (13.1.28) and Eq. (13.1.3) for $T = 0.8$, $Z = Z_n = 0.1n$, $n = 1,\ldots,10$ (points); solution of (13.1.30), using the "experimental" parameter values (Table 13.2) (line)

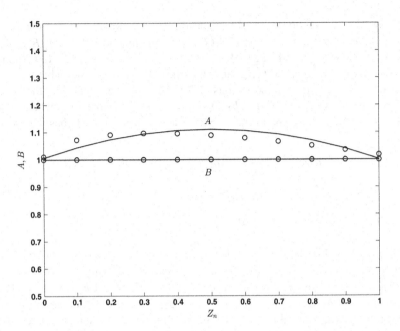

Fig. 13.6 Functions $A(T, Z)$, $B(T, Z)$ in (13.1.3) for $T = 0.8$, $Z = Z_n = 0.1n$, $n = 1, \ldots, 10$ (points); approximations (13.1.16) (lines)

Table 13.2 Model parameter values a_1, a_2, a_t, b_1, b_2, b_t, K (chemical adsorption)		Theoretical values	Experimental values
	a_1	0,4224	0,4643
	a_2	−0,4268	−0,3535
	a_t	0,0068	0,0071
	b_1	$-2,3695 \times 10^{-5}$	–
	b_2	$1,8337 \times 10^{-5}$	–
	b_t	−0,0001	–
	K	1	1,1263

least-squares functions (13.1.27). The obtained (experimental) values of a_1, a_2, a_t, K are presented in Table 13.2. They are used for the solution of (13.1.30), and the result is presented in Fig. 13.5 (line).

The parameter values of a_1, a_2, a_t, in Tables 13.1, 13.2, are equal because they are related to the radial non-uniformity of the velocity only.

13.1.7   Influence of the Model Parameters

In the case of physical adsorption, the influence of the model parameters in the model (13.1.6), (13.1.7), where the experimental values of the parameters a_1, a_2, a_t, K (Table 13.1) are used, will be presented in the cases:

1. $K_1 = 2$, $K_2 = 10$, $\theta = 0.01$, $\omega = 1$ (Fig. 13.7).
2. $K_1 = 1$, $K_2 = 20$, $\theta = 0.01$, $\omega = 1$ (Fig. 13.8).
3. $K_1 = 1$, $K_2 = 10$, $\theta = 0.03$, $\omega = 1$ (Fig. 13.9).
4. $K_1 = 1$, $K_2 = 10$, $\theta = 0.01$, $\omega = 2$ (Fig. 13.10).

The solution of (13.1.6) in these conditions and Eq. (13.1.3) permits to obtain concentration $C(T, R, 1)$, the average concentrations $\overline{C}(T_m, 1)$, $T_m = 0.1m$, $m = 1, 2, \ldots, 10$, and the artificial experimental data (13.1.26) (the points on Figs. 13.7–13.10). The lines on Figs. 13.7–13.10 are the solutions of (13.1.17), (13.1.18), where the parameter values in the cases (1–4) are used, together with the experimental values of the parameters a_1, a_2, a_t, K (Table 13.1).

In the case of chemical adsorption, the influence of the model parameters in the model (13.1.30), where the experimental values of the parameters a_1, a_2, a_t, K (Table 13.2) are used, will be presented in the case $K_3 = 2$, $\theta = 0.01$, $\omega = 1$ (Fig. 13.11).

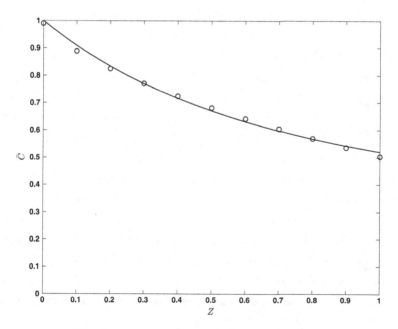

Fig. 13.7 Function $\overline{C}(T, 1)$ as solution of (13.1.17), (13.1.18) (line); artificial experimental data (13.1.26) (points) in the case 1 ($K_1 = 2$)

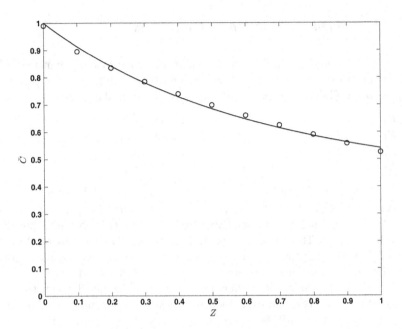

Fig. 13.8 Function $\overline{C}(T, 1)$ as solution of (13.1.17), (13.1.18) (line); artificial experimental data (13.1.26) (points) in the case 2 ($K_2 = 20$)

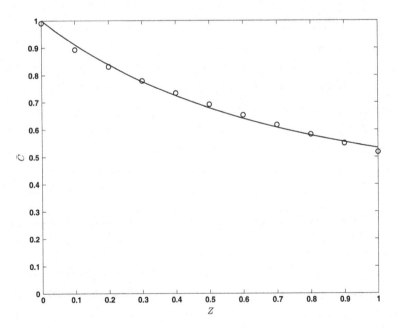

Fig. 13.9 Function $\overline{C}(T, 1)$ as solution of (13.1.17), (13.1.18) (line); artificial experimental data (13.1.26) (points) in the case 3 ($\theta = 0.03$)

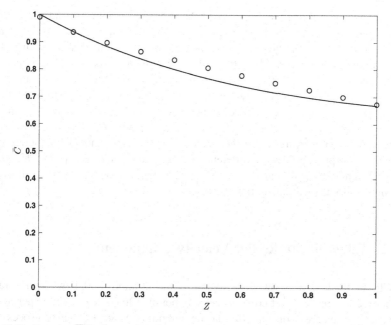

Fig. 13.10 Function $\overline{C}(T, 1)$ as solution of (13.1.17), (13.1.18) (line); artificial experimental data (13.1.26) (points) in the case 4 ($\omega = 2$)

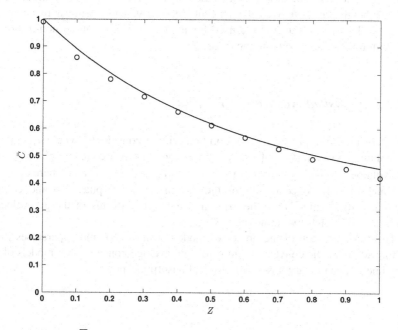

Fig. 13.11 Function $\overline{C}(T, 1)$ as solution of (13.1.30) (line); artificial experimental data (13.1.26) (points) in the chemical adsorption case $K_3 = 2$

The presented numerical analysis of the industrial column for adsorption processes shows that average-concentration model, where the radial velocity component is equal to zero (in the cases of a constant velocity radial non-uniformity along the column height), is possible to be used in the cases of an axial modification of the radial non-uniformity of the axial velocity component. The use of experimental data, for the average concentration at the column end and for different times, for a concrete adsorption process, permits to obtain the model parameters (a_1, a_2, a_t, K), related to the radial non-uniformity of the velocity. These parameter values permit to use the average-concentration model for modeling of different adsorption processes (different values of the parameters K_1, K_2, K_3, θ, ω, i.e., different values of the column height, average velocity, reagent diffusivity, physical and chemical adsorption rate constant, etc.).

13.2 Effect of the Radial Velocity Component

In Chaps. 10–12 was shown that average-concentration model, where the radial velocity component is not equal to zero, is possible to be used. The experimental data, for the average concentration at the column end, for a concrete process and column, permit to obtain the model parameters, related to the radial non-uniformity of the velocity. These parameter values permit to use the average-concentration model for modeling of different processes in the cases of different values of the column height, average velocity, reagent diffusivity, and chemical reaction rate constant. This possibility will be used for modeling of a longtime non-stationary adsorption process in column apparatus.

13.2.1 Physical Adsorption Process

For convenience, a physical gas adsorption will be considered, where c_{11}, c_{13} are the volume concentrations of the AC in gas and solid phases (capillaries volume). The volume concentration of the AS in the solid phase is $c_{23} \leq c_{23}^0$, where c_{23}^0 is the maximal (initial) concentration of the AS in the solid phase. In the column $u_1(r, z)$, $v_1(r, z)$ (m s^{-1}) are the axial and radial components of the gas velocity, while the solid phase is immobile ($u_3(r, z) \equiv 0$).

The new approach of the processes modeling in the column apparatuses [5–7] permits to create the convection–diffusion and average-concentration models of the longtime non-stationary gas–solid physical adsorption process:

$$u_1 \frac{\partial c_{11}}{\partial z} + v_1 \frac{\partial c_{11}}{\partial r} = D_{11} \left(\frac{\partial^2 c_{11}}{\partial z^2} + \frac{1}{r} \frac{\partial c_{11}}{\partial r} + \frac{\partial^2 c_{11}}{\partial r^2} \right) - k_0(c_{11} - c_{13});$$

$$\frac{dc_{13}}{dt} = k_0(c_{11} - c_{13}) - qk_1 c_{13} \frac{c_{23}}{c_{23}^0} + k_2 c_{23}^0 \left(1 - \frac{c_{23}}{c_{23}^0} \right);$$

$$\frac{dc_{23}}{dt} = -qk_1 c_{13} \frac{c_{23}}{c_{23}^0} + k_2 c_{23}^0 \left(1 - \frac{c_{23}}{c_{23}^0} \right); \tag{13.2.1}$$

$$t = 0, \quad c_{13} \equiv 0, \quad c_{23} \equiv c_{23}^0;$$

$$r = 0, \quad \frac{\partial c_{11}}{\partial r} \equiv 0; \quad r = r_0, \quad \frac{\partial c_{11}}{\partial r} \equiv 0;$$

$$z = 0, \quad c_{11} \equiv c_{11}^0, \quad u_1^0 c_{11}^0 \equiv u_1(r) c_{11}^0 - D_{11} \left(\frac{\partial c_{11}}{\partial z} \right)_{z=0}.$$

$$\frac{\partial u_1}{\partial z} + \frac{\partial v_1}{\partial r} + \frac{v_1}{r} = 0; \tag{13.2.2}$$

$$r = r_0, \quad v_1(r_0, z) \equiv 0, \quad z = 0, \quad u_1 = u_1(r, 0).$$

$$\alpha(t, z) \bar{u}_1 \frac{d\bar{c}_{11}}{dz} + [\beta(t, z) + \gamma(t, z)] \bar{u}_1 \bar{c}_{11} = D_{11} \frac{d^2 \bar{c}_{11}}{dz^2} - k_0(\bar{c}_{11} - \bar{c}_{13});$$

$$\frac{d\bar{c}_{13}}{dt} = k_0(\bar{c}_{11} - \bar{c}_{13}) - qk_1 \delta(t, z) \bar{c}_{13} \frac{\bar{c}_{23}}{c_{23}^0} + k_2 c_{23}^0 \left(1 - \frac{\bar{c}_{23}}{c_{23}^0} \right);$$

$$\frac{d\bar{c}_{23}}{dt} = -qk_1 \delta(t, z) \bar{c}_{13} \frac{\bar{c}_{23}}{c_{23}^0} + k_2 c_{23}^0 \left(1 - \frac{\bar{c}_{23}}{c_{23}^0} \right); \tag{13.2.3}$$

$$t = 0, \quad \bar{c}_{13} \equiv 0, \quad \bar{c}_{23} \equiv c_{23}^0; \quad z = 0, \quad \bar{c}_{11} \equiv c_{11}^0, \quad \left(\frac{\partial \bar{c}_{11}}{\partial z} \right)_{z=0} \equiv 0;$$

where

$$\alpha(t, z) = \frac{2}{r_0^2} \int_0^{r_0} r U \tilde{c}_{11}(t, r, z) dr, \quad \beta(t, z) = \frac{2}{r_0^2} \int_0^{r_0} r U \frac{\partial \tilde{c}_{11}}{\partial z} dr,$$

$$\gamma(t, z) = \frac{2}{r_0^2} \int_0^{r_0} r V \frac{\partial \tilde{c}_{11}}{\partial r} dr, \quad \delta(t, z) = \frac{2}{r_0^2} \int_0^{r_0} r \tilde{c}_{13}(t, r, z) \tilde{c}_{23}(t, r, z) dr. \tag{13.2.4}$$

In (13.2.3) are used the average velocity and concentrations:

$$\bar{u}_1 = \frac{2}{r_0^2} \int_0^{r_0} r u_1(r,z) dr = u_1^0, \quad \bar{c}_{11}(t,z) = \frac{2}{r_0^2} \int_0^{r_0} r c_{11}(t,r,z) dr,$$

$$\bar{c}_{13}(t,z) = \frac{2}{r_0^2} \int_0^{r_0} r c_{13}(t,r,z) dr, \quad \bar{c}_{23}(t,z) = \frac{2}{r_0^2} \int_0^{r_0} r c_{23}(t,r,z) dr \tag{13.2.5}$$

and the radial non-uniformities of the velocity components and concentrations:

$$U = \frac{u_1}{u_1^0}, \quad V = \frac{v_1}{\varepsilon u_1^0}, \quad \varepsilon = \frac{r_0}{l}, \quad \tilde{c}_{11} = \frac{c_{11}}{\bar{c}_{11}}, \quad \tilde{c}_{13} = \frac{c_{13}}{\bar{c}_{13}}, \quad \tilde{c}_{23} = \frac{c_{23}}{\bar{c}_{23}}, \tag{13.2.6}$$

where l is the column active zone height.

In (13.2.1), (13.2.3) D_{11} (m^2 s^{-1}) is the diffusivity of AC in the gas phase. The concentration of the adsorbed substance (AC) in the solid phase is $(c_{23}^0 - c_{23})$, i.e., equal to the related part of AS in the solid phase.

The using of dimensionless (generalized) variables [5–7] permits to make a theoretical analysis of the models (13.2.1), (13.2.3), where, as characteristic scales, are used the inlet velocity and concentrations, characteristic time (t_0) and column parameters (r_0, l):

$$T = \frac{t}{t^0}, \quad R = \frac{r}{r_0}, \quad Z = \frac{z}{l}, \quad U = \frac{u_1}{u_1^0}, \quad V = \frac{v_1}{\varepsilon u_1^0},$$

$$C = \frac{c_{11}}{c_{11}^0}, \quad C_0 = \frac{c_{23}}{c_{23}^0}, \quad C_1 = \frac{c_{13}}{c_{11}^0}, \quad \overline{C} = \frac{\bar{c}_{11}}{c_{11}^0}, \quad \overline{C}_0 = \frac{\bar{c}_{23}}{c_{23}^0}, \quad \overline{C}_1 = \frac{\bar{c}_{13}}{c_{11}^0}. \tag{13.2.7}$$

If we put (13.2.7) in (13.2.1, 13.2.3), the models in generalized variables have the forms:

$$U\frac{\partial C}{\partial Z} + V\frac{\partial C}{\partial R} = Fo\left(\varepsilon^2 \frac{\partial^2 C}{\partial Z^2} + \frac{1}{R}\frac{\partial C}{\partial R} + \frac{\partial^2 C}{\partial R^2}\right) - K(C - C_1);$$

$$\frac{dC_1}{dT} = \omega K(C - C_1) - K_1 C_0 C_1 + K_2 \theta^{-1}(1 - C_0);$$

$$\frac{dC_0}{dT} = -K_1 \theta C_0 C_1 + K_2(1 - C_0); \tag{13.2.8}$$

$$T = 0, \quad C_0 \equiv 1, \quad C_1 \equiv 0; \quad R = 0, \quad \frac{\partial C}{\partial R} \equiv 0; \quad R = 1, \quad \frac{\partial C}{\partial R} \equiv 0;$$

$$Z = 0, \quad C \equiv 1, \quad 1 \equiv U(R) - Pe^{-1}\left(\frac{\partial C}{\partial Z}\right)_{Z=0}.$$

$$A(T,Z)\frac{d\overline{C}}{dZ} + [B(T,Z)+G(T,Z)]\overline{C} = Pe^{-1}\frac{d^2\overline{C}}{dZ^2} - K(\overline{C}-\overline{C}_1);$$

$$\frac{d\overline{C}_1}{dT} = \omega K(\overline{C}-\overline{C}_1) - \Delta(T,Z)K_1\overline{C}_0\overline{C}_1 + K_2\theta^{-1}(1-\overline{C}_0);$$

$$\frac{d\overline{C}_0}{dT} = -\Delta(T,Z)K_1\theta\overline{C}_0\overline{C}_1 + K_2(1-\overline{C}_0);$$
(13.2.9)

$$T=0, \quad \overline{C}_0 \equiv 1, \quad \overline{C}_1 \equiv 0; \quad Z=0, \quad \overline{C} \equiv 1, \quad \left(\frac{\partial\overline{C}}{\partial Z}\right)_{Z=0} \equiv 0.$$

In (13.2.8), (13.2.9), T is a parameter in $C(T,R,Z)$, $\overline{C}(T,Z)$, while (R,Z) are parameters in $C_0(T,R,Z)$, $C_1(T,R,Z)$ and in (13.2.9) (Z) is parameter in $\overline{C}_0(T,Z)$, $\overline{C}_1(T,Z)$. In (13.2.8), (13.2.9), the parameters are:

$$Fo = \frac{D_{11}l}{u_1^0 r_0^2}, \quad Pe = \frac{u_1^0 l}{D_{11}}, \quad \theta = \frac{c_{11}^0}{c_{23}^0},$$

$$K = \frac{k_0 l}{u_1^0}, \quad K_1 = k_1 t^0 q, \quad K_2 = k_2 t^0, \quad \omega = \frac{u_1^0 t^0}{l}.$$
(13.2.10)

In (13.2.9) are used the functions:

$$\overline{C}(T,Z) = 2\int_0^1 RC(T,R,Z)dR, \quad \widetilde{C}(T,R,Z) = \frac{C(T,R,Z)}{\overline{C}(T,Z)},$$

$$\overline{C}_0(T,Z) = 2\int_0^1 RC_0(T,R,Z)dR, \quad \widetilde{C}_0(T,R,Z) = \frac{C_0(T,R,Z)}{\overline{C}_0(T,Z)},$$

$$\overline{C}_1(T,Z) = 2\int_0^1 RC_1(T,R,Z)dR, \quad \widetilde{C}_1(T,R,Z) = \frac{C_1(T,R,Z)}{\overline{C}_1(T,Z)},$$
(13.2.11)

$$A(T,Z) = 2\int_0^1 RU(R,Z)\widetilde{C}(T,R,Z)dR,$$

$$B(T,Z) = 2\int_0^1 RU(R,Z)\frac{\partial\widetilde{C}}{\partial Z}dR,$$

$$G(T,Z) = 2\int_0^1 RV(R,Z)\frac{\partial\widetilde{C}}{\partial R}dR, \quad \Delta(Z) = 2\int_0^1 R\widetilde{C}_0(Z)\widetilde{C}_1(Z)dR.$$

In industrial conditions $Fo \sim 10^{-6}$, $Pe^{-1} < 10^{-6}$ and the models (13.2.8), (13.2.9) have the forms:

$$U\frac{\partial C}{\partial Z} + V\frac{\partial C}{\partial R} = -K(C - C_1);$$

$$\frac{dC_1}{dT} = \omega K(C - C_1) - K_1 C_0 C_1 + K_2 \theta^{-1}(1 - C_0);$$

$$\frac{dC_0}{dT} = -K_1 \theta C_0 C_1 + K_2(1 - C_0);$$

$$T = 0, \quad C_0 \equiv 1, \quad C_1 \equiv 0; \quad R = 1, \quad C \equiv 0; \quad Z = 0, \quad C \equiv 1.$$

(13.2.12)

$$A(T,Z)\frac{d\overline{C}}{dZ} + [B(T,Z) + G(T,Z)]\overline{C} = -K(\overline{C} - \overline{C}_1);$$

$$\frac{d\overline{C}_1}{dT} = \omega K(\overline{C} - \overline{C}_1) - A(T,Z)K_1\overline{C}_0\overline{C}_1 + K_2\theta^{-1}(1 - \overline{C}_0);$$

$$\frac{d\overline{C}_0}{dT} = -A(T,Z)K_1\theta\overline{C}_0\overline{C}_1 + K_2(1 - \overline{C}_0);$$

$$T = 0, \quad \overline{C}_0 \equiv 1, \quad \overline{C}_1 \equiv 0; \quad Z = 0, \quad \overline{C} \equiv 1.$$

(13.2.13)

13.2.2 Axial and Radial Velocities

The theoretical analysis of the change in the radial non-uniformity of the axial velocity component (effect of the radial velocity component) in a column can be made by an appropriate hydrodynamic model, where the average velocity at the cross section of the column is a constant (inlet average axial velocity component), while the maximal velocity (and as a result the radial non-uniformity of the axial velocity component, too) decreases along the column height. In generalized variables (13.2.7), this model has the form:

$$U = (2 - 0.4Z) - 2(1 - 0.4Z)R^2, \quad V = 0.2(R - R^3),$$

$$\overline{U} = 2\int_0^1 RUdR = 1.$$

(13.2.14)

From Chap. 10 is seen that $V < 0.1$ and must be presented with the help of a small parameter $0.1 = \alpha \ll 1$, i.e.,

$$V = \alpha V_0, \quad V_0 = 0.2(R - R^3)\alpha^{-1}$$

(13.2.15)

As a result, the problem (10.2.9) has the form:

$$U\frac{\partial C}{\partial Z} + \alpha V_0 \frac{\partial C}{\partial R} = -K(C - C_1);$$
$$\frac{dC_1}{dT} = \omega K(C - C_1) - K_1 C_0 C_1 + K_2 \theta^{-1}(1 - C_0);$$
$$\frac{dC_0}{dT} = -K_1 \theta C_0 C_1 + K_2 (1 - C_0);$$
$$T = 0, \quad C_0 \equiv 1, \quad C_1 \equiv 0; \quad R = 1, \quad C \equiv 0; \quad Z = 0, \quad C \equiv 1.$$

(13.2.16)

In (13.2.16), $0.1 = \alpha \ll 1$ is a small parameter and (13.2.16) must be solved by the perturbation method (see Chap. 7), i.e., the concentrations must be presented as

$$C(R, Z) = C^0(R, Z) + \alpha C^1(R, Z),$$
$$C_0(R, Z) = C_0^0(R, Z) + \alpha C_0^1(R, Z),$$
$$C_1(R, Z) = C_1^0(R, Z) + \alpha C_1^1(R, Z).$$

(13.2.17)

The (13.2.17) in (13.2.16) leads to the next set of problems

$$U\frac{\partial C^0}{\partial Z} = -K(C^0 - C_1^0);$$
$$\frac{dC_1^0}{dT} = \omega K(C^0 - C_1^0) - K_1 C_0^0 C_1^0 + K_2 \theta^{-1}(1 - C_0^0);$$
$$\frac{dC_0^0}{dT} = -K_1 \theta C_0^0 C_1^0 + K_2(1 - C_0^0);$$
$$T = 0, \quad C_0^0 \equiv 1, \quad C_1^0 \equiv 0; \quad R = 1, \quad C^0 \equiv 0; \quad Z = 0, \quad C^0 \equiv 1.$$

(13.2.18)

$$U\frac{\partial C^1}{\partial Z} = -K(C^1 - C_1^1) - V_0\frac{\partial C^0}{\partial R};$$
$$\frac{dC_1^1}{dT} = \omega K(C^1 - C_1^1) - K_1 C_0^1 C_1^0 - K_1 C_0^0 C_1^1 + K_2 \theta^{-1}(1 - C_0^1);$$
$$\frac{dC_0^1}{dT} = -K_1 \theta C_0^1 C_1^0 - K_1 \theta C_0^0 C_1^1 + K_2(1 - C_0^1);$$
$$T = 0, \quad C_0^1 \equiv 0, \quad C_1^1 \equiv 0; \quad R = 1, \quad C^1 \equiv 0; \quad Z = 0, \quad C^1 \equiv 0.$$

(13.2.19)

The solution of (13.2.18), (13.2.19) is possible to be obtained, using cubic spline interpolations for $\frac{\partial C^0}{\partial R}$, C_0^0, C_1^0 and the algorithms in Chap. 3.

The solution of (13.2.18), (13.2.19), using (13.2.11), permits to obtain the average concentrations $\overline{C}(T_m, 1)$, $T_m = 0.1m$, $m = 1, 2, \ldots, 10$ ("theoretical" values) and functions $A(Z_n)$, $B(Z_n)$, $G(Z_n)$, $\Delta(Z_n)$, $Z_n = 0.1(n + 1)$, $n = 0, 1, \ldots, 9$ in (13.2.9), which are possible to be presented as appropriate approximations (see Chap. 10).

13.2.3 Chemical Adsorption Process

The use of the convection-type (3.2.25) and average-concentration (6.2.18) models in the cases of the chemical adsorption, and the models (13.2.12), (13.2.13), leads to:

$$U\frac{\partial C}{\partial Z} + \alpha V_0 \frac{\partial C}{\partial R} = -K(C - C_1);$$

$$\frac{dC_1}{dT} = \omega K(C - C_1) - K_3 C_0 C_1; \qquad \frac{dC_0}{dT} = -K_3 \theta C_0 C_1; \qquad (13.2.20)$$

$$T = 0, \quad C_0 \equiv 1, \quad C_1 \equiv 0; \quad R = 1, \quad \frac{\partial C}{\partial R} \equiv 0; \quad Z = 0, \quad C \equiv 1.$$

$$A(T,Z)\frac{d\overline{C}}{dZ} + [B(T,Z) + G(T,Z)]\overline{C} = -K(\overline{C} - \overline{C}_1);$$

$$\frac{d\overline{C}_1}{dT} = \omega K(\overline{C} - \overline{C}_1) - BK_3\overline{C}_0\overline{C}_1; \qquad \frac{d\overline{C}_0}{dT} = -BK_3\theta\overline{C}\,\overline{C}_0\overline{C}_1; \qquad (13.2.21)$$

$$T = 0, \quad \overline{C}_0 \equiv 1, \quad \overline{C}_1 \equiv 0; \quad Z = 0, \quad \overline{C} \equiv 1.$$

In (13.2.20), (13.2.21) $K_3 = kt^0$.

The presented numerical analysis of the industrial column for adsorption processes shows that average-concentration model, where the radial velocity component is not equal to zero, is possible to be used in the cases of an axial modification of the radial non-uniformity of the axial velocity component. The use of experimental data, for the average concentration at the column end and for different times, for a concrete adsorption process, permits to obtain the model parameters, related to the radial non-uniformity of the velocity. These parameter values permit to use the average-concentration model for modeling of different adsorption processes (different values of the model parameters, i.e., different values of the column height, average velocity, reagent diffusivity, physical and chemical adsorption rate constant, etc.).

References

1. Boyadjiev B, Boyadjiev C (2017) New models of industrial column chemical reactors. Bul Chem Commun 49(3):706–710
2. Boyadjiev B, Boyadjiev C (2017) New models of industrial column absorbers. 1. Co-current absorption processes. Bul Chem Commun 49(3):711–719
3. Boyadjiev B, Boyadjiev C (2017) New models of industrial column absorbers. 1. Counter-current absorption processes. Bul Chem Commun 49(3):720–728
4. Boyadjiev B, Boyadjiev C (2018) A new approach for modeling of industrial absorption columns. J Eng Thermophys 27(1)

5. Boyadjiev C, Doichinova M, Boyadjiev B, Popova-Krumova P (2016) Modeling of column apparatus processes. Springer, Berlin, Heidelberg
6. Boyadjiev C, Boyadjiev B, Popova-Krumova P, Doichinova M (2015) An innovative approach for adsorption column modeling. Chem Eng Technol 38(4):675–682
7. Boyadjiev B, Doichinova M, Boyadjiev C (2015) Computer modelling of column apparatuses. 2. Multistep modeling approach. J Eng Thermophys 24(4):362–370

Chapter 14
Industrial Column Catalytic Reactors

In Chaps. 10–13 were presented convection-type and average-concentration models of chemical [1], co-current absorption [2], countercurrent absorption [3], and non-stationary adsorption [4] processes in industrial column apparatuses, where the radial velocity component is not equal to zero in the cases of an axial modification of the axial velocity radial non-uniformity along the column height. This problem will be solved in the cases of the catalytic reactions in gas–solid systems (physical and chemical adsorption mechanisms) in the industrial column apparatuses [5].

14.1 Effect of the Axial Modification of the Radial Non-uniformity of the Axial Velocity Component

14.1.1 Physical Adsorption Mechanism

In Chaps. 3–6 are presented the convection–diffusion and average-concentration type models. In the case of physical adsorption mechanism, the model (3.3.5 and 3.3.6) of the catalytic process has the form:

$$U(R)\frac{\partial C_{11}}{\partial Z} = Fo_{11}\left(\varepsilon\frac{\partial^2 C_{11}}{\partial Z^2} + \frac{1}{R}\frac{\partial C_{11}}{\partial R} + \frac{\partial^2 C_{11}}{\partial R^2}\right) - K_{01}(C_{11} - C_{13});$$

$$U(R)\frac{\partial C_{21}}{\partial Z} = Fo_{21}\left(\varepsilon\frac{\partial^2 C_{21}}{\partial Z^2} + \frac{1}{R}\frac{\partial C_{21}}{\partial R} + \frac{\partial^2 C_{21}}{\partial R^2}\right) - K_{02}(C_{21} - C_{23});$$

$$R = 0, \quad \frac{\partial C_{i1}}{\partial R} \equiv 0; \quad R = 1, \quad \frac{\partial C_{i1}}{\partial R} \equiv 0;$$

$$Z = 0, \quad C_{i1} \equiv 1, \quad 1 \equiv U(R) - Pe_{i1}^{-1}\left(\frac{\partial C_{i1}}{\partial Z}\right)_{Z=0}; \quad i = 1, 2,$$

(14.1.1)

© Springer International Publishing AG, part of Springer Nature 2018
C. Boyadjiev et al., *Modeling of Column Apparatus Processes,*
Heat and Mass Transfer, https://doi.org/10.1007/978-3-319-89966-4_14

where the solution of (14.1.1) depends on the two functions:

$$C_{13} = \frac{C_{11} + K_1(1 - C_{33})}{1 + K_2 C_{33}}, \quad C_{23} = \frac{C_{21}}{1 + K_3(1 - C_{33})} \tag{14.1.2}$$

and C_{33} is the solution of the cubic equation:

$$
\begin{aligned}
&\omega_3 (C_{33})^3 + \omega_2 (C_{33})^2 + \omega_1 C_{33} + \omega_0 = 0, \\
&\omega_3 = K_3 (K_1 K_4 - K_2 K_5), \\
&\omega_2 = K_5 (K_2 + 2K_2 K_3 - K_3) - K_4 (K_1 + 2K_1 K_3 + K_3 C_{11}) + K_2 C_{21}, \\
&\omega_1 = K_4 (C_{11} + K_1)(1 + K_3) + K_5(1 + 2K_3 - K_2 - K_2 K_3) + (1 - K_2)C_{21}, \\
&\omega_0 = -C_{21} - K_3 K_5 - K_5.
\end{aligned}
\tag{14.1.3}
$$

where as a solution of (14.1.3) $0 \leq C_{33} \leq 1$ is used.

The average-concentration model (6.3.23), (6.3.24) in this case is:

$$
\begin{aligned}
&A_1 \frac{d\overline{C}_{11}}{dZ} + \frac{dA_1}{dZ} \overline{C}_{11} = Pe_1^{-1} \frac{d^2 \overline{C}_{11}}{dZ^2} - K_{01}(\overline{C}_{11} - \overline{C}_{13}); \\
&A_2 \frac{d\overline{C}_{21}}{dZ} + \frac{dA_2}{dZ} \overline{C}_{21} = Pe_2^{-1} \frac{d^2 \overline{C}_{21}}{dZ^2} - K_{02}(\overline{C}_{21} - \overline{C}_{23});
\end{aligned}
\tag{14.1.4}
$$

$$Z = 0, \quad \overline{C}_{11} = 1, \quad \left(\frac{d\overline{C}_{11}}{dZ}\right)_{Z=0} = 0, \quad \overline{C}_{21} = 1, \quad \left(\frac{d\overline{C}_{21}}{dZ}\right)_{Z=0} = 0,$$

where the solution depends on the two functions:

$$\overline{C}_{13} = \frac{\overline{C}_{11} + K_1(1 - \overline{C}_{33})}{1 + BK_2 \overline{C}_{33}}, \quad \overline{C}_{23} = \frac{\overline{C}_{21}}{1 + K_3(1 - G\overline{C}_{33})} \tag{14.1.5}$$

and $\overline{C}_{33}$ is the solution of the cubic equation:

$$
\begin{aligned}
&\bar{\omega}_3 (\overline{C}_{33})^3 + \bar{\omega}_2 (\overline{C}_{33})^2 + \bar{\omega}_1 \overline{C}_{33} + \bar{\omega}_0 = 0, \\
&\bar{\omega}_3 = BGK_3 (K_1 K_4 - K_2 K_5), \\
&\bar{\omega}_2 = K_5 (BK_2 + 2BK_2 K_3 - GK_3) - \\
&\quad - K_4 (BK_1 + BK_1 K_3 + BGK_1 K_3 + BGK_3 \overline{C}_{11}) + BGK_2 \overline{C}_{21}, \\
&\bar{\omega}_1 = BK_4 (\overline{C}_{11} + K_1)(1 + K_3) + \\
&\quad + K_5(1 + K_3 + GK_3 - BK_2 - BK_2 K_3) + (G - BK_2)\overline{C}_{21}, \\
&\bar{\omega}_0 = -\overline{C}_{21} - K_3 K_5 - K_5
\end{aligned}
\tag{14.1.6}
$$

As a solution of (14.1.6) $0 \leq \overline{C}_{33} \leq 1$ is used.

The parameters in (14.1.1)–(14.1.6) are:

$$K_{0i} = \frac{k_{0i}l}{u_1^0}, \quad Fo_{i1} = \frac{D_{i0}l}{u_1^0 r_0^2}, \quad Pe_{i1} = \frac{u_1^0 l}{D_{i0}}, \quad \varepsilon = \frac{r_0^2}{l^2} = Fo_{i1}^{-1} Pe_{i1}^{-1},$$

$$K_1 = \frac{k_2\, c_{33}^0}{k_{01}\, c_{11}^0}, \quad K_2 = \frac{b_0 k_1}{k_{01}}, \quad K_3 = \frac{k_{23} c_{33}^0}{k_{02}}, \quad K_4 = \frac{b_0 k_1}{k_{23} c_{21}^0}\frac{c_{11}^0}{c_{33}^0}, \tag{14.1.7}$$

$$K_5 = \frac{k_2}{k_{23} c_{21}^0}, \quad i = 1, 2.$$

The new functions in (14.1.4)–(14.1.6) have the forms:

$$A_i(Z) = \alpha_i(lZ) = \alpha_i(z) = 2\int_0^1 RU(R)\frac{C_{i1}(R,Z)}{\overline{C}_{i1}(Z)}dR, \quad i = 1, 2,$$

$$B(Z) = \beta(lZ) = \beta(z) = 2\int_0^1 R\frac{C_{13}(R,Z)}{\overline{C}_{13}(Z)}\frac{C_{33}(R,Z)}{\overline{C}_{33}(Z)}dR,$$

$$G(Z) = \gamma(lZ) = \gamma(z) = 2\int_0^1 R\frac{C_{23}(R,Z)}{\overline{C}_{23}(Z)}\frac{C_{33}(R,Z)}{\overline{C}_{33}(Z)}dR,$$

$$\overline{C}_{11}(Z) = 2\int_0^1 RC_{11}(R,Z)dR, \quad \overline{C}_{21}(Z) = 2\int_0^1 RC_{21}(R,Z)dR, \tag{14.1.8}$$

$$\overline{C}_{13}(Z) = 2\int_0^1 RC_{13}(R,Z)dR, \quad \overline{C}_{23}(Z) = 2\int_0^1 RC_{23}(R,Z)dR,$$

$$\overline{C}_{33}(Z) = 2\int_0^1 RC_{33}(R,Z)dR.$$

In the industrial conditions, it is possible to use the next approximations of the models (14.1.1)–(14.1.8):

$$0 = Fo_{i1} \le 10^{-2}, \quad 0 = Pe_{i1} \le 10^{-2}, \quad i = 1, 2 \tag{14.1.9}$$

and as a result, the models (14.1.1)–(14.1.6) have the convective forms:

$$U(R)\frac{dC_{11}}{dZ} = -K_{01}(C_{11} - C_{13}); \quad U(R)\frac{dC_{21}}{dZ} = -K_{02}(C_{21} - C_{23}); \tag{14.1.10}$$

$$Z = 0, \quad C_{i0} \equiv 1; \quad i = 1, 2.$$

$$A_1 \frac{d\overline{C}_{11}}{dZ} + \frac{dA_1}{dZ} \overline{C}_{11} = -K_{01} \left(\overline{C}_{11} - \overline{C}_{13} \right);$$

$$A_2 \frac{d\overline{C}_{21}}{dZ} + \frac{dA_2}{dZ} \overline{C}_{21} = -K_{02} \left(\overline{C}_{21} - \overline{C}_{23} \right); \qquad (14.1.11)$$

$$Z = 0, \quad \overline{C}_{11} \equiv 1, \quad \overline{C}_{21} \equiv 1.$$

In (14.1.1)–(14.1.6) the functions $C_{13}, C_{23}, C_{33}, \overline{C}_{13}, \overline{C}_{23}, \overline{C}_{33}$ are the solutions of (14.1.2), (14.1.3), (14.1.5), (14.1.6), where $B(Z) \equiv 1$, $G(Z) \equiv 1$, $0 \leq C_{33} \leq 1$ and $0 \leq \overline{C}_{33} \leq 1$.

14.1.2 Modeling of the Industrial Column Catalytic Reactors

The effect of the axial modification of the radial non-uniformity of the axial velocity component will be analyzed, using the hydrodynamic model (13.1.5), where the average velocity at the cross section of the column is a constant, while the maximal velocity (and as a result, the radial non-uniformity of the axial velocity component too) decreases along the column height. As a result, the convection-type models (14.1.10), (14.1.2), (14.1.3) have the forms:

$$U_n(R, Z_n) \frac{dC_{11n}}{dZ_n} = -K_{01}(C_{11n} - C_{13n});$$

$$U_n(R, Z_n) \frac{dC_{21n}}{dZ_n} = -K_{02}(C_{21n} - C_{23n}); \qquad (14.1.12)$$

$$0 \leq R \leq 1, \quad 0.1n \leq Z_n \leq 0.1(n+1), \quad n = 0, 1, \ldots, 9,$$

where R is a parameter. The solution of (14.1.12) depends on the two functions:

$$C_{13n} = \frac{C_{11n} + K_1(1 - C_{33n})}{1 + K_2 C_{33n}}, \quad C_{23n} = \frac{C_{21n}}{1 + K_3(1 - C_{33n})}, \quad n = 0, 1, \ldots, 9,$$

$$\qquad (14.1.13)$$

where C_{33n} is the solution of the cubic equation:

$$\omega_{3n}(C_{33n})^3 + \omega_{2n}(C_{33n})^2 + \omega_{1n}C_{33n} + \omega_{0n} = 0,$$

$$\omega_{3n} = K_3(K_2 K_5 - K_1 K_4),$$

$$\omega_{2n} = K_5(K_3 - K_2 - 2K_2 K_3) + K_4(K_1 + 2K_1 K_3 + K_3 C_{11n}) - K_2 C_{21n},$$

$$\omega_{1n} = K_5(K_2 - 1 - 2K_3 + K_2 K_3) - K_4(K_1 + C_{11n})(1 + K_3) + (K_2 - 1)C_{21n},$$

$$\omega_{0n} = C_{21n} + K_3 K_5 + K_5$$

$$\qquad (14.1.14)$$

and as a solution of (14.1.14) $0 \leq C_{33n} \leq 1$ must be used.

The boundary (input) conditions of (4.1.12), (4.1.14) are:

$$Z_0 = 0, \quad C_{110}(R, Z_0) \equiv 1, \quad C_{210}(R, Z_0) \equiv 1,$$
$$C_{130}(R, Z_0) \equiv 0, \quad C_{230}(R, Z_0) \equiv 0, \quad C_{330}(R, Z_0) \equiv 1$$

(14.1.15)

and

$$Z_n = 0.1n, \quad C_{11n}(R, Z_n) = C_{11(n-1)}(R, Z_n),$$
$$C_{21n}(R, Z_n) = C_{21(n-1)}(R, Z_n), \quad C_{13n}(R, Z_n) = C_{13(n-1)}(R, Z_n),$$
$$C_{23n}(R, Z_n) = C_{23(n-1)}(R, Z_n), \quad C_{33n}(R, Z_n) = C_{33(n-1)}(R, Z_n),$$
$$n = 1, \ldots, 9.$$

(14.1.16)

14.1.3 Model Equations Solution

A solution of the problem (14.1.12)–(14.1.16) will be obtained for the case:

$$K_{0i} = 1, \quad i = 1, 2, \quad K_1 = 2.5, \quad K_2 = 1, \quad K_3 = 1, \quad K_4 = 0.5, \quad K_5 = 1$$

(14.1.17)

as the next matrix forms:

$$C_{11n(\zeta)}(R, Z_n) = \left\| C_{11n(\rho\zeta)} \right\|, \quad C_{21n(\zeta)}(R, Z_n) = \left\| C_{21n(\rho\zeta)} \right\|,$$
$$C_{13n(\zeta)}(R, Z_n) = \left\| C_{13n(\rho\zeta)} \right\|, \quad C_{23n(\zeta)}(R, Z_n) = \left\| C_{23n(\rho\zeta)} \right\|,$$
$$C_{33n(\zeta)}(R, Z_n) = \left\| C_{33n(\rho\zeta)} \right\|; \quad R = \frac{\rho - 1}{\rho^0 - 1}, \quad \rho = 1, 2, \ldots, \rho^0, \quad \rho^0 = 101;$$
$$Z_n = 0.1n + 0.01\zeta, \quad n = 0, 1, \ldots, 9, \quad \zeta = 0, 1, \ldots, 10.$$

(14.1.18)

A multistep approach will be used, where $n = 0, 1, \ldots, 9$, $\zeta = 0, 1, \ldots, 10$ are the step numbers.

As a first step, $(n = 0)$ will be used:

$$C_{110(\zeta)}(R, Z_0) = \left\| C_{110(\rho\zeta)} \right\|, \quad C_{210(\zeta)}(R, Z_0) = \left\| C_{210(\rho\zeta)} \right\|$$
$$C_{130(\zeta)}(R, Z_0) = \left\| C_{130(\rho\zeta)} \right\| \equiv 0, \quad C_{230(\zeta)}(R, Z_0) = \left\| C_{230(\rho\zeta)} \right\| \equiv 0, \quad (14.1.19)$$
$$C_{330(\zeta)}(R, Z_0) = \left\| C_{330(\rho\zeta)} \right\| \equiv 1, \quad 0 \leq Z_0 \leq 0.1, \quad \zeta = 1, 2, \ldots, 10.$$

The calculation procedure starts with $\zeta = 0$, i.e.,

$$C_{110(0)}(R, Z_0) = \left\|C_{110(\rho 0)}\right\|, \quad C_{210(0)}(R, Z_0) = \left\|C_{210(\rho 0)}\right\|$$
$$C_{130(0)}(R, Z_0) = \left\|C_{130(\rho 0)}\right\| \equiv 0, \quad C_{230(0)}(R, Z_0) = \left\|C_{230(\rho 0)}\right\| \equiv 0, \quad (14.1.20)$$
$$C_{330(0)}(R, Z_0) = \left\|C_{330(\rho 0)}\right\| \equiv 1, \quad 0 \le Z_0 \le 0.1,$$

where $C_{110(0)}(R, Z_0)$ and $C_{210(0)}(R, Z_0)$ are solution of the problem:

$$U_0(R, Z_0)\frac{dC_{110(0)}}{dZ_0} = -K_{01}C_{110(0)};$$

$$U_0(R, Z_0)\frac{dC_{210(0)}}{dZ_0} = -K_{02}C_{210(0)}; \quad 0 \le Z_0 \le 0.1; \quad (14.1.21)$$

$$Z_0 = 0, \quad C_{110(0)}(R, Z_0) \equiv 1, \quad C_{210(0)}(R, Z_0) \equiv 1.$$

From the solution of (14.1.21), it is possible to be obtained:

$$Z_0 = 0.01\zeta, \quad \zeta = 1, \quad C_{110(0)}(R, 0.01) = \widehat{C}_{110(1)}(R),$$
$$C_{210(0)}(R, 0.01) = \widehat{C}_{210(1)}(R). \quad (14.1.22)$$

The next steps ($\zeta = 1, 2, \ldots, 10$) are the consistent calculations of the elements of the matrix forms:

$$C_{330(\zeta)}(R, Z_0) = \left\|C_{330(\rho\zeta)}\right\|, \quad C_{130(\zeta)}(R, Z_0) = \left\|C_{130(\rho\zeta)}\right\|,$$
$$C_{230(\zeta)}(R, Z_0) = \left\|C_{230(\rho\zeta)}\right\|, \quad C_{110(\zeta)}(R, Z_0) = \left\|C_{110(\rho\zeta)}\right\|, \quad (14.1.23)$$
$$C_{210(\zeta)}(R, Z_0) = \left\|C_{210(\rho\zeta)}\right\|, \quad \zeta = 1, 2, \ldots, 10,$$

after consistently solving the problems (14.1.14), (14.1.13), (14.1.12) for $n = 0$:

$$\omega_{30(\zeta)}\left(C_{330(\zeta)}\right)^3 + \omega_{20(\zeta)}\left(C_{330(\zeta)}\right)^2 + \omega_{10(\zeta)}\widehat{C}_{330(\zeta)} + \omega_{00(\zeta)} = 0,$$
$$\omega_{30(\zeta)} = K_3(K_2K_5 - K_1K_4),$$
$$\omega_{20(\zeta)} = K_5(K_3 - K_2 - 2K_2K_3) + K_4\left(K_1 + 2K_1K_3 + K_3\widehat{C}_{110(\zeta)}\right)$$
$$\quad - K_2\widehat{C}_{210(\zeta)}, \quad (14.1.24)$$
$$\omega_{10(\zeta)} = K_5(K_2 - 1 - 2K_3 + K_2K_3) - K_4\left(K_1 + \widehat{C}_{110(\zeta)}\right)(1 + K_3)$$
$$\quad + (K_2 - 1)\widehat{C}_{210(\zeta)},$$
$$\omega_{00(\zeta)} = \widehat{C}_{210(\zeta)} + K_3K_5 + K_5.$$

$$C_{130(\zeta)} = \frac{\widehat{C}_{110(\zeta)} + K_1\left(1 - C_{330(\zeta)}\right)}{1 + K_2C_{330(\zeta)}}, \quad C_{230(\zeta)} = \frac{\widehat{C}_{210(\zeta)}}{1 + K_3\left(1 - C_{330(\zeta)}\right)}. \quad (14.1.25)$$

$$U_0(R, Z_0)\frac{dC_{110(\zeta)}}{dZ_0} = -K_{01}\left(C_{110(\zeta)} - C_{130(\zeta)}\right);$$

$$U_0(R, Z_0)\frac{dC_{210(\zeta)}}{dZ_0} = -K_{02}\left(C_{210(\zeta)} - C_{230(\zeta)}\right); \qquad (14.1.26)$$

$$0.01\zeta \le Z_0 \le 0.1, \quad \zeta = 1, 2, \ldots, 10.$$

For the solution of (14.1.24)–(14.1.25) have to be used

$$\widehat{C}_{110(\zeta)}(R) = C_{110(\zeta-1)}(R, 0.01\zeta), \quad \widehat{C}_{210(\zeta)}(R) = C_{210(\zeta-1)}(R, 0.01\zeta),$$

$$\zeta = 1, 2, \ldots, 10. \qquad (14.1.27)$$

The procedures (14.1.24)–(14.1.27) for $\zeta = 1, 2, \ldots, 10$ are repeated for $n = 1, 2, \ldots, 9$. The zero steps $\zeta = 0$ are:

$$\begin{aligned}
C_{11n(0)}(R, Z_n) &= \left\|C_{11n(\rho 0)}\right\| = \left\|C_{11(n-1)(\rho 10)}\right\|, \\
C_{21n(0)}(R, Z_n) &= \left\|C_{21n(\rho 0)}\right\| = \left\|C_{21(n-1)(\rho 10)}\right\|, \\
C_{13n(0)}(R, Z_n) &= \left\|C_{13n(\rho 0)}\right\| = \left\|C_{13(n-1)(\rho 10)}\right\|, \\
C_{23n(0)}(R, Z_n) &= \left\|C_{23n(\rho 0)}\right\| = \left\|C_{23(n-1)(\rho 10)}\right\|, \\
C_{33n(0)}(R, Z_n) &= \left\|C_{33n(\rho 0)}\right\| = \left\|C_{33(n-1)(\rho 10)}\right\|, \\
Z_n &= 0.1n, \quad n = 1, 2, \ldots, 9.
\end{aligned} \qquad (14.1.28)$$

The calculation procedures (14.1.24)–(14.1.28) for $\zeta = 1, 2, \ldots, 10$ and different $n = 1, 2, \ldots, 9$ permit to be obtained five matrixes as solution of the model Eqs. (14.1.12)–(14.1.16):

$$C_{11}(R, Z) = \sum_{n=0}^{9}\left\|C_{11n(\rho\zeta)}\right\|, \quad C_{21}(R, Z) = \sum_{n=0}^{9}\left\|C_{21n(\rho\zeta)}\right\|,$$

$$C_{13}(R, Z) = \sum_{n=0}^{9}\left\|C_{13n(\rho\zeta)}\right\|, \quad C_{23}(R, Z) = \sum_{n=0}^{9}\left\|C_{23n(\rho\zeta)}\right\|,$$

$$C_{33}(R, Z) = \sum_{n=0}^{9}\left\|C_{33n(\rho\zeta)}\right\|; \qquad (14.1.29)$$

$$R = \frac{\rho - 1}{\rho^0 - 1}, \quad \rho = 1, 2, \ldots, \rho^0, \quad \rho^0 = 101;$$

$$Z = Z_n = 0.1n + 0.01\zeta, \quad n = 0, 1, \ldots, 9, \quad \zeta = 1, 2, \ldots, 10.$$

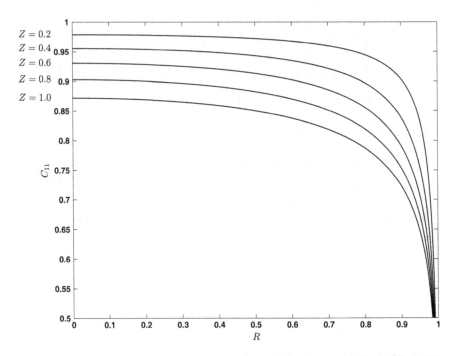

Fig. 14.1 Radial distribution of the concentration $C_{11}(R, Z)$, $Z = 0.2(1)$, $0.4(2)$, $0.6(3)$, $0.8(4)$, $1.0(5)$

The solutions (14.1.29) of the model Eqs. (14.1.12)–(14.1.16) permit to be presented in Figs. 14.1 and 14.2 the concentration distributions of AC for different Z in the case (14.1.17).

The obtained concentrations as solutions of (14.1.12)–(14.1.16) in the case (14.1.17) permit to be obtained average concentrations $\overline{C}_{11}(Z)$, $\overline{C}_{21}(Z)$, $\overline{C}_{13}(Z)$, $\overline{C}_{23}(Z)$, $\overline{C}_{33}(Z)$ and functions $A_i(Z)$, $i = 1, 2$, $B(Z)$, $G(Z)$ in (14.1.8). The obtained results show that $B(Z) \equiv 1$, $G(Z) \equiv 1$, practically. The functions $\overline{C}_{11}(Z)$, $\overline{C}_{21}(Z)$, $A_i(Z)$, $i = 1, 2$ are presented as "theoretical values" for different $Z_n = 0.1(n + 1)$, $n = 0, 1, \ldots, 9$ (points) in Figs. 14.3 and 14.4.

From Fig. 14.4, it is seen that the functions $A_i(Z)$, $i = 1, 2$, are possible to be presented as quadratic approximations (the lines in Fig. 14.4):

$$A_i(Z) = a_{0i} + a_{1i}Z + a_{2i}Z^2, \quad i = 1, 2 \tag{14.1.30}$$

and the approximations ("theoretical") parameter values are presented in Table 14.1, where $K_{0i} = 1$, $i = 1, 2$ are "theoretical" values too.

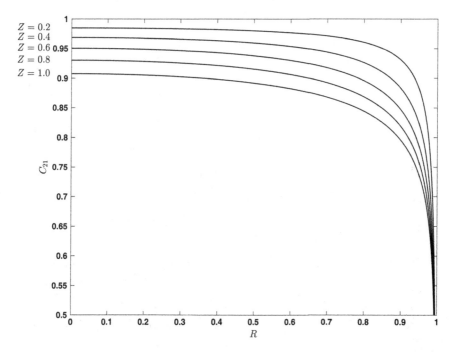

Fig. 14.2 Radial distribution of the concentration $C_{21}(R, Z)$, $Z = 0.2(1)$, $0.4(2)$, $0.6(3)$, $0.8(4)$, $1.0(5)$

From (14.1.11), (14.1.30), it is possible to be obtained the final form of the average-concentration model:

$$\left(a_{01} + a_{11}Z + a_{21}Z^2\right)\frac{d\overline{C}_{11}}{dZ} + (a_{11} + 2a_{21}Z)\overline{C}_{11} = -K_{01}\left(\overline{C}_{11} - \overline{C}_{13}\right);$$

$$\left(a_{02} + a_{12}Z + a_{22}Z^2\right)\frac{d\overline{C}_{21}}{dZ} + (a_{12} + 2a_{22}Z)\overline{C}_{21} = -K_{02}\left(\overline{C}_{21} - \overline{C}_{23}\right); \qquad (14.1.31)$$

$$Z = 0, \quad \overline{C}_{11} \equiv 1, \quad \overline{C}_{21} \equiv 1,$$

where the functions $\overline{C}_{13}$, $\overline{C}_{23}$ are obtained in (14.1.5), (14.1.6), where $B(Z) \equiv 1$, $G(Z) \equiv 1$, and the model parameters a_{0i}, a_{1i}, a_{2i}, K_{0i}, $i = 1, 2$ must be obtained, using experimental data.

The solution of (14.1.5), (14.1.6), (14.1.31) must be obtained as five vector forms:

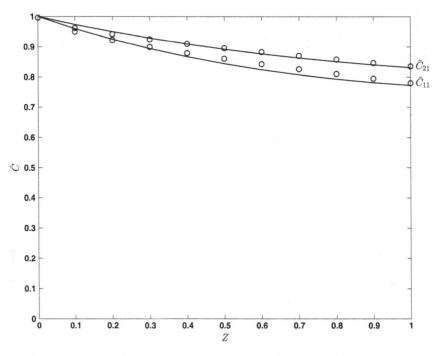

Fig. 14.3 "Theoretical values" of the average concentration $\overline{C}_{11}(Z)$, $\overline{C}_{21}(Z)$ for different $Z = 0.1(n+1)$, $n = 0, 1, \ldots, 9$ (points); $\overline{C}_{11}(Z)$, $\overline{C}_{21}(Z)$ as a solution of (14.1.31) (lines), using the "experimental" parameter values in Table 14.1

$$\overline{C}_{11(\varsigma)}(Z) = \left|\overline{C}_{11(\varsigma)}\right|, \quad \overline{C}_{21(\varsigma)}(Z) = \left|\overline{C}_{21(\varsigma)}\right|, \quad \overline{C}_{13(\varsigma)}(Z) = \left|\overline{C}_{13(\varsigma)}\right|,$$

$$\overline{C}_{23(\varsigma)}(Z) = \left|\overline{C}_{23(\varsigma)}\right|, \quad \overline{C}_{33(\varsigma)}(Z) = \left|\overline{C}_{33(\varsigma)}\right|, \tag{14.1.32}$$

$$Z = \frac{\varsigma - 1}{\varsigma^0 - 1}, \quad \varsigma = 1, 2, \ldots, \varsigma^0, \quad \varsigma^0 = 101.$$

A multistep approach will be used, where $\varsigma = 1, 2, \ldots, \varsigma^0$ are the step numbers. As a first step, $(\varsigma = 1)$ will be used:

$$\overline{C}_{11(1)}(Z) = \left|\overline{C}_{11(1)}\right|, \quad \overline{C}_{21(1)}(Z) = \left|\overline{C}_{21(1)}\right|,$$

$$\overline{C}_{13(1)}(Z) = \left|\overline{C}_{13(1)}\right| \equiv 0, \quad \overline{C}_{23(1)}(Z) = \left|\overline{C}_{23(1)}\right| \equiv 0, \tag{14.1.33}$$

$$\overline{C}_{33(1)}(Z) = \left|\overline{C}_{33(1)}\right| \equiv 1, \quad 0 \leq Z \leq 1.$$

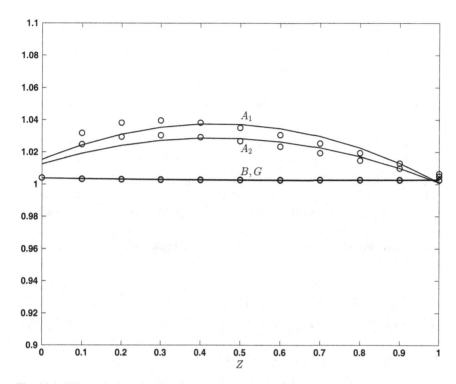

Fig. 14.4 "Theoretical values" of the functions $A_i(Z)$, $i = 1, 2$, for different $Z_n = 0.1(n + 1)$, $n = 0, 1, \ldots, 9$ (points); quadratic approximations (14.1.30) (lines), using the "theoretical" parameter values in Table 14.1

Table 14.1 Model parameter values a_{0i}, a_{1i}, a_{2i}, K_{0i}, $i = 1, 2$, b_0, b_1, b_2, g_0, g_1, g_2 (physical adsorption mechanism)	"Theoretical" values	"Experimental" values
	$a_{01} = 1.0153$	$a_{01} = 0.8026$
	$a_{11} = 0.1013$	$a_{11} = 0.1041$
	$a_{21} = -0.1154$	$a_{21} = -0.1199$
	$a_{02} = 1.0125$	$a_{02} = 0.9239$
	$a_{12} = 0.0747$	$a_{12} = 0.0833$
	$a_{22} = -0.0862$	$a_{22} = -0.0716$
	$K_{01} = 1$	$K_{01} = 1.1429$
	$K_{02} = 1$	$K_{02} = 1.1636$
	$b_0 = 1.0038$	
	$b_1 = -0.0039$	
	$b_2 = 0.0027$	
	$g_0 = 1.0038$	
	$g_1 = -0.0031$	
	$g_2 = 0.0022$	

where $\overline{C}_{11(1)}(Z)$, $\overline{C}_{21(1)}(Z)$ are solution of the problem:

$$
\left(a_{01} + a_{11}Z + a_{21}Z^2\right)\frac{d\overline{C}_{11(1)}}{dZ} + \left(a_{11} + 2a_{21}Z\right)\overline{C}_{11(1)} = -K_{01}\overline{C}_{11(1)};
$$

$$
\left(a_{02} + a_{12}Z + a_{22}Z^2\right)\frac{d\overline{C}_{21(1)}}{dZ} + \left(a_{12} + 2a_{22}Z\right)\overline{C}_{21(1)} = -K_{02}\overline{C}_{21(1)};\ 0 \leq Z \leq 1;
$$

$$
Z = 0, \quad \overline{C}_{11(1)} \equiv 1, \quad \overline{C}_{21(1)} \equiv 1.
$$

$$(14.1.34)$$

The solution of (14.1.34) permits to be obtained:

$$
Z = 0.01, \quad \overline{C}_{11(1)}(0.01) = \widehat{C}_{11(2)}, \quad \overline{C}_{21(1)}(0.01) = \widehat{C}_{21(2)}. \tag{14.1.35}
$$

The next steps $(\varsigma = 2, 3, \ldots, \varsigma^0)$ are the consistent calculations of the elements of the vector forms:

$$
\overline{C}_{33(\varsigma)}(Z) = \left|\overline{C}_{33(\varsigma)}\right|, \quad \overline{C}_{13(\varsigma)}(Z) = \left|\overline{C}_{13(\varsigma)}\right|, \quad \overline{C}_{23(\varsigma)}(Z) = \left|\overline{C}_{23(\varsigma)}\right|,
$$

$$
\overline{C}_{11(\varsigma)}(Z) = \left|\overline{C}_{11(\varsigma)}\right|, \quad \overline{C}_{21(\varsigma)}(Z) = \left|\overline{C}_{21(\varsigma)}\right|, \tag{14.1.36}
$$

$$
Z = \frac{\varsigma - 1}{\varsigma^0 - 1}, \quad \varsigma = 1, 2, \ldots, \varsigma^0, \quad \varsigma^0 = 101.
$$

after consistently solving the problems (14.1.5), (14.1.6), (14.1.31):

$$
\bar{\omega}_3\left(\overline{C}_{33(\varsigma)}\right)^3 + \bar{\omega}_2\left(\overline{C}_{33(\varsigma)}\right)^2 + \bar{\omega}_1\overline{C}_{33(\varsigma)} + \bar{\omega}_0 = 0,
$$

$$
\bar{\omega}_3 = K_3(K_2K_5 - K_1K_4),
$$

$$
\bar{\omega}_2 = K_5(-K_2 - K_2K_3 - K_2K_3 + K_3)
$$

$$
\quad\quad + K_4\left(K_1 + K_1K_3 + K_1K_3 + K_3\overline{C}_{11(\varsigma)}\right) + K_2\overline{C}_{21}(\varsigma), \tag{14.1.37}
$$

$$
\bar{\omega}_1 = K_5(K_2 - 1 - K_3 - K_3 + K_2K_3)
$$

$$
\quad\quad - K_4\left(K_1 + \overline{C}_{11}\right)(1 + K_3) - \overline{C}_{21(\varsigma)} + K_2\overline{C}_{21(\varsigma)},
$$

$$
\bar{\omega}_0 = \overline{C}_{21(\varsigma)} + K_3K_5 + K_5
$$

$$
\overline{C}_{13(\zeta)} = \frac{\overline{C}_{11(\varsigma)} + K_1\left(1 - \overline{C}_{33(\varsigma)}\right)}{1 + K_2\overline{C}_{33(\varsigma)}}, \quad \overline{C}_{23(\varsigma)} = \frac{\overline{C}_{21(\varsigma)}}{1 + K_3\left(1 - \overline{C}_{33(\varsigma)}\right)}, \tag{14.1.38}
$$

$$\left(a_{01} + a_{11}Z + a_{21}Z^2\right)\frac{d\overline{C}_{11(\varsigma)}}{dZ} + (a_{11} + 2a_{21}Z)\overline{C}_{11(\varsigma)} = -K_{01}\left(\overline{C}_{11(\varsigma)} - \overline{C}_{13(\varsigma)}\right);$$

$$\left(a_{02} + a_{12}Z + a_{22}Z^2\right)\frac{d\overline{C}_{21(\varsigma)}}{dZ} + (a_{12} + 2a_{22}Z)\overline{C}_{21(\varsigma)} = -K_{02}\left(\overline{C}_{21(\varsigma)} - \overline{C}_{23(\varsigma)}\right);$$

$$Z = 0.01(\varsigma - 1), \quad \overline{C}_{11(\varsigma)} = \hat{C}_{11(\varsigma)}, \quad \overline{C}_{21(\varsigma)} = \hat{C}_{21(\varsigma)}.$$

$$(14.1.39)$$

The calculation procedures (14.1.33)–(14.1.39) for $\varsigma = 1, 2, \ldots, \varsigma_0$ permit to be obtained five vectors as solution of the model Eqs. (14.1.5), (14.1.6), (14.1.31):

$$C_{11}(Z) = \sum_{\varsigma=1}^{\varsigma_0}\left|C_{11(\varsigma)}\right|, \quad C_{21}(Z) = \sum_{\varsigma=1}^{\varsigma_0}\left|C_{21(\varsigma)}\right|, \quad C_{13}(Z) = \sum_{\varsigma=1}^{\varsigma_0}\left|C_{13(\varsigma)}\right|,$$

$$C_{23}(Z) = \sum_{\varsigma=1}^{\varsigma_0}\left|C_{23(\varsigma)}\right|, \quad C_{33}(Z) = \sum_{\varsigma=1}^{\varsigma_0}\left|C_{33(\varsigma)}\right|;$$

$$Z = 0.01\varsigma, \quad \varsigma = 1, 2, \ldots, \varsigma_0.$$

$$(14.1.40)$$

14.1.4 Parameter Identification

The obtained "theoretical" values of the average concentrations $\overline{C}_{11}(Z)$, $\overline{C}_{21}(Z)$ for different $Z_n = 0.1(n+1)$, $n = 0, 1, \ldots, 9$ (points in Fig. 14.3) permit to be obtained the artificial experimental data at the column end $Z = 1$:

$$\overline{C}_{11\,\exp}^m(1) = (0.95 + 0.1S_m)\overline{C}_{11}(1), \quad \overline{C}_{21\,\exp}^m(1) = (0.95 + 0.1S_m)\overline{C}_{21}(1),$$

$$m = 1, \ldots 10,$$

$$(14.1.41)$$

where $0 \leq S_m \leq 1$, $m = 1, \ldots, 10$ are obtained by a generator of random numbers. The artificial experimental data (14.1.41) are used for the parameters $(a_{0i}, a_{1i}, a_{2i}, K_{0i}, i = 1, 2)$ identification in the average-concentration model (14.1.5), (14.1.6), (14.1.31) by the minimization of the least-squares function:

$$Q(a_{0i}, a_{1i}, a_{2i}, K_{0i}) = \sum_{m=1}^{10}\left[\overline{C}_{11}(1, a_{0i}, a_{1i}, a_{1i}, K_{0i}) - \overline{C}_{11\,\exp}^m(1)\right]^2$$

$$+ \sum_{m=1}^{10}\left[\overline{C}_{21}(1, a_{0i}, a_{1i}, a_{2i}, K_{0i}) - \overline{C}_{21\,\exp}^m(1)\right]^2, \quad i = 1, 2.$$

$$(14.1.42)$$

The obtained ("experimental") values of a_{0i}, a_{1i}, a_{2i}, K_{0i}, $i = 1, 2$ are presented in Table 14.1. They are used for the calculation of the functions $\overline{C}_{11}(Z)$, $\overline{C}_{21}(Z)$ in (14.1.31), and the results are presented in Fig. 14.3 (lines).

The results in Fig. 14.3 show that the experimental data obtained from the column end are useful for the parameter identifications.

14.1.5 Chemical Adsorption Mechanism

The difference between the physical and chemical adsorption mechanisms (in the stationary case) is that the gas–solid interphase mass transfer rate of the first reagent is equal to the chemical reaction between this reagent and AS in the solid phase (catalyst capillaries).

14.1.6 Convection-Type and Average-Concentration Models

The use of model Eqs. (3.3.9), (3.3.19), (3.3.21), (6.3.23), (6.3.25), (6.3.26) leads to the convection-type and average-concentration models in the case of chemical adsorption mechanism:

$$U(R)\frac{dC_{11}}{dZ} = -K_{01}(C_{11} - C_{13}); \quad U(R)\frac{dC_{21}}{dZ} = -K_{02}(C_{21} - C_{23});$$
$$Z = 0, \quad C_{i0} \equiv 1; \quad i = 1, 2. \tag{14.1.43}$$

$$C_{13} = \frac{C_{11}}{1 + K_1 C_{33}}, \quad C_{23} = \frac{C_{21}}{1 + K_2(1 - C_{33})},$$
$$(C_{21}K_1 - C_{11}K_2K_3)(C_{33})^2$$
$$+ (C_{21} + C_{11}K_3 + C_{11}K_2K_3 - C_{21}K_1)C_{33} - C_{21} = 0, \tag{14.1.44}$$

$$A_1\frac{d\overline{C}_{11}}{dZ} + \frac{dA_1}{dZ}\overline{C}_{11} = -K_{01}(\overline{C}_{11} - \overline{C}_{13});$$
$$A_2\frac{d\overline{C}_{21}}{dZ} + \frac{dA_2}{dZ}\overline{C}_{21} = -K_{02}(\overline{C}_{21} - \overline{C}_{23}); \tag{14.1.45}$$
$$Z = 0, \quad \overline{C}_{11} \equiv 1, \quad \overline{C}_{21} \equiv 1.$$

$$\overline{C}_{13} = \frac{\overline{C}_{11}}{1 + BK_1\overline{C}_{33}}, \quad \overline{C}_{23} = \frac{\overline{C}_{21}}{1 + K_2(1 - G\overline{C}_{33})},$$
$$BG(\overline{C}_{21}K_1 - \overline{C}_{11}K_2K_3)(\overline{C}_{33})^2$$
$$+ (G\overline{C}_{21} + B\overline{C}_{11}K_3 + B\overline{C}_{11}K_2K_3 - B\overline{C}_{21}K_1)\overline{C}_{33} - \overline{C}_{21} = 0. \tag{14.1.46}$$

In (14.1.43)–(14.1.46) are used the parameters:

$$K_1 = \frac{k_{13}c_{33}^0}{k_{01}}, \quad K_2 = \frac{k_{23}c_{33}^0}{k_{02}}, \quad K_3 = \frac{k_{13}c_{11}^0}{k_{23}c_{21}^0}. \tag{14.1.47}$$

As solutions of the equations in (14.1.44), (14.1.46) must be used $0 \leq C_{33} \leq 1$ and $0 \leq \overline{C}_{33} \leq 1$.

14.1.7   Industrial Column Modeling

If put (13.1.5) in (14.1.43), the convective model has the form:

$$U_n(R, Z_n)\frac{dC_{11n}}{dZ_n} = -K_{01}(C_{11n} - C_{13n});$$

$$U_n(R, Z_n)\frac{dC_{21n}}{dZ_n} = -K_{02}(C_{21n} - C_{23n}); \tag{14.1.48}$$

$$0 \leq R \leq 1, 0.1n \leq Z_n \leq 0.1(n+1), \quad n = 0, 1, \dots, 9,$$

where R is a parameter. The solution of (14.1.48) depends on the two functions:

$$C_{13n} = \frac{C_{11n}}{1 + K_1 C_{33n}}, \quad C_{23n} = \frac{C_{21n}}{1 + K_2(1 - C_{33n})}, \tag{14.1.49}$$

where C_{33n} is the solution of the equation:

$$\begin{aligned}(C_{21n}K_1 - C_{11n}K_2K_3)(C_{33n})^2 \\ + (C_{21n} + C_{11n}K_3 + C_{11n}K_2K_3 - C_{21n}K_1)C_{33n} - C_{21n} = 0,\end{aligned} \tag{14.1.50}$$

and for solving of (14.1.45) $0 \leq C_{33n} \leq 1$ has to be used.

The boundary (input) conditions of (14.1.48)–(14.1.50) are:

$$\begin{aligned}Z_0 = 0, \quad C_{110}(R, Z_0) &\equiv 1, \quad C_{210}(R, Z_0) \equiv 1, \\ C_{130}(R, Z_0) &\equiv 0, \quad C_{230}(R, Z_0) \equiv 0, \quad C_{330}(R, Z_0) \equiv 1\end{aligned} \tag{14.1.51}$$

and

$$Z_n = 0.1n, \quad C_{11n}(R, Z_n) = C_{11(n-1)}(R, Z_n),$$
$$C_{21n}(R, Z_n) = C_{21(n-1)}(R, Z_n),$$
$$C_{13n}(R, Z_n) = C_{13(n-1)}(R, Z_n),$$
$$C_{23n}(R, Z_n) = C_{23(n-1)}(R, Z_n),$$
$$C_{33n}(R, Z_n) = C_{33(n-1)}(R, Z_n), \quad n = 1, \dots, 9. \tag{14.1.52}$$

14.1.8 Convective Model Equations Solution

A solution of the problem (14.1.48)–(14.1.52) will be obtained for the case:

$$K_{0i} = 1, \quad i = 1, 2, \quad K_1 = 1, \quad K_2 = 0.5, \quad K_3 = 1, \tag{14.1.53}$$

as the matrix forms (14.1.18), using the calculation procedure (14.1.18)–(14.1.29), where (14.1.24), (14.1.25) must be replaced by (14.1.50), (14.1.49).

The solutions of the model Eqs. (14.1.48)–(14.1.52) permit to be presented in Figs. 14.5 and 14.6 the concentration distributions of AC for different Z in the case (14.1.53).

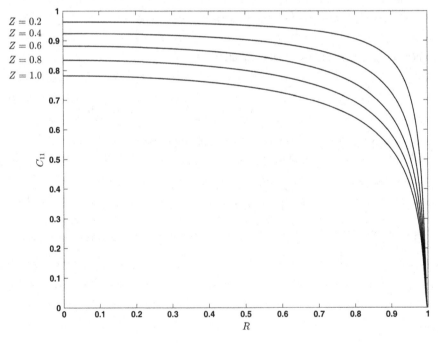

Fig. 14.5 Radial distribution of the concentration $C_{11}(R, Z)$, $Z = 0.2(1)$, $0.4(2)$, $0.6(3)$, $0.8(4)$, $1.0(5)$

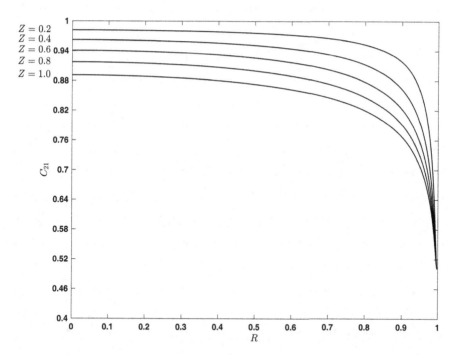

Fig. 14.6 Radial distribution of the concentration $C_{21}(R, Z)$, $Z = 0.2(1)$, $0.4(2)$, $0.6(3)$, $0.8(4)$, $1.0(5)$

The obtained concentrations as solutions of (14.1.48)–(14.1.52) in the case (14.1.53) permit to be obtained average concentrations $\overline{C}_{11}(Z)$, $\overline{C}_{21}(Z)$, $\overline{C}_{13}(Z)$, $\overline{C}_{23}(Z)$, $\overline{C}_{33}(Z)$ and functions $A_i(Z)$, $i = 1, 2$, $B(Z)$, $G(Z)$ in (14.1.8). The obtained results show that $B(Z) \equiv 1$, $G(Z) \equiv 1$, practically. The functions $\overline{C}_{11}(Z)$, $\overline{C}_{21}(Z)$, $A_i(Z)$, $i = 1, 2$ are presented as "theoretical values" for different $Z_n = 0.1(n + 1)$, $n = 0, 1, \ldots, 9$ (points) in Figs. 14.7 and 14.8.

From Fig. 14.8, it is seen that the functions $A_i(Z)$, $i = 1, 2$ are possible to be presented as quadratic approximations (14.1.30) (the lines in Fig. 14.8) and the approximations ("theoretical") parameter values are presented in Table 14.2, where $K_{0i} = 1$, $i = 1, 2$ are "theoretical" values too.

The final form of the average-concentration model is (14.1.31), (14.1.46) and the model parameters a_{0i}, a_{1i}, a_{2i}, K_{0i}, $i = 1, 2$ must be obtained, using experimental data.

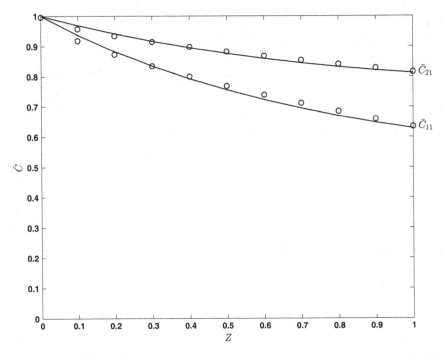

Fig. 14.7 "Theoretical values" of the average concentration $\overline{C}_{11}(Z)$, $\overline{C}_{21}(Z)$ for different $Z_n = 0.1(n+1)$, $n = 0, 1, \ldots, 9$ (points); $\overline{C}_{11}(Z)$, $\overline{C}_{21}(Z)$ as solution of (14.1.31), (14.1.46) (lines), using the "experimental" parameter values in Table 14.2

14.1.9　Average-Concentration Model Equations Solution

The solution of (14.1.31), (14.1.46) must be obtained as five vector forms (14.1.32), using the calculation procedure (14.1.33)–(14.1.40), where (14.1.37), (14.1.38) are replaced by (14.1.50), (14.1.49).

14.1.10　Parameter Identification

The obtained "theoretical" values of the average concentrations $\overline{C}_{i1n}(Z_n)$, $Z_n = 0.1$ $(n+1)$, $n = 0, 1, \ldots, 9$, $i = 1, 2$ (the points in Fig. 14.7) permit to be obtained the artificial experimental data (14.1.36) at the column end $Z = 1$.

The artificial experimental data (14.1.41) are used for the parameter $(a_{0i}, a_{1i}, a_{2i}, K_{0i}, i = 1, 2)$ identification in the average-concentration model (14.1.31), (14.1.46) by the minimization of the least-squares function (14.1.42). The obtained ("experimental") values of a_{0i}, a_{1i}, a_{2i}, K_{0i}, $i = 1, 2$ are presented in

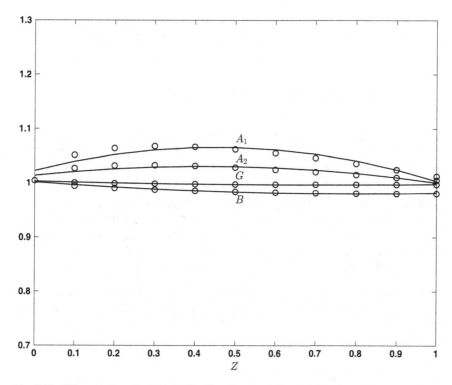

Fig. 14.8 "Theoretical values" of the functions $A_i(Z)$, $i = 1, 2$, for different $Z_n = 0.1$ $(n+1)$, $n = 0, 1, \ldots, 9$ (points); quadratic approximations (14.1.30) (lines), using the "theoretical" parameter values in Table 14.2

Table 14.2 Model parameter values (chemical adsorption mechanism)	"Theoretical" values	"Experimental" values
	$a_{01} = 1.0223$	$a_{01} = 0.9006$
	$a_{11} = 0.1903$	$a_{11} = 0.1882$
	$a_{21} = -0.2089$	$a_{21} = -0.197$
	$a_{02} = 1.0135$	$a_{02} = 0.9096$
	$a_{12} = 0.0792$	$a_{12} = 0.0895$
	$a_{22} = -0.0921$	$a_{22} = -0.0863$
	$K_{01} = 1$	$K_{01} = 1.1664$
	$K_{02} = 1$	$K_{02} = 1.1595$
	$b_0 = 1.001$	
	$b_1 = -0.05367$	
	$b_2 = 0.0344$	
	$g_0 = 1.0028$	
	$g_1 = -0.01805$	
	$g_2 = 0.0132$	

Table 14.2. They are used for the calculation of the functions $\overline{C}_{11}(Z)$, $\overline{C}_{21}(Z)$ in (14.1.31), (14.1.46), and the results are presented in Fig. 14.7 (lines).

14.1.11   Influence of the Model Parameters

In the case of physical adsorption mechanism, the influence of the model parameters K_1, K_2, K_3, K_4, K_5 in the model (14.1.31), (14.1.5), (14.1.6), where the experimental values of the parameters a_{0i}, a_{1i}, a_{2i}, K_{0i}, $i = 1, 2$ (Table 14.1) are used, will be presented in the cases:

1. $K_1 = 2$, $K_2 = 1$, $K_3 = 1.5$, $K_4 = 0.5$, $K_5 = 1$ (Fig. 14.9).
2. $K_1 = 2.5$, $K_2 = 1.5$, $K_3 = 1.5$, $K_4 = 0.5$, $K_5 = 1$ (Fig. 14.10).
3. $K_1 = 2.5$, $K_2 = 1$, $K_3 = 1$, $K_4 = 0.5$, $K_5 = 1$ (Fig. 14.11).
4. $K_1 = 2.5$, $K_2 = 1$, $K_3 = 1.5$, $K_4 = 1$, $K_5 = 1$ (Fig. 14.12).
5. $K_1 = 2.5$, $K_2 = 1$, $K_3 = 1.5$, $K_4 = 0.5$, $K_5 = 1.5$ (Fig. 14.13).

The solutions of (14.1.12)–(14.1.16) and (14.1.8) permit to be obtained the "theoretical" values of the average concentrations $\overline{C}_{i1n}(Z_n)$, $Z_n = 0.1(n+1)$,

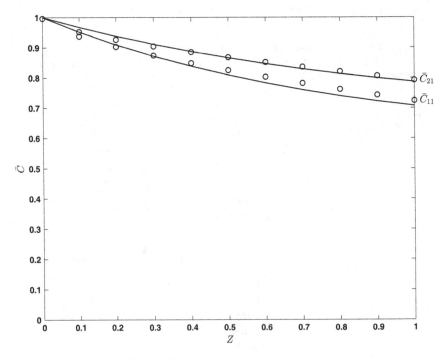

Fig. 14.9 Effect of K_1: $\overline{C}_{11}(Z)$, $\overline{C}_{21}(Z)$ as solution of (14.1.31), (14.1.5), (14.1.6) (lines); "theoretical" values $\overline{C}_{11n}(Z_n)$, $\overline{C}_{21n}(Z_n)$, $Z_n = 0.1(n+1)$, $n = 0, 1, \ldots, 9$ (points)

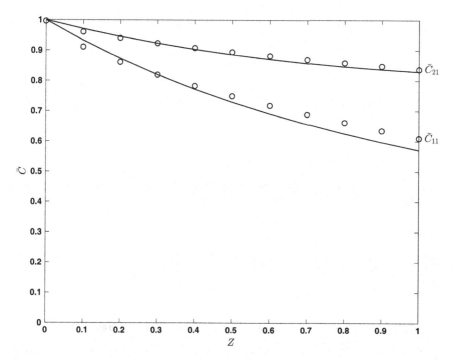

Fig. 14.10 Effect of K_2: $\overline{C}_{11}(Z)$, $\overline{C}_{21}(Z)$ as solution of (14.1.31), (14.1.5), (14.1.6) (lines); "theoretical" values $\overline{C}_{11n}(Z_n)$, $\overline{C}_{21n}(Z_n)$, $Z_n = 0.1(n+1)$, $n = 0, 1, \ldots, 9$ (points)

$n = 0, 1, \ldots, 9$, $i = 1, 2$, which are presented (the points) in Figs. 14.9, 14.10, 14.11, 14.12 and 14.13.

In the case of chemical adsorption mechanism, the influence of the model parameters K_1, K_2, K_3 in the model (14.1.31), (14.1.46), where the experimental values of the parameters a_{0i}, a_{1i}, a_{2i}, K_{0i}, $i = 1, 2$ (Table 14.2) are used, will be presented in the cases:

1. $K_1 = 1.5$, $K_2 = 0.5$, $K_3 = 1$ (Fig. 14.14).
2. $K_1 = 1$, $K_2 = 1$, $K_3 = 1$ (Fig. 14.15).
3. $K_1 = 1$, $K_2 = 0.5$, $K_3 = 1, 5$ (Fig. 14.16).

The solutions of (14.1.48)–(14.1.52) and (14.1.8) in the cases 1–3 permit to be obtained "theoretical" values of the average concentrations $\overline{C}_{i1n}(Z_n)$, $Z_n = 0.1(n+1)$, $n = 0, 1, \ldots, 9$, $i = 1, 2$ (the points in Figs. 14.14, 14.15, and 14.16), which are compared with $\overline{C}_{11}(Z)$, $\overline{C}_{21}(Z)$ as solution of (14.1.31), (14.1.46) (lines), using the "experimental" parameter values in Table 14.2.

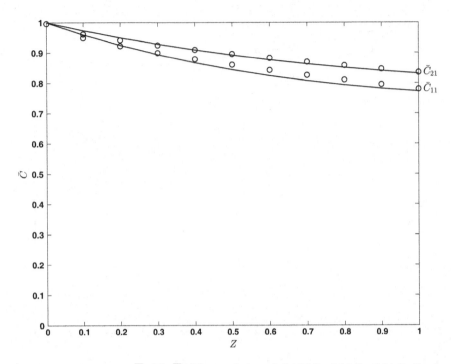

Fig. 14.11 Effect of K_3: $\overline{C}_{11}(Z)$, $\overline{C}_{21}(Z)$ as solution of (14.1.31), (14.1.5), (14.1.6) (lines); "theoretical" values $\overline{C}_{11n}(Z_n)$, $\overline{C}_{21n}(Z_n)$, $Z_n = 0.1(n+1)$, $n = 0, 1, \ldots, 9$ (points)

The numerical analysis of the catalytic processes modeling in column apparatuses, in the cases of physical and chemical adsorption mechanisms, shows that average-concentration model, where the radial velocity component is equal to zero (in the cases of a constant velocity radial non-uniformity along the column height), is possible to be used in the cases of an axial modification of the radial non-uniformity of the axial velocity component. The use of experimental data, for the average concentration at the column end, for a concrete process and column, permits to be obtained the model parameters, related to the radial non-uniformity of the velocity. These parameter values permit to be used the average-concentration model for modeling of different processes.

14.2   Effect of the Radial Velocity Component

In Chaps. 10–13, it was shown that average-concentration model, where the radial velocity component is not equal to zero, is possible to be used. This possibility will be used for modeling of the catalytic processes in column apparatuses too. For convenience, a gas–solid process will be considered.

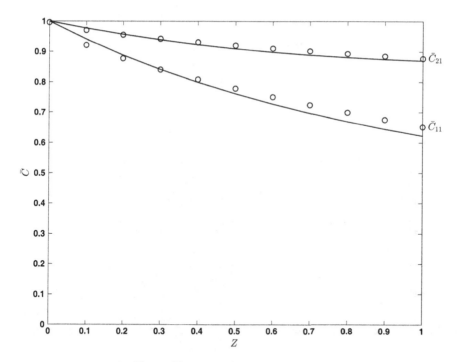

Fig. 14.12 Effect of K_4: $\overline{C}_{11}(Z)$, $\overline{C}_{21}(Z)$ as solution of (14.1.31), (14.1.5), (14.1.6) (lines); "theoretical" values $\overline{C}_{11n}(Z_n)$, $\overline{C}_{21n}(Z_n)$, $Z_n = 0.1(n+1)$, $n = 0, 1, \ldots, 9$ (points)

The catalytic processes will be considered as three-component ($i = 1, 2, 3$), two-phase ($j = 1, 3$) gas–solid interphase mass transfer processes in column apparatus, accompanied by adsorption and chemical reaction at the gas–solid interphase. In these conditions, the convection–diffusion model has the form:

$$u_j \frac{\partial c_{ij}}{\partial z} + v_j \frac{\partial c_{ij}}{\partial r} = D_{ij} \left(\frac{\partial^2 c_{ij}}{\partial z^2} + \frac{1}{r} \frac{\partial c_{ij}}{\partial r} + \frac{\partial^2 c_{ij}}{\partial r^2} \right) \pm Q_{ij}(c_{ij});$$

$$r = 0, \quad \frac{\partial c_{ij}}{\partial r} = 0; \quad r = r_0, \quad \frac{\partial c_{ij}}{\partial r} = 0; \tag{14.2.1}$$

$$z = 0, \quad c_{ij} = c_{ij}^0, \quad u_j^0 c_{ij}^0 = u_j c_{ij}^0 - D_{ij} \frac{\partial c_{ij}}{\partial z}; \quad i = 1, 2, 3, \quad j = 1, 2.$$

$$\frac{\partial u_j}{\partial z} + \frac{\partial v_j}{\partial r} + \frac{v_j}{r} = 0; \quad r = r_0, \quad v_j(r_0, z) \equiv 0, \quad z = 0, \quad u_j = u_j(r, 0), \tag{14.2.2}$$

where $u_j = u_j(r, z)$, $v_j = v_j(r, z)$, $j = 1, 3$ (m s^{-1}) are the axial and radial velocity components in the gas and solid phases. Practically, solid phase (catalyst) is immobile, i.e., $u_3 = v_3 \equiv 0$.

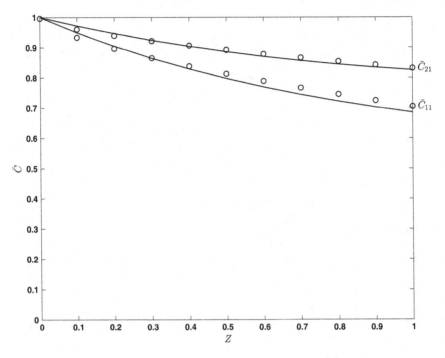

Fig. 14.13 Effect of K_5: $\overline{C}_{11}(Z)$, $\overline{C}_{21}(Z)$ as solution of (14.1.31), (14.1.5), (14.1.6) (lines); "theoretical" values $\overline{C}_{11n}(Z_n)$, $\overline{C}_{21n}(Z_n)$, $Z_n = 0.1(n+1)$, $n = 0, 1, \ldots, 9$ (points)

The terms Q_{ij}, $i = 1, 2, 3$, $j = 1, 3$ (kg $-$ mol m^{-3} s^{-1}) are the reaction rates in the phases of the interphase mass transfer, physical or chemical adsorption, and catalytic chemical reaction, presented as volume reactions; i.e., they are volume sources or sinks in the phase part of the column volumes and participate in the mass balance in the phase volumes (no in the column volume).

14.2.1   Physical Adsorption Mechanism

The convection–diffusion model of the catalytic process, in the case of physical adsorption mechanism and $v \neq 0$, is possible to be obtained from (3.3.3) and (14.0.1):

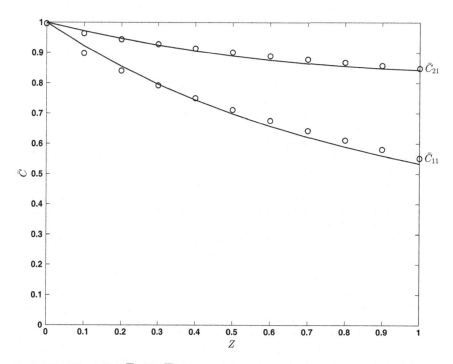

Fig. 14.14 Effect of K_1: $\overline{C}_{11}(Z)$, $\overline{C}_{21}(Z)$ as solution of (14.1.31), (14.1.46) (lines); "theoretical" values $\overline{C}_{11n}(Z_n)$, $\overline{C}_{21n}(Z_n)$, $Z_n = 0.1(n+1)$, $n = 0,1,\ldots,9$ (points)

$$u_1 \frac{\partial c_{11}}{\partial z} + v_1 \frac{\partial c_{11}}{\partial r} = D_{11}\left(\frac{\partial^2 c_{11}}{\partial z^2} + \frac{1}{r}\frac{\partial c_{11}}{\partial r} + \frac{\partial^2 c_{11}}{\partial r^2}\right) - k_{01}(c_{11} - c_{13});$$

$$u_1 \frac{\partial c_{21}}{\partial z} + v_1 \frac{\partial c_{21}}{\partial r} = D_{21}\left(\frac{\partial^2 c_{21}}{\partial z^2} + \frac{1}{r}\frac{\partial c_{21}}{\partial r} + \frac{\partial^2 c_{21}}{\partial r^2}\right) - k_{02}(c_{21} - c_{23});$$

$$r = 0, \quad \frac{\partial c_{11}}{\partial r} = \frac{\partial c_{21}}{\partial r} \equiv 0; \quad r = r_0, \quad \frac{\partial c_{11}}{\partial r} = \frac{\partial c_{21}}{\partial r} \equiv 0; \qquad (14.2.3)$$

$$z = 0, \quad c_{11} \equiv c_{11}^0, \quad u_1^0 c_{11}^0 \equiv u_1(r)c_{11}^0 - D_{11}\left(\frac{\partial c_{11}}{\partial z}\right)_{z=0},$$

$$c_{21} \equiv c_{21}^0, \quad u_1^0 c_{21}^0 \equiv u_1(r)c_{21}^0 - D_{21}\left(\frac{\partial c_{21}}{\partial z}\right)_{z=0}.$$

$$k_{01}(c_{11} - c_{13}) - qk_1 c_{13} \frac{c_{33}}{c_{33}^0} + k_2 c_{33}^0 \left(1 - \frac{c_{33}}{c_{33}^0}\right) = 0,$$

$$k_{02}(c_{21} - c_{23}) - k_{23} c_{23}(c_{33}^0 - c_{33}) = 0, \qquad (14.2.4)$$

$$- qk_1 c_{13} \frac{c_{33}}{c_{33}^0} + k_2 c_{33}^0 \left(1 - \frac{c_{33}}{c_{33}^0}\right) + k_{23} c_{23}(c_{33}^0 - c_{33}) = 0.$$

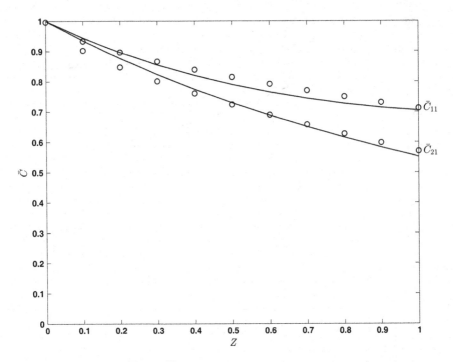

Fig. 14.15 Effect of K_2: $\overline{C}_{11}(Z)$, $\overline{C}_{21}(Z)$ as solution of (14.1.31), (14.1.46) (lines); "theoretical" values $\overline{C}_{11n}(Z_n)$, $\overline{C}_{21n}(Z_n)$, $Z_n = 0.1(n+1)$, $n = 0, 1, \dots, 9$ (points)

The introduction of the average values of the velocity and concentration in the convection–diffusion model (14.2.3), (14.2.4) leads to the average-concentration model:

$$\alpha_1(z)\bar{u}_1 \frac{d\bar{c}_{11}}{dz} + [\varphi_1(z) + \delta_1(z)]\bar{u}_1\bar{c}_{11} = D_{11}\frac{d^2\bar{c}_{11}}{dz^2} - k_{01}(\bar{c}_{11} - \bar{c}_{13});$$

$$\alpha_2(z)\bar{u}_1 \frac{d\bar{c}_{21}}{dz} + [\varphi_2(z) + \delta_2(t,z)]\bar{u}_1\bar{c}_{21} = D_{21}\frac{d^2\bar{c}_{21}}{dz^2} - k_{02}(\bar{c}_{21} - \bar{c}_{23}); \quad (14.2.5)$$

$$z = 0, \quad \bar{c}_{11} \equiv c_{11}^0, \quad \left(\frac{d\bar{c}_{11}}{dz}\right)_{z=0} \equiv 0, \quad \bar{c}_{21} \equiv c_{21}^0, \quad \left(\frac{d\bar{c}_{21}}{dz}\right)_{z=0} \equiv 0.$$

$$k_{01}(\bar{c}_{11} - \bar{c}_{13}) - q\beta(z)k_1\bar{c}_{13}\frac{\bar{c}_{33}}{c_{33}^0} + k_2c_{33}^0\left(1 - \frac{\bar{c}_{33}}{c_{33}^0}\right) = 0;$$

$$k_{02}(\bar{c}_{21} - \bar{c}_{23}) - k_{23}\bar{c}_{23}c_{33}^0 + \gamma(z)k_{23}\bar{c}_{23}\bar{c}_{33} = 0; \quad (14.2.6)$$

$$- q\beta(z)k_1\bar{c}_{13}\frac{\bar{c}_{33}}{c_{33}^0} + k_2c_{33}^0\left(1 - \frac{\bar{c}_{33}}{c_{33}^0}\right) + k_{23}\bar{c}_{23}c_{33}^0 - \gamma(z)k_{23}\bar{c}_{23}\bar{c}_{33} = 0,$$

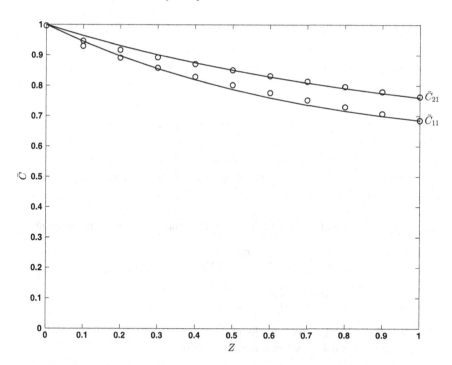

Fig. 14.16 Effect of K_3: $\overline{C}_{11}(Z)$, $\overline{C}_{21}(Z)$ as solution of (14.1.31), (14.1.46) (lines); "theoretical" values $\overline{C}_{11n}(Z_n)$, $\overline{C}_{21n}(Z_n)$, $Z_n = 0.1(n+1)$, $n = 0, 1, \ldots, 9$ (points)

where

$$\alpha_i(z) = \frac{2}{r_0^2} \int\limits_0^{r_0} r U \tilde{c}_{i1}(r, z) \, dr, \quad \varphi_i(z) = \frac{2}{r_0^2} \int\limits_0^{r_0} r U \frac{\partial \tilde{c}_{i1}}{\partial z} dr,$$

$$\delta_i(z) = \frac{2}{r_0^2} \int\limits_0^{r_0} r V \frac{\partial \tilde{c}_{i1}}{\partial r} dr, \quad i = 1, 2, \tag{14.2.7}$$

$$\beta(z) = \frac{2}{r_0^2} \int\limits_0^{r_0} r \tilde{c}_{13}(r, z) \tilde{c}_{33}(r, z) \, dr, \quad \gamma(z) = \frac{2}{r_0^2} \int\limits_0^{r_0} r \tilde{c}_{23}(r, z) \, \tilde{c}_{33}(r, z) \, dr.$$

In (14.2.5)–(14.2.7), the average velocity and concentrations are used :

$$\bar{u}_1 = \frac{2}{r_0^2}\int_0^{r_0} ru_1(r,z)\,dr = u_1^0, \quad \bar{c}_{11}(z) = \frac{2}{r_0^2}\int_0^{r_0} rc_{11}(r,z)\,dr,$$

$$\bar{c}_{21}(z) = \frac{2}{r_0^2}\int_0^{r_0} rc_{21}(r,z)\,dr, \quad \bar{c}_{13}(z) = \frac{2}{r_0^2}\int_0^{r_0} rc_{13}(r,z)\,dr, \qquad (14.2.8)$$

$$\bar{c}_{23}(z) = \frac{2}{r_0^2}\int_0^{r_0} rc_{23}(r,z)\,dr, \quad \bar{c}_{33}(z) = \frac{2}{r_0^2}\int_0^{r_0} rc_{33}(r,z)\,dr$$

and the radial non-uniformities of the velocity components and concentrations:

$$U = \frac{u_1}{u_1^0}, \quad V = \frac{v_1}{\varepsilon u_1^0}, \quad \varepsilon = \frac{r_0}{l}, \quad \tilde{c}_{11} = \frac{c_{11}}{\bar{c}_{11}}, \quad \tilde{c}_{21} = \frac{c_{21}}{\bar{c}_{21}},$$

$$\tilde{c}_{13} = \frac{c_{13}}{\bar{c}_{13}}, \quad \tilde{c}_{23} = \frac{c_{23}}{\bar{c}_{23}}, \quad \tilde{c}_{33} = \frac{c_{33}}{\bar{c}_{33}}, \qquad (14.2.9)$$

where l is the column active zone height.

The use of dimensionless (generalized) variables:

$$R = \frac{r}{r_0}, \quad Z = \frac{z}{l}, \quad U = \frac{u_1}{u_1^0}, \quad V = \frac{v_1}{\varepsilon u_1^0},$$

$$C_{11} = \frac{c_{11}}{c_{11}^0}, \quad C_{13} = \frac{c_{13}}{c_{11}^0}, \quad C_{21} = \frac{c_{21}}{c_{21}^0}, \quad C_{23} = \frac{c_{23}}{c_{21}^0}, \quad C_{33} = \frac{c_{33}}{c_{33}^0}, \qquad (14.2.10)$$

$$\overline{C}_{11} = \frac{\bar{c}_{11}}{c_{11}^0}, \quad \overline{C}_{13} = \frac{\bar{c}_{13}}{c_{11}^0}, \quad \overline{C}_{21} = \frac{\bar{c}_{21}}{c_{21}^0}, \quad \overline{C}_{23} = \frac{\bar{c}_{23}}{c_{21}^0}, \quad \overline{C}_{33} = \frac{\bar{c}_{33}}{c_{33}^0},$$

leads to:

$$U\frac{\partial C_{11}}{\partial Z} + V\frac{\partial C_{11}}{\partial R} = Fo_{11}\left(\varepsilon^2\frac{\partial^2 C_{11}}{\partial Z^2} + \frac{1}{R}\frac{\partial C_{11}}{\partial R} + \frac{\partial^2 C_{11}}{\partial R^2}\right) - K_{01}(C_{11} - C_{13});$$

$$U\frac{\partial C_{21}}{\partial Z} + V\frac{\partial C_{21}}{\partial R} = Fo_{21}\left(\varepsilon^2\frac{\partial^2 C_{21}}{\partial Z^2} + \frac{1}{R}\frac{\partial C_{21}}{\partial R} + \frac{\partial^2 C_{21}}{\partial R^2}\right) - K_{02}(C_{21} - C_{23});$$

$$R = 0, \quad \frac{\partial C_{i1}}{\partial R} \equiv 0; \quad R = 1, \quad \frac{\partial C_{i1}}{\partial R} \equiv 0;$$

$$Z = 0, \quad C_{i1} \equiv 1, \quad 1 \equiv U(R) - Pe_{i1}^{-1}\left(\frac{\partial C_{i1}}{\partial Z}\right)_{Z=0}; \quad i = 1, 2.$$

$$(14.2.11)$$

$$C_{13} = \frac{C_{11} + K_1(1 - C_{33})}{1 + K_2 C_{33}}, \quad C_{23} = \frac{C_{21}}{1 + K_3(1 - C_{33})};$$

$$\omega_3(C_{33})^3 + \omega_2(C_{33})^2 + \omega_1 C_{33} + \omega_0 = 0,$$

$$\omega_3 = K_3(K_2 K_5 - K_1 K_4),$$

$$\omega_2 = K_5(K_3 - K_2 - 2K_2 K_3) + K_4(K_1 + 2K_1 K_3 + K_3 C_{11}) - K_2 C_{21},$$

$$\omega_1 = K_5(K_2 - 1 - 2K_3 + K_2 K_3) - K_4(K_1 + C_{11})(1 + K_3) + (K_2 - 1)C_{21},$$

$$\omega_0 = C_{21} + K_3 K_5 + K_5;$$

$$(14.2.12)$$

$$A_1(Z)\frac{d\overline{C}_{11}}{dZ} + [\Phi_1(Z) + \Delta_1(Z)]\overline{C}_{11} = Pe^{-1}\frac{d^2\overline{C}_{11}}{dZ^2} - K(\overline{C}_{11} - \overline{C}_{13});$$

$$A_2(Z)\frac{d\overline{C}_{21}}{dZ} + [\Phi_2(Z) + \Delta_2(Z)]\overline{C}_{21} = Pe^{-1}\frac{d^2\overline{C}_{21}}{dZ^2} - K(\overline{C}_{21} - \overline{C}_{23}); \quad (14.2.13)$$

$$Z = 0, \quad \overline{C}_{11} \equiv 1, \quad \overline{C}_{21} \equiv 1, \quad \left(\frac{\partial \overline{C}_{11}}{\partial Z}\right)_{Z=0} \equiv 0, \quad \left(\frac{\partial \overline{C}_{21}}{\partial Z}\right)_{Z=0} \equiv 0.$$

$$\overline{C}_{13} = \frac{\overline{C}_{11} + K_1(1 - \overline{C}_{33})}{1 + BK_2\overline{C}_{33}}, \quad \overline{C}_{23} = \frac{\overline{C}_{21}}{1 + K_3(1 - G\overline{C}_{33})};$$

$$\bar{\omega}_3(\overline{C}_{33})^3 + \bar{\omega}_2(\overline{C}_{33})^2 + \bar{\omega}_1\overline{C}_{33} + \bar{\omega}_0 = 0,$$

$$\bar{\omega}_3 = BGK_3(K_2 K_5 - K_1 K_4),$$

$$\bar{\omega}_2 = K_5(-BK_2 - BK_2 K_3 - BGK_2 K_3 + GK_3)$$

$$\quad + K_4(BK_1 + BK_1 K_3 + BGK_1 K_3 + BGK_3\overline{C}_{11}) + BGK_2\overline{C}_{21}, \quad (14.2.14)$$

$$\bar{\omega}_1 = K_5(BK_2 - 1 - K_3 - GK_3 + BK_2 K_3) - BK_4(K_1 + \overline{C}_{11})(1 + K_3)$$

$$\quad - G\overline{C}_{21} + BK_2\overline{C}_{21},$$

$$\bar{\omega}_0 = \overline{C}_{21} + K_3 K_5 + K_5.$$

The parameters in (14.2.11)–(14.2.14) are:

$$K_{0i} = \frac{k_{0i}l}{u_1^0}, \quad Fo_{i1} = \frac{D_{i0}l}{u_1^0 r_0^2}, \quad Pe_{i1} = \frac{u_1^0 l}{D_{i0}},$$

$$\varepsilon = \frac{r_0^2}{l^2} = Fo_{i1}^{-1}Pe_{i1}^{-1}, \quad i = 1, 2;$$

$$(14.2.15)$$

$$K_1 = \frac{k_2}{k_{01}}\frac{c_{33}^0}{c_{11}^0}, \quad K_2 = \frac{qk_1}{k_{01}}, \quad K_3 = \frac{k_{23}c_{33}^0}{k_{02}},$$

$$K_4 = \frac{qk_1}{k_{23}c_{21}^0}\frac{c_{11}^0}{c_{33}^0}, \quad K_5 = \frac{k_2}{k_{23}c_{21}^0}.$$

The new functions in (14.2.13), (14.2.14) have the forms:

$$\overline{C}_{11}(Z) = 2 \int_0^1 RC_{11}(R,Z)dR, \quad \overline{C}_{13}(Z) = 2 \int_0^1 RC_{13}(R,Z)dR,$$

$$\overline{C}_{21}(Z) = 2 \int_0^1 RC_{21}(R,Z)dR, \quad \overline{C}_{23}(Z) = 2 \int_0^1 RC_{23}(R,Z)dR,$$

$$\overline{C}_{33}(Z) = 2 \int_0^1 RC_{33}(R,Z)dR, \quad \widetilde{C}_{11}(R,Z) = \frac{C_{11}(R,Z)}{\overline{C}_{11}(Z)},$$

$$\widetilde{C}_{13}(R,Z) = \frac{C_{13}(R,Z)}{\overline{C}_{13}(Z)}, \quad \widetilde{C}_{21}(R,Z) = \frac{C_{21}(R,Z)}{\overline{C}_{21}(Z)},$$

$$\widetilde{C}_{23}(R,Z) = \frac{C_{23}(R,Z)}{\overline{C}_{23}(Z)}, \quad \widetilde{C}_{33}(R,Z) = \frac{C_{33}(R,Z)}{\overline{C}_{33}(Z)}, \tag{14.2.16}$$

$$B(Z) = 2 \int_0^1 R\widetilde{C}_{13}(Z)\widetilde{C}_{33}(Z)dR, \quad G(Z) = 2 \int_0^1 R\widetilde{C}_{23}(Z)\widetilde{C}_{33}(Z)dR,$$

$$A_i(Z) = 2 \int_0^1 RU(R,Z)\widetilde{C}_{i1}(R,Z)dR, \quad \Phi_i(Z) = 2 \int_0^1 RU(R,Z)\frac{\partial \widetilde{C}_{i1}}{\partial Z}dR,$$

$$\Delta_i(Z) = 2 \int_0^1 RV(R,Z)\frac{\partial \widetilde{C}_{i1}}{\partial R}dR, \quad i = 1,2.$$

In the industrial conditions, it is possible to be used the next approximations in the models (14.2.11), (14.2.13):

$$0 = Fo_{i1} \leq 10^{-2}, \quad 0 = Pe_{i1} \leq 10^{-2}, \quad i = 1,2 \tag{14.2.17}$$

and as a result, the models (14.2.11), (14.2.13) have the convective forms:

$$U\frac{\partial C_{11}}{\partial Z} + V\frac{\partial C_{11}}{\partial R} = -K_{01}(C_{11} - C_{13});$$

$$U\frac{\partial C_{21}}{\partial Z} + V\frac{\partial C_{21}}{\partial R} = -K_{02}(C_{21} - C_{23}); \tag{14.2.18}$$

$$R = 1, \quad C_{i1} \equiv 0; \quad Z = 0, \quad C_{i1} \equiv 1; \quad i = 1,2.$$

$$A_1(Z)\frac{d\overline{C}_{11}}{dZ} + [\Phi_1(Z) + \Delta_1(Z)]\overline{C}_{11} = -K_{01}(\overline{C}_{11} - \overline{C}_{13});$$

$$A_2(Z)\frac{d\overline{C}_{21}}{dZ} + [\Phi_2(Z) + \Delta_2(Z)]\overline{C}_{21} = -K_{02}(\overline{C}_{21} - \overline{C}_{23}); \qquad (14.2.19)$$

$$Z = 0, \quad \overline{C}_{11} \equiv 1, \quad \overline{C}_{21} \equiv 1.$$

In (14.2.18), (14.2.19), the functions $C_{13}, C_{23}, C_{33}, \overline{C}_{13}, \overline{C}_{23}, \overline{C}_{33}$ are the solutions of (14.2.12), (14.2.14), where $0 \le C_{33} \le 1$ and $0 \le \overline{C}_{33} \le 1$.

14.2.2 Axial and Radial Velocities

The radial non-uniformity of the axial velocity component in a column apparatus is a result of the fluid hydrodynamics at the column inlet, where it is a maximum and decreases along the column height as a result of the fluid viscosity. As a result, a radial velocity component is initiated. The theoretical analysis of the change in the radial non-uniformity of the axial velocity component (effect of the radial velocity component) in a column can be made by an appropriate hydrodynamic model, where the average velocity at the cross section of the column is a constant, while the maximal velocity (and as a result, the radial non-uniformity of the axial velocity component, too) decreases along the column height. In generalized variables, this model has the form:

$$U = (2 - 0.4Z) - 2(1 - 0.4Z)R^2, \quad V = 0.2(R - R^3),$$

$$\overline{U} = 2\int_0^1 RU dR = 1. \qquad (14.2.20)$$

From Chap. 10, it is seen that $V < 0.1$ and must be presented with the help of a small parameter $0.1 = \alpha \ll 1$, i.e.,

$$V = \alpha V_0, \quad V_0 = 0.2(R - R^3)\alpha^{-1}. \qquad (14.2.21)$$

As a result, Eq. (14.2.18) have the form:

$$U\frac{\partial C_{11}}{\partial Z} + \alpha V_0 \frac{\partial C_{11}}{\partial R} = -K_{01}(C_{11} - C_{13});$$

$$U\frac{\partial C_{21}}{\partial Z} + \alpha V_0 \frac{\partial C_{21}}{\partial R} = -K_{02}(C_{21} - C_{23}); \qquad (14.2.22)$$

$$R = 1, \quad C_{i1} \equiv 0; \quad Z = 0, \quad C_{i1} \equiv 1; \quad i = 1, 2.$$

14.2.3 Convection-Type Model Equations Solution

In (14.2.21), $0.1 = \alpha \ll 1$ is a small parameter and must be used the perturbation method (see Chap. 7); i.e., the concentrations must be presented as

$$
\begin{aligned}
C_{11}(R,Z) &= C_{11}^0(R,Z) + \alpha C_{11}^1(R,Z),\\
C_{21}(R,Z) &= C_{21}^0(R,Z) + \alpha C_{21}^1(R,Z),\\
C_{23}(R,Z) &= C_{23}^0(R,Z) + \alpha C_{23}^1(R,Z),\\
C_{33}(R,Z) &= C_{33}^0(R,Z) + \alpha C_{33}^1(R,Z).
\end{aligned}
\tag{14.2.23}
$$

From the cubic Eq. (14.2.12), it is followed that $C_{33}^1 \equiv 0$.

The introduction of (14.2.23) in (14.2.22), (14.2.12) leads to the next set of problems:

$$
\begin{aligned}
&U\frac{\partial C_{11}^0}{\partial Z} = -K_{01}\left(C_{11}^0 - C_{13}^0\right);\\[4pt]
&U\frac{\partial C_{21}^0}{\partial Z} = -K_{02}\left(C_{21}^0 - C_{23}^0\right);\\[4pt]
&R = 1, \quad C_{il} \equiv 0; \quad Z = 0, \quad C_{i1}^0 \equiv 1; \quad i = 1,2.\\[4pt]
&C_{13}^0 = \frac{C_{11}^0 + K_1\left(1 - C_{33}^0\right)}{1 + K_2 C_{33}^0}, \quad C_{23}^0 = \frac{C_{21}^0}{1 + K_3\left(1 - C_{33}^0\right)};\\[4pt]
&\omega_3\left(C_{33}^0\right)^3 + \omega_2\left(C_{33}^0\right)^2 + \omega_1 C_{33}^0 + \omega_0 = 0,\\[4pt]
&\omega_3 = K_3(K_2 K_5 - K_1 K_4),\\[4pt]
&\omega_2 = K_5(K_3 - K_2 - 2K_2 K_3) + K_4\left(K_1 + 2K_1 K_3 + K_3 C_{11}^0\right) - K_2 C_{21}^0,\\[4pt]
&\omega_1 = K_5(K_2 - 1 - 2K_3 + K_2 K_3) - K_4\left(K_1 + C_{11}^0\right)(1 + K_3) + (K_2 - 1)C_{21}^0,\\[4pt]
&\omega_0 = C_{21}^0 + K_3 K_5 + K_5;
\end{aligned}
$$

$$
\tag{14.2.24}
$$

$$
\begin{aligned}
&U\frac{\partial C_{11}^1}{\partial Z} = -K_{01}\left(C_{11}^1 - C_{13}^1\right) - V_0\frac{\partial C_{11}^0}{\partial R};\\[4pt]
&U\frac{\partial C_{21}^1}{\partial Z} = -K_{02}\left(C_{21}^1 - C_{23}^1\right) - V_0\frac{\partial C_{21}^0}{\partial R};\\[4pt]
&R = 1, \quad C_{il} \equiv 0; \quad Z = 0, \quad C_{i1} \equiv 0; \quad i = 1,2.\\[4pt]
&C_{13}^1 = \frac{C_{11}^1}{1 + K_2 C_{33}^0}, \quad C_{23}^1 = \frac{C_{21}^1}{1 + K_3\left(1 - C_{33}^0\right)}.
\end{aligned}
\tag{14.2.25}
$$

The sequential solution of the problems (14.2.24), (14.2.25) uses cubic spline interpolations for $\frac{\partial C_{11}^0}{\partial R}, \frac{\partial C_{21}^0}{\partial R}, C_{33}^0$.

14.2.4 Chemical Adsorption Mechanism

The difference between the physical and chemical adsorption mechanisms (in the stationary case) is in the models (14.2.4), (14.2.6), (14.2.12), (14.2.14), which must be replaced by (3.3.17), (3.3.19), (6.3.20), (6.3.24). As a result, the convection-type and average-concentration models in the case of chemical adsorption mechanism have the forms:

$$U\frac{\partial C_{11}}{\partial Z} + \alpha V_0 \frac{\partial C_{11}}{\partial R} = -K_{01}(C_{11} - C_{13});$$

$$U\frac{\partial C_{21}}{\partial Z} + \alpha V_0 \frac{\partial C_{21}}{\partial R} = -K_{02}(C_{21} - C_{23}); \tag{14.2.26}$$

$$R = 1, \quad C_{11} = C_{21} \equiv 0; \quad Z = 0, \quad C_{i0} \equiv 1; \quad i = 1, 2.$$

$$C_{13} = \frac{C_{11}}{1 + K_1 C_{33}}, \quad C_{23} = \frac{C_{21}}{1 + K_2(1 - C_{33})},$$

$$(C_{21}K_1 - C_{11}K_2K_3)(C_{33})^2 \tag{14.2.27}$$

$$+ (C_{21} + C_{11}K_3 + C_{11}K_2K_3 - C_{21}K_1)C_{33} - C_{21} = 0,$$

whereas solutions of the equations in (14.2.26), (14.2.27) must be used $0 \le C_{33} \le 1$. In (14.2.27) are used the parameters:

$$K_1 = \frac{k_{13}c_{33}^0}{k_{01}}, \quad K_2 = \frac{k_{23}c_{33}^0}{k_{02}}, \quad K_3 = \frac{k_{13}c_{11}^0}{k_{23}c_{21}^0}. \tag{14.2.28}$$

A new approach for the catalytic processes modeling in industrial column apparatuses, in the cases of physical and chemical adsorption mechanisms, is presented. A numerical analysis shows that convection–diffusion and average-concentration model, where the radial velocity component is not equal to zero, is possible to be used in the cases of an axial modification of the radial non-uniformity of the axial velocity component. The use of experimental data, for the average concentration at the column end, for a concrete process and column, permits to be obtained the model parameters, related to the radial non-uniformity of the velocity. These parameter values permit to be used the average-concentration model for modeling of different processes.

References

1. Boyadjiev B, Chr Boyadjiev (2017) New models of industrial column chemical reactors. Bulg Chem Commun 49(3):706–710
2. Boyadjiev B, Boyadjiev Chr (2017) New models of industrial column absorbers. Co-current absorption processes. Bulg Chem Commun 49(3):711–719

3. Boyadjiev B, Boyadjiev Chr (2017) New models of industrial column absorbers. 2. Counter-current absorption processes. Bulg Chem Commun 49(3):720–728
4. Boyadjiev B, Chr Boyadjiev (2018) A new approach for modeling of industrial adsorption columns. J Eng Thermophys 27(1):82–97 (in press)
5. Boyadjiev B, Boyadjiev Chr (2018) A new approach for modeling of industrial catalytic columns. J Eng Thermophys (in press)
6. Chr Boyadjiev, Doichinova M, Boyadjiev B, Popova-Krumova P (2016) Modeling of column apparatus processes. Springer, Berlin Heidelberg

Part V
Waste Gas Purification in Column Apparatises

New Patents

The solid fuel combustion in the thermal power plants, which use sulfur-rich fuels, poses the problem of sulfur dioxide removal from the waste gases. This problem is complicated by the fact that it is required to purify huge volumes of gas with low SO_2 concentration. The huge gas flow rates require big apparatus size, which is possible to be substantially lowered when the maximal process rate of the gas purification is achieved [1–6].

Practically, the waste gas purification in the thermal power plants uses absorption methods. The SO_2 absorption intensification needs a quantitative description of the process using a suitable mathematical model [7–11], which has to be created on the basis of a qualitative analysis of the process mechanism. The solution of this problem will be presented in the cases of physical and chemical absorption of SO_2 in column apparatuses, using one-phase absorbents as water and water solutions of NaOH, Na_2CO_3, and NH_4OH.

In many practical cases [12, 13], two-phase absorbents are used (as water suspensions of $CaCO_3$ or $Ca(OH)_2$) because they have a low price and big absorption capacity. The presence of the active component in the absorbent, as a solution and solid phase, leads to the introduction of a new process (the dissolution of the solid phase) and to creation of conditions for variations of the absorption mechanism (interphase mass transfer through two interphase surfaces—gas/liquid and liquid/solid). A theoretical analysis of the methods and apparatuses for waste gas purification from SO_2 using two-phase absorbent ($CaCO_3$ suspension) [12, 13] as applied by some companies (Babcock & Wilcox Power Generation Group, Inc., Alstom Power Italy, Idreco-Insigma-Consortium) in the thermal power plants will be presented, too.

The gas absorption is practically realized in column apparatuses, where two-phase gas–liquid flows are different dispersed systems as gas–liquid drops, liquid–gas bubbles or gas–thin liquid films. The absorption rate is related to the interphase mass transfer resistance, which is disturbed between the gas and liquid phases in the column. The absorption rate increase is possible to be realized if the mass transfer resistance in the limiting phase (the phase with the higher resistance) decreases as a result of the intensification of convective transfer, because the role of the diffusion mass transfer is much smaller. The intensification of the convective transfer must be realized in the dispersion medium, because the convective transfer in the dispersion phase (bubbles, drops) is very limited. As a result, the optimal realization of the absorption process, where the interphase mass transfer is limited by the mass transfer in the gas phase, is a gas–liquid drops (film) column, while in the opposite case a liquid–gas bubbles column is optimal.

In the cases of comparable interphase mass transfer resistances in the gas and liquid phases, an intensification of the mass transfer should be realized in two phases. In these conditions, a new patent [14] is proposed, where the process optimization is realized in a two-zone column, where the upper zone of the process is physical absorption in a gas–liquid drops system (intensification of the gas-phase mass transfer), while in the lower zone it is a physical absorption in liquid–gas bubbles system (intensification of the liquid-phase mass transfer).

The method, which uses two-phase absorbents ($CaCO_3$ suspension), has a number of shortcomings. The chemical reaction of $CaCO_3$ with SO_2 lead to CO_2 emission (every molecule of SO_2 absorbed from the air is equivalent to a molecule of CO_2 emitted in the air) and this creates a new ecological problem, because the greenhouse effects of SO_2 and CO_2 are similar. Further disadvantages are the impossibility of the absorbent regeneration and the large quantity of the by-products (gypsum) generated. These problems are solved in a new patent [15], using a two-step process—physical absorption of SO_2 by water and adsorption of SO_2 from the water solution by synthetic anionite particles. The adsorbent regeneration is made by NH_4OH solution. The obtained $(NH_4)_2SO_3$ (NH_4HSO_3) is used (after reaction with HNO_3) for production of concentrated solutions of SO_2 and NH_4NO_3.

Countercurrent absorbers are used in the purification of the large amounts of waste gas from combustion plants, where the gas velocity (as a result and absorbers diameter too) is limited by the rate of the absorbent drops falling in an immobile gas medium, i.e., the gas velocity must be less than the drops velocity (~ 4 [m s^{-1}] practically). This disadvantage is avoided by a new patent [16], where a co-current SO_2 absorption is realized.

References

1. Boyadjiev Chr (2011) On the SO_2 problem in power engineering. In: Proceedings of energy forum, Bulgaria, pp 114–125
2. Boyadjiev Chr (2012) On the SO_2 problem in power engineering. In: Proceedings of Asia-Pacific power and energy engineering conference (APPEEC 2012), China, vol 1
3. Boyadjiev Chr (2012) On the SO_2 problem of solid fuel combustion. In: Proceedings, VIII All-Russian conference with international participation "solid fuel combustion", Novosibirsk
4. Boyadjiev Chr (2014) On the SO_2 problem of solid fuel combustion. Therma Eng 61(9): 691–695
5. Boyadjiev Chr, Doichinova M, Popova-Krumova P, Boyadjiev B (2014) Gas purification from SO_2 in thermal power plants. Chem Eng Techn 37(7):1243–1250

6. Boyadjiev Chr, Doichinova M, Popova P (2012) On the SO_2 problem in power engineering. Trans Academenergo 2:44–65
7. Boyadjiev Chr (2010) Theoretical chemical engineering. Modeling and simulation. Springer, Berlin Heidelberg
8. Boyadjiev Chr (2006) Diffusion models and scale-up. Int J Heat Mass Transfer 49:796–799
9. Boyadjiev Chr (2009) Modeling of column apparatuses. Trans Academenergo 3:7–22
10. Doichinova M, Boyadjiev Chr (2012) On the column apparatuses modeling. Int J Heat Mass Transfer 55:6705–6715
11. Boyadjiev Chr (2013) A new approach for the column apparatuses modeling in chemical engineering. J Pure Appl Math: Adv Appl 10(2):131–150
12. Boyadjiev Chr (2011) Mechanism of gas absorption with two-phase absorbents. Int J Heat Mass Transfer 54:3004–3008
13. Boyadjiev Chr, Popova P, Doichinova M (2011) On the SO_2 problem in power engineering. 2. Two-phase absorbents. In: Proceedings of 15th workshop on transport phenomena in two-phase flow, Bulgaria, pp 104–115
14. Boyadjiev Chr, Boyadjiev B, Doichinova M, Popova-Krumova P (2013) Method and apparatus for gas absorption. Patent application no 111168
15. Boyadjiev Chr, Boyadjiev B, Doichinova M, Popova-Krumova P (2014) Method and apparatus for gas cleaning from sulfur dioxide. Patent application no 111398
16. Boyadjiev Chr, Boyadjiev B, Doichinova M, Popova-Krumova P (2014) Co-current apparatus for gas absorption. Patent application no 111473

Chapter 15
Bizonal Absorption Apparatus

The chemical absorption of average soluble gases (ASG) in the case of slow chemical reaction (e.g., absorption of CO_2 with aqueous solutions of NaOH, where Henry's number in the system CO_2/H_2O is $\chi^{20\,°C} = 1.16$) is possible to be used for waste gas purification. The absorption process intensification has to be realized through intensification of the convective mass transfer in the gas phase (in gas–liquid drops system) and in the liquid phase (in liquid–gas bubbles system). This theoretical result is applied to a new method and bizonal apparatus for gas absorption [1]. In the upper equipment zone, a physical absorption (as a result of the short reaction time, i.e., short existence of the absorbent drops) is realized in a gas–liquid drops system and the big convective transfer in the gas phase leads to decrease of the mass transfer resistances in this phase. In the lower zone, a chemical absorption in a liquid–gas bubbles system takes place and the big convective transfer in the liquid phase lowers the mass transfer resistances in this phase. The large volume of the liquid in this zone causes an increase of the chemical reaction time, and as a result, further decrease of the mass transfer resistance in the liquid phase is realized. In the column tank, the chemical reaction only takes place.

15.1 Absorption Column

The bizonal absorption column [1–3] is shown in Fig. 15.1. The apparatus comprises a cylindrical absorption column 1 fitted at its lower end with a tangential inlet 2 for submission of the waste gas. A horizontal gas distribution plate 3 (bell plate or any other device providing the cleansing of the bubbling gas while passing through the absorbent in the lower part of the column) is placed above the lower part of the absorption column 1 so that between it and the bottom of the absorption column 1, a gas input zone is formed. On the distribution plate 3, vertical gas distribution pipes 4 are installed, the number which depends on the gas flux flow rate. Each of the

© Springer International Publishing AG, part of Springer Nature 2018
C. Boyadjiev et al., *Modeling of Column Apparatus Processes*,
Heat and Mass Transfer, https://doi.org/10.1007/978-3-319-89966-4_15

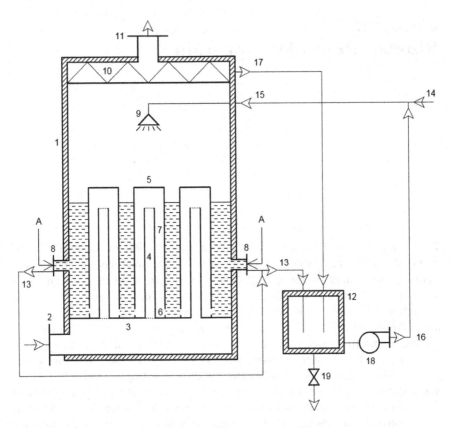

Fig. 15.1 Bizonal absorption column (A-A, 20)

distribution pipes 4 is covered with concentric bubbling cap 5, which lies on plate 3 and has slots 6 at the bottom. Between the distribution pipes 4 and the bubbling caps 5 are formed passages 7, which are open to the pipes 4. The volume of the pipes 4 is filled with absorbent suspension.

The upper part of the absorption column 1 is equipped with sprinkler system 9, which is located above the drop separator 10 and the outlet 11 provided for the purified gas. The exits 8 (of the bubbling area) are connected with the tank 12 through the pipe 13. The level of the absorbent in the bubbling zone is controlled by the turn-cock 20. The tank 12 is provided with the turn-cock 19. Sprinkler system 9 is connected by circulation pipes 15 and 16 and circulation pump 18 with the tank 12. The absorbent, separated in the separator 10, passes through the outlet 17 and enters the tank 12. Item 14 is the absorbent inlet of the absorption system.

A tangential inlet of the gas phase [4, 5] leads to a significant decrease of the velocity radial non-uniformity below the gas distribution plate. This effect increases further when the gas is distributed in the pipes, and as a result, the mass transfer rate in the gas phase increases.

The sprinkler system leads to decrease of the velocity radial non-uniformity in the liquid phase, and as a result, the mass transfer rate in the liquid phase increases too.

15.1.1   Physical Absorption Modeling in the Upper Zone

Let us consider the absorption column [3] in Fig. 15.1 with a diameter D (m), where l_1 (m) is the height of the upper zone between bubbling caps and sprinkler system (gas–liquid drops system zone) and l_2 (m) is the height of the lower (liquid–gas bubbles system) zone between distribution plate and the liquid surface (practically the bubbling caps height). The diameter of the gas distribution pipes 4 is D_0 (m), the diameter of bubbling caps is D_1 (m), and n is their number. The gas and liquid flow rates in the column are Q_G and Q_L (m^3 s^{-1}), the inlet velocities in the gas and liquid phases are u_G^0 and u_L^0 (m s^{-1}), and $\varepsilon_G, \varepsilon_L$ ($\varepsilon_G + \varepsilon_L = 1$) are holdup coefficients, respectively:

$$u_G^0 = \frac{Q_G}{\varepsilon_G F}, \quad u_L^0 = \frac{Q_L}{\varepsilon_L F}, \quad F = \frac{\pi D^2}{4}, \quad \varepsilon_G = \frac{Q_G + Q_L}{Q_G}, \quad \varepsilon_L = \frac{Q_G + Q_L}{Q_L}.$$
(15.1.1)

The absorber working volume W (m^3) in the lower zone is:

$$W = (F - nF_1)l_2, \quad F_1 = \frac{\pi D_1^2}{4}.$$
(15.1.2)

Let us consider the physical absorption in gas–liquid drops system [2] in the upper zone of the column (Fig. 15.1), where the radial non-uniformities of the velocity distributions in the gas and liquid phases are practically absent. This is a result [4, 5] of the tangential inlet of the gas phase in the column, the gas distribution pipes in the middle column zone, and the sprinkler system in the upper column zone. In these conditions, the radial non-uniformities of the concentration distributions in the gas and liquid phases are absent too and it is possible to use average-concentration values over the cross-sectional area of the column.

In the upper zone, a convection–diffusion model [2, 3] is possible to be used for a countercurrent absorption process in systems of two cylindrical coordinates —(z_1, r), (z_2, r), $(z_1 + z_2 = l_1)$, where $\bar{c}_G = \bar{c}_G(z_1)$ and $\bar{c}_L = \bar{c}_L(z_2)$ are the axial distributions of the average SO$_2$ concentration in the gas and liquid phases

$$u_G^0 \frac{d\bar{c}_G}{dz_1} = D_G \frac{d^2\bar{c}_G}{dz_1^2} - k(\bar{c}_G - \chi\bar{c}_L); \quad z_1 = 0, \quad \bar{c}_G = c_G^0, \quad \left(\frac{d\bar{c}_G}{dz_1}\right)_{z_1=0} = 0;$$

$$u_L^0 \frac{d\bar{c}_L}{dz_2} = D_L \frac{d^2\bar{c}_L}{dz_2^2} + k(\bar{c}_G - \chi\bar{c}_L); \quad z_2 = 0, \quad \bar{c}_L = 0, \quad \left(\frac{d\bar{c}_L}{dz_2}\right)_{z_2=0} = 0,$$

$$(15.1.3)$$

where c_G^0 is the inlet concentration of SO_2 in the gas phase, which is equal to the outlet SO_2 concentration in the gas phase from the lower zone $(\bar{c}_1(l_2))$, and χ is Henry's number.

15.1.2   Chemical Absorption Modeling in the Lower Zone

The gas bubbling in the absorbent volume W in the lower zone creates an ideal mixing regime, and as a result, the concentrations (kg mol m^{-3}) of ASG $(\bar{c}_2)$ and chemical reagent $(\bar{c}_3)$ in the liquid phase are constants (as a result of the big liquid volume).

The mass flow rates (kg mol s^{-1}) of ASG at the inlet and outlet of the column are $Q_G c_1^0$ and $Q_G c_G(l_1)$, where c_1^0 and $c_G(l_1)$ are the ASG concentration (kg mol m^3) in the input and output of the gas flow. The difference between them is the ASG absorption rate V_1 (kg mol s^{-1}) in the column:

$$V_1 = Q_G [c_1^0 - \bar{c}_G(l_1)]. \tag{15.1.4}$$

The chemical reaction rate V_2 (kg mol m^{-3} s^{-1}) between ASG and chemical reagent (CR) in the lower zone is possible to be presented as

$$V_2 = k_0\bar{c}_2\bar{c}_3, \tag{15.1.5}$$

where k_0 is the chemical reaction rate constant.

The concentration of ASG $(\bar{c}_2)$ in the liquid phase in the lower zone is a constant if the ASG absorption rate (V_1) in the column is equal to the amount of ASG (V_2W), which reacts chemically with CR in the lower zone, i.e.

$$\bar{c}_2 = \frac{Q_G [c_1^0 - \bar{c}_G(l_1)]}{Wk_0\bar{c}_3}. \tag{15.1.6}$$

The axial distributions $\bar{c}_1(z)$ of the average ASG concentration in the gas phase in the lower zone are possible to be obtained as a solution to the problem:

$$\bar{u}_1 \frac{d\bar{c}_1}{dz} = D_G \frac{d^2\bar{c}_1}{dz^2} - k_1(\bar{c}_1 - \chi\bar{c}_2); \quad z = 0, \quad \bar{c}_1 = c_1^0, \quad \left(\frac{d\bar{c}_1}{dz}\right)_{z=0} = 0; \quad (15.1.7)$$

where k_1 (s^{-1}) is the interphase mass transfer coefficient in the liquid–gas bubbles zone.

The mathematical model of the ASG chemical absorption in the column on Fig. 15.1 is the set of Eqs. (15.1.3), (15.1.6) and (15.1.7), where $c_G^0 = \bar{c}_1(l_2)$. The quantitative description of the process requires the use of generalized (dimensionless) variables in the model.

15.1.3 Generalized (Dimensionless) Variables Model

The maximal values of the variables will be used as scales in the generalized variables:

$$Z = \frac{z}{l_2}, \quad Z_1 = \frac{z_1}{l_1}, \quad Z_2 = \frac{z_2}{l_1},$$

$$C_G = \frac{\bar{c}_G}{c_1^0}, \quad C_L = \frac{\bar{c}_L\chi}{c_1^0}, \quad C_1 = \frac{\bar{c}_1}{c_1^0}, \quad C_G^0 = \frac{c_G^0}{c_1^0}, \quad C_2 = \frac{\bar{c}_2\chi}{c_1^0}. \quad (15.1.8)$$

If (15.1.8) is put into Eqs. (15.1.3), (15.1.6) and (15.1.7), the model of the ASG chemical absorption in a bizonal column takes the form:

$$\frac{dC_G}{dZ_1} = Pe_G^{-1}\frac{d^2C_G}{dZ_1^2} - K_G(C_G - C_L);$$

$$Z_1 = 0, \quad C_G = C_G^0 = C_1(1), \quad \left(\frac{dC_G}{dZ_1}\right)_{Z_1=0} = 0. \quad (15.1.9)$$

$$\frac{dC_L}{dZ_2} = Pe_L^{-1}\frac{d^2C_L}{dZ_2^2} + K_L(C_G - C_L); \quad Z_2 = 0, \quad C_L = 0, \quad \left(\frac{dC_L}{dZ_2}\right)_{Z_2=0} = 0. \quad (15.1.10)$$

$$\frac{dC_1}{dZ} = Pe^{-1}\frac{d^2C_1}{dZ^2} - K(C_1 - C_2); \quad Z = 0, \quad C_1 = 1, \quad \left(\frac{dC_1}{dZ}\right)_{Z=0} = 0. \quad (15.1.11)$$

$$C_2 = \frac{Q_G[1 - C_G(1)]\chi}{Wk_0\bar{c}_3}. \quad (15.1.12)$$

The following dimensionless parameters are used in (15.1.9)–(15.1.12):

$$Pe_G = \frac{u_G^0 l_1}{D_G}, \quad Pe_L = \frac{u_L^0 l_1}{D_L}, \quad Pe = \frac{\bar{u}_1 l_2}{D_G}, \quad K_G = \frac{kl_1}{u_G^0}, \quad K_L = \frac{kl_1 \chi}{u_L^0}, \quad K = \frac{k_1 l_2}{\bar{u}_1}.$$

$$(15.1.13)$$

15.1.4 Industrial Conditions

Let us consider Fig. 15.1 as an industrial absorption column with a diameter $D = 18.2$ (m), where $l_1 = 6.9$ (m) is the height of the upper zone between the bubbling caps and the sprinkler system (gas–liquid drops system zone) and $l_2 = 1$ (m) is the height of the lower (liquid–gas bubbles system) zone between the distribution plate and the liquid surface (practically height of the bubbling caps). The absorption process is characterized by the average gas velocity $u_G^0 = 4.14$ (m s^{-1}), the average liquid drops velocity $u_L^0 = 3.75$ (m s^{-1}) [time of the drops existence 1.8 (s)], the inlet ASG concentration in the gas phase $c_1^0 = 10^{-4}$ (kg mol m^{-3}), the inlet ASG concentration in the liquid phase $\bar{c}_L = 0$ (kg mol m^{-3}), the liquid (m^3 s^{-1})/gas (m^3 s^{-1}) ratio $L/G = 0.02$, Henry's number $\chi = 1.16$, and the diffusivity of ASG in the gas (air) $D_1 = 10^{-5}$ (m^2 s^{-1}) and the liquid (water) $D_2 = 10^{-9}$ (m^2 s^{-1}) phases. The desired absorption degree will be 94%.

From the presented industrial conditions and (15.1.13) follows

$$Pe_G \sim 10^6, \quad Pe_L \sim 10^{10}, \quad Pe \sim 10^6, \quad \frac{K_G}{K_L} = \frac{u_L^0}{u_G^0 \chi} = 0.695 \qquad (15.1.14)$$

and the model (15.1.9)–(15.1.12) has a convective form:

$$\frac{dC_G}{dZ_1} = -K_G(C_G - C_L); \quad Z_1 = 0, \quad C_G = C_G^0 = C_1(1). \qquad (15.1.15)$$

$$\frac{dC_L}{dZ_2} = 1.439 K_G(C_G - C_L); \quad Z_2 = 0, \quad C_L = 0. \qquad (15.1.16)$$

$$\frac{dC_1}{dZ} = -K(C_1 - C_2); \quad Z = 0, \quad C_1 = 1. \qquad (15.1.17)$$

$$C_2 = \frac{Q_G[1 - C_G(1)]\chi}{W k_0 \bar{c}_3}. \qquad (15.1.18)$$

15.2 Algorithm for Model Equations Solution

15.2.1 Upper Zone Model

The following iterative Algorithm I is applied for solving the model equations for the upper zone, where $s = 0, 1, 2\ldots$ is the iterative steps.

The zero step $(s = 0)$ is solving the problem

$$\frac{dC_G^0}{dZ_1} = -K_G C_G^0; \quad Z_1 = 0, \quad C_G^0 \equiv 1 \tag{15.2.1}$$

and the presentation of C_G^0 as a polynomial $C_G^0 = a_0^0 + a_1^0 Z_1 + a_2^0 Z_1^2$.

The step $s = 1, 2, \ldots$ is solving the problem

$$\frac{dC_L^s}{dZ_2} = 1.439 K_G \left[a_0^{(s-1)} + a_1^{(s-1)}(1 - Z_2) + a_2^{(s-1)}(1 - Z_2)^2 - C_L^s \right]; \tag{15.2.2}$$
$$Z_2 = 0, \quad C_L^s \equiv 0,$$

presentation of C_L^s as a polynomial $C_L^s = b_0^s + b_1^s Z_2 + b_2^s Z_2^2$, solving the problem

$$\frac{dC_G^s}{dZ_1} = -K_G \left[C_G^s - b_0^s - b_1^s(1 - Z_1) - b_2^s(1 - Z_1)^2 \right]; \quad Z_1 = 0, \quad C_G^s = 1 \tag{15.2.3}$$

and presentation of C_G^s as a polynomial $C_G^s = a_0^s + a_1^s Z_1 + a_2^s Z_1^2$.

The stop criterion is

$$\int_0^1 \left(C_G^s - C_G^{(s-1)} \right)^2 dZ_1 \leq 10^{-4}. \tag{15.2.4}$$

According to the above-mentioned industrial conditions, the desired absorption degree has to be 94%, i.e., $C_G(1) = 0.06$. The minimization of the least-squares function

$$F(K_G) = [C_G(1) - 0.06]^2 \tag{15.2.5}$$

obtains the model parameter K_G; its value is 3.86. As a result, the model (15.1.15)–(15.1.18) has the form:

$$\frac{dC_G}{dZ_1} = -3.86(C_G - C_L); \quad Z_1 = 0, \quad C_G = C_1(1). \tag{15.2.6}$$

$$\frac{dC_L}{dZ_2} = 5.32(C_G - C_L); \quad Z_2 = 0, \quad C_L = 0. \tag{15.2.7}$$

$$\frac{dC_1}{dZ} = -K(C_1 - C_2); \quad Z = 0, \quad C_1 = 1. \tag{15.2.8}$$

$$C_2 = \frac{Q_G[1 - C_G(1)]\chi}{Wk_0\bar{c}_3}. \tag{15.2.9}$$

The following iterative Algorithm II is applied for solving the model Eqs. (15.2.6)–(15.2.9), where $C_2^0 = 0.5$ is used as a zero iteration step:

1. Determination of $C_1^s(Z)$ as a solution of the problem

$$\frac{dC_1^s}{dZ} = -K\left(C_1^s - C_2^{(s-1)}\right); \quad Z = 0, \quad C_1^s = 1, \tag{15.2.10}$$

   where $s = 1, 2, \ldots$ is the iteration number.
2. Finding $C_1^s(1)$.
3. Solving the problem (15.2.6), (15.2.7) using the Algorithm I (15.2.1)–(15.2.4).
4. Finding $C_G(1)$.
5. Calculation of C_2^s, using (15.2.9).
6. Checking if $\left(C_2^s - C_2^{(s-1)}\right)^2 \le 10^{-4}$?
   Yes—continuation of 8.
   No—continuation of 7.
7. Calculation of $C_0^s = \frac{C_2^s + C_2^{(s-1)}}{2}$ setting $C_2^s = C_0^s$, $s = s + 1$ and continuation as 1.
8. Stop.

15.2.2 Numerical Results

The presented Algorithms I and II are used for solving the model Eqs. (15.2.6)–(15.2.9), where $K_G = 3.86$, $K = 1$. As a result, the value of $C_2^{s_0}$ is 0.509, where s_0 is the number of the last iteration. The obtained concentration distributions are shown in Figs. 15.2, 15.3, and 15.4.

The values of the gas outlet (inlet) concentration in the lower (upper) zone of column—$C_G^0 = 0.689$, $C_G(1) = 0.0414$—are obtained from the model equations' solution.

Considering the set value of the desired outlet gas concentration in the column $C_G\left(Z_1^{(0)}\right) = 0.06$ with solving the model equations, the interphase mass transfer coefficient in the upper zone of the column and the required height of the upper part of the column are determined as follows: $l_1 = 6.9Z_1^{(0)} = 5.87$ (m), $Z_1^{(0)} = 0.85$.

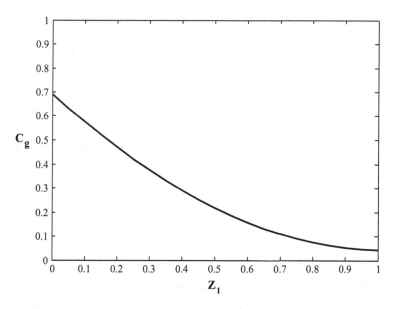

Fig. 15.2 Concentration of ASG in the gas phase (upper zone) ($K_G = 3.86$, $K = 1$, $C_2^{(i_0)} = 0.509$)

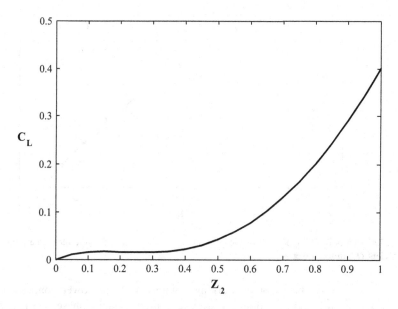

Fig. 15.3 Concentration of ASG in the liquid phase (upper zone) ($K_G = 3.86$, $K = 1$, $C_2^{(i_0)} = 0.509$)

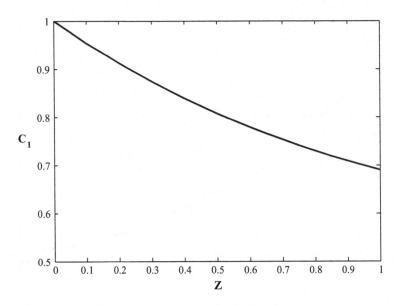

Fig. 15.4 Concentration of ASG in the gas phase (lower zone) ($K_G = 3.86$, $K = 1$, $C_2^{(i_0)} = 0.509$)

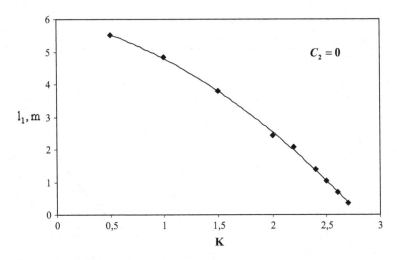

Fig. 15.5 Height of the upper part of the column (l_1) as function of the interphase mass transfer coefficient (K) in the lower zone

The increase of the absorber working volume W in the lower zone leads to decreasing of the concentration of ASG ($\bar{c}_2$) in the liquid phase (15.1.6) and practically $C_2 = 0$. In these conditions, the height of the upper part of the column (l_1) decreases, when the interphase mass transfer coefficient (K) in the lower zone increases. These results are shown in Fig. 15.5.

The results obtained show that the increase of the absorber working volume W of the column leads to decrease of the inlet concentration of ASG in the upper zone. In these conditions, the height of the upper part of the column (l_1) decreases.

From Fig. 15.5, it is seen that the increase of the intensity in the bubble (lower) zone of the column (interphase mass transfer coefficient K) can result in a significant reduction of the height in the upper part of the apparatus and thus the overall height of the apparatus.

The proposed patent [1] introduces a second (liquid–gas bubbles) absorption zone, and as a result, the absorption degree increases or the column height decreases.

References

1. Boyadjiev C, Boyadjiev B, Doichinova M, Popova-Krumova P (2013) Method and apparatus for gas absorption. Patent application No 111168
2. Boyadjiev C, Doichinova M, Popova-Krumova P, Boyadjiev B (2015) On the gas purification from low SO_2 concentration. 1. Absorption processes modeling. Int J Eng Res 4(9):550–557
3. Boyadjiev C, Doichinova M, Popova-Krumova P, Boyadjiev B (2015) On the gas purification from low SO_2 concentration. 2. Two-steps chemical absorption modeling. Int J Eng Res 4 (9):558–563
4. Boyadjiev C, Doichinova M, Popova-Krumova P, Boyadjiev B (2013) Column reactor for chemical processes. Utility model, BG 1776 U1 17.06.2013
5. Boyadjiev C, Doichinova M, Popova-Krumova P, Boyadjiev B (2014) Intensive column apparatus for chemical reactions. OALib J 1(3):1–9

Chapter 16
Absorption–Adsorption Method

Different companies (Babcock & Wilcox Power Generation Group, Inc., Alstom Power Italy, Idreco-Insigma-Consortium) propose methods and apparatuses for waste gas purification from SO_2 using two-phase absorbent ($CaCO_3$ suspension). The adsorption (absorption) of SO_2 on materials derived from natural carbonates [1–3] has the drawback of waste accumulation. The basic problem of the carbonate absorbents is that its chemical reaction with SO_2 lead to CO_2 emission (every molecule of SO_2 absorbed from the air is equivalent to a molecule of CO_2 emitted in the air), because the ecological problems (greenhouse effects) of SO_2 and CO_2 are similar. The large quantity of by-products is a problem, too. Another drawback of these methods is the impossibility for regeneration of the absorbents.

The theoretical analysis [4–10] of the method and apparatus for waste gas purification from SO_2 using two-phase absorbent ($CaCO_3$ suspension) shows that the process in the absorption column in the gas–liquid drops flow is practically physical absorption as a result of the low concentration of the dissolved $CaCO_3$ and SO_2 in the drops and its brief existence in the gas–liquid dispersion. An increase of the process efficiency is proposed in the patent [11], where an absorption column with two absorption zones (lower liquid–gas bubbles zone and upper gas–liquid drops zone) is used. The effect is the possibility to increase the absorption degree or to lower the column height.

The use of synthetic anionites (basic anion-exchange resins—R–OH form of Amberlite, Duolite, Kastel, Varion, Wofatit) as adsorbents [12–14] for gas purification from SO_2 provides possibilities for adsorbent regeneration. The chemical reaction of SO_2 with the synthetic anionites in gas–solid systems can be presented by the stoichiometric equation

$$SO_2 + R-OH = R-HSO_3. \qquad (16.0.1)$$

After saturation of the synthetic anionite particles with sulfur dioxide, the adsorbent regeneration is possible to be carried out with water solution of NH_4OH:

© Springer International Publishing AG, part of Springer Nature 2018
C. Boyadjiev et al., *Modeling of Column Apparatus Processes,*
Heat and Mass Transfer, https://doi.org/10.1007/978-3-319-89966-4_16

$$R-HSO_3 + 2NH_4OH = R-OH + (NH_4)_2SO_3 \text{ (or } NH_4HSO_3). \qquad (16.0.2)$$

16.1 Two-Step Absorption–Adsorption Method

The main disadvantages of the $CaCO_3$ suspension method (CO_2 emission, gypsum accumulation and inability to regenerate the adsorbent) are handled by the patent [15] (see Fig. 16.1), where the waste gas purification is realized in two steps, i.e., physical absorption of SO_2 with water

$$H_2O + SO_2 = HSO_3^- + H^+ \qquad (16.1.1)$$

and chemical adsorption of HSO_3^- from the water solution by synthetic anionite particles:

$$HSO_3^- + H^+ + R-OH = R-HSO_3 + H_2O. \qquad (16.1.2)$$

The adsorbent regeneration is made with NH_4OH solution [see (16.0.2)]. The obtained $(NH_4)_2SO_3$ (NH_4HSO_3) is used (after reaction with HNO_3) for production of concentrated SO_2 (gas) and NH_4NO_3 (solution).

In Fig. 16.1, the gas enters in the middle part of the column 2 through the inlet 1, gets in contact with the absorbent (H_2O) in the upper part of the column, where a gas–liquid drops countercurrent flow is formed, passes through the liquid drops separator 4, and out through the outlet 5. The absorbent enters the column through the inlets 11 and spreader 3, absorbs the SO_2 from the gas in the middle part of the column, collects at the bottom of the column, and goes out through the exit 6. After that, the absorbent ($H_2O + SO_2$) passes through one of the valves 7 and enters in one of the adsorbers 8, where SO_2 is adsorbed on the particles of synthetic anionite and the regenerated absorbent (H_2O) is returned by pumps 9 and valves 10 to the absorption column through the inlets 11.

The absorbent, separated in the separator 4, passes through the outlets 12 and valve 13 and enters the adsorbers 8.

After the synthetic anionite particles have been saturated with sulfur dioxide in the adsorbers 8, the adsorbent regeneration is realized by means of water solution of NH_4OH, which is obtained in the NH_3 absorber 14 and through the pump 15 and the valves 16 enters the adsorbers 8. The water solution of $(NH_4)_2SO_3$ obtained in 8 (after adsorbent regeneration) enters through the pumps 9 and valves 17 in the reactor 18, where it reacts with HNO_3 (coming from the tank 19). The obtained concentrated SO_2 (gas) and NH_4NO_3 solution goes out of the system through the outlet 21. The pump 20 allows the solution of $(NH_4)_2SO_3$ to be re-circulated in order to achieve higher concentration. The absorber 2 is possible to be replaced by a bi-zonal absorber (see Fig. 15.1).

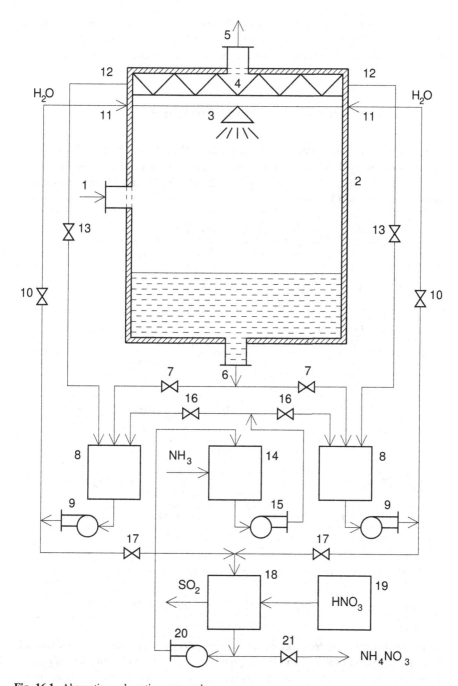

Fig. 16.1 Absorption–adsorption approach

The main processes in the absorption–adsorption method are the physical absorption of SO_2 by H_2O in a countercurrent gas–liquid drops system and a chemical adsorption of SO_2 by synthetic anionite particles in a liquid–solid system.

16.2   Absorption and Adsorption Modeling

The modeling of a non-stationary (as a result of the adsorbent saturation) absorption–adsorption cycle of the method for gas purification from SO_2 uses a combination of the physical absorption model (3.1.2) and (3.1.3) and chemical adsorption model (3.2.18) and (3.2.19):

$$\frac{\partial c_1}{\partial t} + u_1 \frac{\partial c_1}{\partial z_1} = D_1 \left(\frac{\partial^2 c_1}{\partial z_1^2} + \frac{1}{r_1} \frac{\partial c_1}{\partial r_1} + \frac{\partial^2 c_1}{\partial r_1^2} \right) - k_0 (c_1 - \chi c_2);$$

$$\frac{\partial c_2}{\partial t} + u_2 \frac{\partial c_2}{\partial z_2} = D_2 \left(\frac{\partial^2 c_2}{\partial z_2^2} + \frac{1}{r_1} \frac{\partial c_2}{\partial r_1} + \frac{\partial^2 c_2}{\partial r_1^2} \right) + k_0 (c_1 - \chi c_2);$$

$$t = 0, \quad c_1 \equiv c_1^0, \quad c_2 \equiv 0;$$

$$r_1 = 0, \quad \frac{\partial c_1}{\partial r_1} = \frac{\partial c_2}{\partial r_1} \equiv 0; \quad r_1 = r_{10}, \quad \frac{\partial c_1}{\partial r_1} = \frac{\partial c_2}{\partial r_1} \equiv 0;$$

$$z_1 = 0, \, c_1(r_1, 0) \equiv c_1^0, \quad u_1^0 c_1^0 \equiv u_1(r_1)c_1^0 - D_1 \left(\frac{\partial c_1}{\partial z_1} \right)_{z_1=0};$$

$$z_2 = 0, \, c_2(r_1, 0) \equiv \bar{c}_{12}(l_2), \quad u_2^0 \bar{c}_{12}(l_2) \equiv u_2(r_1)\bar{c}_{12}(l_2) - D_2 \left(\frac{\partial c_2}{\partial z_2} \right)_{z_2=0}.$$

$$(16.2.1)$$

$$\frac{\partial c_{12}}{\partial t} + u \frac{\partial c_{12}}{\partial z} = D_{12} \left(\frac{\partial^2 c_{12}}{\partial z^2} + \frac{1}{r} \frac{\partial c_{12}}{\partial r_2} + \frac{\partial^2 c_{12}}{\partial r_2^2} \right) - k_1 (c_{12} - c_{13});$$

$$\frac{dc_{13}}{dt} = k_1 (c_{12} - c_{13}) - k c_{13} c_{23}; \qquad \frac{dc_{23}}{dt} = -k c_{13} c_{23};$$

$$t = 0, \quad c_{12} \equiv 0, \quad c_{13} \equiv 0, \quad c_{23} \equiv c_{23}^0; \qquad\qquad (16.2.2)$$

$$r_2 = 0, \quad \frac{\partial c_{12}}{\partial r_2} \equiv 0; \quad r_2 = r_{20}, \quad \frac{\partial c_{12}}{\partial r_2} \equiv 0;$$

$$z = 0, \quad c_{12}(r_2, 0) \equiv \bar{c}_2(l_1), \quad u^0 \bar{c}_2(l_1) \equiv u(r_2)\bar{c}_2(l_1) - D_2 \left(\frac{\partial c_{12}}{\partial z} \right)_{z=0}.$$

In the absorber model (16.2.1), c_1, c_2, D_1, D_2 are the concentrations and diffusivities of SO_2 in the gas and liquid phases, u_1, u_2—the velocities in the gas and liquid phases, r_1, l_1—the radius and height of the working zone of the column, t—the time. In the adsorber model (16.2.2), c_{12}, D_{12} are the concentration and diffusivity of SO_2 in the liquid phase in the adsorber, c_{13}, c_{23}—the concentrations of

SO_2 and active sides in the adsorbent, u—the velocity in the liquid phases, r_2, l_2— the radius and height of the working zone of the column, t—the time. In the absorption–adsorption cycle, the average outlet concentration of SO_2 in the liquid phase of the absorber is the inlet concentration of SO_2 in the liquid phase of the adsorber $[c_{12}(r_2, 0) \equiv \bar{c}_2(l_1)]$, while the average outlet concentration of SO_2 in the liquid phase of the adsorber is the inlet concentration of SO_2 in the liquid phase of the absorber $[(c_2(r_1, 0) \equiv \bar{c}_{12}(l_2)]$.

16.2.1 Generalized Analysis

The use of dimensionless (generalized) variables [16, 17] allows to make a qualitative analysis of the models (16.2.1) and (16.2.2), whereas characteristic scales are used the average velocity, the inlet and initial concentrations, the characteristic time t_0 (saturation time of the adsorbent), and the column's dimensions (r_1, r_2, l_1, l_2):

$$T = \frac{t}{t_0}, \quad R_1 = \frac{r_1}{r_{10}}, \quad R_2 = \frac{r_2}{r_{20}}, \quad Z = \frac{z}{l_2}, \quad Z_1 = \frac{z_1}{l_1}, \quad Z_2 = \frac{z_2}{l_1},$$

$$U = \frac{u}{u^0}, \quad U_1 = \frac{u_1}{u_1^0}, \quad U_2 = \frac{u_2}{u_2^0}, \quad C_1 = \frac{c_1}{c_1^0}, \quad C_2 = \frac{c_2 \chi}{c_1^0}, \quad (16.2.3)$$

$$C_{12} = \frac{c_{12}\chi}{c_1^0}, \quad C_{13} = \frac{c_{13}\chi}{c_1^0}, \quad C_{23} = \frac{c_{23}}{c_{23}^0}.$$

When (16.2.3) is put into (16.2.1) and (16.2.2), the models in generalized variables take the form:

$$\gamma_1 \frac{\partial C_1}{\partial T} + U_1 \frac{\partial C_1}{\partial Z_1} = Fo_1 \left(\beta_1 \frac{\partial^2 C_1}{\partial Z_1^2} + \frac{1}{R_1} \frac{\partial C_1}{\partial R_1} + \frac{\partial^2 C_1}{\partial R_1^2} \right) - K_1(C_1 - C_2);$$

$$\gamma_2 \frac{\partial C_2}{\partial T} + U_2 \frac{\partial C_2}{\partial Z_2} = Fo_2 \left(\beta_1 \frac{\partial^2 C_2}{\partial Z_2^2} + \frac{1}{R_1} \frac{\partial C_2}{\partial R_1} + \frac{\partial^2 C_2}{\partial R_1^2} \right) + K_2(C_1 - C_2);$$

$$T = 0, \quad C_1 \equiv 1, \quad C_2 \equiv 0;$$

$$R_1 = 0, \quad \frac{\partial C_1}{\partial R_1} = \frac{\partial C_2}{\partial R_1} \equiv 0; \quad R_1 = 1, \quad \frac{\partial C_1}{\partial R_1} = \frac{\partial C_2}{\partial R_1} \equiv 0;$$

$$Z_1 = 0, \quad C_1(R_1, 0) \equiv 1, \quad 1 \equiv U_1(R_1) - Pe_1^{-1} \left(\frac{\partial C_1}{\partial Z_1} \right)_{Z_1=0};$$

$$Z_2 = 0, \quad C_2(R_1, 0) \equiv \overline{C}_{12}(1), \quad 1 \equiv U_2(R_1) - \frac{Pe_2^{-1}}{\overline{C}_{12}(1)} D_2 \left(\frac{\partial C_2}{\partial Z_2} \right)_{Z_2=0}.$$

$$(16.2.4)$$

$$\gamma_0 \frac{\partial C_{12}}{\partial T} + U \frac{\partial C_{12}}{\partial Z} = Fo_0 \left(\beta_0 \frac{\partial^2 C_{12}}{\partial Z^2} + \frac{1}{R_2} \frac{\partial C_{12}}{\partial R_2} + \frac{\partial^2 C_{12}}{\partial R_2^2} \right) - K_0 (C_{12} - C_{13});$$

$$\frac{dC_{13}}{dT} = K_3 (C_{12} - C_{13}) - K_4 C_{13} C_{23} = 0; \qquad \frac{dC_{23}}{dT} = -K_5 C_{13} C_{23};$$

$$T = 0, \quad C_{12} \equiv 0, \quad C_{13} \equiv 0, \quad C_{23} \equiv 1;$$

$$R_2 = 0, \quad \frac{\partial C_{12}}{\partial R_2} \equiv 0; \quad R_2 = 1, \quad \frac{\partial C_{12}}{\partial R_2} \equiv 0;$$

$$Z = 0, \quad C_{12}(R_2, 0) \equiv \overline{C}_2(1), \quad 1 \equiv U(R_2) - \frac{Pe_0^{-1}}{\overline{C}_2(1)} \left(\frac{\partial C_{12}}{\partial Z} \right)_{Z=0}.$$

$$(16.2.5)$$

The following parameters are used in (16.2.4) and (16.2.5):

$$K_0 = \frac{k_1 l_2}{u^0}, \quad K_1 = \frac{k_0 l_1}{u_1^0}, \quad K_2 = \frac{k_0 l_1 \chi}{u_2^0},$$

$$K_3 = k_1 t_0, \quad K_4 = k t_0 c_{23}^0, \quad K_5 = k t_0 \frac{c_1^0}{\chi},$$

$$\gamma_0 = \frac{l_2}{t_0 u^0}, \quad \gamma_1 = \frac{l_1}{t_0 u_1^0}, \quad \gamma_2 = \frac{l_1}{t_0 u_2^0}, \quad \beta_0 = \frac{r_{20}^2}{l_2^2}, \quad \beta_1 = \frac{r_{10}^2}{l_1^2}, \qquad (16.2.6)$$

$$Fo_0 = \frac{D_2 l_2}{u^0 r_{20}^2}, \quad Fo_1 = \frac{D_1 l_1}{u_1^0 r_{10}^2}, \quad Fo_1 = \frac{D_2 l_1}{u_2^0 r_{10}^2},$$

$$\overline{C}_2(1) = 2 \int_0^1 R_1 C_2(R_1, 1) \, dR_1, \quad \overline{C}_{12}(1) = 2 \int_0^1 R_2 C_{12}(R_2, 1) \, dR_2.$$

For lengthy processes $(0 = \gamma_0 \sim \gamma_1 \sim \gamma_2 \leq 10^{-2})$, high columns $(0 = \beta_0 \sim \beta_1 \leq 10^{-2})$, and typical fluid velocities $(0 = Fo_0 \sim Fo_1 \sim Fo_2 \leq 10^{-2})$, the model has the form:

$$U_1 \frac{dC_1}{dZ_1} = -K_1 (C_1 - C_2); \quad Z_1 = 0, \ C_1(R_1, 0) \equiv 1.$$
$$\qquad\qquad\qquad\qquad\qquad\qquad\qquad\qquad\qquad (16.2.7)$$
$$U_2 \frac{dC_2}{dZ_2} = K_2 (C_1 - C_2); \quad Z_2 = 0, \ C_2(R_1, 0) \equiv \overline{C}_{12}(1).$$

$$U \frac{dC_{12}}{dZ} = -K_0 (C_{12} - C_{13}); \quad Z = 0, \quad C_{12}(R_2, 0) \equiv \overline{C}_2(1).$$
$$\frac{dC_{13}}{dT} = K_3 (C_{12} - C_{13}) - K_4 C_{13} C_{23} = 0; \quad \frac{dC_{23}}{dT} = -K_5 C_{13} C_{23}; \qquad (16.2.8)$$
$$T = 0, \quad C_{13} \equiv 0, \quad C_{23} \equiv 1.$$

16.3 Average-Concentration Model

The presented models (16.2.7) and (16.2.8) show that in the practical cases convective type of models has to be used:

$$u_1 \frac{dc_1}{dz_1} = -k_0(c_1 - \chi c_2); \quad z_1 = 0, \; c_1(t, r_1, 0) \equiv c_1^0;$$

$$u_2 \frac{dc_2}{dz_2} = k_0(c_1 - \chi c_2); \quad z_2 = 0, \; c_2(t, r_1, 0) \equiv \bar{c}_{12}(t, l_2). \tag{16.3.1}$$

$$u \frac{dc_{12}}{dz} = -k_1(c_{12} - c_{13}); \quad z = 0, \quad c_{12}(t, r_2, 0) \equiv \bar{c}_2(t, l_1);$$

$$\frac{dc_{13}}{dt} = k_1(c_{12} - c_{13}) - k \, c_{13}c_{23} = 0; \quad \frac{dc_{23}}{dt} = -k \, c_{13}c_{23}; \tag{16.3.2}$$

$$t = 0, \quad c_{13}(0, r_2, z) \equiv 0, \quad c_{23}(0, r_2, z) \equiv c_{23}^0.$$

The average values of the velocities and concentrations in the column's cross-sectional area can be obtained [16, 17] using the expressions:

$$\bar{u} = \frac{2}{r_{20}^2} \int_0^{r_{20}} r_2 u(r_2) dr_2 = u^0, \quad \bar{u}_1 = \frac{2}{r_{10}^2} \int_0^{r_{10}} r_1 u_1(r_1) dr_1 = u_1^0,$$

$$\bar{u}_2 = \frac{2}{r_{10}^2} \int_0^{r_{10}} r_1 u_2(r_1) dr_1 = u_2^0, \quad \bar{c}_1(t, z_1) = \frac{2}{r_{10}^2} \int_0^{r_{10}} r_1 c_1(t, r_1, z_1) dr_1,$$

$$\bar{c}_2(t, z_2) = \frac{2}{r_{10}^2} \int_0^{r_{10}} r_1 c_2(t, r_1, z_2) dr_1, \quad \bar{c}_{12}(t, z) = \frac{2}{r_{20}^2} \int_0^{r_{20}} r_2 c_{12}(t, r_2, z) dr_2,$$

$$c_{13}(t, z) = \frac{2}{r_{20}^2} \int_0^{r_{20}} r_2 c_{13}(t, r_2, z) dr_2, \quad \bar{c}_{23}(t, z) = \frac{2}{r_{20}^2} \int_0^{r_{20}} r_2 c_{23}(t, r_2, z) dr_2. \tag{16.3.3}$$

The velocitie and concentration distributions in (16.3.1) and (16.3.2) can be presented with the help of the average functions (16.3.3):

$$u(r_2) = \bar{u}\tilde{u}(r_2), \quad u_1(r_1) = \bar{u}_1\tilde{u}_1(r_1), \quad u_2(r_1) = \bar{u}_2\tilde{u}_2(r_1),$$

$$c_1(t, r_1, z_1) = \bar{c}_1(t, z_1)\tilde{c}_1(t, r_1, z_1), \quad c_2(t, r_1, z_2) = \bar{c}_2(t, z_2)\tilde{c}_1(t, r_1, z_2),$$

$$c_{12}(t, r_2, z) = \bar{c}_{12}(t, z)\tilde{c}_{12}(t, r_2, z), \quad c_{13}(t, r_2, z) = \bar{c}_{13}(t, z)\tilde{c}_{13}(t, r_2, z), \tag{16.3.4}$$

$$c_{23}(t, r_2, z) = \bar{c}_{23}(t, z)\tilde{c}_{23}(t, r_2, z).$$

Here $\tilde{u}(r_2)$, $\tilde{u}_1(r_1)$, $\tilde{u}_2(r_1)$, $\tilde{c}_1(t, r_1, z_1)$, $\tilde{c}_2(t, r_1, z_2)$, $\tilde{c}_{12}(t, r_2, z)$, $\tilde{c}_{13}(t, r_2, z)$, $\tilde{c}_{23}(t, r_2, z)$ present the radial non-uniformity of the velocity and the concentration distributions, satisfying the conditions:

$$\frac{2}{r_{20}^2} \int_0^{r_{20}} r_2 \tilde{u}(r_2) dr_2 = 1, \quad \frac{2}{r_{10}^2} \int_0^{r_{10}} r_1 \tilde{u}_1(r_1) dr_1 = 1, \quad \frac{2}{r_{10}^2} \int_0^{r_{10}} r_1 \tilde{u}_2(r_1) dr_1 = 1,$$

$$\int_0^{r_{10}} r_1 \tilde{c}_1(t, r_1, z_1) dr_1 = 1, \quad \frac{2}{r_{10}^2} \int_0^{r_{10}} r_1 \tilde{c}_2(t, r_1, z_2) dr_1 = 1,$$

$$\frac{2}{r_{20}^2} \int_0^{r_{20}} r_2 \tilde{c}_{12}(t, r_2, z) dr_2 = 1, \quad \frac{2}{r_{20}^2} \int_0^{r_{20}} r_2 \tilde{c}_{13}(t, r_2, z) dr_2 = 1,$$

$$\frac{2}{r_{20}^2} \int_0^{r_{20}} r_2 \tilde{c}_{23}(t, r_2, z) dr_2 = 1.$$

$$(16.3.5)$$

The use of averaging procedure (II.6–II.10) leads to:

$$\alpha_1 \bar{u}_1 \frac{d\bar{c}_1}{dz_1} + \frac{d\alpha_1}{dz_1} \bar{u}_1 \bar{c}_1 = -k_0(\bar{c}_1 - \chi \bar{c}_2); \quad z_1 = 0, \ \bar{c}_1(t, 0) \equiv c_1^0;$$

$$\alpha_2 \bar{u}_2 \frac{d\bar{c}_2}{dz_2} + \frac{d\alpha_2}{dz_2} \bar{u}_2 \bar{c}_2 = k_0(\bar{c}_1 - \chi \bar{c}_2); \quad z_2 = 0, \ \bar{c}_2(t, 0) \equiv \bar{c}_{12}(t, l_2).$$

$$(16.3.6)$$

$$\alpha \bar{u} \frac{d\bar{c}_{12}}{dz} + \frac{d\alpha}{dz} \bar{u} \bar{c}_{12} = -k_1(\bar{c}_{12} - \bar{c}_{13}); \quad z = 0, \quad \bar{c}_{12}(t, 0) \equiv \bar{c}_2(t, l_1);$$

$$\frac{d\bar{c}_{13}}{dt} = k_1(\bar{c}_{12} - \bar{c}_{13}) - \beta k \bar{c}_{13} \bar{c}_{23} = 0; \quad \frac{d\bar{c}_{23}}{dt} = -\beta k \bar{c}_{13} \bar{c}_{23};$$

$$t = 0, \quad \bar{c}_{13}(0, z) \equiv 0, \quad \bar{c}_{23}(0, z) \equiv c_{23}^0.$$

$$(16.3.7)$$

The following functions are used in (16.3.6) and (16.3.7):

$$\alpha(t, z) = \frac{2}{r_{20}^2} \int_0^{r_{20}} r_2 \tilde{u}(r_2) \tilde{c}_{12}(t, r_2, z) dr_2,$$

$$\alpha_1(t, z_1) = \frac{2}{r_{10}^2} \int_0^{r_{10}} r_1 \tilde{u}_1(r_1) \tilde{c}_1(t, r_1, z_1) dr_1,$$

$$\alpha_2(t, z_2) = \frac{2}{r_{10}^2} \int_0^{r_{10}} r_1 \tilde{u}_2(r_1) \tilde{c}_2(t, r_1, z_2) dr_1,$$

$$\beta(t, z) = \frac{2}{r_{20}^2} \int_0^{r_{20}} r_2 \tilde{c}_{13}(t, r_2, z) \tilde{c}_{23}(t, r_2, z) dr_2. \tag{16.3.8}$$

16.3.1 Generalized Analysis

The use of the dimensionless (generalized) variables

$$T = \frac{t}{t_0}, \quad Z = \frac{z}{l_2}, \quad Z_1 = \frac{z_1}{l_1}, \quad Z_2 = \frac{z_2}{l_1},$$

$$\overline{C}_1 = \frac{\bar{c}_1}{c_1^0}, \quad \overline{C}_2 = \frac{\bar{c}_2 \chi}{c_1^0}, \quad \overline{C}_{12} = \frac{\bar{c}_{12} \chi}{c_1^0}, \quad \overline{C}_{13} = \frac{\bar{c}_{13} \chi}{c_1^0}, \quad \overline{C}_{23} = \frac{\bar{c}_{23}}{c_{23}^0}. \tag{16.3.9}$$

leads to

$$A_1 \frac{d\overline{C}_1}{dZ_1} + \frac{dA_1}{dZ_1} \overline{C}_1 = -K_1(\overline{C}_1 - \overline{C}_2); \quad Z_1 = 0, \overline{C}_1(T, 0) \equiv 1;$$

$$A_2 \frac{d\overline{C}_2}{dZ_2} + \frac{dA_2}{dZ_2} \overline{C}_2 = K_2(\overline{C}_1 - \overline{C}_2); \quad Z_2 = 0, \overline{C}_2(T, 0) \equiv \overline{C}_{12}(T, 1). \tag{16.3.10}$$

$$A \frac{d\overline{C}_{12}}{dZ} + \frac{dA}{dZ} \overline{C}_{12} = -K_0(\overline{C}_{12} - \overline{C}_{13}); \quad Z = 0, \quad \overline{C}_{12}(T, 0) \equiv \overline{C}_2(T, 1);$$

$$\frac{d\overline{C}_{13}}{dT} = K_3(\overline{C}_{12} - \overline{C}_{13}) - BK_4\overline{C}_{13}\overline{C}_{23} = 0; \quad \frac{d\overline{C}_{23}}{dT} = -BK_5\overline{C}_{13}\overline{C}_{23};$$

$$T = 0, \quad \overline{C}_{13}(0, Z) \equiv 0, \quad \overline{C}_{23}(0, Z) \equiv 1. \tag{16.3.11}$$

The following functions are used in (16.3.10) and (16.3.11):

$$A(T, Z) = \alpha(t_0 T, l_2 Z) = \alpha(t, z) = 2 \int_0^1 RU(R_2) \frac{C_{12}(T, R_2, Z)}{\overline{C}_{12}(T, Z)} dR_2$$

$$A_1(T, Z_1) = \alpha_1(t_0 T, l_1 Z_1) = \alpha_1(t, z_1) = 2 \int_0^1 R_1 U_1(R_1) \frac{C_1(T, R_1, Z_1)}{\overline{C}_1(T, Z_1)} dR_1$$

$$A_2(T, Z_2) = \alpha_2(t_0 T, l_1 Z_2) = \alpha_2(t, z_2) = 2 \int_0^1 R_1 U_2(R_1) \frac{C_2(T, R_1, Z_2)}{\overline{C}_2(T, Z_2)} dR_1$$

$$B(T, Z) = \beta(t_0 T, l_2 Z) = \beta(t, z) = 2 \int_0^1 R_2 \frac{C_{13}(T, R_2, Z)}{\overline{C}_{13}(T, Z)} \frac{C_{23}(T, R_2, Z)}{\overline{C}_{23}(T, Z)} dR_2,$$

$$\overline{C}_1(T, Z_1) = 2 \int_0^1 R_1 C_1(T, R_1, Z_1) dR_1, \quad \overline{C}_2(T, Z_2) = 2 \int_0^1 R_1 C_2(T, R_1, Z_2) dR_1,$$

$$\overline{C}_{13}(T, Z) = 2 \int_0^1 R_2 C_{13}(T, R_2, Z) dR_2, \quad \overline{C}_{23}(T, Z) = 2 \int_0^1 R_2 C_{23}(T, R_2, Z) dR_2.$$

$$(16.3.12)$$

In Chap. 5, it was shown that $B(T, Z) \equiv 1$ and $A(T, Z)$, $A_1(T, Z_1)$, $A_2(T, Z_2)$ can be presented as linear approximations:

$$A = 1 + a_z Z + a_t T, \quad A_1 = 1 + a_z^1 Z_1 + a_t^1 T, \quad A_2 = 1 + a_z^2 Z_2 + a_t^2 T. \quad (16.3.13)$$

As a result, the model of the absorption–desorption process has the form:

$$\left(1 + a_z^1 Z_1 + a_t^1 T\right) \frac{d\overline{C}_1}{dZ_1} + a_z^1 \overline{C}_1 = -K_1\left(\overline{C}_1 - \overline{C}_2\right);$$

$$Z_1 = 0, \quad \overline{C}_1(T, 0) \equiv 1;$$

$$\left(1 + a_z^2 Z_2 + a_t^2 T\right) \frac{d\overline{C}_2}{dZ_2} + a_z^2 \overline{C}_2 = K_2\left(\overline{C}_1 - \overline{C}_2\right);$$

$$Z_2 = 0, \quad \overline{C}_2(T, 0) \equiv \overline{C}_{12}(T, 1).$$

$$(16.3.14)$$

$$\left(1 + a_z Z + a_t T\right) \frac{d\overline{C}_{12}}{dZ} + a_z \overline{C}_{12} = -K_0\left(\overline{C}_{12} - \overline{C}_{13}\right);$$

$$Z = 0, \quad \overline{C}_{12}(T, 0) \equiv \overline{C}_2(T, 1);$$

$$\frac{d\overline{C}_{13}}{dT} = K_3\left(\overline{C}_{12} - \overline{C}_{13}\right) - K_4 \overline{C}_{13} \overline{C}_{23} = 0;$$

$$\frac{d\overline{C}_{23}}{dT} = -K_5 \overline{C}_{13} \overline{C}_{23};$$

$$(16.3.15)$$

$$T = 0, \quad \overline{C}_{13}(0, Z) \equiv 0, \quad \overline{C}_{23}(0, Z) \equiv 1.$$

16.3.2 Algorithm of the Solution

The solution of (16.3.14) and (16.3.15) can be obtained as five matrix forms:

$$\overline{C}_1(T,Z) = \|C_{(1)\tau\zeta}\|, \quad \overline{C}_2(T,Z) = \|C_{(2)\tau\zeta}\|, \quad \overline{C}_{12}(T,Z) = \|C_{(12)\tau\zeta}\|,$$
$$\overline{C}_{13}(T,Z) = \|C_{(13)\tau\zeta}\|, \quad \overline{C}_{23}(T,Z) = \|C_{(23)\tau\zeta}\|;$$
$$0 \leq T \leq 1, \quad T = \frac{\tau - 1}{\tau^0 - 1}, \quad \tau = 1, 2, \ldots, \tau^0;$$
$$0 \leq Z \leq 1, \quad Z = \frac{\zeta - 1}{\zeta^0 - 1}, \quad \zeta = 1, 2, \ldots, \zeta^0; \quad \tau^0 = \zeta^0.$$

$$(16.3.16)$$

A multi-step approach is possible to be used. At each step, the problems (16.3.14) and (16.3.15) have to be solved consecutively, where T is a parameter in (16.3.14), Z is a parameter in (16.3.15), $\overline{C}_2^{(s)}(T,0) \equiv \overline{C}_{12}^{(s-1)}(T,1)$, $\overline{C}_{12}^{(0)}(T,1) \equiv 0$, where the superscript values $(s = 0, 1, 2, \ldots)$ are the step numbers.

16.3.3 Parameter Identification

The availability of experimental data for the SO_2 concentrations in the gas and liquid phases at the absorber and adsorber outlets $\overline{C}_1^{exp}(T_n, 1)$, $\overline{C}_{12}^{exp}(T_n, 1)$, $T_n = 0.05n$, $n = 1, 2, \ldots, 20$ permits to use the next algorithm for the parameter identification in the model (16.3.14) and (16.3.15):

1. Put $a_z = a_t = a_z^1 = a_t^1 = a_z^2 = a_t^2 = 0$ in (16.3.14) and (16.3.15) and minimize the least squares functions:

$$F_1(K_1, K_2) = \sum_{n=1}^{20} \left[\overline{C}_1(T_n, 1) - \overline{C}_1^{exp}(T_n, 1)\right]^2,$$
$$(16.3.17)$$
$$F_2(K_0, K_3, K_4, K_5) = \sum_{n=1}^{20} \left[\overline{C}_{12}(T_n, 1) - \overline{C}_{12}^{exp}(T_n, 1)\right]^2,$$

where $\overline{C}_1(T_n, 1)$, $\overline{C}_{12}(T_n, 1)$ are obtained as a solution of (16.3.14) and (16.3.15) for $T_n = 0.05n$, $n = 1, 2, \ldots, 20$.
2. Enter the obtained parameter values $(K_p, p = 0, 1, \ldots, 5)$ in (16.3.14) and (16.3.15) and minimize the least squares functions:

$$F_3\left(a_z^1, a_t^1, a_z^2, a_t^2\right) = \sum_{n=1}^{20} \left[\overline{C}_1(T_n, 1) - \overline{C}_1^{\exp}(T_n, 1)\right]^2,$$

$$\hspace{9cm} (16.3.18)$$

$$F_4(a_z, a_t) = \sum_{n=1}^{20} \left[\overline{C}_{12}(T_n, 1) - \overline{C}_{12}^{\exp}(T_n, 1)\right]^2,$$

where $\overline{C}_1(T_n, 1)$, $\overline{C}_{12}(T_n, 1)$ are obtained as a solution of (16.3.14) and (16.3.15) for $T_n = 0.05n$, $n = 1, 2, \ldots, 20$.

3. Enter the calculated values of the parameters a_z, a_t, a_z^1, a_t^1, a_z^2, a_t^2 in (16.3.14) and (16.3.15) and minimize the least squares functions (16.3.17), etc.

The proposed patent [15] makes it possible to create a waste-free technology for waste gas purification from sulfur dioxide by means of regenerable absorbent and adsorbent. The proposed method [18] permits to use the absorption columns, where the $CaCO_3$ suspensions are used.

16.4　An Absorption–Adsorption Apparatus

In the proposed absorption–adsorption method for waste gas purification from sulfur dioxide, the absorption is realized in the countercurrent absorber (where practical gas velocity does not exceed 5 m s^{-1}) and the adsorption is carried out in a fixed bed adsorber. The efficiency of the process can be increased if the absorption is realized in co-current flows, and the adsorption takes place in the flexible adsorbent. For this, it can use a new absorption–adsorption apparatus with bubbling plates. The bubbling of the gas at a plate through a layer of aqueous suspension of synthetic anionite allows an increasing of the gas velocity (reduction of the diameter of the absorption column), elimination of the adsorption column, and carrying out the adsorption in a flexible adsorbent. The movement of the gas between the plates leads to mixing in the gas phase, which increases the absorption rate because the absorption of sulfur dioxide in water is limited by mass transfer in the gas phase.

The new absorption–adsorption apparatus with bubble plates [19, 20] is shown in Fig. 16.2. The gas enters tangentially into the column 1 through the inlet 2, passes through the distribution pipes 5, concentric bubble caps 6 of the plates 4, and exits the column through the outlet 3. The aqueous suspension of synthetic anionite enters the column 1 via the valve 11 and the pipes 7 and creates of plates 4 layer with a certain thickness. After saturation of the adsorbent with sulfur dioxide, the aqueous suspension is output from the column through the pipes 8 and valves 11 and enters in the system 9 for the regeneration of the adsorbent. The suspension of the regenerated adsorbent is removed from 9 by pump 10 is returned to the plates 4 in the column 1.

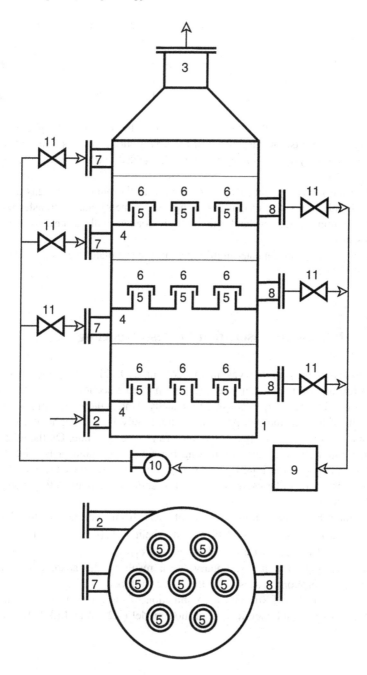

Fig. 16.2 Absorption–adsorption plates column

The operation of the absorption–adsorption apparatus is the following cyclic scheme:

1. Supply all the plates with the necessary amount (volume) of aqueous suspension of synthetic anionite.
2. Start the absorption–adsorption process and monitor the SO_2 concentration of the exit 3.
3. When the increasing the SO_2 concentration at the outlet of the gas 3 exceeds permissible limits, the suspension of the first (bottom) plate is transferred in the regeneration system 9, and the plate is loaded with a new (regenerated) suspension.
4. When the increasing the SO_2 concentration at the outlet of the gas 3 exceeds permissible limits, the suspension of the second (next) plate is transferred in the regeneration system 9, and the plate is loaded with a new (regenerated) suspension.
5. The procedures are repeated until reaching the top plate, then starts again from the first plate.

16.5 Absorption–Adsorption Process Modeling

Tangentially, supplying the gas in the column [19, 21, 22] reduces the radial non-uniformity of the velocity at the column cross-sectional area and the gas velocity is a constant ($u_1 =$ const.), practically. Under these conditions, the concentration of SO_2 in the gas phase is changed only in the height of the column (($c_{11} = c_{11}(t, x)$), which increases the rate of mass transfer rate. On the other hand, the gas bubbling creates an ideal mixing regime in the liquid phase and the concentration of SO_2 in the liquid phase is $c_{12} = c_{12}(t)$. The concentration of SO_2 in the solid phase (capillaries) is c_{13}, and the concentration of active sites in the adsorbent is c_{23}.

The interphase mass transfer rate of SO_2 from the gas to the liquid is $k_0(c_{11} - \chi c_{12})$, while the liquid to the adsorbent is $k_1(c_{12} - c_{13})$. The chemical reaction rate of the SO_2 with the adsorbent is $k c_{13} c_{23}$.

The modeling of a non-stationary (as a result of the adsorbent saturation) absorption–adsorption process on the plate number $n(n = 1, \ldots, N)$, for gas purification from SO_2 [20], uses a combination of the physical absorption model (3.1.2) and (3.1.3) and chemical adsorption model (3.2.18) and (3.2.19), i.e.,

$$\frac{\partial c_{11}^{(n)}}{\partial t} + u_1 \frac{\partial c_{11}^{(n)}}{\partial z} = D_1 \frac{\partial^2 c_{11}^{(n)}}{\partial z^2} - k_0 \left(c_{11}^{(n)} - \chi c_{12}^{(n)} \right);$$

$$\frac{dc_{13}^{(n)}}{dt} = k_1 \left(c_{12}^{(n)} - c_{13}^{(n)} \right) - k c_{13}^{(n)} c_{23}^{(n)}; \quad \frac{dc_{23}^{(n)}}{dt} = -k c_{13}^{(n)} c_{23}^{(n)};$$

$$t = 0, \quad c_{11}^{(n)} \equiv c_{11}^{0}, \quad c_{13}^{(n)} \equiv 0, \quad c_{23}^{(n)} \equiv c_{23}^{0}; \tag{16.5.1}$$

$$z = 0, \, c_{11}^{(n)}(0) \equiv c_{11}^{(n-1)}(l), \quad 0 \equiv \left(\frac{\partial c_{11}^{(n)}}{\partial z} \right)_{z=0}, \quad c_{11}^{(1)}(0) \equiv c_{11}^{0};$$

$$n = 1, \ldots, N.$$

In (16.5.1), $z = 0$ is the entrance to the gas input of the plate number $n(n = 1, \ldots, N)$, l is the distance between the plates, D_1 is the diffusivity of SO_2 in the gas phase.

The concentration of SO_2 in the water of each plate is determined by the amount of absorbed SO_2 (W_1) and its distribution between liquid phase (W_2), the solid phase (capillaries) (W_3^1) and adsorbed on the surface of the capillaries is (W_3^2):

$$W_1 = V_1 \left[c_{11}^{(n)}(0) - c_{11}^{(n)}(l) \right], \quad W_2 = V_2 c_{12}^{(n)},$$
$$W_3^1 = V_3 c_{13}^{(n)}, \quad W_3^2 = V_3 \left(c_{23}^{0} - c_{23}^{(n)} \right), \tag{16.5.2}$$

i.e.,

$$c_{12}^{(n)} = \frac{V_1 \left[c_{11}^{(n)}(0) - c_{11}^{(n)}(l) \right] - V_3 c_{13}^{(n)} - V_3 \left(c_{23}^{0} - c_{23}^{(n)} \right)}{V_2}, \tag{16.5.3}$$

16.5.1 Generalized Analysis

The use of dimensionless (generalized) variables [16, 17] allows to make a qualitative analysis of the models (16.5.1) and (16.5.3), whereas characteristic scales are used the inlet and initial concentrations, the characteristic time t_0 (saturation time of the adsorbent) and the distance between the plates l:

$$T = \frac{t}{t_0}, \quad Z = \frac{z}{l}, \quad C_{11} = \frac{c_{11}}{c_{11}^0},$$
$$C_{12} = \frac{c_{12}\chi}{c_{11}^0}, \quad C_{13} = \frac{c_{13}\chi}{c_{11}^0}, \quad C_{23} = \frac{c_{23}}{c_{23}^0}. \tag{16.5.4}$$

When (16.5.4) is put into (16.5.1) and (16.5.3), the model in generalized variables takes the form:

$$\gamma \frac{\partial C_{11}^{(n)}}{\partial T} + \frac{\partial C_{11}^{(n)}}{\partial Z} = Pe^{-1} \frac{d^2 C_{11}^{(n)}}{dZ^2} - K_0 \left(C_{11}^{(n)} - C_{12}^{(n)} \right);$$

$$\frac{dC_{13}^{(n)}}{dT} = K_1 \left(C_{12}^{(n)} - C_{13}^{(n)} \right) - KC_{13}^{(n)} C_{23}^{(n)}; \quad \frac{dC_{23}^{(n)}}{dT} = -K\alpha^{-1} C_{13}^{(n)} C_{23}^{(n)};$$

$$T = 0, \quad C_{11}^{(n)} \equiv 1, \quad C_{13}^{(n)} \equiv 0, \quad C_{23}^{(n)} \equiv 1. \tag{16.5.5}$$

$$Z = 0, \quad C_{11}^{(n)}(0) \equiv C_{11}^{(n-1)}(1), \quad 0 \equiv \left(\frac{dC_{11}^{(n)}}{dZ} \right)_{Z=0}, \quad C_{11}^{(1)}(0) \equiv 1;$$

$$n = 1, \dots, N.$$

$$C_{12}^{(n)} = \frac{\chi V_1 \left[C_{11}^{(n)}(0) - C_{11}^{(n)}(l) \right] - V_3 C_{13}^{(n)} - \alpha V_3 \left(1 - C_{23}^{(n)} \right)}{V_2}, \quad n = 1, \dots, N. \tag{16.5.6}$$

The following parameters are used in (16.5.5) and (16.5.6):

$$Pe = \frac{u_1 l}{D_1}, \quad K = k t_0 c_{23}^0, \quad K_0 = \frac{k_0 l}{u_1}, \quad K_1 = k_1 t_0,$$

$$\alpha = \frac{\chi c_{23}^0}{c_{11}^0}, \quad \gamma = \frac{l}{t_0 u_1}. \tag{16.5.7}$$

Practically, $0 = \gamma < 10^{-2}$, $0 = Pe^{-1} < 10^{-2}$ and as a result from (16.5.5) follows:

$$\frac{dC_{11}^{(n)}}{dZ} = -K_0 \left(C_{11}^{(n)} - C_{12}^{(n)} \right); \tag{16.5.8}$$

$$Z = 0, \quad C_{11}^{(n)}(0) \equiv C_{11}^{(n-1)}(1), \quad C_{11}^{(1)}(0) \equiv 1; \quad n = 1, \dots, N.$$

$$\frac{dC_{13}^{(n)}}{dT} = K_1 \left(C_{12}^{(n)} - C_{13}^{(n)} \right) - KC_{13}^{(n)} C_{23}^{(n)}; \quad \frac{dC_{23}^{(n)}}{dT} = -K\alpha C_{13}^{(n)} C_{23}^{(n)}; \tag{16.5.9}$$

$$T = 0, \quad C_{13}^{(n)} \equiv 0, \quad C_{23}^{(n)} \equiv 1.$$

The solution of the equations of the model (16.5.6), (16.5.8), and (16.5.9) uses a two-stage algorithm. In the first stage must be solved the equations for $n = 1$. In the second stage must be applied consistently for every plate the algorithm for $n = 1$.

16.5.2 Algorithm of the Solution

1. Put $n = 1$.
2. Put $C_{12}^{(1)} = X_i = 0.1i$, $i = 1, \ldots, 10$.
3. The solution of (16.5.8) leads to $C_{11}^{(1i)} = C_{11}^{(1)}(Z, X_i)$, $C_{11}^{(1)}(0, X_i)$, $C_{11}^{(1)}(1, X_i)$, $i = 1, \ldots, 10$.
4. The solution of (16.5.9) leads to $C_{13}^{(1i)} = C_{13}^{(1)}(T, X_i)$, $C_{23}^{(1i)} = C_{23}^{(1)}(T, X_i)$, $i = 1, \ldots, 10$.
5. The solutions in 3 and 4 must be introduced in (16.5.6) and as a result is obtained $C_{12}^{(1i)} = C_{12}^{(1)}(T, X_i)$, $i = 1, \ldots, 10$.
6. Put $T = T_i = 0.1j$, $j = 1, \ldots, 10$ in $C_{12}^{(1i)} = C_{12}^{(1)}(T, X_i)$, $i = 1, \ldots, 10$ an as a result is obtained $\overline{C}_{12}^{(1i)} = C_{12}^{(1)}(T_j, X_i)$, $i = 1, \ldots, 10$, $j = 1, .., 10$.
7. A polynomial approximation $P_{12}^{(1j)}(T_j, X)$, $j = 1, .., 10$ of $\overline{C}_{12}^{(1i)} = C_{12}^{(1)}(T_j, X_i)$, $i = 1, \ldots, 10$, $j = 1, .., 10$ with respect to X must be obtained.
8. The solutions of the equations $X = P_{12}^{(1j)}(T_j, X)$, $j = 1, .., 10$ with respect to X permits to be obtained X_j, $j = 1, .., 10$ and the obtained solutions must be denoted as $X_j = C_{12}^{(1)}(T_j)$, $j = 1, \ldots, 10$.
9. A polynomial approximation of $C_{12}^{(1)}(T_j)$, $j = 1, .., 10$ with respect to T permit to be obtained $C_{12}^{(1)}(T) = C_{12}^{(1)}(T_j)$, $j = 1, \ldots, 10$.
10. The introducing of $C_{12}^{(1)}(T)$ in (16.5.8) and (16.5.9) permits to be obtained its solutions for $n = 1$.
11. The obtained solution of (16.5.8) $C_{11}^{(1)} = C_{11}^{(1)}(T, Z)$ permits to be obtained $C_{11}^{(1)} = C_{11}^{(1)}(T, 1)$ and as a result to be used the algorithm 1–11 for consistent solutions of the equations set (16.5.6) and (16.5.8) for $n = 2, \ldots, N$.

16.5.3 Parameter Identification

The parameters in the model (16.5.6), (16.5.8), and (16.5.9), which are subjected to experimental determination are K, K_0, K_1. They may be obtained from experimental data of the SO_2 concentration at the gas outlet from the first plate $C_{11\,exp}^{(1j)} = C_{11}^{(1)}(T_j, 1)$, $T_i = 0.1j$, $j = 1, \ldots, 10$, where $T = 1$ is the time for the full saturation of the adsorbent on the first plate. For this purpose must be minimized the function of the least squares with respect to K, K_0, K_1:

$$F(K, K_0, K_1) = \sum_{j=1}^{10} \left[C_{11}^{(1)}(T_j, 1) - \overline{C}_{11\,exp}^{(1j)} \right]^2. \tag{16.5.10}$$

The proposed utility model [19] uses an absorption–adsorption column and makes it possible to create a waste-free technology for waste gas purification from sulfur dioxide by means of regenerable adsorbent, where the system for the regeneration of the adsorbent is similar to the regeneration system in the patent [15]. The efficiency of the processes is increased in an absorption–adsorption apparatus, where the absorption is realized in co-current flows and the adsorption takes place in the flexible adsorbent. For this is proposed a new absorption–adsorption column apparatus with bubbling plates. A mathematical model of the absorption–adsorption process is presented, too.

References

1. Bruce KR (1989) Comparative SO_2 reactivity of CaO derived from $CaCO_3$ and $Ca(OH)_2$. AIChE J 35(1):37
2. Fahlenkamp H (1985) Ewicklungstendenzen der Rauchgasentschwerfelungstechnik auf Kalksteinbasis. Vortagsveroff Haus TechEssen no. 490:9
3. Jozewicz W, Kirchgessner DA (1989) Activation and reactivity of novel calcium-based sorbents for dry SO_2—control in boilers. Power Technol 58:221
4. Boyadjiev C (2011) Mechanism of gas absorption with two-phase absorbents. Int J Heat Mass Transfer 54:3004–3008
5. Boyadjiev C (2011) On the SO_2 problem in power engineering. In: Proceedings, energy forum, Bulgaria, p 114–125
6. Boyadjiev C (2012) On the SO_2 problem in power engineering. In: Proceedings, Asia-Pacific power and energy engineering conference (APPEEC 2012), China, vol 1
7. Boyadjiev C (2012) On the SO_2 problem of solid fuel combustion. In: Proceedings, VIII All-Russian conference with international participation "solid fuel combustion", Novosibirsk
8. Boyadjiev C, Doichinova M, Popova P (2011) On the SO_2 problem in power engineering. 1. Gas absorption. In: Proceedings, 15th workshop on transport phenomena in two-phase flow, 94–103, Sunny Beach Resort, Bulgaria, pp 17–22
9. Boyadjiev C, Popova P, Doichinova M (2011) On the SO_2 problem in power engineering. 2. Two-phase absorbents. Proceedings, 15th workshop on transport phenomena in two-phase flow, Bulgaria, pp 104–115
10. Boyadjiev C, Doichinova M, Popova P (2012) On the SO_2 problem in power engineering. Trans Academenergo 1:44–65
11. Boyadjiev C, Boyadjiev B, Doichinova M, Popova-Krumova P (2013) Method and apparatus for gas absorption. Patent application No 111168
12. Pantofchieva L, Boyadjiev C (1995) Adsorption of sulphur dioxide by synthetic anion exchangers. Bulg Chem Comm 28:780
13. Boyadjiev C, Pantofchieva L, Hristov J (2000) Sulphur dioxide adsorption in a fixed bed of a synthetic anionite. Theor Found Chem Eng 34(2):141
14. Hristov J, Boyadjiev C, Pantofchieva L (2000) Sulphur dioxide adsorption in a magnetically stabilized bed of a synthetic anionite. Theor Found Chem Eng 34(5):489
15. Boyadjiev C, Boyadjiev B, Doichinova M, Popova-Krumova P (2014) Method and apparatus for gas cleaning from sulfur dioxide. Patent application No 111398
16. Boyadjiev C (2010) Theoretical chemical engineering. modeling and simulation. Springer, Berlin
17. Doichinova M, Boyadjiev C (2012) On the column apparatuses modeling. Int J Heat Mass Transfer 55:6705–6715

18. Boyadjiev C, Boyadjiev B (2017) An absorption-adsorption method for gases purification from SO_2 in power plants. Recent Innov in Chem Eng (in press)
19. Boyadjiev C, Boyadjiev B (2015) Absorption-adsorption apparatus for gas cleaning from sulfur dioxide. Utility model BG 2196:U1
20. Boyadjiev C, Boyadjiev B (2017) An absorption-adsorption apparatus for gases purification from SO_2 in power plants. Open Access Libr J 4(5):e3546
21. Boyadjiev C, Boyadjiev B (2013) Column reactor for chemical processes. Utility model BG 1776:U1
22. Boyadjiev C, Doichinova M, Popova-Krumova P, Boyadjiev B (2014) Intensive column apparatus for chemical reactions. Open Access Libr J 1(3):1–9

Chapter 17
Co-current Apparatus

Countercurrent absorbers are used for purification of the large amounts of waste gases emitted from the combustion plants. The gas velocity (and as a result the absorbers diameter, too) is limited by the rate of the absorbent drops fall in an immobile gas medium, i.e., the gas velocity must be less than the drops velocity ($\sim$ 4 (m s^{-1}) practically).

Many companies (Babcock & Wilcox Power Generation Group, Inc., Alstom Power Italy, Idreco-Insigma-Consortium) offer facilities for purification of waste gases from SO$_2$ in thermal power plants, which use countercurrent absorbers, where the gas velocity is 3.98 (m s^{-1}) (Alstom Power Italy) or 4.14 (m s^{-1}) (Idreco-Insigma-Consortium).

The use of co-current absorption apparatus of Venturi type [1, 2] for gas purification from sulfur dioxide is associated with very large hydraulic losses.

A new approach [3–5] for the column apparatuses modeling is applied for the SO$_2$ absorption process. On its base new patents have been proposed [6, 7] for purification of the emitted waste gases from SO$_2$, applying a countercurrent absorption column with two absorption zones—lower liquid–gas bubbles zone and an upper gas–liquid drops zone [6], which could be also combined with an adsorption column filled with a synthetic anionite [7].

Wherever the gas velocity is not limited, there is a possibility to decrease the column diameter of the industrial co-current absorbers. In a new patent [8], a co-current apparatus for absorption of average and highly soluble gases is proposed, which can also be used for removing SO$_2$ from the waste gases.

17.1 Co-current Absorber

A new patent is proposed [8], where the gas velocity can be increased 5–6 times (15–25 (m s^{-1})), and the diameter of the column to fall more than twice.

© Springer International Publishing AG, part of Springer Nature 2018
C. Boyadjiev et al., *Modeling of Column Apparatus Processes*,
Heat and Mass Transfer, https://doi.org/10.1007/978-3-319-89966-4_17

A co-current absorption apparatus for average soluble gas (ASG), according the patent [8], is presented in Figs. 17.1, 17.2, and 17.3. The apparatus represents a cylindrical absorption column 1. The gas enters axially into the upper end of the

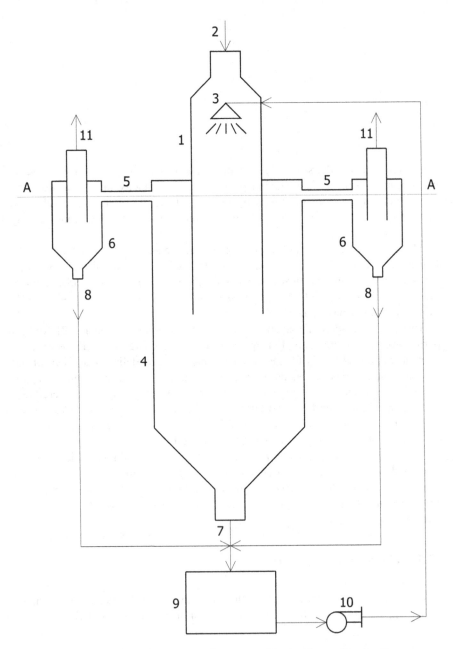

Fig. 17.1 Co-current absorption apparatus from gas purification from sulfur dioxide

Fig. 17.2 Co-current
absorption apparatus from gas
purification from sulfur
dioxide. The pipes 5 are
arranged radially

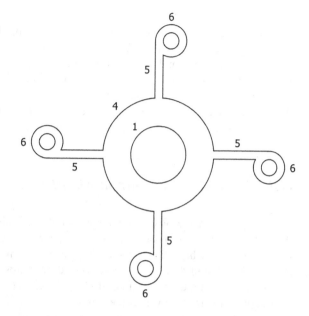

Fig. 17.3 Co-current
absorption apparatus from gas
purification from sulfur
dioxide. The pipes 5 are
arranged tangentially

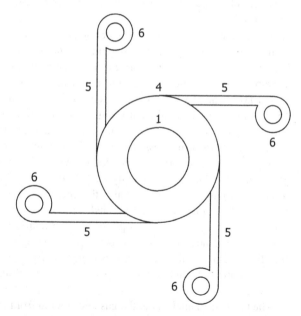

column 2 and passes through a system 3 of one or several rows of sprinklers of the
absorbent (not shown). In the lower part of the absorption, column 1 is covered by
the main droplet precipitator 4, which via hole 12 and the tube 7 is connected to the
apparatus 9 for the chemical treatment or regeneration of the absorbent. The upper

part of the main droplet precipitator 4 is connected by a pipes 5 with four (several) additional droplet precipitators 6, which (in this example) are of a cyclone type. The pipes 5 are arranged radially (Fig. 17.2) or tangentially (Fig. 17.3) relative to the main droplet precipitator 4. Any additional droplet precipitator 6 has outlets for the gas 11 and the absorbent 8. The main droplet precipitator 4 and the additional droplet precipitators 6 are connected via pipes 7 and 8 with the apparatus 9 for the chemical treatment or regeneration of the absorbent. The apparatus 9 is connected with the system sprinklers 3 through the pump 10.

17.1.1   Use of the Co-current Absorber

The gas enters in the absorption column 1 through the inlet 2 with a velocity $W = 15 - 25$ (m s^{-1}), passes through the sprinklers system 3 for the absorbent, which provides a high liquid/gas ratio (about 0.02 (m^3) liquid/1 (m^3) gas). The resulting gas–liquid drops dispersion flows downwards in the absorption column, enters into the main drops separator 4 (where the main separation of the gas-liquid dispersion is realized), reverses the direction of its movement, passes through the tubes 5, enters the additional drops separators 6, and leaves the apparatus through the tubes 11.

The diameter of the pipes 5 is defined by $d = Dn^{-0.5}$ (m), where D (m) is the diameter of the absorption column and n is the number of the tubes 5. The working height of the absorber may vary in the range 5–10 m depending on the sprinklers system 3 and the ratio between the liquid and the gas flow rate, which is usually about 0.02. The normal gas velocities in the co-current column are in the range 15–25 m s^{-1} that is significantly higher compared with the velocity in a counter-current column (~ 4 (m s^{-1})), which allows the diameter of the co-current column to be at least two times less than that of the countercurrent column.

The co-current absorber is intended for absorption of medium and highly soluble gases, because the big gas velocity leads to decrease of the interphase mass transfer resistance in the gas phase, while the increase of the liquid/gas ratio leads to decrease of the interphase mass transfer resistance in the liquid phase (see 2.1.7). The co-current gas absorber is possible to be realized at the end of vertical waste gas pipeline if additionally absorbent sprinklers system is mounted to the tube 1 (see Fig. 17.1), and the equipment is equipped with gas velocity inverter serving as a main drops separator 4 and a system (battery) of subsidiary drops separators 6 (cyclones).

The theoretical analysis of the gas purification from low SO_2 concentration in the cases of two-phase absorbent method [9, 10] and absorption–adsorption method [11] shows that the main process is practically countercurrent physical absorption of SO_2 by H_2O [9]. A co-current model [12] will be presented here below.

17.2   Convection–Diffusion Type of Model

The mathematical model of the physical absorption of ASG in a co-current absorption column will be created in the approximations of the mechanics of continua, where the base of the model will be the convection–diffusion equation [3–5].

Let ε_1 and ε_2 are the gas and liquid parts in the column volume $(\varepsilon_1 + \varepsilon_2 = 1)$, i.e., the gas and liquid holdup coefficients. If c_{11} (c_{12}) is the concentration (kg mol m^{-3}) of the ASG in the gas (liquid) phase, the mass sink (source) in the medium elementary volume is equal to the rate of the interphase mass transfer $k(c_{11} - \chi c_{12})$. As a result, the convection–diffusion equations in a column apparatus have the form:

$$
\begin{aligned}
u_1 \frac{\partial c_{11}}{\partial z} &= D_{11}\left(\frac{\partial^2 c_{11}}{\partial z^2} + \frac{1}{r}\frac{\partial c_{11}}{\partial r} + \frac{\partial^2 c_{11}}{\partial r^2}\right) - k(c_{11} - \chi c_{12}), \\
u_2 \frac{\partial c_{12}}{\partial z} &= D_{12}\left(\frac{\partial^2 c_{12}}{\partial z^2} + \frac{1}{r}\frac{\partial c_{12}}{\partial r} + \frac{\partial^2 c_{12}}{\partial r^2}\right) + k(c_{11} - \chi c_{12}),
\end{aligned}
\tag{17.2.1}
$$

where $u_1(r), u_2(r)$ are the velocity distributions in the gas and liquid (symmetric with respect to the longitudinal coordinate z), $c_{11}(z, r), c_{12}(z, r)$ and D_{11}, D_{12} are the concentration distributions and the diffusivities of SO_2 in the gas and liquid, k-interphase mass transfer coefficient.

Let us consider the co-current gas–liquid drops absorption process in a column with radius r_0 and working zone height l. The boundary conditions [2] of (17.2.1) have the form:

$$
\begin{aligned}
r = 0, \quad &\frac{\partial c_{11}}{\partial r} = \frac{\partial c_{12}}{\partial r} \equiv 0; \quad r = r_0, \quad \frac{\partial c_{11}}{\partial r} = \frac{\partial c_{12}}{\partial r} \equiv 0; \\
z = 0, \quad &c_{11}(r, 0) \equiv c_{11}^0, \quad u_1^0 c_{11}^0 \equiv u_1(r)c_{11}^0 - D_{11}\left(\frac{\partial c_{11}}{\partial z}\right)_{z=0}, \\
&c_{12}(r, 0) \equiv c_{12}^0, \quad u_2^0 c_{12}^0 \equiv u_2(r)c_{12}^0 - D_{12}\left(\frac{\partial c_{12}}{\partial z}\right)_{z=0},
\end{aligned}
\tag{17.2.2}
$$

where $u_1^0, u_2^0, c_{11}^0, c_{12}^0$ are the inlet (average) velocities and concentrations in the gas and liquid phases. Practically $c_{12}^0 = 0$ and the liquid (m^3 s^{-1})/gas (m^3 s^{-1}) ratio is $L/G = 0.02$ ($\varepsilon_1 = 0.98, \varepsilon_2 = 0.02$), Henry's number $\chi = 1.25$, the diffusivity of ASG in the gas (air) is $D_{11} = 10^{-5}$ (m^2 s^{-1}) and in the liquid (water) $D_{12} = 10^{-9}$ [m^2 s^{-1}]. The desired absorption degree is set to 94%.

17.2.1 Generalized Analysis

The use of dimensionless (generalized) variables [2] permits to perform a qualitative analysis of the model (17.2.1) and (17.2.2), whereas characteristic scales are used the following formula for the average velocities, the inlet concentrations, and the column characteristics:

$$R = \frac{r}{r_0}, \quad Z = \frac{z}{l}, \quad U_1 = \frac{u_1}{u_1^0}, \quad U_2 = \frac{u_2}{u_2^0}, \quad C_{11} = \frac{c_{11}}{c_{11}^0}, \quad C_{12} = \frac{c_{12}\chi}{c_{11}^0}. \quad (17.2.3)$$

If (17.2.3) is substituted into (17.2.1) and (17.2.2), the model in generalized variables takes the form:

$$U_1(R)\frac{\partial C_{11}}{\partial Z} = Fo_{11}\left(\frac{r_0^2}{l^2}\frac{\partial^2 C_{11}}{\partial Z^2} + \frac{1}{R}\frac{\partial C_{11}}{\partial R} + \frac{\partial^2 C_{11}}{\partial R^2}\right) - \frac{kl}{u_1^0}(C_{11} - C_{12});$$

$$U_2(R)\frac{\partial C_{12}}{\partial Z} = Fo_{12}\left(\frac{r_0^2}{l^2}\frac{\partial^2 C_{12}}{\partial Z^2} + \frac{1}{R}\frac{\partial C_{12}}{\partial R} + \frac{\partial^2 C_{12}}{\partial R^2}\right) + \frac{kl\chi}{u_2^0}(C_{11} - C_{12});$$

$$R = 0, \quad \frac{\partial C_{11}}{\partial R} = \frac{\partial C_{12}}{\partial R} \equiv 0; \quad R = 1, \quad \frac{\partial C_{11}}{\partial R} = \frac{\partial C_{12}}{\partial R} \equiv 0; \quad (17.2.4)$$

$$Z = 0, \quad C_{11}(R,0) \equiv 1, \quad 1 \equiv U_1(R) - Pe_{11}^{-1}\left(\frac{\partial C_{11}}{\partial Z}\right)_{Z=0},$$

$$C_{12}(R,0) \equiv 0, \quad \left(\frac{\partial C_{12}}{\partial Z}\right)_{Z=0} \equiv 0,$$

where

$$Fo_{11} = \frac{D_{11}l}{u_1^0 r_0^2}, \quad Fo_{12} = \frac{D_{12}l}{u_2^0 r_0^2}, \quad Pe_{11} = \frac{u_1^0 l}{D_{11}}, \quad Pe_{12} = \frac{u_2^0 l}{D_{12}} \quad (17.2.5)$$

are the Fourier and the Peclet numbers.

The very small values of the Fourier number and the very big values of the Peclet number, which result from the big inlet (average) velocities $\left(u_1^0, u_2^0\right)$, show that the diffusion mass transfer is negligible in comparison with the convection mass transfer and the model (17.2.4) is possible to be presented in zero approximation with respect to the small parameters $\left(0 = Fo_{11} \sim Fo_{12} \sim Pe_{11}^{-1} \sim Pe_{12}^{-1} \leq 10^{-2}\right)$. As a result, the model (17.2.4) has the convective form:

$$U_1(R)\frac{dC_{11}}{dZ} = -K_1(C_{11} - C_{12}); \quad U_2(R)\frac{dC_{12}}{dZ} = K_2(C_{11} - C_{12});$$
$$Z = 0, \quad C_{11}(R,0) \equiv 1, \quad C_{12}(R,0) \equiv 0, \quad (17.2.6)$$

where

$$K_1 = \frac{kl}{u_1^0}, \quad K_2 = \frac{kl\chi}{u_2^0}, \quad \frac{K_1}{K_2} = \frac{u_2^0}{\chi u_1^0} = 0.8$$

$$(\chi = 1.25, \quad u_1^0 \approx u_2^0)$$

(17.2.7)

and R is a parameter in the functions $C_{11}(R, Z), C_{12}(R, Z)$.

17.2.2   Concentration Distributions

Let us consider an example, the parabolic velocity distributions (Poiseuille flow) in the gas and liquid phases:

$$u_1 = 2u_1^0\left(1 - \frac{r^2}{r_o^2}\right), \quad u_2 = 2u_2^0\left(1 - \frac{r^2}{r_o^2}\right), \quad U_1 = U_2 = 2(1 - R^2). \quad (17.2.8)$$

For this case the model of the physical absorption in a co-current column has the form:

$$2(1 - R^2)\frac{dC_{11}}{dZ} = -0.8K_2(C_{11} - C_{12});$$

$$2(1 - R^2)\frac{dC_{12}}{dZ} = K_2(C_{11} - C_{12}); \quad (17.2.9)$$

$$Z = 0, \quad C_{11}(R, 0) \equiv 1, \quad C_{12}(R, 0) \equiv 0.$$

The solutions of (17.2.9) obtained for different values of the parameter $R(0 \le R \le 1)$ and the functions $C_{11}(R, Z), C_{12}(R, Z)$ in the case $K_2 = 1.0, Z = 0.2, 0.5, 1.0$ are shown on Figs. 17.4 and 17.5.

Fig. 17.4 Concentration distribution $C_{11}(R, Z)$

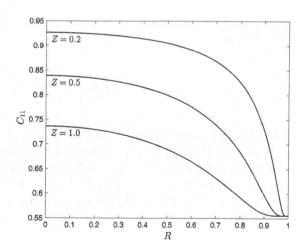

Fig. 17.5 Concentration
distribution $C_{12}(R,Z)$

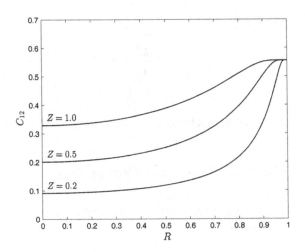

17.2.3 Absorption Degree

The absorption degree in the column G can be obtained as a difference between the average values of the inlet and outlet convective (ASG) mass fluxes through the column:

$$G = 1 - \frac{J}{u_1^0 c_{11}^0}, \qquad (17.2.10)$$

where J is the average outlet convective mass flux through the column:

$$J = \frac{2}{r_0^2} \int_0^{r_0} rI(r)\,dr, \quad I(r) = u_1(r)c_{11}(r,l). \qquad (17.2.11)$$

The use of the dimensionless variables (17.2.3) and the velocity distribution (17.2.8) leads to

$$J = 4u_1^0 c_{11}^0 \int_0^1 R(1 - R^2)C_{11}(R,1)\,dR, \qquad (17.2.12)$$

where $C_{11}(R,1)$ is the solution of (17.2.9) for $Z = 1$.

From (17.2.12) and (17.2.9), it is seen that the absorption degree in the column G is related to the interphase mass transfer coefficient k through the dimensionless parameters K_1 and K_2 in (17.2.9). For the case $\varepsilon_1 = 0.98, \varepsilon_2 = 0.02, \chi = 1.25, u_1^0 = u_2^0 = 20$ (m s^{-1}), $l = 10$ (m), the dimensionless parameters are $K_1 = 0.8K_2$ and if $K_2 = 1$ the values of K_1 is $K_1 = 0.8$ and the absorption degree is $G = 0.667$.

The absorption degree depends on the interphase mass transfer. The absorption degree 0.94% is possible to be obtained for $K_2 = 0.0864$ ($K_1 = 0.0691$). These values of K_1, K_2 can be obtained with different combinations of values of the parameters $l, \varepsilon_1, \varepsilon_2, u_1^0, u_2^0$.

The average outlet concentration $\bar{C}_{11}(1)$ of ASG in the gas phase and the absorption degree are:

$$\bar{C}_{11}(1) = 2 \int_0^1 RC_{11}(R,1)dR, \quad G = 1 - \bar{C}_{11}(1). \tag{17.2.13}$$

17.3 Average-Concentration Model

In the diffusion type model (17.2.1), the velocity distributions in the phases cannot be obtained, because the equation of the interphase surface is not possible to be obtained. The problem can be avoided, if the average values of the velocity and concentration over the cross-sectional area of the column are used. From (17.2.6) follows that a convection type of model has to be used:

$$u_1 \frac{dc_{11}}{dz} = -k(c_{11} - \chi c_{12}); \quad u_2 \frac{dc_{12}}{dz} = k(c_{11} - \chi c_{12});$$

$$z = 0, \quad c_{11}(r,0) \equiv c_{11}^0, \quad c_{12}(r,0) \equiv 0. \tag{17.3.1}$$

The average values of the velocity and concentration in the column's cross-sectional area are possible to be obtained (see Part II) using the expressions

$$\bar{u}_1 = \frac{2}{r_0^2} \int_0^{r_0} ru_1(r)dr = u_1^0, \quad \bar{u}_2 = \frac{2}{r_0^2} \int_0^{r_0} ru_2(r)dr = u_2^0,$$

$$\bar{c}_{11}(z) = \frac{2}{r_0^2} \int_0^{r_0} rc_{11}(r,z)dr, \quad \bar{c}_{12}(z) = \frac{2}{r_0^2} \int_0^{r_0} rc_{12}(r,z)dr. \tag{17.3.2}$$

The velocity and concentration distributions in (17.3.1) can be represented with the help of the average functions (17.3.2):

$$u_1(r) = \bar{u}_1 \tilde{u}_1(r), \quad u_2(r) = \bar{u}_2 \tilde{u}_2(r),$$

$$c_{11}(r,z) = \bar{c}_{11}(z)\tilde{c}_{11}(r,z), \quad c_{12}(r,z) = \bar{c}_{12}(z)\tilde{c}_{12}(r,z), \tag{17.3.3}$$

where $\tilde{u}_1(r), \tilde{u}_2(r), \tilde{c}_{11}(r, z), \tilde{c}_{12}(r, z)$ represent the radial non-uniformity of both the velocity and the concentration distributions, satisfying the conditions:

$$\frac{2}{r_0^2} \int_0^{r_0} r\tilde{u}_1(r)dr = 1, \quad \frac{2}{r_0^2} \int_0^{r_0} r\tilde{u}_2(r)dr = 1,$$

$$\frac{2}{r_0^2} \int_0^{r_0} r\tilde{c}_{11}(r, z)dr = 1, \quad \frac{2}{r_0^2} \int_0^{r_0} r\tilde{c}_{12}(r, z)dr = 1. \tag{17.3.4}$$

The average-concentration model may be obtained [3] when putting (17.3.3) into (17.3.1) then multiplying by r and integrating with respect to r over the interval $[0, r_0]$. The result is:

$$\alpha_1(z)\bar{u}_1 \frac{d\bar{c}_{11}}{dz} + \frac{d\alpha_1}{dz} \bar{u}_1 \bar{c}_{11} = -k(\bar{c}_{11} - \chi\bar{c}_{12});$$

$$\alpha_2(z)\bar{u}_2 \frac{d\bar{c}_{12}}{dz} + \frac{d\alpha_2}{dz} \bar{u}_2 \bar{c}_{12} = k(\bar{c}_{11} - \chi\bar{c}_{12}); \tag{17.3.5}$$

$$z = 0, \quad \bar{c}_{11}(r, 0) \equiv c_{11}^0, \quad \bar{c}_{12}(r, 0) \equiv 0.$$

where

$$\alpha_1 = \frac{2}{r_0^2} \int_0^{r_0} r\tilde{u}_1 \tilde{c}_{11}dr, \quad \alpha_2 = \frac{2}{r_0^2} \int_0^{r_0} r\tilde{u}_2 \tilde{c}_{12}dr. \tag{17.3.6}$$

The use of the generalized variables

$$z = lZ, \quad \bar{c}_1 = c_{11}^0 \bar{C}_{11}, \quad \bar{c}_{12} = \frac{c_{11}^0}{\chi} \bar{C}_{12}, \tag{17.3.7}$$

leads to

$$A_1(Z) \frac{d\bar{C}_{11}}{dZ} + \frac{dA_1}{dZ} \bar{C}_{11} = -K_1(\bar{C}_{11} - \bar{C}_{12});$$

$$A_2(Z) \frac{d\bar{C}_{12}}{dZ} + \frac{dA_2}{dZ} \bar{C}_{12} = K_2(\bar{C}_{11} - \bar{C}_{12}); \tag{17.3.8}$$

$$Z = 0, \quad \bar{C}_{11} \equiv 1, \quad \bar{C}_{12} \equiv 0.$$

where

$$\alpha_1(z) = \alpha_1(lZ) = A_1(Z), \quad \alpha_2(z) = \alpha_2(lZ) = A_2(Z). \tag{17.3.9}$$

From (17.2.8) follows

$$\tilde{u}_1 = \tilde{u} = 2\left(1 - R^2\right) \tag{17.3.10}$$

and from (17.3.6), (17.3.7), (17.3.9) it is possible to obtain:

$$A_1(Z) = 4 \int_0^1 R\left(1 - R^2\right) \frac{C_{11}(R, Z)}{\bar{C}_{11}(Z)} \, dR, \quad \bar{C}_{11}(Z) = 2 \int_0^1 R C_{11}(R, Z) \, dR,$$

$$A_2(Z) = 4 \int_0^1 R\left(1 - R^2\right) \frac{C_{12}(R, Z)}{\bar{C}_{12}(Z)} \, dR, \quad \bar{C}_{12}(Z) = 2 \int_0^1 R C_{12}(R, Z) \, dR.$$

$$\tag{17.3.11}$$

The functions $A_1(Z), A_2(Z)$ of (17.3.11) in the case $K_1 = 0.8, K_2 = 1.0$ are obtained and presented on Fig. 17.6, where it can be seen that the linear approximations is possible to be used:

$$A_1 = a_{10} + a_{11}Z, \quad A_2 = a_{20} + a_{21}Z. \tag{17.3.12}$$

The obtained "theoretical" values of $a_{10}, a_{11}, a_{20}, a_{21}$ and $K_1 = 0.8, K_2 = 1.0$ are presented in the Table 17.1.

As a result the model (17.3.8) has the form:

$$\left(a_{10} + a_{11}Z\right) \frac{d\bar{C}_{11}}{dZ} + a_{11}\bar{C}_{11} = -0.8K_2(\bar{C}_{11} - \bar{C}_{12});$$

$$\left(a_{20} + a_{21}Z\right) \frac{d\bar{C}_{12}}{dZ} + a_{21}\bar{C}_{12} = K_2(\bar{C}_{11} - \bar{C}_{12}); \quad Z = 0, \quad \bar{C}_{11} \equiv 1, \quad \bar{C}_{12} \equiv 0.$$

$$\tag{17.3.13}$$

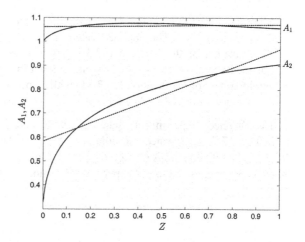

Fig. 17.6 Functions $A_1(Z), A_2(Z)$ (lines) and its linear approximation (dashed lines)

Table 17.1 Model parameter values

Parameters	K_1	K_2	a_{10}	a_{11}	a_{20}	a_{21}
"Theoretical" values	0.8	1	1.0617	0.0105	0.5834	0.3853
"Experimental" values	1.3376	1.672	0.986	0.0097	0.9873	0.2507

17.3.1 Parameter Identification

The obtained average concentrations $\bar{C}_{11}(Z), \bar{C}_{12}(Z)$ (17.3.11) permit to obtain "artificial experimental data" for $Z = 0.1$:

$$\bar{C}^m_{11\,exp}(Z) = (0.95 + 0.1S_m)\bar{C}_{11}(Z),$$
$$\bar{C}^m_{12\,exp}(Z) = (0.95 + 0.1S_m)\bar{C}_{12}(Z), \quad m = 1,\ldots 10, \quad Z = 0.1, \tag{17.3.14}$$

where $0 \le S_m \le 1, m = 1, \ldots, 10$ are obtained using a generator of random numbers. The obtained "artificial experimental data" (17.3.14) are used for illustration of the parameter identification in the average concentrations models (17.3.13) by minimization of the least-squares function Q:

$$Q(K_2, a_{10}, a_{11}, a_{20}, a_{21}) = \sum_{m=1}^{10} \left[\bar{C}_{11}(Z, K_2, a_{10}, a_{11}, a_{20}, a_{21}) - \bar{C}^m_{11\,exp}(Z) \right]^2$$
$$+ \sum_{m=1}^{10} \left[\bar{C}_{12}(Z, K_2, a_{10}, a_{11}, a_{20}, a_{21}) - \bar{C}^m_{12\,exp}(Z) \right]^2, \quad Z = 0.1 \tag{17.3.15}$$

where the values of $\bar{C}_{11}(Z, K_2, a_{10}, a_{11}, a_{20}, a_{21})$ and $\bar{C}_{12}(Z, K_2, a_{10}, a_{11}, a_{20}, a_{21})$ are obtained as solutions of (17.3.13) for $Z = 0.1$.

The parameters $K_2, a_{10}, a_{11}, a_{20}, a_{21}$, in the model (17.3.13) are possible to be obtained using the following algorithm:

1. Minimization of the function (17.3.15) with respect to K_2, where $a_{10} = 1, a_{11} = 0, a_{20} = 1, a_{21} = 0$. The obtained value of K_2 must be replaced in (17.3.13).
2. Minimization of the function (17.3.15) with respect to $a_{10}, a_{11}, a_{20}, a_{21}$. The obtained values of $a_{10}, a_{11}, a_{20}, a_{21}$ are entered in (17.3.13).
3. Minimization of the function (17.3.15) with respect to K_2. The obtained value of K_2 is replaced in (17.3.13), etc.

The obtained "experimental" parameter values are presented in Table 17.1.

In Fig. 17.7 is presented the comparison of the "artificial experimental data" (17.3.14) and the functions $\bar{C}_{11}(Z), \bar{C}_{12}(Z)$ as a solution of the model Eq. (17.3.13), whereas parameters are used "experimental" values in Table 17.1.

Fig. 17.7 Comparison of the model (17.3.13) and the "artificial experimental data" (17.3.14)

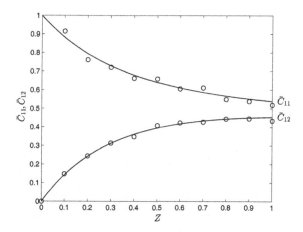

References

1. Elenkov D, Boyadjiev C (1967) Hydrodynamics and mass transfer in a nozzles Venturi absorber. II. Absorption of sulfur dioxide in water and aqueous solutions of surfactants. Int Chem Eng 7(2):191

2. Еленков Д, Хр Бояджиев, Ив Кръстев, Бояджиев Л (1966) Абсорбция на серен двуокис с разтвори на натриев карбонат в струест абсорбер тип Вентури. Известия на ИОНХ-БАН 4:153–166

3. Chr Boyadjiev (2010) Theoretical chemical engineering. Modeling and simulation. Springer, Berlin

4. Doichinova M, Boyadjiev C (2012) On the column apparatuses modeling. Int J Heat Mass Transfer 55:6705–6715

5. Chr Boyadjiev (2013) A new approach for the column apparatuses modeling in chemical engineering. J Pure Appl Math Adv Appl 10(2):131–150

6. Boyadjiev C, Boyadjiev B, Doichinova M, Popova-Krumova P (2013) Method and apparatus for gas absorption. Patent application № 111168

7. Boyadjiev C, Boyadjiev B, Doichinova M, Popova-Krumova P (2014) Method and apparatus for gas cleaning from sulfur dioxide. Patent application № 111398

8. Boyadjiev C, Boyadjiev B, Doichinova M, Popova-Krumova P (2014) Co-current apparatus for gas absorption. Patent application № 111473

9. Boyadjiev C, Doichinova M, Popova-Krumova P, Boyadjiev B (2015) On the gas purification from low SO_2 concentration. 1. Absorption methods modeling. Int J Eng Res 4(9):550–557

10. Boyadjiev C, Doichinova M, Popova-Krumova P, Boyadjiev B (2015) On the gas purification from low SO_2 concentration. 2. Two-steps chemical absorption modeling. Int J Eng Res 4(9):558–563

11. Boyadjiev C, Boyadjiev B, Doichinova M, Popova-Krumova P (2015) An absorption-adsorption method for gas purification from SO_2 in power plants. OALib J (in press)

12. Boyadjiev B, Boyadjiev C, Doichinova M, Popova-Krumova P (2015) Co-current apparatus for gas purification from SO_2. Recent Innov Chem Eng 8(1):25–29

Part VI
Book Conclusions

Chapter 18
Conclusion

The column apparatuses are the main devices for separation and chemical processes realization in chemical, power, biotechnological, and other industries. They are different types as plate columns, packed bed columns, bubble columns, trickle columns, catalyst bed columns, etc.

The processes in column apparatuses (except for the plate columns) are realized in one, two, or three phases. The gas phase moves among the columns as jets and bubbles. The liquid phase present in the column is as droplets, films, and jets. The solid phase forms are packed beds, catalyst particles, or slurries ($CaCO_3/H_2O$ suspension).

In the book are presented mathematical models of the complexes of elementary processes in column apparatuses, which are mathematical structures, where the mathematical operators are mathematical descriptions of the elementary processes; i.e., the models express a full correspondence between physical effects and mathematical operators.

The complex processes in the column apparatuses are a combination of hydrodynamic processes, convective and diffusive mass (heat) transfer processes and chemical reactions of the reagents (components of the phases).

The fundamental problem in the column apparatuses modeling is the complicated hydrodynamic behavior of the flows in the columns, where the velocity distributions and interphase boundaries are unknown.

The column apparatuses are possible to be modeled using a new approach on the basis of the physical approximations of the mechanics of continua, where the mathematical point (in the phase volume or on the surface between the phases) is equivalent to a small (elementary) physical volume, which is sufficiently small with respect to the apparatus volume, but at the same time sufficiently large with respect to the intermolecular volumes in the medium.

The physical elementary volumes are presented as mathematical points in a cylindrical coordinate system. The concentrations (kg mol m^{-3}) of the reagents (components of the phases) are the quantities of the reagents (kg mol) in 1 m^3 of the phase volumes.

© Springer International Publishing AG, part of Springer Nature 2018
C. Boyadjiev et al., *Modeling of Column Apparatus Processes*,
Heat and Mass Transfer, https://doi.org/10.1007/978-3-319-89966-4_18

The homogeneous chemical reactions and interphase mass transfer are presented as volume reactions (kg mol m^{-3} s^{-1}) in the phases; i.e., volume sources or sinks in the phase volumes in the column.

The volume reactions lead to different values of the reagent (substance) concentrations in the elementary volumes and as a result, two mass transfer effects exist —convective transfer (caused by the fluid motion) and diffusion transfer (caused by the concentration gradient).

The convective transfer in column apparatus is caused by a laminar or turbulent (as a result of large-scale turbulent pulsations) flow in the small (elementary) volume.

The diffusive transfer is molecular or turbulent (caused by small-scale turbulent pulsations).

The mathematical models of the processes in the column apparatuses, in the physical approximations of the mechanics of continua, represent the mass balances in the elementary phase volumes (phase parts in the elementary column volume) between the convective transfers, the diffusive transfers, and the volume mass sources (sinks) (as a result of the homogeneous chemical reactions and different heterogeneous reactions as absorption, adsorption, and catalytic reactions). They are convection–diffusion type and average-concentration type, where the most important feature is the presentation of the heterogeneous reactions such as volume sources (sinks).

The convection–diffusion-type models permit a qualitative analysis of the processes (models) to be made in order to obtain the main, small, and slight physical effects (mathematical operators) and to discard the slight effects (operators). As a result, the process mechanism identification becomes possible. These models permit to determine the mass transfer resistances in the gas and liquid phases and to find the optimal dispersion system in the gas absorption (gas–liquid drops or liquid–gas bubbles). The convection–diffusion model is a base of the average-concentration models, which allow a quantitative analysis of the processes in column apparatuses.

The convection–diffusion models are possible to be used for qualitative analysis only, because the velocity distribution functions are unknown and cannot be obtained. The problem can be avoided by the average-concentration type of models if the average values of the velocity and concentration over the cross-sectional area of the column are used; i.e., the medium elementary volume (in the physical approximations of the mechanics of continua) will be equivalent to a small cylinder with column radius and a height, which is sufficiently small with respect to the column height and at the same time sufficiently large with respect to the inter-molecular distances in the medium.

The model parameters in the average-concentration type of models are related with the radial non-uniformity of the velocity distribution and show the influence of the column radius (scale-up effect) on the mass transfer kinetics. These parameters are possible to be obtained using experimental data. In the cases of a constant radial non-uniformity of the velocity distribution, the output value of the average concentration of a short (10–20% from the real column height) column is sufficient for determination of the model parameters. In the cases of a variable radial non-uniformity of the velocity distribution along the column height, the values of the average concentrations at the column end permit to be obtained the model parameters. The obtained values of the model parameters are possible to be used in the cases of chemical reactions and mass transfer of different substances in the column.

The convection–diffusion and average-concentration models are used for qualitative and quantitative analyzes of different processes in column apparatuses:

1. Chemical reactions;
2. Physical and chemical absorption;
3. Physical and chemical adsorption;
4. Heterogeneous catalytic chemical reactions.

In many cases, the computer modeling of the processes in column apparatuses, made on the base of the new approach, using the convection–diffusion-type model and average-concentration type model, does not allow a direct use of the MATLAB program. In these cases, it is necessary to create combinations of appropriate algorithms:

1. Appropriate combination of MATLAB and perturbations method;
2. Solving the equations set in different coordinate systems;
3. Multi-step algorithm and MATLAB.

The new approach for modeling the column apparatuses allow the solving of some practical problems, in connection with the purification of gases from sulfur dioxide, by the proposing of several patents and utility models, and modeling of the processes:

1. A new patent for the gas purification from average soluble gases is proposed, where the process optimization is realized in a two-zone column, wherein the upper zone the process is physical absorption in a gas–liquid drops system (intensification of the gas phase mass transfer), while in the lower zone it is a physical absorption in liquid–gas bubbles system (intensification of the liquid phase mass transfer);
2. A new patent, using a two-step process—physical absorption of SO_2 by water and adsorption of SO_2 from the water solution by synthetic anionite particles, is proposed. The adsorbent regeneration is made by NH_4OH solution. The obtained $(NH_4)_2SO_3$ (NH_4HSO_3) is used (after reaction with HNO_3) for production of concentrated SO_2 and NH_4NO_3 solutions;
3. A new patent is presented, using a co-current absorber, where the gas velocity increase and the column diameter decrease.
4. A new utility model of an absorption–adsorption apparatus for gas cleaning from sulfur dioxide.
5. A new utility model of a column reactor for chemical processes.

The presented book is a theoretical base for qualitative and quantitative analyzes of different separation and chemical processes in chemical, power, biotechnological, and other industries.

The presented theoretical analysis in the book shows that the mechanics of continua cannot be used to model the chemical, absorption, adsorption, and catalytic processes in the column apparatuses because fluid velocities in the phases and interphase boundaries are unknown and cannot be determined. These problems are overcome by replacing the surface (heterogeneous) phase boundary reactions with equivalent volume (homogeneous) reactions, and the unknown phase velocities are replaced by the average velocity of the cross section of the column. This leads to parameters in the models that must be determined by experimental data.

The theoretical analysis of the process models in the column apparatuses (in the approximations of the mechanics of continua) shows that the radial non-uniformity of the axial component of the velocity is the reason for the reduction of the efficiency of the processes and the appearance of the experimentally determinable parameters. The book shows that the tangential introduction of the gases (liquids) into the column minimizes the radial non-uniformity of the axial component of the velocity and cancels the values of the experimentally determinable parameters. The only parameters that remain to be determined are the interphase mass transfer coefficients (as volume factors), but they do not depend on the diameter of the column and can be determined by experimental data obtained on a model (small diameter) column.

Printed in the United States
By Bookmasters